T0188909

Communications
in Computer and Information Science   1869

Rationale

The CCIS series is devoted to the publication of proceedings of computer science conferences. Its aim is to efficiently disseminate original research results in informatics in printed and electronic form. While the focus is on publication of peer-reviewed full papers presenting mature work, inclusion of reviewed short papers reporting on work in progress is welcome, too. Besides globally relevant meetings with internationally representative program committees guaranteeing a strict peer-reviewing and paper selection process, conferences run by societies or of high regional or national relevance are also considered for publication.

Topics

The topical scope of CCIS spans the entire spectrum of informatics ranging from foundational topics in the theory of computing to information and communications science and technology and a broad variety of interdisciplinary application fields.

Information for Volume Editors and Authors

Publication in CCIS is free of charge. No royalties are paid, however, we offer registered conference participants temporary free access to the online version of the conference proceedings on SpringerLink (http://link.springer.com) by means of an http referrer from the conference website and/or a number of complimentary printed copies, as specified in the official acceptance email of the event.

CCIS proceedings can be published in time for distribution at conferences or as post-proceedings, and delivered in the form of printed books and/or electronically as USBs and/or e-content licenses for accessing proceedings at SpringerLink. Furthermore, CCIS proceedings are included in the CCIS electronic book series hosted in the SpringerLink digital library at http://link.springer.com/bookseries/7899. Conferences publishing in CCIS are allowed to use Online Conference Service (OCS) for managing the whole proceedings lifecycle (from submission and reviewing to preparing for publication) free of charge.

Publication process

The language of publication is exclusively English. Authors publishing in CCIS have to sign the Springer CCIS copyright transfer form, however, they are free to use their material published in CCIS for substantially changed, more elaborate subsequent publications elsewhere. For the preparation of the camera-ready papers/files, authors have to strictly adhere to the Springer CCIS Authors' Instructions and are strongly encouraged to use the CCIS LaTeX style files or templates.

Abstracting/Indexing

CCIS is abstracted/indexed in DBLP, Google Scholar, EI-Compendex, Mathematical Reviews, SCImago, Scopus. CCIS volumes are also submitted for the inclusion in ISI Proceedings.

How to start

To start the evaluation of your proposal for inclusion in the CCIS series, please send an e-mail to ccis@springer.com.

Haijun Zhang · Yinggen Ke · Zhou Wu ·
Tianyong Hao · Zhao Zhang · Weizhi Meng ·
Yuanyuan Mu
Editors

International Conference on Neural Computing for Advanced Applications

4th International Conference, NCAA 2023
Hefei, China, July 7–9, 2023
Proceedings, Part I

Springer

Editors
Haijun Zhang 🆔
Harbin Institute of Technology
Shenzhen, China

Zhou Wu 🆔
Chongqing University
Chongqing, China

Zhao Zhang 🆔
Hefei University of Technology
Hefei, China

Yuanyuan Mu 🆔
Chaohu University
Hefei, China

Yinggen Ke 🆔
Chaohu University
Hefei, China

Tianyong Hao 🆔
South China Normal University
Guangzhou, China

Weizhi Meng 🆔
Technical University of Denmark
Kongens Lyngby, Denmark

ISSN 1865-0929   ISSN 1865-0937 (electronic)
Communications in Computer and Information Science
ISBN 978-981-99-5843-6   ISBN 978-981-99-5844-3 (eBook)
https://doi.org/10.1007/978-981-99-5844-3

This Springer imprint is published by the registered company Springer Nature Singapore Pte Ltd.
The registered company address is: 152 Beach Road, #21-01/04 Gateway East, Singapore 189721, Singapore

Paper in this product is recyclable.

Preface

Neural computing and Artificial Intelligence (AI) have become hot topics in recent years. To promote multi-disciplinary development and application of neural computing, a series of NCAA conferences was initiated on the theme of *"make the academic more practical"*, providing an open platform for academic discussions, industrial showcases, and basic training tutorials. This volume contains the papers accepted for this year's International Conference on Neural Computing for Advanced Applications (NCAA 2023). NCAA 2023 was organized by Chaohu University and co-organized by Shandong Jianzhu University and Wuhan Textile University, and it was supported by Springer and Cancer Innovation. After the effects of COVID-19, the mainstream part of NCAA 2023 was turned into a hybrid event, with mainly offline participants, in which people could freely connect to keynote speeches and presentations face to face.

NCAA 2023 received 211 submissions, of which 84 high-quality papers were selected for publication in this volume after double-blind peer review, leading to an acceptance rate of just under 40%. These papers were categorized into 13 technical tracks: *Neural network (NN) theory, NN-based control systems, neuro-system integration and engineering applications; Machine learning and deep learning for data mining and data-driven applications; Computational intelligence, nature-inspired optimizers, and their engineering applications; Neuro/fuzzy systems, multi-agent control, decision making, and their applications in smart construction and manufacturing; Deep learning-driven pattern recognition, computer vision and its industrial applications; Natural language processing, knowledge graphs, recommender systems, and their applications; Neural computing-based fault diagnosis and forecasting, prognostic management, and cyber-physical system security; Sequence learning for spreading dynamics, forecasting, and intelligent techniques against epidemic spreading; Multimodal deep learning for representation, fusion, and applications; Neural computing-driven edge intelligence, machine learning for mobile systems, pervasive computing, and intelligent transportation systems; Applications of data mining, machine learning and neural computing in language studies; Computational intelligent fault diagnosis and fault-tolerant control, and their engineering applications; Other Neural computing-related applications.*

The authors of each paper in this volume have reported their novel results of computing theory or applications. The volume cannot cover all aspects of neural computing and advanced applications, but may still inspire insightful thoughts for readers. We hope that more secrets of AI will be unveiled and academics will drive more practical developments and solutions to real-world applications.

July 2023

Haijun Zhang
Yinggen Ke
Zhou Wu
Tianyong Hao
Zhao Zhang
Weizhi Meng
Yuanyuan Mu

Organization

Honorary Chairs

John MacIntyre   University of Sunderland, UK
Tommy W. S. Chow   City University of Hong Kong, China

General Co-chairs

Haijun Zhang   Harbin Institute of Technology, China
Yinggen Ke   Chaohu University, China
Zhou Wu   Chongqing University, China

Program Co-chairs

Tianyong Hao   South China Normal University, China
Zhao Zhang   Hefei University of Technology, China
Weizhi Meng   Technical University of Denmark, Denmark

Organizing Committee Co-chairs

Rongqi Yu   Chaohu University, China
Yanxia Sun   University of Johannesburg, South Africa
Mingbo Zhao   Donghua University, China

Local Arrangement Co-chairs

Yuanyuan Mu   Chaohu University, China
Peng Yu   Chaohu University, China
Yongfeng Zhang   Jinan University, China

Registration Co-chairs

Yaqing Hou	Dalian University of Technology, China
Jing Zhu	Macau University of Science and Technology, China
Shuqiang Wang	Chinese Academy of Sciences, China
Weiwei Wu	Southeast University, China
Zhili Zhou	Nanjing University of Information Science and Technology, China

Publication Co-chairs

Kai Liu	Chongqing University, China
Yu Wang	Xi'an Jiaotong University, China
Yi Zhang	Fuzhou University, China
Bo Wang	Huazhong University of Science and Technology, China
Xianghua Chu	Shenzhen University, China

Publicity Co-chairs

Fei He	Coventry University, UK
Xiao-Zhi Gao	University of Eastern Finland, Finland
Choujun Zhan	South China Normal University, China
Zenghui Wang	University of South Africa, South Africa
Yimin Yang	University of Western Ontario, Canada
Zili Chen	Hong Kong Polytechnic University, China
Reza Maleklan	Malmö University, Sweden
Sinan Li	University of Sydney, Australia

Sponsor Co-chairs

Wangpeng He	Xidian University, China
Bingyi Liu	Wuhan University of Technology, China
Cuili Yang	Beijing University of Technology, China
Guo Luo	Nanfang College of Sun Yat-sen University, China
Shi Cheng	Shaanxi Normal University, China

Forum Chair

Jingjing Cao   Wuhan University of Technology, China

Tutorial Chair

Jicong Fan   Chinese University of Hong Kong, Shenzhen,
                                      China

Competition Chair

Chengdong Li   Shandong Jianzhu University, China

NCAA Steering Committee Liaison

Jianghong Ma   Harbin Institute of Technology, China

Web Chair

Xinrui Yu   Harbin Institute of Technology, China

Program Committee Members

Dong Yang   City University of Hong Kong, China
Sheng Li   University of Georgia, USA
Jie Qin   Swiss Federal Institute of Technology (ETH),
                                      Switzerland
Xiaojie Jin   Bytedance AI Lab, USA
Zhao Kang   University of Electronic Science and Technology,
                                      China
Xiangyuan Lan   Hong Kong Baptist University, China
Peng Zhou   Anhui University, China
Chang Tang   China University of Geosciences, China
Dan Guo   Hefei University of Technology, China
Li Zhang   Soochow University, China
Xiaohang Jin   Zhejiang University of Technology, China
Wei Huang   Zhejiang University of Technology, China

Chao Chen	Chongqing University, China
Jing Zhu	Macau University of Science and Technology, China
Weizhi Meng	Technical University of Denmark, Denmark
Wei Wang	Dalian Ocean University, China
Jian Tang	Beijing University of Technology, China
Heng Yue	Northeastern University, China
Yimin Yang	University of Western Ontario, Canada
Jianghong Ma	Harbin Institute of Technology, China
Jicong Fan	Chinese University of Hong Kong (Shenzhen), China
Xin Zhang	Tianjin Normal University, China
Xiaolei Lu	City University of Hong Kong, China
Penglin Dai	Southwest Jiaotong University, China
Liang Feng	Chongqing University, China
Xiao Zhang	South-Central University for Nationalities, China
Bingyi Liu	Wuhan University of Technology, China
Cheng Zhan	Southwest University, China
Qiaolin Pu	Chongqing University of Posts and Telecommunications, China
Hao Li	Hong Kong Baptist University, China
Junhua Wang	Nanjing University of Aeronautics and Astronautics, China
Yu Wang	Xi'an Jiaotong University, China
BinQiang Chen	Xiamen University, China
Wangpeng He	Xidian University, China
Jing Yuan	University of Shanghai for Science and Technology, China
Huiming Jiang	University of Shanghai for Science and Technology, China
Yizhen Peng	Chongqing University, China
Jiayi Ma	Wuhan University, China
Yuan Gao	Tencent AI Lab, China
Xuesong Tang	Donghua University, China
Weijian Kong	Donghua University, China
Zhili Zhou	Nanjing University of Information Science and Technology, China
Yang Lou	City University of Hong Kong, China
Chao Zhang	Shanxi University, China
Yanhui Zhai	Shanxi University, China
Wenxi Liu	Fuzhou University, China
Kan Yang	University of Memphis, USA
Fei Guo	Tianjin University, China

Wenjuan Cui	Chinese Academy of Sciences, China
Wenjun Shen	Shantou University, China
Mengying Zhao	Shandong University, China
Shuqiang Wang	Chinese Academy of Sciences, China
Yanyan Shen	Chinese Academy of Sciences, China
Haitao Wang	China National Institute of Standardization, China
Yuheng Jia	City University of Hong Kong, China
Chengrun Yang	Cornell University, USA
Lijun Ding	Cornell University, USA
Zenghui Wang	University of South Africa, South Africa
Xianming Ye	University of Pretoria, South Africa
Yanxia Sun	University of Johannesburg, South Africa
Reza Maleklan	Malmö University, Sweden
Xiaozhi Gao	University of Eastern Finland, Finland
Jerry Lin	Western Norway University of Applied Sciences, Norway
Xin Huang	Hong Kong Baptist University, China
Xiaowen Chu	Hong Kong Baptist University, China
Hongtian Chen	University of Alberta, Canada
Gautam Srivastava	Brandon University, Canada
Bay Vo	Ho Chi Minh City University of Technology, Vietnam
Xiuli Zhu	University of Alberta, Canada
Rage Uday Kiran	University of Aizu, Japan
Matin Pirouz Nia	California State University Fresno, USA
Vicente Garcia Diaz	University of Oviedo, Spain
Youcef Djenouri	University of South-Eastern Norway, Norway
Jonathan Wu	University of Windsor, Canada
Yihua Hu	University of York, UK
Saptarshi Sengupta	Murray State University, USA
Wenxiu Xie	City University of Hong Kong, China
Christine Ji	University of Sydney, Australia
Jun Yan	Yidu Cloud, China
Jian Hu	Yidu Cloud, China
Alessandro Bile	Sapienza University of Rome, Italy
Jingjing Cao	Wuhan University of Technology, China
Shi Cheng	Shaanxi Normal University, China
Xianghua Chu	Shenzhen University, China
Valentina Colla	Scuola Superiore S. Anna, Italy
Mohammad Hosein Fazaeli	Amirkabir University of Technology, Iran
Vikas Gupta	LNM Institute of Information Technology, Jaipur
Tianyong Hao	South China Normal University, China

Hongdou He	Yanshan University, China
Wangpeng He	Xidian University, China
Yaqing Hou	Dalian University of Technology, China
Essam Halim Houssein	Minia University, Egypt
Wenkai Hu	China University of Geosciences, China
Lei Huang	Ocean University of China, China
Weijie Huang	University of Hefei, China
Zhong Ji	Tianjin University, China
Qiang Jia	Jiangsu University, China
Yang Kai	Yunnan Minzu University, China
Andreas Kanavos	Ionian University, Greece
Zhao Kang	Southern Illinois University Carbondale, USA
Zouaidia Khouloud	Badji Mokhtar Annaba University, Algeria
Chunshan Li	Harbin Institute of Technology, China
Dongyu Li	Beihang University, China
Kai Liu	Chongqing University, China
Xiaofan Liu	City University of Hong Kong, China
Javier Parra Arnau	Karlsruhe Institute of Technology, Germany
Santwana Sagnika	Kalinga Institute of Industrial Technology, India
Atriya Sen	Rensselaer Polytechnic Institute, USA
Ning Sun	Nankai University, China
Shaoxin Sun	Chongqing University, China
Ankit Thakkar	Nirma University, India
Ye Wang	Chongqing University of Posts and Telecommunications, China
Yong Wang	Sun Yat-sen University, China
Zhanshan Wang	Northeastern University, China
Quanwang Wu	Chongqing University, China
Xiangjun Wu	Henan University, China
Xingtang Wu	Beihang University, China
Zhou Wu	Chongqing University, China
Wun-She Yap	Universiti Tunku Abdul Rahman, Malaysia
Rocco Zaccagnino	University of Salerno, Italy
Kamal Z. Zamli	Universiti Malaysia Pahang, Malaysia
Choujun Zhan	South China Normal University, China
Haijun Zhang	Harbin Institute of Technology, Shenzhen, China
Menghua Zhang	Jinan University, China
Zhao Zhang	Hefei University of Technology, China
Mingbo Zhao	Donghua University, China
Dongliang Zhou	Harbin Institute of Technology, Shenzhen, China
Guo Luo	Nanfang College of Sun Yat-sen University, China
Chengdong Li	Shandong Jianzhu University, China

Yongfeng Zhang   Jinan University, China
Kai Yang   Wuhan Textile University, China
Li Dong   Shenzhen Technology University, China

Contents – Part I

Computational Intelligence, Nature-Inspired Optimizers, and Their Engineering Applications

Contents – Part II

**Natural Language Processing, Knowledge Graphs, Recommender
Systems, and Their Applications**

Neural Computing-Based Fault Diagnosis and Forecasting, Prognostic Management, and Cyber-Physical System Security

Sequence Learning for Spreading Dynamics, Forecasting, and Intelligent Techniques Against Epidemic Spreading (2)

Applications of Data Mining, Machine Learning and Neural Computing in Language Studies

Computational Intelligent Fault Diagnosis and Fault-Tolerant Control, and Their Engineering Applications

Other Neural Computing-Related Topics

Neural Network (NN) Theory, NN-Based Control Systems, Neuro-System Integration and Engineering Applications

ESN-Based Control of Bending Pneumatic Muscle with Asymmetric and Rate-Dependent Hysteresis

Hongge Ru, Jian Huang$^{(\boxtimes)}$, and Bo Wang

Huazhong University of Science and Technology, Wuhan 430074, China
{rhg_hust,huang_jan,wb8517}@hust.edu.cn

Abstract. Soft bending pneumatic muscles (SBPMs) suffer from imprecise control due to the complicated nonlinearity, including the intrinsic rate-dependent and asymmetric hysteresis. In this work, we designed and fabricated a fiber-reinforced soft-bending pneumatic muscle (FSBPM) with high bending efficiency. A real-time visual feedback system was applied to recognize the bending angle of the FSBPM. To tackle the hysteresis problem of the FSBPM, we introduced an inverse hysteresis compensation method (IHCM) for the FSBPM, which combined the inverse hysteresis compensation with the feedback control strategy. The inverse hysteresis model was directly approximated by an echo state network (ESN). Both fixed frequency and variable frequency trajectory tracking experimental results show that compared with the traditional PID control, the proposed method effectively improves the tracking performance.

Keywords: Bending pneumatic muscle · Hysteresis · Echo state network

1 Introduction

Similar to biological muscles in structure, pneumatic muscles are light in weight, with a high output-force-weight ratio and inherent compliance. Usually consisting of a rubber tube and braid mesh wrapped around, they can either extend or contract as the inner pressure changes [1], which depends on the elasticity of the rubber and the reinforcement pattern of the fiber-wrapping. Pneumatic muscles have been applied in various fields [2,3]. To implement bending and twisting motions, the SBPMs are designed based on the specific shape of chambers and different reinforcement ways of the fabric respectively or together [4,5].

In the past few years, safe human-robot interaction has received increasing attention, leading to the fast-growing research interests in SBPMs. However, imprecise sensing and complicated system modeling, which are caused by compliance, make the motion control of the SBPMs difficult [6,7]. Meanwhile, the viscoelasticity of the material of the SBPMs is much more complicated, resulting in severe nonlinear characteristics [6]. Thus, precise modeling and trajectory

© The Author(s), under exclusive license to Springer Nature Singapore Pte Ltd. 2023
H. Zhang et al. (Eds.): NCAA 2023, CCIS 1869, pp. 3–17, 2023.
https://doi.org/10.1007/978-981-99-5844-3_1

tracking control of SBPMs remains a main challenging problem. [8] proposed a finite-element method model for bending in free space. A static model of a flexible pneumatic bending joint was described in [9]. Zheng et al. presented a 3D dynamic model for an arm inspired by octopus anatomy [10]. Several typical octopus arm motions were simulated. She et al. proposed an analysis model which could be applied to predict the shape and displacement of the actuator but was inadequate for motion control due to the lack of dynamics [11]. There are also several empirical models of soft bending robotics [12], whose differential equations are too complicated to be applied in real-time motion control. To the best of our knowledge, there still are limitations for existing mechanisms or empirical models of the SBPMs to be applied in motion control. In this case, data-driven-based control strategies are considered to be the solution to the trajectory tracking problem of the FSBPM.

In recent years, much attention has been paid to hysteresis in many mechanical systems. Hysteresis causes a delayed system response and makes the motion control challenging. To address the hysteresis problem, phenomenological models like the Bouc-Wen (BW) model [13], Duhem model [14], Preisach model [15], PI model [16], Maxwell model [17] and their modified versions [18] have been well studied to characterize hysteresis. A physical model was proposed to fit the magnetic hysteresis curves, using the equations of the Jiles-Atherton model [19]. Considering that feedforward controllers can compensate for hysteresis, the inversion of the hysteresis model is typically applied in a cascade control structure for hysteresis compensation [20,21]. Rate-dependent and asymmetric, the hysteresis of the SBPMs can not be characterized by the general hysteresis model. Although modifications of those models can handle complex hysteresis for different systems, the parameter identification process is complicated and time-consuming, resulting in inconvenience in application. Data-driven models based on artificial neural networks (ANNs) have advantages, e.g., strong capacities of self-learning and self-adapting, and are potential solutions to the motion control of SBPMs. Xu et al. presented a data-driven model of length-pressure hysteresis of a linear PM based on Gaussian mixture models [22]. In [23], back-propagation ANNs were introduced into a modeling method to compensate for the hysteresis behavior of a piezoelectric scanner. Zhang et al. proposed a D-extended un-parallel PI model combined with a deep learning-based convolution neural network [24]. Mirikitani et al. introduced an applicable network for the nonlinear model [25]. The recurrent neural networks (RNNs) and the proposed approach contributed to better generalization and stable numerical performance in the time-series modeling problems. Based on these researches, a feedback control strategy combined with the ESN-based inverse hysteresis compensation method has been proposed for the motion control of the FSBPM.

The ESN, a typical RNN, is characterized by an internal layer called the dynamic reservoir (DR), which is composed of a large number of neurons sparsely connected to each other. An ESN can uniquely maps the temporal past input to the echo states, thus mapping inputs into high-dimensional space and reserve past information. Different from normal RNNs, the ESN only adjusts the set of output weights leading from the internal nodes to the output nodes, which reduces computational complexity significantly due to the sparse connections in the DR [26]. Thus, the ESN is a promising data-driven model method to handle the rate-dependent and asymmetric hysteresis. Consequently, the ESN is chosen to approximate the inversion of the hysteresis model for the FSBPM in our work.

Although the feedforward control scheme is customarily implemented alone, a combination of feedforward and feedback is believed to show better performance in reducing the tracking error related to modeling imperfection, complex dynamics, and uncertainties [21]. The feedforward with feedback control strategies has been proven to be effective [22,27]. In this paper, a data-driven inverse hysteresis model based on the ESN combined with a PID controller was designed for the FSBPM motion control problem.

The main contributions of this paper are: (i) a bending pneumatic muscle with high bending efficiency has been fabricated, and the bending angle of the FSBPM has been recognized in real time by a visual sensor under a mosaic background; (ii) a direct inverse hysteresis model is proposed based on the ESN to handle the asymmetric and rate-dependent hysteresis of the FSBPM; (iii)the inverse hysteresis compensation method has been validated effective through series of experiments.

The rest of this paper is organized as follows. Section 2 gives an introduction of the FSBPM platform. Section 3 includes the hysteresis of the FSBPM, the feedforward compensation based on the inversion of the hysteresis, as well as the feedback control strategy combined with the feedforward compensation. In Sect. 4, a series of experiments is presented, including the motion control experimental results at vary frequencies. In Sect. 5, we can safely arrive at the conclusion that the proposed strategy combining the inverse hysteresis compensation method with the PID controller has satisfactory performance and can effectively mitigate the hysteresis problem.

2   The FSBPM

The existing design and fabrication of soft bending pneumatic muscles are mainly divided into two categories: existing elastomer and silicone rubber molding. In our previous research [28], the fabrication process of the FSBPM and the visual feedback system have been described in detail. In this section, brief introduction to related study is given.

2.1   Structure of the FSBPM

The FSBPM designed in this paper includes the main body and the fiber-glass, as is shown in Fig. 1. The actuator is symmetrical along the cross-section. Chambers are cubic in shape, same in size, and equidistantly arranged. According to [29], the actuators with the rectangle cross profile have higher bending efficiency and are easier to carry out compared with those with the other three shapes. A piece of fiber-glass is embedded at the bottom of the actuator, which is immutable when pressurized but soft.

Fig. 1. The fiber-reinforced soft bending pneumatic muscle and the cross section view

2.2   The Asymmetric Rate-Dependent Hysteresis of the FSBPM

When inflated with the air, the chamber will expand. The bottom layer is nonstretchable because of the fiber-glass, resulting in the bending action of the FSBPM.

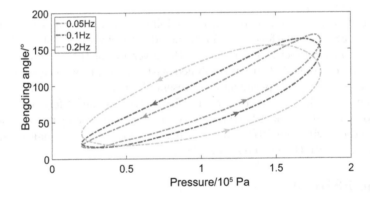

Fig. 2. The open-loop hysteresis loop at the $u(t) = (0.8sin(2\pi ft - \pi/2) + 1)$ excitation

With the help of a series of open-loop excitations for the FSBPM, we have found the hysteresis of the FSBPM is asymmetric and rate-dependent. The general bending angle-pressure hysteresis loops of the FSBPM with different constant excitation frequencies are shown in Fig. 2, where u is the inner pressure. The frequencies(f) are 0.05 Hz, 0.1 Hz and 0.2 Hz respectively. The hysteresis loops with the variable frequency (linearly change from 0.2 Hz to 0.05 Hz) are shown in Fig. 3. The amplitude shrinks, and the loop contracts when the frequency decreases. The right edge of the loop indicates the bending/inflation process when the pressure increases, and the left edge indicates the straightening/deflation process when the pressure decreases. As a result, two neural networks are separately adopted to approximate the inverse hysteresis of the FSBPM.

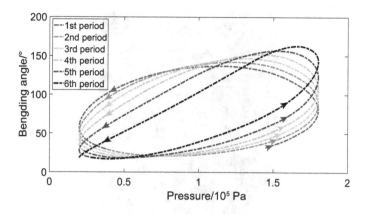

Fig. 3. The open-loop hysteresis loop at the $u(t) = (0.8sin(2\pi(0.05 + 0.005(30 - t)) \cdot (30 - t) - \pi/2) + 1)$ excitation

2.3   Mathematical Description of the FSBPM

Supposing that the body of the FSBPM is approximated to an arc, the state of the FSBPM can be described by the bending angle, which is the central angle corresponding to the arc. As is shown in Fig. 4, the inlet of the FSBPM is fixed on a mosaic background plate. The reference point is located at the beginning of the arc, while the black distal is located at the end. The bending angle is defined as θ_3. In the coordinate system, we have

$$\theta_3 = 2 \cdot \theta_1 = 2 \cdot arctan(\frac{x_c - x_0}{y_c - y_0}) \tag{1}$$

In this case, we only need to locate the center of the black distal, *i.e.*, (x_c, y_c).

3   Feedback Control Strategy Combined with the Feedforward Compensation

In our research, the direct approximation of the inverse hysteresis model is proposed based on the ESN. Then, a feedback control strategy combined with the feedforward compensation is proposed for the motion control of the FSBPM.

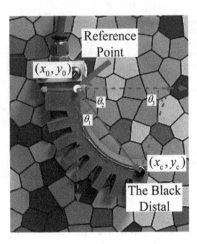

Fig. 4. Principle of visual feedback

3.1   Inversion of the Hysteresis Based on ESN

The echo state network (ESN) was firstly proposed by Herbert Jaeger [30], which usually employs analog neurons (typically linear), sigmoid or leaky integrator units, and simple sparsely-connected graphs as network topologies. A typical structure of the ESN includes an input unit, a recurrent neural network called the reservoir, and a linear readout output layer that maps states of the reservoir to the output of the ESN. The learning algorithms (typically ridge regression algorithm) of the ESN are computationally efficient and easy to use. The supervised ESN training is carried out by updating states of the reservoir and computing the output weights with the following formulas:

$$\mathbf{s}^T(t+1) = (1-\gamma)f(\mathbf{W}\mathbf{s}^T(t) + \mathbf{W}_{\text{in}}\mathbf{u}^T(t)) + \gamma\mathbf{s}^T(t) \in \mathbb{R}^{N \times 1} \qquad (2)$$

$$\mathbf{s_a}^T = [\mathbf{s}, \mathbf{u}]^T \in \mathbb{R}^{(N+K)\times 1} \qquad (3)$$

$$\mathbf{S} = [\mathbf{s_a}^T(1), \dots, \mathbf{s_a}^T(t), \dots, \mathbf{s_a}^T(N_1)] \in \mathbb{R}^{(N+K)\times N_1} \qquad (4)$$

$$\mathbf{Y}_{\text{target}} = [u(1), \dots, u(t), \dots, u(N_1)] \qquad (5)$$

$$\mathbf{W}_{\text{out}} = \mathbf{Y}_{\text{target}}\mathbf{S}^T\left(\mathbf{S}\mathbf{S}^T + \lambda\mathbf{I}\right)^{-1} \in \mathbb{R}^{1\times(N+K)} \qquad (6)$$

where $\mathbf{s}(t)$ represents the state of the reservoir at time t. The initial state of the reservoir is usually set as zero. $\mathbf{W} \in \mathbb{R}^{N \times N}$ is the weight matrix of the reservoir. $\mathbf{u} \in \mathbb{R}^{1 \times K}$ is the input vector, where K is the number of inputs. γ is the leaky rate ranging from 0 to 1. $\mathbf{W}_{\text{in}} \in \mathbb{R}^{N \times K}$ denotes the input weight matrix. $\mathbf{s_a}$ means the dataset of the augmented state $\mathbf{s}$. $\mathbf{S}$ is the time series of

$\mathbf{s_a}$, where N_1 indicates the length of the training set. $\mathbf{Y}_{target} \in \mathbb{R}^{1 \times N_1}$ is the dataset of the desired output of the ESN, which is the teacher matrix. $\mathbf{W}_{out}$ is the weight matrix of output layer. λ is generally set as a small constant. Both $\mathbf{W}$ and $\mathbf{W}_{in}$ are initialized randomly. To ensure the echo state property, $\mathbf{W}$ has to be contractive, which means the spectral radius of $\mathbf{W}$, $\rho(\mathbf{W})$, must be less than 1. In this paper, we calculate the output weights by applying a ridge regression algorithm shown in Eq. 6. Figure 5 shows the scheme for the training of the ESN as the hysteresis compensation. The activation function is chosen as a hyperbolic tangent:

$$f(\cdot) \triangleq \tanh(\cdot) \tag{7}$$

After obtaining the output weights of the ESN, the state update is the same as Eq. 2. The output equation is written as:

$$y(t) = \mathbf{W}_{out}[\mathbf{s}(t), \mathbf{u}(t)]^T \tag{8}$$

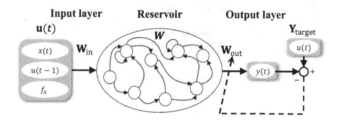

Fig. 5. Block diagram of training ESN for the hysteresis compensation

To approximate the inverse of hysteresis of the FSBPM based on the ESN, a sinusoidal signal with a specific frequency (the same as the excitation frequency) is input to collect the open-loop response of the FSBPM. The output and the input of the FSBPM at the last sampling time ($x(t)$ and $u(t-1)$) are taken as the input of the ESN. The frequency of the sinusoidal signal is also included because of the rate-dependent hysteresis character. The corresponding input, $u(t)$, is defined as the teaching signal of the ESN, as is shown in Fig. 5. Consequently, $\mathbf{u}$ is defined as:

$$\mathbf{u}(t) = [x(t), u(t-1), f_x] \tag{9}$$

where f_x is the frequency of the sinusoidal signal. In this way, the trained ESN characterizes the inverse of the hysteresis, and is applied in the feedforward control strategy as a inverse hysteresis compensator. The training process of the ESN is shown in Algorithm 1.

Based on the proposed inverse hysteresis model, the general open-loop control structure is shown in Fig. 6. The reference signal $x_d(t)$ substitute one of the

---

Algorithm 1. The Training Process of the ESN

1: Randomly initialize the matrices, $\mathbf{W}$, $\mathbf{W}_{in}$.
2: Scaling down the spectral radius of $\mathbf{W}$ to ensure $\rho(\mathbf{W}) < 1$.
3: Update the reservoir states $\mathbf{s}$ according to Equation 2.
4: Collect the augmented states $\mathbf{s}$, the input and the output of the FSBPM for the training set.
5: Calculate $\mathbf{W}_{out}$ by the ridge regression algorithm.

---

input of the ESN, $x(t)$, which results in $x \rightarrow x_d$ under ideal condition. However, the tracking accuracy cannot be guaranteed, considering inaccurate offline identification and unknown disturbances. To alleviate these issues, a closed-loop control strategy is combined with the feedforward compensation.

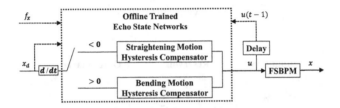

Fig. 6. Block diagram of feedforward compensation

3.2   Feedback Control Strategy Based on Feedforward Compensation

Inverse hysteresis model based control strategies have been reviewed in [21], showing that the tracking performance can be improved in the presence of unexpected disturbances and dynamics of mechanical systems with the feedback scheme. In this study, we combine the feedforward compensation with a PID controller to improve the trajectory tracking accuracy. The proposed control strategy, inverse hysteresis compensation method (for simplicity, it is abbreviated as the IHCM), is shown in Fig. 7. The FFC refers to the feedforward compensation shown in Fig. 6. x_d represents the desired trajectory. x is the actual bending angle of the FSBPM. A general Kalman filter is employed for real-time filtering. e is the tracking error introduced into the particle swarm optimization (PSO) algorithm to optimize the PID parameters.

Based on this control strategy and the visual feedback system, we have conducted a series of experiments to verify the effectiveness of the proposed methodology.

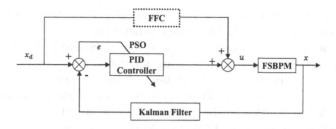

Fig. 7. The Feedback Control Strategy Based on Feedforward Compensation

4 Experiments

enlargethispage12pt In this section, the proposed control strategy is applied to the trajectory tracking control of the FSBPM. The controller is implemented in MATLAB, version 2013b. The visual feedback algorithm is realized by OpenCV.

Fig. 8. The experimental platform

4.1 Experimental Platform

Figure 8 shows the experimental platform. The visual feedback based FSBPM platform consists of a high-speed camera which can work at a maximum frame rate of 110 fps, an electromagnetic proportion valve, an xPC target system, an air pump, a mosaic background for visual recognition, an aluminum frame, and a fiber-reinforced soft bending pneumatic muscle. The FSBPM is fixed on the acrylic board, and the direction of curving is parallel to the board.

Raw images are captured by the high-speed camera and transmitted to the PC, where real-time visual recognition is accomplished. Then, based on the UDP protocol, visual recognition results are sent to the controller in the xPC system, which is designed in Simulink. Through a NI-6025 signal acquisition board, the valve is enabled to drive the FSBPM. Related information on the primary devices is shown in Table 1.

Table 1. The List of Main Devices

Equipment name	Type	Performance index
High-speed camera	MVC610DAM-GE110	Max width: 659*494
		Max frame rate: 110 fps
		Sampling Resolution: 10 bits
Data acquisition card	NI PCI-6025E	16 AI and 2 AO
		32 digital I/O buses
		Sampling rate: 200 KS/s
Electromagnetic valve	ITV1030-211BS	Input: 0–5 V
		Output: 0.005–0.5 MPa
Air compressor	Denair DW35	Regulating range: 0–0.8 MPa
		Input: AC 220 V 50 HZ
Power supply	Q-120DE	Input: AC 220 ±10 V
		Output: ±12 V, ±24 V

4.2 Feedback Control Experiments

For the fixed frequency experiments, the desired trajectories are sinusoidal signals of the same amplitude but different frequencies:

$$x_d = A sin(2\pi f_x t - \frac{\pi}{2}) + B \tag{10}$$

where A is set to 70, and B is 80 according to the workspace of the designed actuator. The frequency f_x is set as 0.1 and 0.2 separately to test the effectiveness of the proposed method. The control sampling time of the valve is set as 1 ms. Meanwhile, the most widely applied traditional PID controller is chosen as a comparative item to verify the superiorities of the IHCM. The maximum absolute error (MAE) and the integral of absolute error (IAE) are introduced as the performance evaluation:

$$\mathbf{MAE} = Max(|x(k) - x_d(k)|_{k=k_0}^{n}) \tag{11}$$

$$\mathbf{IAE} = \frac{1}{n} \sum_{k=k_0}^{n} (|x(k) - x_d(k)|_{k=k_0}^{n}) \tag{12}$$

where k is the total number of single experimental data, and the value of k_0 depends.

For the trajectory tracking experiments, the ESN-based inverse hysteresis compensator is first offline-trained as mentioned above. Then the PSO algorithm is implemented to optimize the PID parameters in both methods. The fitness function of the PSO algorithm can be written as:

$$fit = b\mathbf{MAE} + (1 - b)\mathbf{IAE} \tag{13}$$

where fit represents the fitness, and the optimization goal is to find the minimum of it. b is constant and set to 0.2. Since we focus on the steady-state performance, k_0 is set to the start of the second period.

Based on the experimental setup above, the trajectory tracking experiments are conducted to compare the performance of the IHCM with that of the PID controller. The experimental results are shown in Fig. 9 and Fig. 10. The corresponding evaluation indexes are shown in Table 2. Figure 11 shows the states of the FSBPM with ten images in a control period (5 s) with frequency $f_x = 0.2$ Hz.

To further investigate the proposed control strategy's performance, experiments under variable frequencies of the reference trajectories are conducted.

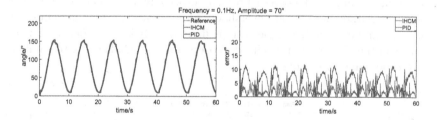

Fig. 9. The tracking performance when f_x is 0.1

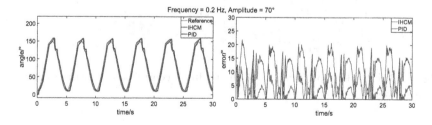

Fig. 10. The tracking performance when f_x is 0.2

Table 2. MAE and IAE of Different Control Method

Frequency	Error	PID	**IHCM**	Improvement
0.1 Hz	MAE	11.8202	**10.4321**	11.74%
	IAE	6.1183	**2.1789**	64.38%
0.2 Hz	MAE	21.7541	**13.8399**	36.38%
	IAE	10.5179	**4.6776**	55.52%

Fig. 11. The states of the FSBPM in a control period (5 s) with frequency $f_x = 0.2$ Hz

Figure 12 and Table 3 show the experimental results, which indicate the effectiveness and reliability of the proposed method. The IHCM shows better performance in tackling the hysteresis problem, especially at the peaks and troughs of the wave. Although there are overshoots and vibrations, the maximum overshoot to maximum bending angle ratio is limited to 10%. Considering that the resonance of the viscoelasticity is hard to overcome, it is challenging to eliminate the overshoots and vibrations.

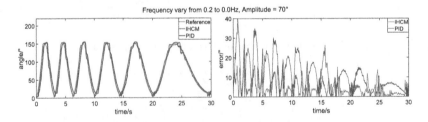

Fig. 12. The tracking performance of the feedback control based on the feedforward compensation

Table 3. MAE and IAE for the variable frequency experiments

Error	PID	**IHCM**	Improvement
MAE	39.6837	**24.8154**	37.46%
IAE	11.1752	**4.5904**	58.92%

5 Conclusions

In this paper, we propose a feedback control strategy combined with the feedforward compensation for a fiber-reinforced soft bending pneumatic muscle based on the visual feedback. The fabricated FSBPM is lightweight, inherently compliant, and capable of implementing curving actions. According to the hysteresis characteristic of the FSBPM, the ESN-based inverse hysteresis compensation method is proposed to realize the trajectory tracking of the FSBPM. Compared with traditional PID control, the performance of the trajectory tracking and the hysteresis has been significantly improved, whereas overshoots and vibrations resulting from the resonance of the material and imprecise approximation can not be eliminated. An adaptive online approximation algorithm may improve the modeling accuracy, while the mechanical design is equally important.

In the future, intelligent soft sensors will be crucial to improvements in the control performance and applications of the SBPMs. Efficient methods or precise modeling that can describe the nonlinear characteristics like hysteresis are worth further investigation.

Acknowledgements. This work is supported by the National Natural Science Foundation of China under Grant U1913207 and by the Program for HUST Academic Frontier Youth Team. The authors would like to thank the support from these foundations.

References

1. Morin, A.H.: Elastic diaphragm. U.S. patent no. 2642091
2. Irshaidat, M., Soufian, M., Al-Ibadi, A., Nefti-Meziani, S.: A novel elbow pneumatic muscle actuator for exoskeleton arm in post-stroke rehabilitation. In: 2019 2nd IEEE International Conference on Soft Robotics (RoboSoft). IEEE (2019). https://doi.org/10.1109/robosoft.2019.8722813
3. Zhou, W., Li, Y.: Modeling and analysis of soft pneumatic actuator with symmetrical chambers used for bionic robotic fish. Soft Rob. **7**(2), 168–178 (2020). https://doi.org/10.1089/soro.2018.0087
4. Al-Fahaam, H., Davis, S., Nefti-Meziani, S.: Power assistive and rehabilitation wearable robot based on pneumatic soft actuators. In: 2016 21st international conference on methods and models in automation and robotics (MMAR), pp. 472–477. IEEE (2016)
5. Hao, Y., et al.: Universal soft pneumatic robotic gripper with variable effective length. In: 2016 35th Chinese Control Conference (CCC), pp. 6109–6114. IEEE (2016)

6. Al-Fahaam, H., Davis, S., Nefti-Meziani, S.: The design and mathematical modelling of novel extensor bending pneumatic artificial muscles (ebpams) for soft exoskeletons. Rob. Auton. Syst. **99**, 63–74 (2018)
7. Chen, W., Xiong, C., Liu, C., Li, P., Chen, Y.: Fabrication and dynamic modeling of bidirectional bending soft actuator integrated with optical waveguide curvature sensor. Soft Rob. **6**(4), 495–506 (2019)
8. Polygerinos, P., et al.: Modeling of soft fiber-reinforced bending actuators. IEEE Trans. Rob. **31**(3), 778–789 (2015)
9. Zhang, L., Bao, G., Yang, Q., Ruan, J., Qi, L.: Static model of flexible pneumatic bending joint. In: 2006 9th International Conference on Control, Automation, Robotics and Vision, pp. 1–5. IEEE (2006)
10. Zheng, T., Branson, D.T., Guglielmino, E., Caldwell, D.G.: A 3D dynamic model for continuum robots inspired by an octopus arm. In: 2011 IEEE International Conference on Robotics and Automation, pp. 3652–3657. IEEE (2011)
11. She, Y., Chen, J., Shi, H., Su, H.J.: Modeling and validation of a novel bending actuator for soft robotics applications. Soft Rob. **3**(2), 71–81 (2016)
12. Aggarwal, A.: An improved parameter estimation and comparison for soft tissue constitutive models containing an exponential function. Biomech. Model. Mechanobiol. **16**(4), 1309–1327 (2017)
13. Martins, S.A.M., Aguirre, L.A.: Sufficient conditions for rate-independent hysteresis in autoregressive identified models. Mech. Syst. Signal Process. **75**, 607–617 (2016)
14. Chen, P., Bai, X.X., Qian, L.J., Choi, S.B.: An approach for hysteresis modeling based on shape function and memory mechanism. IEEE/ASME Trans. Mechatron. **23**(3), 1270–1278 (2018)
15. Nguyen, P.B., Choi, S.B., Song, B.K.: Development of a novel diagonal-weighted preisach model for rate-independent hysteresis. Proc. Inst. Mech. Engineers Part C: J. Mech. Eng. Sci. **231**(5), 961–976 (2017)
16. Xie, S., Mei, J., Liu, H., Wang, Y.: Hysteresis modeling and trajectory tracking control of the pneumatic muscle actuator using modified prandtl-ishlinskii model. Mech. Mach. Theory **120**, 213–224 (2018)
17. Liu, Y., Du, D., Qi, N., Zhao, J.: A distributed parameter maxwell-slip model for the hysteresis in piezoelectric actuators. IEEE Trans. Ind. Electron. **66**(9), 7150–7158 (2018)
18. Xie, S.L., Liu, H.T., Mei, J.P., Gu, G.Y.: Modeling and compensation of asymmetric hysteresis for pneumatic artificial muscles with a modified generalized prandtl-ishlinskii model. Mechatronics **52**, 49–57 (2018)
19. Pop, N., Caltun, O.: Jiles-atherton magnetic hysteresis parameters identification. Acta Physica Polonica, A. **120**(3), 491–496 (2011)
20. Shakiba, S., Ourak, M., Vander Poorten, E., Ayati, M., Yousefi-Koma, A.: Modeling and compensation of asymmetric rate-dependent hysteresis of a miniature pneumatic artificial muscle-based catheter. Mech. Syst. Signal Process. **154**, 107532 (2021)
21. Sabarianand, D., Karthikeyan, P., Muthuramalingam, T.: A review on control strategies for compensation of hysteresis and creep on piezoelectric actuators based micro systems. Mech. Syst. Signal Process. **140**, 106634 (2020)
22. Xu, J., Xiao, M., Ding, Y.: Modeling and compensation of hysteresis for pneumatic artificial muscles based on gaussian mixture models. Sci. China Technol. Sci. **62**(7), 1094–1102 (2019)

23. Wu, Y., Fang, Y., Ren, X., Lu, H.: Back propagation neural networks based hysteresis modeling and compensation for a piezoelectric scanner. In: 2016 IEEE International Conference on Manipulation, Manufacturing and Measurement on the Nanoscale (3M-NANO), pp. 119–124. IEEE (2016)
24. Zhang, Y., Gao, J., Yang, H., Hao, L.: A novel hysteresis modelling method with improved generalization capability for pneumatic artificial muscles. Smart Mater. Struct. **28**(10), 105014 (2019)
25. Mirikitani, D.T., Nikolaev, N.: Recursive bayesian recurrent neural networks for time-series modeling. IEEE Trans. Neural Netw. **21**(2), 262–274 (2009)
26. Huang, J., Qian, J., Liu, L., Wang, Y., Xiong, C., Ri, S.: Echo state network based predictive control with particle swarm optimization for pneumatic muscle actuator. J. Franklin Inst. **353**(12), 2761–2782 (2016)
27. Zhang, X., et al.: Decentralized adaptive neural approximated inverse control for a class of large-scale nonlinear hysteretic systems with time delays. IEEE Trans. Syst, Man Cybern. Syst. **49**(12), 2424–2437 (2018)
28. Ru, H., Huang, J., Chen, W., Xiong, C.: Modeling and identification of rate-dependent and asymmetric hysteresis of soft bending pneumatic actuator based on evolutionary firefly algorithm. Mech. Mach. Theory **181**, 105169 (2023). https://doi.org/10.1016/j.mechmachtheory.2022.105169
29. Hu, W., Mutlu, R., Li, W., Alici, G.: A structural optimisation method for a soft pneumatic actuator. Robotics **7**(2), 24 (2018)
30. Jaeger, H.: The "echo state" approach to analysing and training recurrent neural networks-with an erratum note. Bonn, Germany: German Natl. Res. Center Inf. Technol. GMD Tech. Rep. **148**(34), 13 (2001)

Image Reconstruction and Recognition of Optical Flow Based on Local Feature Extraction Mechanism of Visual Cortex

Wanyan Lin, Hao Yi, and Xiumin Li[✉]

College of Automation, Chongqing University, Chongqing 400030, China
xmli@cqu.edu.cn

Abstract. Neurons in the medial superior temporal (MSTd) region of the visual cortex of the brain can efficiently recognize the firing patterns from the neurons in the MT region. The process is similar to sparse coding in non-negative matrix decomposition (NMF), and the modular recognition of images can be achieved through synaptic plasticity learning rules. In this paper, a spiking neural network model based on spike-timing-dependent plasticity (STDP) is built to simulate the processing process of the visual cortex region MT-MSTD. Our results show that STDP can perform similar functions as NMF, i.e. generating sparse and linear superimposed output based on local features, so as to accurately reconstruct input stimuli for image reconstruction. Finally, support vector machine is used to achieve image recognition of optical flow inputs in eight directions. Compared with NMF, local feature extraction using STDP does not need to retrain the decision layer during the testing procedure of new optical flow samples, contributing to more efficient recognition of optical flow images.

Keywords: MT-MSTd · STDP · SNN · Optical flow

1 Introduction

In recent years, a wave of artificial intelligence has swept the world, and major industrial countries around the world have seized the historical opportunity of artificial intelligence as a national strategy. Artificial intelligence has been commonly applied in numerous significant fields, including security, big data, voice recognition, virtual reality, automated tasks, etc. It should be underscored that the current achievements in AI are essentially breakthroughs in machine learning, such as relying on deep learning and a large number of samples to train networks that allow better adaptation to data and enable classification predictions. The use of spiking neural network (SNN) to simulate the image processing mechanism of visual cortex is important to improve the classification effect of SNN in image recognition, while revealing the efficient operation mechanism of visual cortex. In addition, sparse coding explains that the visual cortex completes

H. Zhang et al. (Eds.): NCAA 2023, CCIS 1869, pp. 18–32, 2023.
https://doi.org/10.1007/978-981-99-5844-3_2

iterative processing of information with fewer neurons [1], which can encode and process information efficiently.

SNNs are commonly combined with traditionally effective algorithms for good results in applications, yet there is no evidence to support most of them by brain science so far. In 2016, Samsung Advanced Institute of Technology proposed an error backpropagation mechanism that treats neuronal membrane potentials as differentiable signals and temporal discontinuities as noise, which allows the potential of spike to be exploited directly [2]. Meanwhile, in view of the great success of deep learning, some researchers combined spiking neural networks with deep convolutional networks or traditional probabilistic models to build SNNs according to the architecture of deep convolutional networks and transformed the original model parameters into those of SNNs, resulting in high accuracy in image classification recognition. Kheradpisheh SR. proposed a STDP-based rule-based spiking deep convolutional neural network (SDNN) [3] that extracts image features and trains multilayer parameters with the help of multilayer convolutional operations. Fang Tao et al. applied the network parameters from CNN training to the transformed deep SNN, which substantially improvs the classification effect of SNN [4]. Although abundant SNN learning algorithms have been proposed in the existing literature, how information is encoded in biological nervous systems is still unresolved. The disadvantages of the current SNN algorithm include four main aspects: i) slow training speed of learning algorithms; ii) limited application areas; iii) limited ability to solve complex problems; and iiii) the structure of the network is single and the utilization of the information contained in the images is low; the network structure is single and the utilization of the information contained in the image is not high.

In the field of computer vision, algorithms designed by borrowing from biological vision have been widely utilized in the existing literature. In complex optical flow stimuli obtained from the retina, neurons in the MSTd region tend to respond to stimuli with translational [5], rotational, and small-angle deformation components. Numerous experiments have shown that neurons in the medial superior temporal dorsal subdorsal (MSTd) region of macaques possess continuous visual response selectivity to a wide range of motion stimuli, in which one of the most common properties is orientation selectivity [6]. Beyeler M proposed a hypothesis that MSTd efficiently encodes a large range of continuous visual optical flow signals from neurons in the middle temporal region (MT) with a descent process similar to that of NMF. These properties not only scientifically explain the frequent non-intuitive response of MSTd neurons (often non-intuitive response), but also reveal the relationship between motion perception and efficient coding in visual cortex [7]. The image processing process in the brain is characterized by modular recognition, which greatly improves the efficiency of image recognition and robustness to displacement, rotation, noise interference, and so on. Recently, a spiking neural network model based on evolved spike time-dependent plasticity and homomorphic synaptic scaling (STDP-H) learning rules for MSTd was proposed in the paper [8]. They demonstrate that the SNN model learns a compressed and effective representation of input patterns similar to those emerging from NMF, resulting in a receptive field similar to that observed in monkeys for MSTd. This SNN model indicates that the STDP-H observed in the nervous system may perform a similar function to NMF with sparsity constraints, which provides a testbed for a mechanistic theory of how MSTd can efficiently encode complex visuomotor patterns to support robust self-motion perception.

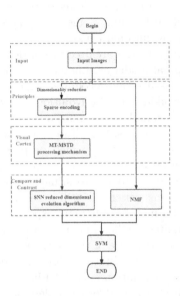

Fig. 1. Experimental algorithm process. The efficient coding process from MT to MSTd in the motor cortex is simulated to achieve sparse coding of the input image and to extract low-dimensional features, thus improving the classification performance.

Inspired by all above observations, in this work, a new SNN computation based on visual cortex image processing mechanisms is constructed by combining interdisciplinary research in neuroscience and artificial intelligence, where SNNs are currently ineffective for image recognition applications or need to be combined with traditional classification methods that are not biologically feasible. The proposed method consists of four main aspects: image dynamic preprocessing, primary visual cortex feature extraction, module recognition, and decision layer design model. The non-negative matrix decomposition algorithm (NMF) and SNN are used for modular recognition of images respectively, and the high-dimensional data are stored in the MT layer to the weights of the MSTd layer in the form of sparse coding, which not only reduces the number of nodes of the network in the layer by modular recognition of images, but also improves the learning efficiency of the decision layer weights training in the later stage, and moreover, the computational complexity is significantly reduced while ensuring the computational accuracy. In addition, for NMF, if new samples are added, it is necessary to re-decompose all the data again, while SNN is able to obtain the discharge response H of the samples individually once the connection weights W are trained by STDP, i.e., there is no need to recompute all samples. Our results show that such SNN model is capable of achieving feature extraction and sparse coding of images and improving the performance of spiking neural networks in image processing, which is vitally important for understanding the visual system of living organisms [9].

The research framework of this paper is shown in Fig. 1. The sparse encoding model contains MT group and MSTd group. And the process of MT-MSTd processing mechanism is similar to the non-negative matrix decomposition NMF. First, the optical flow data from the motor cortex MT is generated. Besides, the readings are encoded as poisson frequencies into discharge sequences as the input of MT group neurons in the SNN model, the connection weights from MT to MSTd are used as the decomposed W matrix, and the discharge frequencies of MSTd group are used as the decomposed H matrix. To enable the $W \times H$ matrix to reconstruct the original optical flow data, Evolutionary Algorithms (EA) are used to adjust the network parameters and evaluate the reconstruction effect at each iteration so that the network can be optimized automatically. Finally, image reconstruction and support vector machine (SVM) classification are used to verify the effectiveness of sparse coding.

2   Methods

2.1   MT Optical Flow Stimulation

Optical flow refers to the movement of patterns in an image and contains information about speed and direction. First, the MT model is used to generate a 15 × 15 pixel visual optical flow. This paper uses Longuet H. [10] proposed visual motion field model on the retina , which assumes that the observer (a point) translates and rotates in the 3D world [11], the points $\overrightarrow{P} = [X, Y, Z]^t$ in the 3D world are projected onto the 2D plane $\overrightarrow{p} = [x, y]^t = f/Z[X, Y]^t$ (that is, the retina). The observer's visual focal length is f and the motion of a three-dimensional point is described as a vector $\dot{\vec{p}} = [\dot{x}, \dot{y}]^t$ in a two-dimensional plane, with a direction of $\tan^{-1}(\dot{y}/\dot{x})$, and a velocity of $\|\dot{\vec{x}}\|$. The vector $\dot{\vec{p}}$ is the sum of the movement of the two parts, the translation vector $\dot{\vec{x}}_T = [\dot{x}_T, \dot{y}_T]^t$ and the rotation vector $\dot{\vec{x}}_R = [\dot{x}_R, \dot{y}_R]^t$:

$$\begin{bmatrix} \dot{x} \\ \dot{y} \end{bmatrix} = \begin{bmatrix} \dot{x}_T \\ \dot{y}_T \end{bmatrix} + \begin{bmatrix} \dot{x}_R \\ \dot{y}_R \end{bmatrix} \tag{1}$$

The translation vector is related to the linear velocity $\vec{v} = [v_x, v_y, v_z]^t$ of the three-dimensional point relative to the observation camera, and the rotation vector is related to the angular velocity $\vec{\omega} = [\omega_x, \omega_y, \omega_z]^t$ of the three-dimensional point relative to the observation camera.

$$\begin{bmatrix} \dot{x}_T \\ \dot{y}_T \end{bmatrix} = \frac{1}{Z} \begin{bmatrix} -f & 0 & x \\ 0 & -f & y \end{bmatrix} \begin{bmatrix} v_x \\ v_y \\ v_z \end{bmatrix} \tag{2}$$

$$\begin{bmatrix} \dot{x}_R \\ \dot{y}_R \end{bmatrix} = \frac{1}{f} \begin{bmatrix} xy & -(f^2 + x^2) & fy \\ (f^2 + y^2) & -xy & -fx \end{bmatrix} \begin{bmatrix} \omega_x \\ \omega_y \\ \omega_z \end{bmatrix} \tag{3}$$

In the simulation, let $f = 0.01, x, y \in [-0.01, 0.01]$, camera angle tilt $30°$ (horizontal line down) to simulate natural view sampled 6000 optical flows moving towards the ground plane and back plane. The linear velocity magnitude of the 3D points is set to a comfortable walking speed $\|\vec{v}\| = \{0.5, 1, 1.5\}m/s$ for humans, the angular velocity magnitude of the 3D points is set to the eye movement speed $\vec{\omega} = \{0, \pm 5, \pm 10\}°/s$ when a person begins to gaze, and the direction of motion is sampled uniformly from all possible 3D directions, with the ground plane and back plane at a distance $d = \{2, 4, 8, 16, 32\}m$ from the observer. In the MT stage [10], each optical flow field is processed by a unit similar to the MT selection function, which selects optical flow in a specific direction θ_{pref} and speed ρ_{pref} at a specific location (x, y). The unit model is:

$$r_{MT}(x, y; \theta_{pref}, \rho_{pref}) = d_{MT}(x, y; \theta_{pref}) s_{MT}(x, y; \rho_{pref}) \tag{4}$$

where d_{MT} is the response of the model element to the direction, and s_{MT} is the response of the model element to the speed. d_{MT} is calculated by computing the difference between the specified spatial position direction $\theta(x, y)$ and the selected direction θ_{pref} of motion:

$$d_{MT}(x, y; \theta_{pref}) = \exp(\sigma_\theta(\cos(\theta(x, y) - \theta_{pref}) - 1)) \tag{5}$$

where, the bandwidth parameter σ_θ is 3, and the resulting tuning width is about $90°$. s_{MT} is a logarithmic gaussian function [12] that is calculated by computing the difference between the spatial position velocity $\rho(x, y)$ and the selected motion velocity ρ_{pref}:

$$s_{MT}(x, y; \rho_{pref}) = \exp(-\frac{\log^2(\frac{\rho(x,y)+s_0}{\rho_{pref}+s_0})}{2\sigma_\rho^2}) \tag{6}$$

In the formula, bandwidth parameters $\sigma_\rho = 1.16$ and speed compensation parameters $s_0 = 0.33$ are selected to prevent the error of log function. The selected parameters of this simulation are $\rho_{pref} = 2°/s$, $\theta_{pref} = \{45, 90, 135, 180, 235, 280, 325, 360\}°/s$. Therefore, the generated model units are $15 \times 15 \times 8 = 1800$, i.e., the original MT optical flow data is 1800 dimensions. In this paper, 6000 samples V of MT optical flow data were generated by computer, where a single sample V_i of dimensional size 1800 contains the response to a set of 8 directions and a velocity.

Both the single sample V_i and the decomposed matrix W contain responses to eight directions and one velocity, the response value is the length of the optical flow vector. Therefore, in the drawing process, the information of 8 directions and speeds is superimposed on a 15 × 15 grid, that is, the vector is transformed into x and y coordinates for superposition, and the coordinates of the endpoint of the vector are obtained. As shown in Fig. 2. Use the MATLAB quiver function to draw the optical flow vector graph.

2.2 Image Reconstruction Using NMF Algorithm

Principle of non-negative matrix decomposition NMF calculation:

For an M-dimensional random vector v, N observations are made, and these observations are noted as $v_i, i = 1, 2, ..., N$. Taking $V = [v_1, v_2, ..., v_N]$, the non-negative matrix decomposition is the decomposition of the original data V into a non-negative $M \times K$-dimensional basis matrix $W = [W_1, W_2, ..., W_N]$ and a $K \times N$-dimensional coefficient matrix $H = [H_1, H_2, ..., H_N]$, as much as possible so that $V \approx WH$, the matrix is represented as $V_i \approx \sum_{j=1}^{K} W_j H_i$. The basis matrix W contains different local features of the sample, and the coefficient matrix H is used to superposition different local features into the original data for reconstruction. In practice, the dimension K affects the effect after $W \times H$ reconstruction, and the best parameters can only be selected by multiple experiments [13].

Fig. 2. The velocity plot of MT optic flow. Five optical flow velocity diagrams are drawn to simulate the projection of 3D coordinates on the retina when the observer moves in the 3D world. It can be seen that the corresponding projection is the projection of points in the 3D world on the retina when the observer moves back, rotates, and translates.

This paper uses an improved NMF algorithm, alternating least-squares algorithm (ALS). The key feature of NMF is its ability to recognize patterns, which decompose data into linear combinations of different local features. $W \times H$ is always approximately equal to V. The decomposed basis under different conditions is uncertain. Least squares non-negative matrix decomposition LS-NMF is an improved NMF algorithm, LS-NMF integrates basis uncertainty into NMF update rules and retains the advantages of NMF, significantly improving the performance of NMF techniques [14].

Firstly, NMF is used to process MT optical flow. NMF uses the standard NNMF function provided by MATLAB for dimensionality reduction. NNMF uses the alternating least squares algorithm to minimize the root mean square residual:

$$D = \frac{\|V - WH\|}{FS} \tag{7}$$

Where, V is MT input matrix, MT each vector is a column of V, F is the number of rows of V, S is the number of columns of V. According to several experiments, the decomposition dimension K is set to 64. Figure 3 shows the decomposition and reconstruction of optical flow by NMF. The reconstructed samples are very close to the original samples, indicating that the original 1800-dimensional information is compressed to 64 dimensions very successfully. It can be seen that the direction of optical flow in each square lattice in the basis matrix W is different, and this form of distribution indicates that the features are separated and contain the local features of the original data, which are reconstructed by multiplying the features in W with the coefficient matrix H. Therefore, the coefficient matrix H contains the weight of each local feature and can represent the original data. In addition, in order to verify the validity of this sparse coding, support vector machine SVM classification will be used later to verify it.

2.3   Image Reconstruction Using the SNN Model

MT optical flow decomposition has been successfully achieved by using NMF, and now the decomposition process similar to NMF is implemented by using SNN based on the STDP learning. The SNN model is shown in Fig. 4. Firstly, the MT model is used to generate a visual optical flow, and the flow value is encoded as a poisson discharge sequence and transmitted to the MT group. Then run the SNN and an evolutionary algorithm is used to ensure that the connection weights W and MST discharge frequency H can reconstruct the image. Finally, the discharge frequency H is classified by SVM to verify the effectiveness of SNN decomposition.

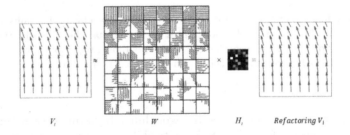

Fig. 3. NMF is applied to decompose and reconstruct the optical flow of MT. The dimension $K = 64$, and the decomposed W is 1800×64, and H is 64×6000. The drawing method of single sample V_i and W has been explained above. H_i is gray value, and the darker the color, the greater the corresponding value. The reconstructed sample is very close to the original sample, indicating that the original 1800-dimensional information is successfully compressed to 64 dimensions.

SNN Model. The neural network is built by using CARLsim, and the neural model uniformly adopts the Izhikevich model. First, the data V is read, and each column vector of V has 1800 information. Therefore, the MT neuron group of 1800 neurons is set, corresponding to each input one by one. The input value is taken as the average parameter, and the pulse train conforming to natural physiology is generated through the poisson process. The larger the value is, the more dense the encoded pulse train is; the smaller the value is, the more sparse the encoded pulse train is. The MST group had 64 neurons, corresponding to dimension K, and the inhibition group had 512 Inh neurons.

The architecture of MT group is $15 \times 15 \times 8$, the architecture of MST group is $8 \times 8 \times 1$, and the architecture of Inh group is $8 \times 8 \times 8$. The connection from MT to MST is gaussian excitatory plastic connection, the connection from MT to Inh is random excitatory plastic connection, and the connection from Inh to MST is random inhibitory plastic connection. The topology can enhance the network's ability to process spatio-temporal information in a more biologically feasible way. The number of synapses of neurons is determined according to the number of established connections, the connection weight of the MT-MSTd group is the decomposed W, and the neuronal firing frequency of the MSTd group is the decomposed H.

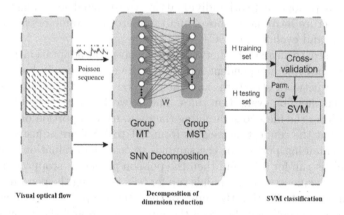

Fig. 4. SNN structure. The architecture of the MT group was $15 \times 15 \times 8$, the MST group was $8 \times 8 \times 1$ and the Inh group was $8 \times 8 \times 8$. MT to MST is a gauss excitatory plastic connection, MT to Inh is a random excitatory plastic connection, and Inh to MST is a random inhibitory plastic connection. In the decision layer, a support vector machine (SVM) is used for image classification.

Network Parameter Adjustment. After obtaining the connection weight W of MT-MSTd and the discharge frequency H of MST group, it is necessary to train the weight W in order to reconstruct the original data, so that the weight W can contain the local characteristics of MT optical flow. In this paper, an evolutionary algorithm is adopted to manage network parameters, and the similarity between the reconstructed $W \times H$ data and the original data is evaluated at each iteration until a high similarity is found so that data reconstruction can be completed.

When using CARLsim to build a network, the firing characteristics of neurons and the tuning of connections are affected by many parameters and therefore require a parameter tuning interface, PTI, to manage them. One of them is ECJ [15], a software interface written in JAVA [16], capable of running evolutionary algorithms EA, also called evolutionary algorithms, including genetic algorithms, evolutionary programming, evolutionary planning and evolutionary strategies, etc. EA uses the biological concept, selection → crossover → variation, to evolve the optimal solution to the problem through N generations of computation. For this procedure, the selected individuals are the given parameters, and the adapted parameters are the variant individuals, and the optimal parameters are obtained after several iterations managed by PTI.

In the above steps, only the ECJ configuration file needs to be written and ECJ automatically executes all steps except the evaluation function of CARLsim. Most of the runtime is spent on CARLsim running on C++/CUDA [17], and ECJ mainly passes parameters and calls CARLsim, thus incurring negligible overhead. communication between the ECJ process and the simulator process it starts is automatic via standard UNIX input/output streams [18]. Among them, the independent network individuals are constructed according to the SNN network structure introduced above, and there are 18 parameters for automatic management and evolution of ECJ.

ECJ is used to invoke the network and evolve the adaptation parameters. At runtime, multiple independent individuals are automatically built according to the above network structure, and then the best adaptation parameters are selected from them by evaluation rules. The network is run by first reading the input data from the file and scrambling the data, using a portion of the samples trained so that the connection weights from MT to MST are adjusted, and then using the remaining portion of the samples to test the evaluation parameters. Each sample is run for 0.5s, and then the poisson process is stopped to generate the discharge sequence and the null is run for 0.5s, allowing the neuron voltage to drop down without affecting the input of the later samples. The evaluation rule is the correlation coefficient between the input samples and the reconstructed.

$$fitness = \frac{\sum_m \sum_n (A_{mn} - \bar{A})(B_{mn} - \bar{B})}{\sqrt{\sum_m \sum_n (A_{mn} - \bar{A}) \sum_m \sum_n (B_{mn} - \bar{B})}} \tag{8}$$

Where $\bar{A}$ and $\bar{B}$ are the column average values of matrix A and matrix B respectively. A is the test sample of input data, and B is the product of MT to MST connection weight matrix W and neuron discharge frequency H of MST group.

The adjustment range of ECJ configuration parameters is shown in Table. 1. Each iteration consists of 15 individual networks, and fitness can reach 72.66% after 100 iterations. The 18 parameters corresponding to the network with the highest fitness are selected as the adaptive parameters.

The effect of SNN reconstructed optical flow is shown in Fig. 5. The optical flow and W plotting superimpose the information on 8 directions and one velocity on top of a 15 × 15 square grid, and H is a grayscale map containing the weight

Table 1. The range of parameters

Parameters	Range
Enhanced and suppressive STDP parameters	[−0.0002,0.004]
Enhanced and suppressive STDP parameters	[5,100]
Dynamically balance the target frequency	[5,20]
MT-MST and MT-Inh connection weights	[0.01,0.5]
Inh-MST connection weights	[0.001,0.5]
MT-MST gaussian connection radius	[0.01,7.5]

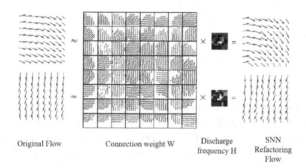

Fig. 5. SNN reconstructs the optic flows.

information of local features. Compared with the above NMF, the reconstruction effect is relatively a little worse, but basically restores the MT optical flow data. And the distribution of W illustrates that the features are separated, and the next comparison can be based on H for classification.

3   Results

To compare the effectiveness of SNN and NMF, support vector machine SVM is used for classification. SVM was originally designed for binary classification, and the common schemes used to deal with multiple classification are: direct method, one-to-many method, one-to-one method, and hierarchical SVMs method [19]. In this paper, we use LIBSVM [20] developed by Professor Chih-Jen Lin of National Taiwan University, which provides interfaces to run on multiple platforms and can solve problems such as classification, regression and distribution estimation. The steps of SVC (Support Vector Classification) classification process are as follows: First, the data set and labels are prepared, the images are compressed into a vector dimension p, and the input matrix is the number of samples n × dimension p. Immediately after, the data are divided into training and test sets and normalized. Subsequently, cross-validation is used to select the best parameters c and g. c is the penalty coefficient and the default is 1; g is the value of γ in the kernel function and the default value is the inverse of the number of features. Then, the best parameters c and g are used to build our SVM model. Finally, the obtained model is used for testing and prediction. As shown in Fig. 6.

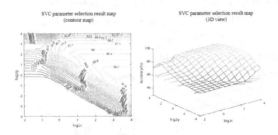

Fig. 6. Cross-validation of SVC.

The optimal parameters c and g are selected for cross-validation, a contour map, and 3D map are drawn, and the local optimal is found through the grid search method. The range of parameter c is $[log2 - 2, log24]$, the range of parameter g is $[log2 - 4, log24]$, and the grid step size is $log20.5$. This is the cross-validation of NMF dimension reduction data, the best c is 11.31, g is 0.71, and the validation accuracy is 92.14%. Then, the model using this parameter is used to predict classification.

To test the effect of SNN reconstruction, eight classifications were selected from a sample of 6,000 MT optical streams that had previously been successfully generated. Flow represents the discharge mode of MT. In addition to translation, there are also rotation and mixing modes. We only select translation mode and select the sample with smaller vector variance after stacking 8 directions in 15 × 15 squares. The selected directions are 8 directions increasing by 45 °C from 0 ° to 360 °. As shown in Fig. 7. According to formula(5), the 225 vectors of the selected sample should meet the angle variance of less than 0.2 and be divided into 8 classes according to the pointing direction after superposition. Each class selected 120, a total of 960 samples.

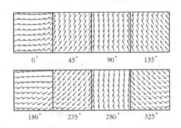

Fig. 7. Optical flow samples with eight different directions.

Some samples were randomly used for training (including those in rotation and mixed mode), mainly to adjust the synaptic weights. Using 960 selected samples as input, the discharge H of the MST group neurons was obtained. H matrix is 960×64 matrix, H is classified by SVM, 800 samples are used to train SVM, and 160 samples are tested. As shown in Fig. 8.

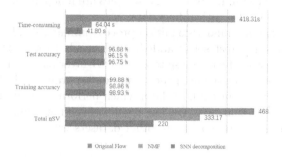

Fig. 8. Image recognition of optical flows based on local feature extraction of NMF or SNN.

The original flow data, NMF-decomposed H and SNN-decomposed discharge frequency were compared on SVM classification, and both NMF-decomposed H and SNN-decomposed discharge frequency H were 64-dimensional. All the data are high in training accuracy and test accuracy, indicating that SVM can effectively differentiate the data. However, the original optical flow data has the largest dimensionality and requires the most total support vectors (Total nSV) and cross-validation time consuming, NMF is the second. SNN decomposition has the same dimensionality K as NMF, but requires less total support vectors and consumes the least time, and performs better than NMF on optical flow data.

Table 2. The before and after of decomposed training set and test set separately.

Handling method	One-time processing of the data set Correlation coefficient with the original data	Separately process the correlation coefficient with the original data	One-time processing SVM classification accuracy	Treat SVM classification accuracy separately
NMF	86.85%	47.68%	96.15%	0%
SNN	67.86%	65.19%	96.75%	93.50%

In addition, when NMF is used, all data sets are decomposed once and then divided into training sets and test sets. Now the training set and test set are decomposed separately, the correlation and classification accuracy of the decomposed matrix and the original data are calculated, and the SNN is compared. As shown in Table 2:

It can be seen that NMF needs to retrain the new testing data before SVM learning. In the case of a separate dimension reduction, the SVM verification accuracy is high, but the accuracy of the tests directly to be 0, which means that the data of the test set and huge difference between the training set and the base matrix W does not contain the local characteristics of the test set. For SNN, the samples can obtain a new response H after the network forward conduction after the training of W. The samples are independent, so SNN is also effective for the new MT optical flow data.

In summary, the modular recognition process similar to NMF is achieved by using SNN, and if the test set is added, NMF needs to retrain W each time, while SNN can classify the response to the new H layer based on the conduction of the multilayer network signal of the network after training W. SNN currently deals with vector type optical flow data, which performs better than NMF, but is limited by the fact that SNN for The results of solving complex problems are not accurate enough to be applied to other datasets yet.

4   Conclusion

4.1   Summary

A biologically feasible SNN model with improved image classification was built by drawing on the image processing mechanisms of the visual cortex. The main research and results of this paper are as follows: motion information in vision goes through MT neurons for specific direction and speed responses, and MSTd area neurons can efficiently encode the recognition of motion stimuli from MT. In this paper, a SNN model is built to implement this process. By using an evolutionary algorithm to automatically optimize the SNN network parameters, the SNN model achieves a modular recognition similar to NMF, and then the effectiveness of this SNN model is verified by image reconstruction and SVM classification.

In decomposing the data using SNN, the optical flow of MT model was firstly generated, and then the SNN was built using CARLsim simulator, and the network parameters were automatically adjusted using evolutionary algorithm. The connection weight of the MT group to the MSTd group in the network is equivalent to W in NMF, and the discharge frequency of the MSTd group is equivalent to H in NMF. The experiments prove that the reconstruction of the MT optical flow using SNN is effective, and the SVM classification accuracy reaches 96.75%, which achieves modular identification. More importantly, the dimensionality reduction using SNN does not need to include all MT optical flow samples and can be processed quickly. After processing the optical flow data using SNN, it improves its performance in SVM classification and, at the same time, solves the problem that NMF needs to re-decompose all the data to get W when adding data.

4.2 Outlook

In this paper, we study spiking neural network, which theoretically has stronger computational performance and lower power consumption, however, the application of spiking neural network in intelligent computing is still relatively small, and there are shortcomings and areas for improvement in this paper.

Firstly, this paper only achieves the effect of processing MT optical flow data, and it needs further research to achieve such effect on other image data sets. Secondly, the MT to MSTd SNN model processes the generated MT optical flow data, which is a vector type and mainly responds to the direction and velocity information of motion. Limited by the time encoding, SNN is not accurate enough in solving the problem, it can not yet reconstruct and accurately restore the image pixel values, and its mechanism is worthy of continued research. Finally, among the supervised learning methods of SNN, in image recognition, the method of training the parameters by first building a deep neural network and then converting it to SNN is highly accurate, but cannot explain the meaning of the conversion. While the unsupervised STDP learning rules are related to memory (synaptic storage), which is more in line with biological properties, and currently STDP is mainly used for unsupervised learning, and for complex tasks a combination of local plus global learning is used, but the classification accuracy is still lower than that of the transformed method, so it is crucial to develop the learning algorithm for SNN.

Ackonwledgments. This paper is supported by STI 2030-Major Project 2021ZD 0201300. The authors would like to give sincere appreciations to Prof. Jeffrey Krichmar and Dr. Kexin Chen for their insightful suggestions and discussions for this work.

References

1. Olshausen, B.A., Field, D.J.: Emergence of simple-cell receptive field properties by learning a sparse code for natural images. Nature **381**(6583), 607–609 (1996). https://doi.org/10.1038/381607a0
2. Jin, Y., Zhang, W., Li, P.: Hybrid macro/micro level backpropagation for training deep spiking neural networks. In: Advances in Neural Information Processing Systems, vol. 31 (2018). https://doi.org/10.48550/arXiv.1805.07866
3. Kheradpisheh, S.R., Ganjtabesh, M., Thorpe, S.J., Masquelier, T.: STDP-based spiking deep convolutional neural networks for object recognition. Neural Netw. **99**, 56–67 (2018). https://doi.org/10.1016/j.neunet.2017.12.005
4. Li, J., Hu, W., Yuan, Y., Huo, H., Fang, T.: Bio-inspired deep spiking neural network for image classification. In: Liu, D., Xie, S., Li, Y., Zhao, D., El-Alfy, ES. (eds.) Neural Information Processing: 24th International Conference, ICONIP 2017, Guangzhou, China, 14–18 November 2017, Proceedings, Part II 24. pp. 294–304. Springer, Cham (2017). https://doi.org/10.1007/978-3-319-70096-0_31
5. Graziano, M., Andersen, R.A., Snowden, R.J.: Tuning of MST neurons to spiral motions. J. Neurosci. **14**(1), 54–67 (1994). https://doi.org/10.1523/JNEUROSCI. 14-01-00054.1994

6. Logan, D.J., Duffy, C.J.: Cortical area MSTD combines visual cues to represent 3-d self-movement. Cerebral Cortex **16**(10), 1494–1507 (2006). https://doi.org/10.1093/cercor/bhj082

7. Beyeler, M., Dutt, N., Krichmar, J.L.: 3d visual response properties of MSTD emerge from an efficient, sparse population code. J. Neurosci. **36**(32), 8399–8415 (2016). https://doi.org/10.1523/JNEUROSCI.0396-16.2016

8. Chen, K., Beyeler, M., Krichmar, J.L.: Cortical motion perception emerges from dimensionality reduction with evolved spike-timing-dependent plasticity rules. J. Neurosci. **42**(30), 5882–5898 (2022). https://doi.org/10.1523/JNEUROSCI.0384-22.2022

9. Beyeler, M., Rounds, E.L., Carlson, K.D., Dutt, N., Krichmar, J.L.: Neural correlates of sparse coding and dimensionality reduction. PLoS Comput. Biol. **15**(6), e1006908 (2019). https://doi.org/10.1371/journal.pcbi.1006908

10. Longuet-Higgins, H.C., Prazdny, K.: The interpretation of a moving retinal image. Proc. R. Soc. London. Ser. B. Biol. Sci. **208**(1173), 385–397 (1980). https://doi.org/10.1098/rspb.1980.0057

11. Britten, K.H., van Wezel, R.J.: Electrical microstimulation of cortical area MST biases heading perception in monkeys. Nat. Neurosci. **1**(1), 59–63 (1998). https://doi.org/10.1038/259

12. Nover, H., Anderson, C.H., DeAngelis, G.C.: A logarithmic, scale-invariant representation of speed in macaque middle temporal area accounts for speed discrimination performance. J. Neurosci. **25**(43), 10049–10060 (2005). https://doi.org/10.1523/JNEUROSCI.1661-05.2005

13. Bidaut, G., Ochs, M.F.: Clutrfree: cluster tree visualization and interpretation. Bioinformatics **20**(16), 2869–2871 (2004). https://doi.org/10.1093/bioinformatics/bth307

14. Wang, G., Kossenkov, A.V., Ochs, M.F.: Ls-NMF: a modified non-negative matrix factorization algorithm utilizing uncertainty estimates. BMC Bioinf. **7**(1), 1–10 (2006). https://doi.org/10.1186/1471-2105-7-175

15. Luke, S.: Ecj then and now. In: Proceedings of the Genetic and Evolutionary Computation Conference Companion, pp. 1223–1230 (2017). https://doi.org/10.1145/3067695.3082467

16. Tang, Z., Zhang, X., Zhang, S.: Robust perceptual image hashing based on ring partition and NMF. IEEE Trans. Knowl. Data Eng. **26**(3), 711–724 (2013). https://doi.org/10.1109/TKDE.2013.45

17. Niedermeier, L., et al.: Carlsim 6: an open source library for large-scale, biologically detailed spiking neural network simulation. In: 2022 International Joint Conference on Neural Networks (IJCNN), pp. 1–10. IEEE (2022). https://doi.org/10.1109/IJCNN55064.2022.9892644

18. Carlson, K.D., Richert, M., Dutt, N., Krichmar, J.L.: Biologically plausible models of homeostasis and STDP: stability and learning in spiking neural networks. In: The 2013 International Joint Conference on Neural Networks (IJCNN), pp. 1–8. IEEE (2013). https://doi.org/10.1109/IJCNN.2013.6706961

19. Carlson, K.D., Nageswaran, J.M., Dutt, N., Krichmar, J.L.: An efficient automated parameter tuning framework for spiking neural networks. Front. Neurosci. **8**, 10 (2014). https://doi.org/10.3389/fnins.2014.00010

20. Chang, C.C., Lin, C.J.: Libsvm: a library for support vector machines. ACM Trans. Intell. Syst. Technol. (TIST) **2**(3), 1–27 (2011). https://doi.org/10.1145/1961189.1961199

Conditional Diffusion Model-Based Data Augmentation for Alzheimer's Prediction

Weiheng Yao[1,2], Yanyan Shen[1], Fred Nicolls[3], and Shu-Qiang Wang[1(✉)] (iD)

[1] Shenzhen Institutes of Advanced Technology, Chinese Academy of Sciences,
Shenzhen 518055, China
sq.wang@siat.ac.cn

[2] Southern University of Science and Technology, Shenzhen 518000, China

[3] University of Cape Town, Cape Town 7701, South Africa
Fred.Nicolls@uct.ac.za

Abstract. Brain imaging plays a crucial role in the study and diagnosis of Alzheimer's disease. However, obtaining brain imaging data is challenging due to the uneven quality of images and the need to consider patient privacy. Consequently, the available data sets are often small, which can limit the effectiveness of analyses and the generalizability of findings. This study proposes a conditional diffusion model-based method for generating brain images of Alzheimer's disease and mild cognitive impairment. The generated data was evaluated for its classification performance by comparing it with datasets containing different proportions of generated data and other data augmentation methods. The performance was visualized to aid the analysis of the experimental results. The analysis of the experimental results showed that the generated data can be used as a reliable data supplement, as it was shown to be beneficial for the classification of Alzheimer's disease. The proposed method offers a promising approach to generating synthetic data for brain imaging research, particularly in neurodegenerative disease diagnosis and treatment.

Keywords: Data augmentation · Conditional Diffusion Model · Alzheimer's disease · Mild cognitive impairment · Brain Image Computing

1 Introduction

Alzheimer's disease (AD) is a widely prevalent neurodegenerative disease characterized by a progressive decline in cognitive abilities that primarily affects the elderly population. The disease is often associated with a range of symptoms including a diminished ability to perform daily activities, impaired decision-making abilities, and reduced socialization skills such as mobility impairment, aphasia, and agnosia. The economic burden of AD is also substantial, with the global cost of the disease estimated to have surpassed USD 1 trillion and expected to double by the year 2030 [1].

H. Zhang et al. (Eds.): NCAA 2023, CCIS 1869, pp. 33–46, 2023.
https://doi.org/10.1007/978-981-99-5844-3_3

Alzheimer's disease (AD) remains an incurable disease [2], underscoring the importance of early and accurate diagnosis coupled with prompt treatment. In this regard, medical imaging techniques have emerged as a crucial diagnostic tool in providing supportive evidence for the identification of AD. Imaging techniques, such as Magnetic Resonance Imaging (MRI) and Positron Emission Tomography (PET) scans, have demonstrated significant utility in facilitating a comprehensive understanding of the structure and function of the brain [3,4], thereby enabling clinicians to identify pathological features such as neuronal degeneration and brain damage indicative of AD. Medical imaging techniques can play a significant role in the early detection and monitoring of Alzheimer's disease, given their ability to assess functional alterations in areas such as brain metabolism and nerve conduction. For instance, positron emission tomography (PET) scans have been shown to detect the extent of beta-amyloid deposits, a major physiological marker of Alzheimer's disease [4], in the brains of affected individuals. By facilitating neuroimaging, physicians can evaluate the severity of the disease and recommend appropriate treatment options, thereby relieving the patient's physical pain and slowing the progression of the disease. In clinical practice, an accurate and efficient analysis of the disease stage, coupled with timely interventions, is essential for optimal patient outcomes.

The increasing availability of medical images and the rapid advancement of medical technology have propelled machine learning [5–7] to become a crucial tool in medical image computing. This is owing to the complexity of medical image processing, which requires analyzing and processing massive amounts of data to extract valuable information. Machine learning techniques have proved beneficial in this regard, facilitating swift and accurate data analysis, detection of diseases and abnormalities, and improving the precision and efficiency of diagnosis. Some commonly applied machine learning approaches in medical image processing include support vector machines, artificial neural networks, and deep learning algorithms. These methods have been widely used in various areas of medical image processing, such as automatic assessment of bone maturity using X-ray films [8,9], automatic identification of cervical spondylosis using DTI images [10], and automatic diagnosis of Alzheimer's disease and mild cognitive impairment using MR images [11–13]. In the future, as machine learning technology continues to develop and improve, it will play an even more important role in the field of medical image processing. The growing prevalence of Alzheimer's disease in the elderly necessitates early diagnosis and treatment. Neuroimaging is widely used in AD research, with generative models playing a key role. These models learn data distributions to generate new data and improve the accuracy and robustness of algorithms by enhancing dataset diversity and reducing overfitting risk.

Related Work. Generative models are being increasingly employed for a variety of tasks in the field of medical imaging, such as brain image alignment, superresolution generation, image segmentation, cross-modality generation, and classification detection. Generative adversarial networks [14], which can be seen as variational inference [15] based generative model, and diffusion models [16] are widely recognized as prominent generative models within the current research landscape. It's crucial to align the images accurately to study the brain's structure and function effectively. Generative models can be used to align these images by learning a mapping between them [17–19]. It can also generate high-resolution images from low-resolution ones, enhancing the details and improving the quality of the images [20, 21]. Different structures and regions can be learned by the generative model, which in turn enables the segmentation of medical images [22–24]. Cross-modality generation involves using generative models to generate images from one modality to another [25–27]. By learning to distinguish between normal and abnormal images based on specific criteria, the generative model can also be used for classification detection [28–30].

Deep learning methods often require large amounts of data to train models to prevent overfitting, however, applying deep learning to medical image analysis suffers from limited training data and class imbalance, and acquiring well-annotated medical data is both expensive and time-consuming. To address this problem, various data enhancement methods have been proposed. Basic enhancement techniques include those that pan, rotate, flip, crop, add noise, etc. to the image to generate an enhanced image. These basic enhancement techniques are the most widely used, with 93 out of 149 studies employing them [31], but they often fail to produce realistic and diverse medical image data. Generative models, on the other hand, can learn features and patterns from a small amount of medical image data, and then generate new and diverse medical image data, thus expanding the original dataset and helping to improve the diagnostic accuracy and efficiency of doctors, as well as the quality and usability of medical image data. Generative models have a GAN-based generative approach, which generates hard-to-fail images through adversarial learning of generators and discriminators. It can effectively solve the medical image sample imbalance problem [32] and generate high-quality images [33–35]. Diffusion models have been particularly popular in recent years, with the advantage of adding random noise to the original data step by step and then learning its inverse process to generate data from the noise, with advantages such as stable training and better quality of the generated samples, and are also widely used in medical image enhancement tasks [36–38]. We propose a novel based on the conditional diffusion model approach to augment brain image data for individuals with normal cognition (CN), mild cognitive impairment (MCI), and Alzheimer's disease (AD). Our proposed method is evaluated by assessing the ability of the model to learn the distinctive features of the real image data using performance metrics such as classification accuracy. The results preserve the critical features required for the accurate classification of the different cognitive states.

2   Method

2.1   Overview

We propose a method for generating brain images for Alzheimer's disease and mild cognitive impairment based on a conditional diffusion model. The method involves training a neural network on a large dataset of brain images to learn the

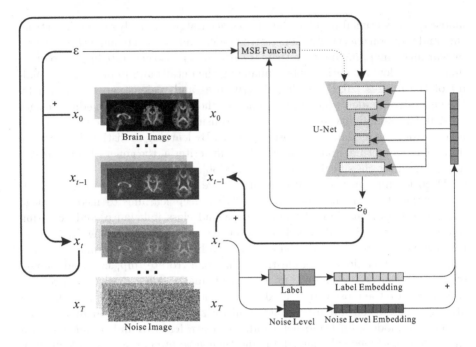

Fig. 1. The thick line in the figure indicates the data stream, the thin line indicates the computational stream in U-Net, and the dashed line indicates the feedback effect of the loss function. Noise level is t and Label is the category l of the input brain image. Only the flow is shown in the figure, and the specific coefficients are ignored.

underlying diffusion process. The trained network can then be used to generate new brain images that capture the characteristic features of Alzheimer's disease and mild cognitive impairment. Our study leverages the power of the Conditional Diffusion-based Probabilistic Model to generate enhanced brain images for data augmentation. The Conditional DDPM framework allows for the incorporation of external conditions into the generation process. Specifically, in our investigation, we introduce disease status labels, namely "CN" for the cognitively normal group, "MCI" for mild cognitive impairment, and "AD" for Alzheimer's disease, as the conditioning variables in the model training. By doing so, we can obtain brain image data that corresponds to each of the disease statuses.

The framework diagram of the model we use is shown in Fig. 1, where the FDP part operates as shown in Sect. 2.2 Eq. (2). the prediction network ϵ_θ of the RDP part is implemented by the U-Net structure. Each layer of the U-Net convolution block has a residual block; the noise level t and the label l are summed and input to the U-Net network after embedding encoding.

2.2 Diffusion Probabilistic Model

The diffusion probability model generates data points x from random noise x_T in two main steps, namely constructing a forward diffusion process (FDP) which obtains diffusion data points x_t at any noise level (time step) t, and training a reverse denoising process (RDP) which generates realistic samples using Markov chains.

For the constructed forward diffusion process (FDP), we add noise gradually starting from x_0. When the noise level is t, the recurrence relationship between images with different noise levels can be expressed by Eq. (1):

$$x_t = \sqrt{\alpha_t}x_{t-1} + \sqrt{\beta_t}\varepsilon_t, \tag{1}$$

where α_t and β_t satisfy $\alpha_t + \beta_t = 1$ and β_t takes a small value close to 0, which is ensured by using a large total number of steps T. The variation of each step of noise added to the original image is relatively small so that clearer results can be obtained.

The formula for x_0 to x_t can be obtained by a recursive formula like Eq. (1) as shown in Eq. (2):

$$x_t = \sqrt{\bar{\alpha}_t}x_0 + \sqrt{1 - \bar{\alpha}_t}\bar{\varepsilon}_t = \sqrt{\bar{\alpha}_t}x_0 + \sqrt{\bar{\beta}_t}\bar{\varepsilon}_t, \tag{2}$$

where $\bar{\alpha}_t = \alpha_0\alpha_1 \cdots \alpha_t$ and $\bar{\beta}_t = 1 - \bar{\alpha}_t$. With the forward diffusion process, it is possible to start deriving the reverse denoising process (RDP) on this basis. The reverse diffusion process is the process from random noise x_T to x_0. Assuming that the reverse recurrence formula is $x_{t-1} = \mu(x_t)$, according to Eq. (1), we can obtain the reverse process $\mu(x_t)$ in the form shown in Eq. (3).

$$\mu(x_t) = \frac{1}{\sqrt{\alpha_t}}\left(x_t - \sqrt{\beta_t}\epsilon(x_t, t)\right) \tag{3}$$

The reverse denoising process requires that x_{t-1} and $\mu(x_t)$ be infinitely close to each other, i.e., the mean square error between them is minimized: the formula $\mathbb{E}\left\|x_{t-1} - \mu(x_t)\right\|^2$ is minimized, and the form obtained after deforming Eq. (1) into this equation is shown in Eq. (4).

$$\mathbb{E}\left\|x_{t-1} - \mu(x_t)\right\|^2 = \frac{\beta_t}{\alpha_t}\mathbb{E}\left\|\varepsilon_t - \epsilon(x_t, t)\right\|^2 \tag{4}$$

The solution objective is reduced to minimize the formula $\mathbb{E}\left\|\varepsilon_t - \epsilon(x_t, t)\right\|^2$. At this point, the equation contains x_t and ε_t two uncertain unknown quantities, which need to be further simplified. Substituting Eq. (2) into Eq. (4) yields the following equation.

$$\mathbb{E}\left\|\varepsilon_t - \epsilon(x_t, t)\right\|^2 = \mathbb{E}\left\|\varepsilon_t - \epsilon\left(\sqrt{\bar{\alpha}_t}x_0 + \sqrt{\alpha_t}\sqrt{\bar{\beta}_{t-1}}\bar{\varepsilon}_{t-1} + \sqrt{\beta_t}\varepsilon_t, t\right)\right\|^2 \tag{5}$$

Adding the noise selects the standard normal distribution $\bar{\varepsilon}_{t-1}, \varepsilon_t \sim \mathcal{N}(0, I)$, so according to $\bar{\alpha}_t = \alpha_0\alpha_1 \cdots \alpha_t$ and $\alpha_t + \beta_t = 1, \sqrt{\alpha_t}\sqrt{\bar{\beta}_{t-1}}\bar{\varepsilon}_{t-1} + \sqrt{\beta_t}\varepsilon_t$ can be expressed as a single random variable $\sqrt{\bar{\beta}_t}\varepsilon$, at which point ε_t can be expressed by ε and $\bar{\varepsilon}_{t-1}$ representations.

But ε and $\bar{\varepsilon}_{t-1}$ are linearly related, introducing the linearly independent $\sqrt{\bar{\beta}_t}\omega = \sqrt{\bar{\beta}_t}\bar{\varepsilon}_{t-1} - \sqrt{\alpha_t}\sqrt{\bar{\beta}_{t-1}}\varepsilon_t$, with $\sqrt{\bar{\beta}_t}\varepsilon = \sqrt{\alpha_t}\sqrt{\bar{\beta}_{t-1}}\bar{\varepsilon}_{t-1} + \sqrt{\bar{\beta}_t}\varepsilon_t$ associates, we can express ε_t as

$$\varepsilon_t = \frac{\sqrt{\bar{\beta}_t}\varepsilon - \sqrt{\alpha_t}\sqrt{\bar{\beta}_{t-1}}\omega}{\sqrt{\bar{\beta}_t}} \tag{6}$$

Substituting Eq. (6) into Eq. (5) and simplifying to obtain the form of our loss function is shown below:

$$\begin{aligned}\mathcal{L}\left(\theta; x_0\right) &= \mathbb{E}\left\|\varepsilon - \frac{\sqrt{\bar{\beta}_t}}{\sqrt{\bar{\beta}_t}}\epsilon\left(\sqrt{\bar{\alpha}_t}x_0 + \sqrt{\bar{\beta}_t}\varepsilon, t\right)\right\|^2 \\ &= \mathbb{E}\left\|\varepsilon - \epsilon_\theta\left(\sqrt{\bar{\alpha}_t}x_0 + \sqrt{\bar{\beta}_t}\varepsilon, t\right)\right\|^2,\end{aligned} \tag{7}$$

where θ denotes the neural network parameters. In the reverse denoising process, according to Eq. (3), we can obtain

$$\begin{aligned}x_{t-1} &= \frac{1}{\sqrt{\alpha_t}}\left(x_t - \frac{\beta_t}{\sqrt{\bar{\beta}_t}}\epsilon_\theta\right) \\ &= \frac{1}{\sqrt{\alpha_t}}\left(x_t - \frac{1-\alpha_t}{\sqrt{1-\bar{\alpha}_t}}\epsilon_\theta\right)\end{aligned} \tag{8}$$

After performing random sampling, it is sufficient to add the noise term $\sqrt{\sigma_t}z, z \sim \mathcal{N}(0, I)$ after Eq. (8). Generally, to keep the FDP and RDP process variance synchronized, take $\sigma_t = \beta_t$. At this point, the complete diffusion probabilistic model is implemented.

2.3 Conditional DDPM

Conditional DDPM is a further extension of DPM. The introduction of conditional control in DPM allows DPM to generate a specific style of image. Its concrete implementation is to introduce conditional control by adding condition c to the noise level t embedding at each step to obtain a new conditional embedding network $\epsilon_\theta(x_t, t, c)$.

In our work, the embedding of disease labels is implemented in a simple way, where the labels are denoted by l and the loss function (7) can be rewritten as

$$\mathcal{L}\left(\theta; x_0\right) = \mathbb{E}\left\|\varepsilon - \epsilon_\theta\left(x_t, t + l\right)\right\|^2, \tag{9}$$

where l is the embedding of the labels as c in the conditional network $\epsilon_\theta(x_t, t, c)$, and l is input to the network from the channel with noise intensity t so that $\epsilon_\theta(x_t, t, c)$ is denoted as $\epsilon_\theta(x_t, t + l)$.

3 Experiments

3.1 Dataset and Experiment Design

Dataset. In this experiment, we used the Alzheimer's Disease Neuroimaging Initiative (ADNI) dataset, which consisted of 362 T1-weighted images displaying tissue morphology and structure. The images were categorized into three groups: the early cognitive impairment (MCI) group had the largest number of samples with 211, the control group (CN) had 87 samples, and the Alzheimer's disease (AD) group had the smallest number of samples with 64. As a result, this experiment involved training with extremely unbalanced samples. Before analysis, the data were preprocessed to obtain slices of the most central regions in three dimensions: the coronal plane, the axial plane, and the sagittal plane.

Implementation Detail. We train on a single Nvidia GeForce RTX 3090 with Adam optimizer, learning rate set to 0.0003, weight decay to 0.001, and training epoch set to 200. The number of sampling steps for the diffusion model is set to 1000. To ensure the robustness of our results, we repeated all experiments three times and took the average.

Metrics. To evaluate the effectiveness of our proposed approach, we employed a variety of classification evaluation metrics, including Accuracy (ACC), Precision (PRE), Recall (REC), Area under the ROC Curve (AUC), and F1-score. These metrics were chosen to provide a comprehensive assessment of the performance of our deep learning model. In addition, we show the generated results and plot the tSNE to demonstrate the distribution of the results, allowing for easy comparison and interpretation of the model's performance.

3.2 Evaluation of Generated Data

In this study, we conducted a comprehensive assessment of the quality of generated data for the classification of Alzheimer's disease (AD) and mild cognitive impairment (MCI). To this end, we utilized six distinct datasets, each comprised of varying combinations of generated and real data. We constructed our classifiers using the downsampling component of the U-Net architecture, combined with a fully connected layer. We evaluated the performance using a series of classification metrics, including accuracy, sensitivity, specificity, the area under the subject working characteristic curve (AUC-ROC, as shown in Fig. 2 (a)), and the F1-score.

Our results (as summarized in Table 1) provide strong evidence for the efficacy of including generated data in the training datasets. Specifically, we observed a significant improvement in the classification metrics for AD and MCI, including accuracy, precision, and recall rates, all of which are essential for reliable and accurate medical diagnosis. Furthermore, the observed improvements in the classification metrics highlight the potential for generated data to mitigate the effects of imbalanced datasets, which are commonly encountered in medical research.

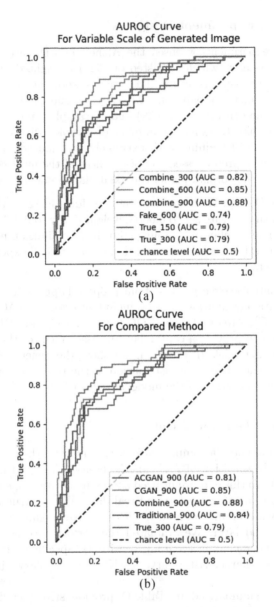

Fig. 2. (a) shows the AUROC curve with different scales of generated data. It is obvious from the graph that the orange line and the green line stand out and have a larger area under the curve. (b) shows the AUROC curve with the comparison method, from which it is obvious that the area under the curve of our method (green line) is better than the other comparison methods with the baseline (Color figure online)

Table 1. Evaluation of classification metrics for AD and MCI diseases on datasets with different amounts of generated and real data

Dataset	# of images		ACC	PRE	REC	F1-score	AUC
	Real	Synthetic					
True_150	150	0	64.52%	68.60%	50.41%	0.5067	0.7856
Combine_300	150	150	63.98%	62.82%	52.61%	0.5337	0.7928
True_300	300	0	67.20%	65.42%	60.31%	0.6079	0.8091
Combine_600	300	300	72.04%	**77.78%**	62.30%	0.6502	0.8420
Combine_900	300	600	**74.73%**	77.28%	**66.52%**	**0.6968**	**0.8590**
Synthetic_600	0	600	63.98%	64.01%	58.41%	0.5834	0.7692

3.3 Evaluation with Compared Methods

In this section, we compare basic augmentation methods and GANs with our method. To ensure a fair comparison, we controlled the number of data obtained from basic enhancement methods to 600, which was consistent with the number of generated data from other comparison methods. The comparison methods for the generative models used were Conditional GAN (CGAN) [39] and Auxiliary Classifier GAN (ACGAN) [40] with the same conditional inputs. To evaluate the performance of the different methods, we used a classifier structure as outlined in Sect. 3.2, and compared various classification metrics. The results of our study are presented in Table 2 and Fig. 2 (b), which show the subject operating characteristic curves (ROCs) for each classification metric. We also draw a radar plot (Fig. 3) to visualize the good and bad classification performance of the different comparison methods.

Table 2. Results of the evaluation of classification metrics in comparison with basic and generative enhancement methods. True_300, which has 300 real samples in Table 1, is still used as the benchmark, and Combine_900 is also consistent with that in Table 1.

Dataset	ACC	PRE	REC	F1-score	AUC
True_300	67.20%	65.42%	60.31%	0.6079	0.8091
Combine_900	**74.73%**	**77.28%**	**66.52%**	**0.6968**	**0.8590**
Traditional_900	68.28%	71.29%	56.18%	0.5712	0.8307
ACGAN	67.74%	67.84%	58.29%	0.5961	0.7950
CGAN	69.89%	72.16%	61.47%	0.6284	0.8361

Based on the results, it is evident that our method outperformed the comparison methods in terms of both accuracy and improvement of sample imbalance.

To present the experimental results clearly and concisely, we employed visualization techniques. The results of our proposed method and the comparison methods, as well as the original image used as a benchmark, are presented in Fig. 4.

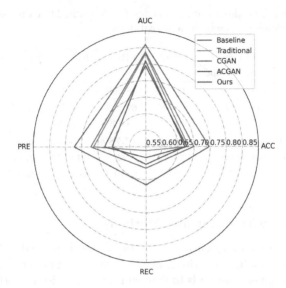

Fig. 3. The radar plot of our method versus the comparison method for classification metrics shows that our method significantly outperforms the comparison method.

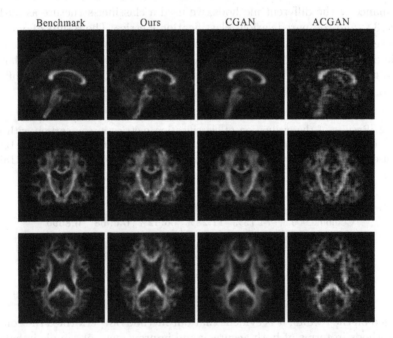

Fig. 4. The visual presentation of the results is shown in Fig. It can be seen that the CGAN results are too smooth and lack details, while the ACGAN results are of poor quality and have a lot of noise. In contrast, our results are of higher quality and have more details.

From the results, we can see that among the comparison methods, the results of ACGAN are of poor quality and noisy. Similarly, the results obtained from CGAN exhibit an over-smoothed appearance, lacking the fine details present in the original image. In contrast, the results generated from our proposed method are remarkably similar to the benchmark image, as they display a high level of detail and exhibit superior image quality.

To visualize the differences between the generated data and the original data distribution, we utilized tSNE plots, as shown in Fig. 5.

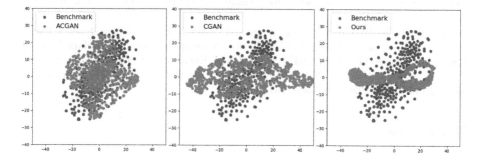

Fig. 5. The tSNE plot of our method versus the comparison method. It can be seen that our generated data has more classification tendency, indicating that the learned features that contribute to classification.

From the tSNE plot, we observed that the original data had unclear classification features, making it challenging to distinguish different classes. Although the ACGAN method provided better coverage of the original data distribution, it only imitated the data and did not capture the underlying disease characteristics. In contrast, our proposed method reflected the classification trend of the original data, and it learned the real features of the disease, as evident from the significant improvement in the classification index.

Therefore, our method not only improves the classification effect but also provides a more accurate representation of the real disease data. The results of our study highlight the importance of understanding the underlying characteristics of medical data when generating synthetic data and demonstrate the effectiveness of our proposed method in improving disease classification.

4 Conclusion

In this research, we propose a new model for generating brain image slices based on a conditional diffusion model that has demonstrated promising results in generating data for specific types of Alzheimer's disease and mild cognitive impairment. The classification performance of our generated data was evaluated through comparison with datasets that had different proportions of generated data and other data enhancement methods. The experimental results showed that our generated data could be used as a reliable data supplement, which significantly improved the classification of Alzheimer's disease. These findings suggest that generative modeling is a viable and effective approach to address data privacy concerns in medical applications. It allows for the generation of synthetic data that can be used to augment existing datasets without violating confidentiality issues. Our study underscores the potential of generative modeling techniques in facilitating data sharing among medical researchers and practitioners, ultimately leading to better diagnoses and treatment of neurodegenerative diseases. Further research is needed to explore the applicability of this approach to other medical domains and to investigate the generalizability of our results.

Acknowledgment. This work was supported by the National Natural Science Foundations of China under Grant 62172403, the Distinguished Young Scholars Fund of Guangdong under Grant 2021B1515020019, the Excellent Young Scholars of Shenzhen under Grant RCYX20200714114641211 and Shenzhen Key Basic Research Project under Grant JCYJ20200109115641762.

References

1. Weidner, W.S., Barbarino, P.: P4-443: the state of the art of dementia research: new frontiers. Alzheimer's Dement. **15**, P1473–P1473 (2019)
2. McKhann, G.M., et al.: The diagnosis of dementia due to alzheimer's disease: Recommendations from the national institute on aging-alzheimer's association workgroups on diagnostic guidelines for alzheimer's disease. Alzheimer's Dement. **7**(3), 263–269 (2011)
3. Jack, C.R., et al.: Medial temporal atrophy on MRI in normal aging and very mild alzheimer's disease. Neurology **49**(3), 786–794 (1997)
4. Nordberg, A.: Pet imaging of amyloid in alzheimer's disease. Lancet Neurol. **3**(9), 519–527 (2004)
5. Zeng, D., Wang, S., Shen, Y., Shi, C.: A GA-based feature selection and parameter optimization for support tucker machine. In: Procedia computer science, 8th International Conference on Advances in Information Technology, vol. 111, pp. 17–23 (2017)
6. Wang, S., Hu, Y., Shen, Y., Li, H.: Classification of diffusion tensor metrics for the diagnosis of a myelopathic cord using machine learning. Int. J. Neural Syst. **28**(02), 1750036 (2018)
7. Wang, S., Chen, Z., You, S., Wang, B., Shen, Y., Lei, B.: Brain stroke lesion segmentation using consistent perception generative adversarial network. Neural Comput. Appl. **34**(11), 8657–8669 (2022)
8. Wang, S., et al.: An ensemble-based densely-connected deep learning system for assessment of skeletal maturity. IEEE Trans. Syst. Man Cybern. Syst. **52**(1), 426–437 (2022)
9. Wang, S., Shen, Y., Zeng, D., Hu, Y.: Bone age assessment using convolutional neural networks. In: 2018 International Conference on Artificial Intelligence and Big Data (ICAIBD), pp. 175–178 (2018)
10. Wang, S.-Q., Li, X., Cui, J.-L., Li, H.-X., Luk, K.D., Hu, Y.: Prediction of myelopathic level in cervical spondylotic myelopathy using diffusion tensor imaging. J. Magn. Reson. Imaging **41**(6), 1682–1688 (2015)

11. Wang, S., Shen, Y., Chen, W., Xiao, T., Hu, J.: Automatic recognition of Mild cognitive impairment from MRI images using expedited convolutional neural networks. In: Lintas, A., Rovetta, S., Verschure, P.F.M.J., Villa, A.E.P. (eds.) ICANN 2017. LNCS, vol. 10613, pp. 373–380. Springer, Cham (2017). https://doi.org/10.1007/978-3-319-68600-4_43

12. Lei, B., Liang, E., Yang, M., Yang, P., Zhou, F., Tan, E.-L., Lei, Y., Liu, C.-M., Wang, T., Xiao, X., et al.: Predicting clinical scores for alzheimer's disease based on joint and deep learning. Exp. Syst. Appl. **187**, 115966 (2022)

13. Wang, S., Wang, H., Shen, Y., Wang, X.: Automatic recognition of mild cognitive impairment and alzheimers disease using ensemble based 3d densely connected convolutional networks. In: 2018 17th IEEE International Conference on Machine Learning and Applications (ICMLA), pp. 517–523 (2018)

14. Goodfellow, I.J., et al.: Generative adversarial nets. In: NIPS (2014)

15. Mo, L., Wang, S.-Q.: A variational approach to nonlinear two-point boundary value problems. Nonlinear Anal. Theor. Methods Appl. **71**(12), 834–838 (2009)

16. Ho, J., Jain, A., Abbeel, P.: Denoising diffusion probabilistic models. Adv. Neural Inf. Process. Syst. **33**, 6840–6851 (2020)

17. Yang, H., Qian, P., Fan, C.: An indirect multimodal image registration and completion method guided by image synthesis. Comput. Math. Methods Med. **2020** (2020). Article ID 2684851

18. Kong, L., Lian, C., Huang, D., Hu, Y., Zhou, Q., et al.: Breaking the dilemma of medical image-to-image translation. Adv. Neural Inf. Process. Syst. **34**, 1964–1978 (2021)

19. Kim, B., Han, I., Ye, J.C.: DiffuseMorph: unsupervised deformable image registration using diffusion model. In: Avidan, S., Brostow, G., Cissé, M., Farinella, G.M., Hassner, T. (eds.) Computer Vision - ECCV 2022. ECCV 2022. LNCS, vol. 13691, pp. 347–364. Springer, Cham (2022). https://doi.org/10.1007/978-3-031-19821-2_20

20. Song, T.-A., Chowdhury, S.R., Yang, F., Dutta, J.: Pet image super-resolution using generative adversarial networks. Neural Netw. **125**, 83–91 (2020)

21. You, S., et al.: Fine perceptive GANs for brain MR image super-resolution in wavelet domain. IEEE Trans. Neural Netw. Learn. Syst. (2022). https://doi.org/10.1109/TNNLS.2022.3153088

22. Ding, Y., et al.: Tostagan: an end-to-end two-stage generative adversarial network for brain tumor segmentation. Neurocomputing **462**, 141–153 (2021)

23. Wolleb, J., Sandkühler, R., Bieder, F., Valmaggia, P., Cattin, P.C.: Diffusion models for implicit image segmentation ensembles, arXiv preprint arXiv: Arxiv-2112.03145 (2021)

24. Pinaya, W.H.L., et al.: Fast unsupervised brain anomaly detection and segmentation with diffusion models. In: MICCAI (2022)

25. Hu, S., Yuan, J., Wang, S.: Cross-modality synthesis from MRI to pet using adversarial u-net with different normalization. In: 2019 International Conference on Medical Imaging Physics and Engineering (ICMIPE), pp. 1–5 (2019)

26. Hu, S., Lei, B., Wang, S., Wang, Y., Feng, Z., Shen, Y.: Bidirectional mapping generative adversarial networks for brain MR to pet synthesis. IEEE Trans. Med. Imaging **41**(1), 145–157 (2021)

27. Conte, G.M., et al.: Generative adversarial networks to synthesize missing t1 and flair MRI sequences for use in a multisequence brain tumor segmentation model. Radiology **299**(2), 313–323 (2021)

28. Yu, W., Lei, B., Ng, M.K., Cheung, A.C., Shen, Y., Wang, S.: Tensorizing GAN with high-order pooling for alzheimer's disease assessment. IEEE Trans. Neural Netw. Learn. Syst. **33**(9), 4945–4959 (2021)
29. Yu, W., et al.: Morphological feature visualization of alzheimer's disease via multi-directional perception GAN. IEEE Trans. Neural Netw. Learn. Syst. **34**, 4401–4415 (2022)
30. Wolleb, J., Bieder, F., Sandkühler, R., Cattin, P.C.: Diffusion models for medical anomaly detection. In: MICCAI (2022)
31. Chlap, P., Min, H., Vandenberg, N., Dowling, J., Holloway, L., Haworth, A.: A review of medical image data augmentation techniques for deep learning applications. J. Med. Imaging Radiat. Oncol. **65**(5), 545–563 (2021)
32. Hu, S., Yu, W., Chen, Z., Wang, S.: Medical image reconstruction using generative adversarial network for alzheimer disease assessment with class-imbalance problem. In: 2020 IEEE 6th International Conference on Computer and Communications (ICCC), pp. 1323–1327 (2020)
33. Shaul, R., David, I., Shitrit, O., Raviv, T.R.: Subsampled brain MRI reconstruction by generative adversarial neural networks. Med. Image Anal. **65**, 101747 (2020)
34. Dar, S.U., Yurt, M., Karacan, L., Erdem, A., Erdem, E., Çukur, T.: Image synthesis in multi-contrast MRI with conditional generative adversarial networks. IEEE Trans. Med. Imaging **38**(10), 2375–2388 (2019)
35. Luo, Y., et al.: Edge-preserving MRI image synthesis via adversarial network with iterative multi-scale fusion. Neurocomputing **452**, 63–77 (2021)
36. Akrout, M., et al.: Diffusion-based data augmentation for skin disease classification: impact across original medical datasets to fully synthetic images, arXiv preprint arXiv:2301.04802 (2023)
37. Pinaya, W.H.L. et al. Brain imaging generation with latent diffusion models. In: Mukhopadhyay, A., Oksuz, I., Engelhardt, S., Zhu, D., Yuan, Y. (eds.) Deep Generative Models. DGM4MICCAI 2022. LNCS, vol. 13609, pp. 117–126. Springer, Cham (2022). https://doi.org/10.1007/978-3-031-18576-2_12
38. Peng, W., Adeli, E., Zhao, Q., Pohl, K.M.: Generating realistic 3d brain MRIS using a conditional diffusion probabilistic model, arXiv preprint arXiv: Arxiv-2212.08034 (2022)
39. Mirza, M., Osindero, S.: Conditional generative adversarial nets, arXiv preprint arXiv: Arxiv-1411.1784 (2014)
40. Odena, A., Olah, C., Shlens, J.: Conditional image synthesis with auxiliary classifier GANs. In: International Conference on Machine Learning (2016)

Design of Dissolved Oxygen Online Controller Based on Adaptive Dynamic Programming Theory

Yingxing Wan$^{(\boxtimes)}$, Cuili Yang, and Yilong Liang

Beijing University of Technology, Chaoyang District, Beijing, China
wanyx@emails.bjut.edu.cn, clyang5@bjut.edu.cn

Abstract. This paper proposes an optimal control solution for regulating the dissolved oxygen (DO) level in wastewater treatment processes (WWTPs). Our method integrates the Echo State Network (ESN) with online Adaptive Dynamic Programming (ADP) to develop an ESN-ADP controller that adeptly handles the dynamic and nonlinear properties of WWTPs. Compared to the traditional Backpropagation (BP) neural network, which may not be optimal for handling the nonlinear and time-varying dynamics of WWTPs, ESN provides better control performance. By utilizing an online learning approach, our algorithm guarantees that the ESN-ADP controller is both adaptive and convergent in the face of changing conditions. Results of experimental tests show that the proposed ESN-ADP controller outperforms other existing control methods in terms of regulatory performance.

Keywords: online learning · Echo state networks · regularized recursive least squares with forgetting · optimal control

1 Introduction

The scarcity of freshwater resources has led to the construction of many WWTP plants for sewage treatment and improving its reuse capabilities [1,2]. However, the WWTP process is highly complex due to significant fluctuations in water discharge and pollutants, resulting in pronounced nonlinear and dynamic behavior [3,4]. Moreover, the extensively adopted sludge bulking method in WWTPs depends on the concentration of dissolved oxygen (DO), which is critical. If the DO concentration is too high, it can expedite the decomposition of organic matter in wastewater, leading to reduced the effectiveness in flocculation and the adsorption capacity for activated sludge. Conversely, low DO concentrations can hinder organic matter degradation by organisms, leading to sludge expansion. Therefore, achieving precise control of DO concentration remains a significant and challenging issue.

© The Author(s), under exclusive license to Springer Nature Singapore Pte Ltd. 2023
H. Zhang et al. (Eds.): NCAA 2023, CCIS 1869, pp. 47–61, 2023.
https://doi.org/10.1007/978-981-99-5844-3_4

Several conventional control methods have been suggested to regulate dis solved oxygen (DO) concentration, which features Proportional-Integral-Derivative (PID) control [5,6], output-based model predictive control [7], and Open-loop control [7]. In [7], To maintain DO, We have developed a control approach that combines both feedforward-feedback and PID control [8,9], where distinct control rules are established based on varying conditions. Although these methods have uncomplicated designs and are user-friendly, they often feature fixed parameters that do not yield satisfactory performance across different operational situations.

Neural network based controllers have recently been proposed to address the challenges posed by nonlinear and dynamic systems. As a case in point, the literature has suggested the use of controllers based on fuzzy neural networks [10,11] to improve DO control, resulting in better control performance compared to PID schemes [10]. However, the fuzzy information processing used in these schemes is relatively simple, which may limit the control precision [11]. Previous studies [12,13] have introduced a model predictive control (MPC) scheme, incorporating neural network based modeling to achieve advanced dissolved oxygen control. However, accurately modeling WWTP can be challenging due to its nonlinear properties.

The field of modern control theory includes a prominent area of research known as adaptive dynamic programming (ADP) control [14]. Adaptive evaluation algorithms, neural networks, and reinforcement learning are all incorporated into this method. A proposed ADP controller for reducing the cumulative tracking error of a wastewater treatment plant was developed in [14]. The resulting control performance exhibited a significant improvement when compared to a traditional PID controller.

The ADP controller relies heavily on the neural network, as it undergoes a parameter adjustment process similar to the learning process of the ADP controller. However, training the neural network can be challenging in WWTPs due to their dynamic and nonlinear nature. Fortunately, researchers have proposed the use of the echo state network (ESN), a type of recurrent neural network, as a potential solution [15]. The ESN employs a large randomly generated hidden layer, referred to as a reservoir, where the output weight vector, which corresponds to the connection weights between the reservoir and output layer, is the only trainable parameter. In contrast, the remaining weights in the network are randomly selected and remain unchanged throughout the learning process.

Designing effective control systems for wastewater treatment processes is a challenging task due to their dynamic and nonlinear nature. In [17], the ESN-ADP controller was proposed as a solution for DO control [18,19]. By combining the ESN and ADP, this control method is able to effectively address the complex characteristics of WWTPs. The ORLS algorithm is used to update the output weights of each ESN model [20]. However, this approach does not fully utilize the recurrent state of the system, and accumulated errors from past data may affect the control performance.

In the context of regulating dissolved oxygen levels in (WWTP) using online control methods, the ESN-ADP controller has shown potential to address the nonlinear and dynamic characteristics of the system. However, there are still limitations to the accuracy of the control, as the recurrent arrived system state and accumulated error from historical data may impact control effectiveness. To improve control accuracy, this study proposes a novel approach that utilizes the regularization technique to enhance the generalization ability and establishes a technique for evaluating iteratively and adaptively method using the forgetting recursive least squares algorithm (FRRLS-ESN-ADP). The proposed algorithm is proven to be stable, and results from its application in WWTP show its effectiveness in achieving improved control accuracy.

In this paper, we present a brief introduction to ADP and ESN in Sect. 2. Section 3 presents the proposed online training algorithm FRRLS-ESN-ADP, which includes the introduction of regularization technique, the controller's evaluation is performed in an iterative and adaptive manner, and the proof of convergence. Section 4 presents the experimental results and performance comparison of the FRRLS-ESN-ADP controller with other existing control methods. Section 5 summarizes the findings and draws conclusions based on the results presented in the previous sections.

2   Preliminary Knowledge

2.1   Optimal Problem Formulation

Simulating wastewater treatment processes in actual WWTPs is challenging due to their nonlinearities and uncertainties in nitrogen removal. To address this, the BSM1, also known as the Benchmark Simulation Model No. 1, is employed as the platform for the system in this study, with defined control test criteria and evaluation procedures. Figure 2 shows a simplified schematic diagram of BSM1 with the ESN-ADP controller. The system consists of a biochemical reaction tank and a secondary sedimentation tank combined, functioning together to treat the wastewater, divided into five units: The first two modules comprise of a sludge-based reactor and anoxic segments, while the remaining units are equipped with aeration systems. The biochemical reaction tank is used to remove most pollutants, and the resulting wastewater is further purified by the secondary sedimentation tank. Achieving effective control of DO is crucial for achieving optimal nitrogen removal efficiency. The coefficient of oxygen transfer in the fifth unit (KLa5) serves as a key parameter for regulating DO (Fig. 1).

The discrete dynamics of WWTP, which are nonlinear in nature, one way to express this is as follows:

$$\boldsymbol{x}(t+1) = \boldsymbol{F}(\boldsymbol{x}(t), \boldsymbol{u}(t)) \tag{1}$$

The state, represented by $\boldsymbol{x}(t) \in \mathbb{R}^n$, is a key factor in control theory, while the control input, denoted by $\boldsymbol{u}(t) \in \mathbb{R}^m$, is an essential variable for manipulating the system, and the function $\boldsymbol{F}(\cdot)$ is continuous and nonlinear. Assuming that

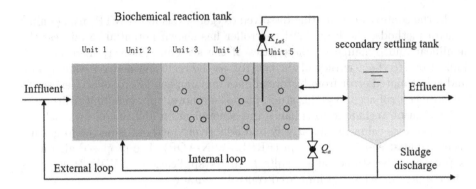

Fig. 1. The main layout of BSM1.

the equilibrium point of system (1) is at $x = 0$ with the control $u = 0$. Suppose Ω is a subset of the n-dimensional Euclidean space $\mathbb{R}^n$ and $\Psi(\Omega)$ is a compact set containing all allowable control laws on Ω. It is assumed that $u(x(t))$ can stabilize the system (1) on the compact set Ω.

Assuming that $J(x(t)) \in \mathbb{R}$ represents the objective function for the optimal regulation problem with a horizon of infinite duration.

$$J(x(t)) = \sum_{i=t}^{\infty} \rho^{i-t} U(x(i), u(x(i))) \qquad (2)$$

where $0 < \rho \leq 1$ represents the discounting factor, $U(x(i), u(x(i))) \in \mathbb{R}$ is the positive definite benefit function which ensures that $U(0, 0) = 0$. The benefit function measures the cost of the control in a single step. The expression (2) is equal to

$$J(x(t)) = U(x(t), u(x(t))) + \rho J(x(t + 1)) \qquad (3)$$

The goal of optimal control is to design a control strategy $u \in \Psi(\Omega)$ that can stabilize the system (1) and minimize the cost function $J(x(t))$. Applying Bellman's principle of optimality, we can obtain the discrete-time HJB equation [?].

$$J^*(x(t)) = \min_{u(x(t))} \{U(x(t), u(x(t))) + \rho J^*(x(t + 1))\} \qquad (4)$$

Using the HJB equation, the optimal control policy $u^*(x(t))$ can be derived as follows:

$$u^*(x(t)) = \arg \min_{u(x(t))} \{U(x(t), u(x(t))) + \rho J^*(x(t + 1))\} \qquad (5)$$

In practice, accurate approximation of the controlled system (1) may not be feasible, which implies that the state $x(k + 1)$ can only be predicted using the current state and control input. This makes it difficult to compute the optimal control variable as presented in Eq. (5). To address this issue, an online approximation algorithm is presented in the following section. The algorithm is designed to generate an effective control strategy using a model-critic-actor architecture based on an ESN and online ADP method.

2.2 Online ESN-ADP Algorithm

In this section, we propose an online ADP algorithm based on Echo State Network (ESN) to approximate the solution of (5). The ESN model is utilized to estimate the nonlinear system states (1), while the critic and actor ESNs are designed to approximate the cost function and control policy, respectively, as shown in Fig. 2. The structure of our ESN-based algorithm is inspired by the DHP algorithm. Note that the two critic networks are the same network and only illustrate the time difference between t and $t+1$. The output weights of the three ESNs are tuned simultaneously to track system trajectories. To implement the proposed ESN-ADP method, an online Recursive Least Squares (RLS) based learning algorithm is used, which incorporates the ℓ_2 regularization technique and forgetting parameter. After the network weight matrices have converged, the optimal control law is computed, and the system is stabilized under the optimal control.

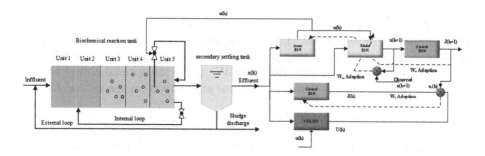

Fig. 2. The schematic diagram of ESN-ADP.

2.3 FRRLS Algorithm for Training ESN-ADP

With known system state $\mathbf{x}(t)$ and control $\mathbf{u}(t)$, the aim of model ESN is to estimate the next state $\mathbf{x}(t+1)$, i.e., the input of model ESN is $[\mathbf{x}(t); \mathbf{u}(t)]$, the network target output is $\mathbf{x}(t+1)$. The model ESN as illustrated in Fig. 3 is consisted by three layers, including the input layer, reservoir and output layer. The number of reservoir nodes is set as h_m. Let $\mathbf{W}_m^{in} \in \mathbb{R}^{(n+m) \times h_m}$, $\mathbf{W}_m^r \in \mathbb{R}^{h_m \times h_m}$, $\mathbf{W}_m \in \mathbb{R}^{h_m \times n}$ stand for the input connection matrix, the internal

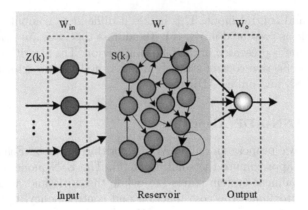

Fig. 3. The architecture of ESN.

weights of reservoir and the output weight matrix, respectively. Then, the input-output updating rule of model ESN is given,

$$s(t+1) = f(\mathbf{W}_m^{in}[\mathbf{x}(t); \mathbf{u}(t)] + \mathbf{W}_m^r s(t)) \tag{6}$$

$$\hat{\mathbf{x}}(t+1) = \mathbf{s}^T(t+1)\mathbf{W}_m(t) \tag{7}$$

where f is the activation function of reservoir, $\mathbf{s} \in \mathbb{R}^{h_m}$ and $\hat{\mathbf{x}}(t+1)$ denote the reservoir state vector and the network output vector, respectively. During training stage of ESN, only $\mathbf{W}_m$ needs to be trained, while $\mathbf{W}_m^{in}$ and $\mathbf{W}_m^r$ are randomly initialized and remain fixed. Let $\mathbf{W}_m^*(t)$ be the optimal output weight matrix at time step t, based on the universal approximation theory, one has

$$\mathbf{x}(t+1) = \mathbf{s}^T(t+1)\mathbf{W}_m^*(t) + \mathbf{v}_m(t+1) \tag{8}$$

where $\mathbf{v}_m(t+1)$ is the reconstruction error. The network estimation error is defined as

$$\tilde{\mathbf{x}}(t+1) = \mathbf{x}(t+1) - \hat{\mathbf{x}}(t+1) \tag{9}$$

To search the optimal value of output weight $\mathbf{W}_m(t)$, the conventional performance measure for ESN online training is to minimize the accumulated network estimation error,

$$\Theta(\mathbf{W}_m(t)) = \sum_{i=0}^{t} \left(\mathbf{x}(i+1) - \mathbf{s}^T(i+1)\mathbf{W}_m(t)\right)^2 \tag{10}$$

In Eq. (10), the timeless of arrived dataset samples is omit. However, in practical applications, the training data is always valid for short time, some outdated date may become a possible source of misleading information, which results in construction error accumulation. To put less weight on outdated date samples, the forgetting factor $0 < \lambda < 1$ is employed. On the other hand, the problem in Eq. (10) can be solved by the online RLS algorithm or its variant. However, if the input signals of ESN are significantly varied over time, their corresponding covariance matrix may become poorly conditioned or ever singular, leading to increased or even unstable estimation error. To reduce the estimation error variance, the ℓ_2 regularization technique is considered. Thus, the performance measure in Eq. (10) is modified as below,

$$\begin{aligned}
\Gamma(\mathbf{W}_m(t)) = {} & \lambda^{t+1}\mathbf{W}_m^T(t)\mathbf{W}_m(t) \\
& + \sum_{i=0}^{t} \lambda^{t-i}\big(\mathbf{x}(i+1) - \mathbf{s}^T(i+1)\mathbf{W}_m(t)\big)^2
\end{aligned} \tag{11}$$

$\mathbf{W}_m^*(t)$ can be computed by minimizing $\Gamma(\mathbf{W}_m(t))$,

$$\mathbf{W}_m^*(t) = \arg\min_{\mathbf{W}_m(t)} \Gamma(\mathbf{W}_m(t)) \tag{12}$$

Now, introduce the diagonal matrix $\Lambda(t+1)$,

$$\Lambda(t+1) = \begin{bmatrix} \lambda^t & & & & \\ & \lambda^{t-1} & & & \\ & & \ddots & & \\ & & & \lambda & \\ & & & & 1 \end{bmatrix} \tag{13}$$

Collect target output into $\mathbf{y}(t)$ and put network states into $\mathbf{H}(t)$,

$$\mathbf{y}(t+1) = \begin{bmatrix} \hat{\mathbf{x}}(1) \\ \hat{\mathbf{x}}(2) \\ \ldots \\ \hat{\mathbf{x}}(t+1) \end{bmatrix} \tag{14}$$

$$\mathbf{H}(t+1) = \begin{bmatrix} \mathbf{s}^T(1) \\ \mathbf{s}^T(2) \\ \ldots \\ \mathbf{s}^T(t+1) \end{bmatrix} \tag{15}$$

Thus, the Eq. (12) is substituted as

$$\begin{aligned}
\mathbf{W}_m(t) = \arg\min_{\mathbf{W}_m(t)} [& \lambda^{t+1}\mathbf{W}_m^T(t)\mathbf{W}_m(t) \\
& + (\mathbf{y}(t+1) - \mathbf{H}(t+1)\mathbf{W}_m(t))^T \, \Lambda(t+1)(\mathbf{y}(t+1) \\
& - \mathbf{H}(t+1)\mathbf{W}_m(t))]
\end{aligned} \tag{16}$$

Actually, the Eq. (16) can be seen as an exponentially weighted least-squares cost function, whose solution is given by

$$\mathbf{W}_m(t) = \left[\lambda^{t+1}\mathbf{I} + \mathbf{H}^T(t+1)\Lambda(t+1)\mathbf{H}(t+1)\right]^{-1} \times \mathbf{H}^T(t+1)\Lambda(t+1)\mathbf{y}(t+1) \tag{17}$$

For the sake of analysis, introduce matrices $\mathbf{P}(t+1)$ and $\mathbf{P}(t)$,

$$\mathbf{P}(t+1) = \left[\lambda^{t+1}\mathbf{I} + \mathbf{H}^T(t+1)\Lambda(t+1)\mathbf{H}(t+1)\right]^{-1} \tag{18}$$

$$\mathbf{P}(t) = \left[\lambda^t\mathbf{I} + \mathbf{H}^T(t)\Lambda(t)\mathbf{H}(t)\right]^{-1} \tag{19}$$

Now, the Eq. (17) can be substituted as

$$\mathbf{W}_m(t) = \mathbf{P}(t+1)\mathbf{H}^T(t+1)\Lambda(t+1)\mathbf{y}(t+1) \tag{20}$$

Since the least squares problem in Eq. (16) is time-updated, $\mathbf{W}_m(t-1)$, the optimal solution at time step $t-1$, can be calculated,

$$\mathbf{W}_m(t-1) = \mathbf{P}(t)\mathbf{H}^T(t)\Lambda(t)\mathbf{y}(t) \tag{21}$$

where $\mathbf{y}(t)$ and $\mathbf{H}(t)$ is the target output and network state matrix at time step $t-1$, respectively. Moreover, these two matrices satisfy

$$\mathbf{y}(t+1) = \begin{bmatrix} \mathbf{y}(t) \\ \hat{\mathbf{x}}(t+1) \end{bmatrix} \tag{22}$$

$$\mathbf{H}(t+1) = \begin{bmatrix} \mathbf{H}(t) \\ \mathbf{s}^T(t+1) \end{bmatrix} \tag{23}$$

The aim of network online training is to obtain the updating rule of $\mathbf{W}_m(t)$. Firstly, let us introduce $\mathbf{K}(t+1) = \mathbf{P}(t+1)^{-1}$,

$$\mathbf{K}(t+1) = \lambda^{t+1}\mathbf{I} + \mathbf{H}^T(t+1)\Lambda(t+1)\mathbf{H}(t+1) \\ = \lambda\mathbf{K}(t) + \mathbf{s}(t+1)\mathbf{s}^T(t+1) \tag{24}$$

Accordingly, one has

$$\mathbf{P}(t+1)^{-1} = \lambda\mathbf{P}(t)^{-1} + \mathbf{s}(t+1)\mathbf{s}^T(t+1) \tag{25}$$

$\mathbf{P}(t+1)$ in Eq. (18) is derived by the Woodbury formula,

$$\mathbf{P}(t+1) = \lambda^{-1}\mathbf{P}(t) - \lambda^{-1}\mathbf{P}(t)\mathbf{s}(t+1) \\ \times \left(\mathbf{I} + \mathbf{s}^T(t+1)\lambda\mathbf{P}(t)\mathbf{s}(t+1)\right)^{-1}\mathbf{s}^T(t+1)\lambda^{-1}\mathbf{P}(t) \tag{26}$$

Then, one obtains

$$\mathbf{H}^T(t+1)\Lambda(t+1)\mathbf{y}(t+1)$$
$$= \mathbf{K}(t+1)\mathbf{W}_m(t-1) - \mathbf{s}(t+1)\mathbf{s}^T(t+1)\mathbf{W}_m(t-1) \qquad (27)$$
$$+ \mathbf{s}(t+1)\hat{\mathbf{x}}(t+1)$$

Based on Eqs. (26) and (27), the updating formulation for $\mathbf{W}_m(t)$ can be derived as

$$\mathbf{W}_m(t) = \mathbf{K}^{-1}(t+1)\mathbf{H}^T(t+1)\Lambda(t+1)\mathbf{y}(t+1)$$
$$= \mathbf{P}(t+1)\mathbf{s}(t+1)\left(\hat{\mathbf{x}}(t+1) - \mathbf{s}^T(t+1)\mathbf{W}_m(t-1)\right) \qquad (28)$$
$$+ \mathbf{W}_m(t-1)$$

By combining Eqs. (26) and (28), it is easily found that $\mathbf{W}_m(t)$ can be recursively updated by $\mathbf{W}_m(t-1)$ and the information of new data $\{\hat{\mathbf{x}}(t+1), \mathbf{s}(t+1)\}$.

Hence, based on the estimated $\mathbf{W}_m(t)$, the system state $\hat{\boldsymbol{x}}(t+1))$ in Eq. (1) can be approximated as

$$\hat{\boldsymbol{x}}(t+1)) = \boldsymbol{F}\left(\hat{\boldsymbol{x}}(t), \boldsymbol{u}(t)\right) = \boldsymbol{s}^T(t+1)\boldsymbol{W}_m(t) \qquad (29)$$

Remark 1. For the matrix inversion in Eq. (17), $O(t^3)$ operations is required, leading to increased storage capacity. By using the recursive calculation process as shown in Eqs. (26) and (28), the complex inverse operation is avoid, which alleviates computational burden to some extent.

Remark 2. The performance measure in Eq. (11) is intuitively appealing. Firstly, the forgetting factor λ is introduced to give more emphasis on new arrived data, which provides adaptation to changing system dynamics. Secondly, the ℓ_2 regularization technique is applied to penalize the value magnitude of weight matrix, such that the network overfitting problem is avoid.

The critic ESN is generated to approximate the cost function $J(\mathbf{x}(t))$, whose estimated value is denoted as $\hat{J}(\mathbf{x}(t))$. For given known input $\mathbf{x}(t)$, the input-output formulation of critic ESN is computed as

$$\mathbf{s}_c(t+1) = f(\mathbf{W}_c^{in}\mathbf{x}(t) + \mathbf{W}_c^r\mathbf{s}_c(t)) \qquad (30)$$

$$\hat{J}(\mathbf{x}(t)) = \mathbf{s}_c^T(t+1)\mathbf{W}_c(t) \qquad (31)$$

where h_c is the nodes number of reservoir, $\boldsymbol{s}_c \in \mathbb{R}^{h_c}$ is the reservoir state vector, $\mathbf{W}_c^{in} \in \mathbb{R}^{n \times h_c}$ is the input-reservoir weight matrix, $\mathbf{W}_c^r \in \mathbb{R}^{h_c \times h_c}$ is the internal weights of reservoir, and $\mathbf{W}_c \in \mathbb{R}^{h_c}$ is the reservoir-output connection weights. The actor ESN is designed to approximate the optimal control policy. Suppose there are h_a nodes in reservoir, the connection weight between input layer and reservoir is denoted as $\mathbf{W}_a^{in} \in \mathbb{R}^{n \times h_a}$, the internal weights of reservoir is set as $\mathbf{W}_a^r \in \mathbb{R}^{h_a \times h_a}$, while the weights between reservoir and output layer is represented as $\mathbf{W}_c \in \mathbb{R}^{h_a \times m}$. The input of actor ESN is $\boldsymbol{x}(t)$, the network output is set as $\hat{\boldsymbol{u}}(\boldsymbol{x}(t))$, which is calculated as

$$\mathbf{s}_a(t) = f(\mathbf{W}_a^{in}\boldsymbol{x}(t) + \mathbf{W}_a^r\mathbf{s}_a(t-1)) \qquad (32)$$

$$\hat{u}(\boldsymbol{x}(t)) = \mathbf{s}_a^T(t)\mathbf{W}_a^*(t-1) \tag{33}$$

where $\boldsymbol{s}_a \in \mathbb{R}^{h_a}$ is the reservoir state vector. Using the same inference process, the online training formula for Critical and Actor networks can be derived:

$$\mathbf{W}_c(t) = \mathbf{W}_c(t-1) \\ + \mathbf{P}_c(t+1)\mathbf{s}_c(t+1)\left(\hat{J}(\mathbf{x}(t)) - \mathbf{s}_c^T(t+1)\mathbf{W}_c(t-1)\right) \tag{34}$$

$$\mathbf{P}_c(t+1) = \lambda^{-1}\mathbf{P}_c(t) - \lambda^{-1}\mathbf{P}_c(t)\mathbf{s}_c(t+1) \\ \times \left(\mathbf{I} + \mathbf{s}_c^T(t+1)\lambda\mathbf{P}_c(t)\mathbf{s}_c(t+1)\right)^{-1}\mathbf{s}_c^T(t+1)\lambda^{-1}\mathbf{P}_c(t) \tag{35}$$

$$\mathbf{W}_a(t) = \mathbf{W}_a(t-1) \\ + \mathbf{P}_a(t+1)\mathbf{s}_a(t+1)\left(\hat{u}(\mathbf{x}(t)) - \mathbf{s}_a^T(t+1)\mathbf{W}_a(t-1)\right) \tag{36}$$

$$\mathbf{P}_a(t+1) = \lambda^{-1}\mathbf{P}_a(t) - \lambda^{-1}\mathbf{P}_a(t)\mathbf{s}_a(t+1) \\ \times \left(\mathbf{I} + \mathbf{s}_a^T(t+1)\lambda\mathbf{P}_a(t)\mathbf{s}_a(t+1)\right)^{-1}\mathbf{s}_a^T(t+1)\lambda^{-1}\mathbf{P}_a(t) \tag{37}$$

3   Convergence of ESN-Based Value Function Approximation

Firstly, the ESN is proved to be available to solve the optimal control problem. Secondly, the convergence of online learning algorithm is studied. Particularly, the upperbounds and lowerbounds of output weight estimation errors are derived.

3.1   Convergence of ESN-Based Value Function Approximation

To carry out the online HDP approach, two ESN are utilized to approximate the value function (4) and control policy (5). In the following, the convergence property of ESN-based value function approximation is discussed. The ESN model has echo state property (ESP). For a selected fixed discount factor $0 < \rho < 1$, two distinct ESN initial states are considered., $\mathbf{s}_1(t)$ and $\mathbf{s}_2(t)$, satisfy the following condition,

$$\| \mathbf{s}_1(t+\Delta t) - \mathbf{s}_2(t+\Delta t) \| \le \rho \| \mathbf{s}_1(t) - \mathbf{s}_2(t) \| \tag{38}$$

Thus, the reservoir state of ESN is asymptotic convergence, which implies $\mathbf{s}(t)$ is bounded.

Theorem 1. *Suppose Assumption 1 holds for a given admissible control $\boldsymbol{u}(\boldsymbol{x}(t))$, then the approximations $\hat{J}(\boldsymbol{x}(t))$ in Eqs. (30) and (31) and $\hat{u}(\boldsymbol{x}(t))$ in Eqs. (32) and (33) converge to their respective limits.*

Proof. Based on Assumption 1, it can be deduced that the states of the model ESN, $s(t)$, are limited. Thus, it follows that the current value of $\hat{J}(\mathbf{x}(t))$ is computed as,

$$\hat{J}(\mathbf{x}(t)) = \mathbf{s}_c^T(t+1)\mathbf{W}_c(t)$$
$$\leq \| \mathbf{s}_c(t+1) \| \| \mathbf{W}_c(t) \| \tag{39}$$

This suggests that the approximation of $J(\mathbf{x}(t))$ using the critic ESN is bounded. Set $s_1(t)$ and $s_2(t)$ be the current state and the desired value, respectively. Thus, $s_2(t) = s_2(t+\Delta t) = J_c$, where J_c is a constant value. Based on Assumption 1, given a constant $0 < \rho_1 < 1$, one has

$$\| \mathbf{s}_c(t+\Delta t) - J_c \| \leq \rho_1 \| \mathbf{s}_c(t) - J_c \| \tag{40}$$

Furthermore, one obtains

$$\| \hat{J}(\mathbf{x}(t+\Delta t)) - J_c \| - \| \hat{J}(\mathbf{x}(t)) - J_c \|$$
$$= \| (\mathbf{s}_c^T(t+\Delta t) - J_c)\mathbf{W}_c(t) \| - \| (\mathbf{s}_c^T(t) - 0)\mathbf{W}_c(t) \| \tag{41}$$
$$\leq \| \mathbf{s}_c(t+\Delta t) - J_c \| \| \mathbf{W}_c(t) \| - \| \mathbf{s}_c(t) - J_c \| \| \mathbf{W}_c(t) \| < 0$$

Thus, the approximated $\hat{J}(\mathbf{x}(t))$ in Eqs. (30) and (31) can eventually converge to a constant value.

By using the similar theoretical analysis from Eqs. (39) to (41), it is easily proved that the approximated $\hat{u}(\boldsymbol{x}(t))$ in Eqs. (32) and (33) is also bounded and eventually converged. □

Lemma 1. *Given the model ESN in (6) and (7) with output weight updating rule in (26) and (27), if their exist constants α, β, and an integer N, which satisfy $0 < \alpha \leq \beta < \infty$, $N \geq h_m$, the following strong persistent excitation (SPE) conditions hold,*

$$\alpha\mathbf{I} \leq \frac{1}{N}\sum_{i=1}^{N}\mathbf{s}(t+i)\mathbf{s}^T(t+i) \leq \beta\mathbf{I}, \quad t > 0 \tag{42}$$

With the forgetting factor $0 < \lambda < 1$, the matrix $\mathbf{P}(t)$ satisfies

$$\frac{\lambda^{N-1}}{1-\lambda}\alpha\mathbf{I} + \lambda^t\left[\mathbf{P}^{-1}(0) - \frac{\alpha}{1-\lambda}\mathbf{I}\right]$$
$$\leq \mathbf{P}^{-1}(t) \leq \frac{N\beta}{1-\lambda}\mathbf{I} + \lambda^t\left[\mathbf{P}^{-1}(0) - \frac{N\beta}{1-\lambda}\mathbf{I}\right], \quad t \geq N \tag{43}$$

Remark 3. The Lemma 1 proves that the ESN is capable of solving the optimal control problem. As the states of ESN gradually converge to the equilibrium point, the obtained output weight matrices of critic ESN and actor ESN are optimal, resulting in the approximated cost function and control policy being optimal as well. This implies that the associated cost function in Eq. (3) attains its minimum value.

4 Experiment and Discussion

In this section, the proposed online ESN-ADP algorithm is applied to realize the optimal control of S_O is BSM1. Data experiment will be given to demonstrate the control performance of the ESN-HDP controller. One is the comparison experiment of the dissolved oxygen stability control of the online ESN-HDP controller based on the FRRLS algorithm; the other is the comparison experiment of the dissolved oxygen tracking control of the online ESN-HDP controller based on the LM algorithm.

For each ESN model, the input weights W^{in} are randomly selected from the range $[-0.1, 0.1]$ without posterior scaling. The reservoir is constructed by the singular value decomposition method, whose singular values are randomly chosen from the range $[0.1, 0.99]$. Furthermore, the sigmoid function $f(x) = 1/(1 + e^{-x})$ is treated as the activation function for reservoir. For model ESN and critic ESN, the initial output weights are selected as $W = W_c = 0.1I$, the initial W_a for actor ESN is set as $0.2I$ which makes sure the initial control is admissible. Some structural parameters of model ESN, critic ESN and actor ESN are shown in Table 1. N is the size of reservoir, Sp is the sparsity of reservoir.

Table 1. Parameter settings of ESN-ADP

Model	Input	Output	N	Sp
Model ESN	$[KL_{a5}(t); S_Ot)]$	$\hat{S}_o(t+1)$	30	0.1
Critic ESN	$S_o(t)$	$\hat{J}(t)$	40	0.1
Actor ESN	$R(t)- S_o(t)$	ΔKL_{a5}	40	0.1

According to the Fig. 4 and Fig. 5, all three controllers can eventually control dissolved oxygen to the set point, but the PID controller has large control

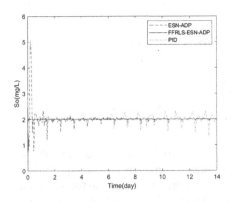

Fig. 4. Comparison of control effect of three controllers

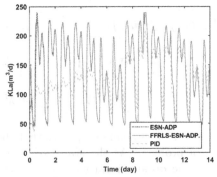

Fig. 5. Curve of control variable

fluctuations and low control accuracy compared to the ESN-ADP controller and the FRRLS-ESN-ADP controller. The ESN-ADP controller has better control performance than the FRRLS-ESN-ADP. At the beginning, the ESN-ADP controller and the FRRLS-ESN-ADP controller fluctuated greatly because the network had insufficient sample size in the early stage of online training. In the later stage of control, the ESN-ADP controller based on FRRLS online algorithm has smaller control fluctuations and higher accuracy for dissolved oxygen.

The Fig. 6 is the change curve of the value function, through the Fig. 6, the value function of the FRRLS-ESN-ADP controller is more stable, which shows that the control smoothness of the FRRLS-ESN-ADP controller has advantages over the ESN-ADP control, but the final consumption is higher than that of the ESN-ADP controller, because a certain amount of computing power will be consumed in the FRRLS algorithm for online update, therefore, the overall consumption will be high. Figures 7, 8 and 9 are the errors change of each network module, and it can be seen through the figure that the error fluctuation of each network module of the ESN-ADP controller is greater than that of the FRRLS-

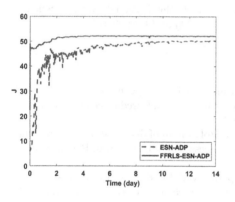

Fig. 6. Value function change curve 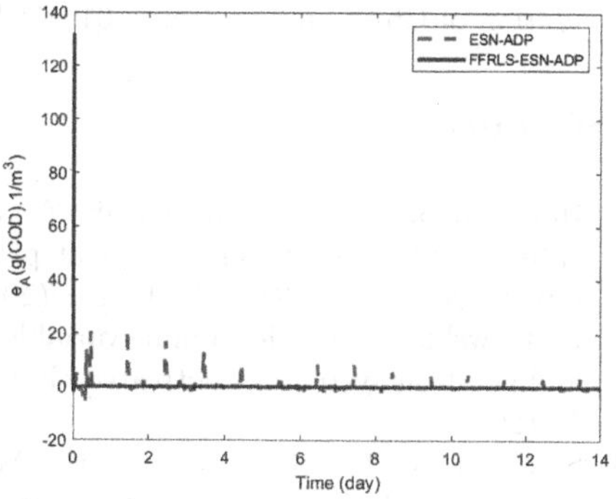   Fig. 7. Errors of Actor networks

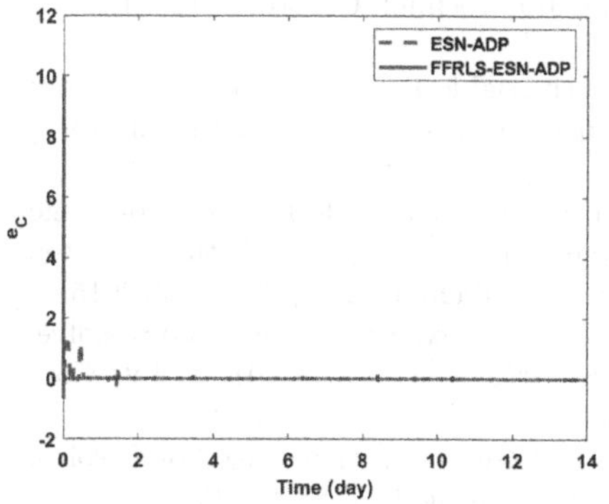

Fig. 8. Errors of Critical network 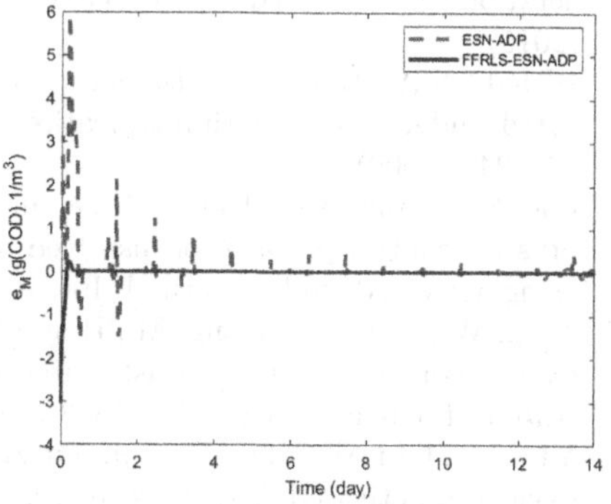   Fig. 9. Errors of Model network

ESN-ADP controller, which shows that the FRRLS algorithm has better anti-interference ability.

5 Conclusion

To address the nonlinear and dynamic characteristics in WWTP, this paper proposes a novel approach that combines ESN and ADP for optimal control of DO. The proposed approach employs the online RLS algorithm to train the output weights of the ESN, and a forgetting factor is incorporated into the cost function to consider the importance of newly arrived data. The resulting ESN-ADP controller outperforms the traditional PID control strategy, achieving more stable and convergent effects. Overall, the combination of ESN and ADP provides a powerful tool for optimizing the DO control in WWTP.

$$\hat{\mathbf{x}}(t + 1) = \mathbf{s}^T(t + 1)\mathbf{W}_m(t) \tag{44}$$

Acknowledgements. This work was supported by the National Key Research and Development Project under Grants 2021ZD0112002, the National Natural Science Foundation of China under Grants 61973010, 62021003 and 61890930-5.

References

1. Holenda, B., Domokos, E., Redey, A., et al.: Dissolved oxygen control of the activated sludge wastewater treatment process using model predictive control. Comput. Chem. Eng. **32**(6), 1270–1278 (2008)
2. Piotrowski, R.: Two-level multivariable control system of dissolved oxygen tracking and aeration system for activated sludge processes. Water Environ. Res. **87**(1), 3–13 (2015)
3. Santin, I., Pedret, C., Vilanova, R.: Applying variable dissolved oxygen set point in a two level hierarchical control structure to a wastewater treatment process. J. Process Control **28**, 40–55 (2015)
4. Piotrowski, R., Skiba, A.: Nonlinear fuzzy control system for dissolved oxygen with aeration system in sequencing batch reactor. Inf. Technol. Control **44**(2), 182–195 (2015)
5. Wahab, N.A., Katebi, R., Balderud, J.: Multivariable PID control design for activated sludge process with nitrification and denitrification. Biochem. Eng. J. **45**(3), 239–248 (2009)
6. Luo, F., Hoang, B.L., Tien, D.N., et al.: Hybrid PI controller design and hedge algebras for control problem of dissolved oxygen in the wastewater treatment system using activated sludge method. Int. Res. J. Eng. Technol. **2**(7), 733–738 (2015)
7. Tang, W., Feng, Q., Wang, M., Hou, Q., Wang, L.: Expert system based dissolved oxygen control in APMP wastewater aerobic treatment process. IEEE Int. Conf. Autom. Logistics **2008**, 1308–1313 (2008)
8. Yang, T., Qiu, W., Ma, Y., et al.: Fuzzy model-based predictive control of dissolved oxygen in activated sludge processes. Neurocomputing **136**, 88–95 (2014)
9. Han, H., Qiao, J.: Nonlinear model-predictive control for industrial processes: an application to wastewater treatment process. IEEE Trans. Ind. Electron. **61**(4), 1970–1982 (2014)

10. Xu, J., Yang, C., Qiao, J.: A novel dissolve oxygen control method based on fuzzy neural network. In: 2017 36th Chinese Control Conference (CCC), pp. 4363–4368 (2017)
11. Fu, W. -T., Qiao, J. -F., Han, G. -T., Meng: Dissolved oxygen control system based on the T-S fuzzy neural network. In: 2015 International Joint Conference on Neural Networks (IJCNN), pp. 1–7 (2015)
12. Werbos, J.: Approximate dynamic programming for real-time control and neural modeling. In: Handbook of Intelligent Control: Neural Fuzzy and Adaptive Approaches. Van Nostrand Reinhold, New York, NY, USA (1992)
13. Yang, Q., Cao, W., Meng, W., Si, J.: Reinforcement-learning-based tracking control of waste water treatment process under realistic system conditions and control performance requirements. IEEE Trans. Syst. Man Cybern. Syst. **52**(8), 5284–5294 (2022)
14. Qiao, J., Wang, Y., Chai, W.: Optimal control based on iterative ADP for wastewater treatment process. J. Beijing Univ. Technol. **44**(2), 200–206 (2018)
15. Al-Tamimi, A., Lewis, F.L., Abu-Khalaf, M.: Discrete-time nonlinear HJB solution using approximate dynamic programming: convergence proof. IEEE Trans. Syst. Man Cybern.-Part B: Cybern. **38**(4), 943–949 (2008)
16. Zhang, H., Liu, C., Su, H., Zhang, K.: Echo state network-based decentralized control of continuous-time nonlinear large-scale interconnected systems. IEEE Trans. Syst. Man Cybern. Syst. **51**(10), 6293–6303 (2021)
17. Bo, Y., Qiao, J.: Heuristic dynamic programming using echo state network for multivariable tracking control of wastewater treatment process. Asian J. Control **17**(5), 1654–1666 (2015)
18. Levant, A.: Exact differentiation of signals with unbounded higher derivatives. In: Proceedings of 45th IEEE Conference on Decision Control, San Diego, CA, USA, pp. 5585–5590 (2006). https://doi.org/10.1109/CDC.2006.377165
19. Fliess, M., Join, C., Sira-Ramirez, H.: Non-linear estimation is easy. Int. J. Model. Ident. Control **4**(1), 12–27 (2008)
20. Ortega, R., Astolfi, A., Bastin, G., Rodriguez, H.: Stabilization of food-chain systems using a port-controlled Hamiltonian description. In: Proceedings of American Control Conference, Chicago, IL, USA, pp. 2245–2249 (2000)

Ascent Guidance for Airbreathing Hypersonic Vehicle Based on Deep Neural Network and Pseudo-spectral Method

Wenzhe Fu, Bo Wang[⊠][iD], Lei Liu, and Yongji Wang

National Key Laboratory of Science and Technology on Multispectral Information Processing, School of Artificial Intelligence and Automation, Huazhong University of Science and Technology, Wuhan 430074, China
{fuwzhust,wb8517,liulei,wangyjch}@hust.edu.cn

Abstract. The hypersonic vehicle has received wide attention because of its high speed and large flight domain. However, the ascent guidance is quite challenging due to its high speed, strong nonlinear dynamics, uncertain parameters, and coupling between the thrust and aerodynamic forces. The traditional model-based guidance laws may have large deviations under uncertain parameters when the model is inaccurate. This paper proposes a data-driven guidance method that consists of deep neural networks and a pseudo-spectral optimization method. The trajectory generation under different uncertain parameters is formulated as an optimal control problem which is solved by the pseudo-spectral method offline. The trajectory database is established based on the vast optimal trajectories and used to train the deep neural network (DNN). Afterward, the ascent guidance law is given by the well-trained DNNs online to lessen the effect of the uncertain parameters and nonlinear characteristics. Meanwhile, a sequential calling strategy is put forward to invoke the multiple DNNs, in which the remaining flight time is designed as the switching variable. The aim is to let the latter networks correct the integral deviation generated by the approximation error of the former networks, and the high-precision terminal states can be achieved. The Mento-Carlo simulations are carried out to give robustness. The results show that the proposed method has the advantage of small terminal deviations and no parameter-tunning.

Keywords: Hypersonic vehicle · Ascent guidance · Deep neural network · Gauss pseudo-spectral method

1 Introduction

Because of the wide range of maneuvering and high speed, hypersonic vehicles have received extensive research and attention [1]. The ascent phase for hypersonic vehicles is quite important because its terminal accuracy has a great impact on the subsequent mission. However, the ascent guidance is quite challenging due

© The Author(s), under exclusive license to Springer Nature Singapore Pte Ltd. 2023
H. Zhang et al. (Eds.): NCAA 2023, CCIS 1869, pp. 62–75, 2023.
https://doi.org/10.1007/978-981-99-5844-3_5

to its high speed, uncertain parameters, strong nonlinear dynamics, and coupling between the thrust and aerodynamic forces.

The ascent guidance method can be divided into three categories [2]. The first method is called nominal trajectory tracking, which is based on the development of stability-related control theory. For example, proportional-integral-derivative (PID) controller [3], finite-time sliding mode (SMC) guidance method [4], and so on. The majority of these methods take advantage of stability. However, these methods may get a large terminal deviation when the dynamic model is inaccurate caused of the uncertain parameters. The second way is the prediction-correction method, which merges the concepts of control and optimization. Such as model predictive control (MPC) [5], adaptive dynamic programming (ADP) [6,7], and so on. The widely used model predictive static programming (MPSP) [8] is one type of variant of MPC, which combines the MPC and the Lagrange multiplier optimization method. They have excellent efficiency in process constraint handling and improving system performance, but bear an expensive online computational burden, while still needing to ensure system stability and optimization feasibility.

The third method is the state-of-the-art intelligent-based guidance method. Benefiting from remarkable advances in computer hardware, a new interest has grown in the literature toward artificial intelligence methods application in the online guidance for hypersonic vehicles in the ascent phase, such as the deep neural network (DNN). Wang et al. [9] proposed an online generation method of ascent trajectory based on feedforward neural networks. The states are used as the input of the networks, and the outputs are the control for endo-atmosphere flight and the costate for vacuum flight. The high precision is guaranteed by optimal control theory when the initial co-states are given by networks. Diao et al. [10] used deep neural networks to predict the nonregular reachable area as the terminal constraints for the boost phase. The boost trajectory is generated by the hp-adaptive pseudospectral method (hpPM). Liu et al. [11] came up with an end-to-end trajectory planning algorithm for multi-stage dual-pulse hypersonic vehicles. The artificial neural network (ANN) is used to fit the range prediction function, and the ascent trajectory is planned by the sequential convex programming (SCP) method.

By summarizing the above literature, we found that there are three common issues about the DNN-based ascent guidance studies. First, the DNN method is widely used in trajectory planning problems as an initial guess for the exact numerical trajectory solver. The exact solution often relies on the efficiency of the numerical solver, and the DNNs have difficulty obtaining accurate results. Second, the DNNs are rarely treated as the guidance controller which copes with uncertain parameters. The network-based tracking guidance method depends on optimal control theory or stability theory. Third, the literature on the application of DNNs in air-breathing hypersonic vehicles is rare. Unlike rocket propellant engines or multi-impulsive engines, which are widely used in powered landing [12,13] or rendezvous mission [14,15], the air-breathing scramjet engine has a strong nonlinear characteristic and a strong coupling between thrust and the aerodynamic force when flying in the endo-atmosphere. The characteristic of nonlinear dynamics and strong coupling makes the application of the DNNs hard in the ascent guidance problem for the air-breathing hypersonic vehicle.

Motivated by the discussions above, the ascent guidance law design for air-breathing hypersonic vehicles based on the DNN method is studied in this paper. To improve the robustness of the ascent guidance method, uncertain parameters are considered in the designing process of the ascent guidance law. The DNN is an effective way to approximate the nonlinear function, therefore, we adopt the DNNs to design the ascent guidance law. However, the approximation error between the ideal controller and the DNNs guidance law will steer the hypersonic vehicle away from the desired trajectory. To obtain satisfactory precision, a continuous invocation strategy is proposed, in which the four well-trained DNNs are called sequentially. The latter networks will correct the integral error induced by the former guidance controller. This process can be viewed as a multi-model controller, which designs the switching rule as the remaining flight time. Finally, an ascent guidance law is obtained. The contributions and distinct features are as follows:

First, the dynamic model is established under the consideration of uncertain parameters. The major uncertain parameters are analyzed and modeled. Second, the optimal control problem under uncertainties is formulated and solved by the pseudo-spectral method offline. The training database is established by sampling the vast trajectories generated by the pseudo-spectral method. Third, several deep feedforward multilayer networks are designed to learn the nonlinear function from the state to the control. Four, to obtain a high-precision result, a continuous invocation strategy is proposed, in which four DNNs are recalled sequentially. The Mento-Carlo simulation results show that the proposed method takes an advantage of small terminal deviations and no parameter-tunning.

2   Dynamic Modeling

The air-breathing hypersonic vehicles fly in the endo-atmosphere. In this paper, the longitudinal motion is considered. Because the flight time of the ascent phase is small, the rotation and eccentricity of the earth are ignored. Then, the point-mass dynamics are as follows:

$$
\begin{cases}
\dot{R} = V \sin \gamma \\
\dot{V} = \dfrac{T \cos \alpha - D}{m} - g \sin \gamma \\
\dot{\gamma} = \dfrac{T \sin \alpha + L}{mV} - \left(\dfrac{g}{V} - \dfrac{V}{R} \right) \cos \gamma \\
\dot{m} = -\dfrac{T}{g_0 I_{sp}},
\end{cases}
\tag{1}
$$

where R is the radius, V is the velocity of the vehicle, γ is the flight path angle, and m is the mass. The above four variable is the state of the vehicle. I_{sp} is the specific impulse. g is the gravity acceleration. The control variable is the angle of attack (AOA) α. All variables above are scalars. L, D and T are lift force, drag force, and thrust force, respectively, which are given by

$$\begin{cases} L = C_L q S_{ref} \\ D = C_D q S_{ref} \\ T = 0.029 \phi_t I_{sp} \rho g_0 V C_T A_e, \end{cases} \quad (2)$$

where q is the dynamic pressure, S_{ref} is the reference area of the aerodynamic force, C_L, C_D, and C_T are the force coefficients, which are functions of the Mach and α. Therefore, they are strongly coupled. ϕ_t is the throttle, ρ is the atmosphere density, and A_e is the capture area of the engine. The detailed parameters are given in the literature [16].

The ascent phase has three major constraints, the dynamic pressure, the heating rate of stagnation point, and the overload.

$$\begin{cases} q = 0.5 \rho V^2 \le q_{max} \\ n = \sqrt{D^2 + L^2}/mg \le n_{max} \\ \dot{Q} = K \sqrt{\rho} V^{3.15} \le \dot{Q}_{max} \end{cases} \quad (3)$$

In an ascent guidance problem for the air-breathing hypersonic vehicle, there are four major uncertain parameters. They are the lift coefficient δC_L, drag coefficient δC_D, the atmosphere density $\delta \rho$, and the thrust coefficient δC_T. The forces in the real dynamic are given by

$$\begin{cases} \widetilde{L} = L(1 + \delta C_L)(1 + \delta \rho) \\ \widetilde{D} = D(1 + \delta C_D)(1 + \delta \rho) \\ \widetilde{T} = L(1 + \delta C_T)(1 + \delta \rho). \end{cases} \quad (4)$$

Thus, the ascent guidance problem for the air-breathing hypersonic vehicle is to find a control angle α online to guide the vehicle to the desired terminal states under uncertainties in Eq. 4.

3   Guidance Law Design

This section gives an ascent guidance method design for hypersonic vehicles. The method consists of a DNN and a Gauss pseudo-spectral method. Due to the effectiveness and efficiency of the pseudo-spectral method in the trajectory planning problem, this method is adopted to generate vast trajectories under uncertainties offline. The trajectory database is established based on the vast optimal trajectories and used to train the DNN. Afterward, the ascent guidance law is given by the well-trained DNN online to lessen the effect of the uncertain parameters.

3.1 Offline Trajectory Database Establishment

The ascent trajectory optimization problem for air-breathing hypersonic vehicle is an optimal control problem, which can be given by

$$P : min \; J = \varphi\left(x_f, u_f\right) + \int_{t_0}^{t_f} L\left(x\left(\tau\right), u\left(\tau\right), \tau\right) d\tau$$
$$s.t. \; \dot{x} = f\left(x, u, \delta p, t\right) \tag{5}$$
$$C\left(x, u, \delta p, t\right) \leq 0$$
$$H\left(x, u, \delta p, t\right) = 0,$$

where t_f, t_0 denote the flight time and the initial time, respectively. u is the control α. $x = [R, V, \gamma, m]^T$ is the state. δp are the uncertain parameters. In the ascent trajectory optimization problem, the cost function usually is to minimize fuel consumption as follows:

$$J = m_0 - m_f \tag{6}$$

The dynamic $\dot{x} = f\left(x, u, p, t\right)$ is given by Eq. 1. The inequality constraint $C\left(x, u, p, t\right) \leq 0$ are given in Eq. 3. To implement the pseudo-spectral optimization method, the time domain is normalized to τ, which is given by

$$\tau = \frac{2t}{t_f - t_0} - \frac{t_f + t_0}{t_f - t_0}, \tag{7}$$

where $\tau \in [-1, 1]$. The Radau pseduo-spectral method is adopted to discrete control and state variables. Let $P_n(\tau)$ be the n-th order Legendre polynomials. The N-th order Legendre-Gauss-Radau (LGR) collocations are given by the root of the $P_{n-1}(\tau) + P_n(\tau)$. Taking the $\tau = -1$ as the discrete node, the control variable and the state variables are given by

$$\begin{cases} x(\tau) \approx \sum_{i=0}^{N} x(\tau_i) L_i(\tau) \\ u(\tau) \approx \sum_{i=0}^{N} u(\tau_i) L_i(\tau), \end{cases} \tag{8}$$

where $L_i(\tau)$ refers to the Lagrange interpolation polynomial, which is given as follows:

$$L_i(\tau) = \prod_{j=0, j \neq i}^{N} \frac{\tau - \tau_j}{\tau_i - \tau_j}, i = 0, 1, \cdots, N \tag{9}$$

Then, the dynamic constrain Eq. 1 is transformed into the constraints at discrete nodes:

$$\sum_{j=0}^{N} D_{ij} x_j - \frac{t_f - t_0}{2} f\left(x_i, u_i, \tau_i, \delta p\right) = 0, \tag{10}$$

where δp_j is the uncertain parameters given in Eq. 4. The $D_{ij} = \dot{L}_j(\tau_j)$ is the differential matrix of Lagrangian interpolation polynomials, $i = 1, 2, \cdots, N$, $j = 0, 1, 2, \cdots, N$. The path constraints are also transformed into the collocation nodes as follows:

$$C(x_i, u_i, \delta p, \tau_i) \leq 0, i = 0, 1, 2, \cdots N \tag{11}$$

Then, the ascent trajectory optimization problems for air-breathing hypersonic vehicles under uncertain parameters are transformed into a sequence of nonlinear programming problems (NLP), which are given by

$$P_1 : min \quad J = m_0 - m_f$$

$$s.t. \quad \sum_{j=0}^{N} D_{ij}x_j - \frac{t_f - t_0}{2}f(x_i, u_i, \tau_i, \delta p) = 0 \tag{12}$$

$$C(x_i, u_i, \delta p, \tau_i) \leq 0,$$

where $i = 0, 1, 2, \cdots N$. The uncertain parameters δp are generated under the uniform distribution. Due to its computational efficiency and high accuracy, the well-known GPOPS toolbox [17] is used to solve this NLP problem offline. Then, the trajectories under uncertain parameters are used to establish the trajectory database.

3.2 DNN Structure and Training

The multi-layer forward network has been shown to be a suitable tool for coping with the initial state deviations in the literature [18], which studies the trajectory optimization problem for the hypersonic vehicle gliding phase. Thus, we adopt the multi-layer forward network to study its applicability to the ascent guidance problem. The proposed DNN is fully connected with one input layer, multiple hidden layers, and one output layer. It has seven hidden layers, and each layer has several neural nodes. The input of the network has 10 variables. The input state consists of three parts, the distance between the current state and desired terminal state, the errors between the normal trajectory and the trajectories under uncertain parameters, and the uncertain parameters. They are as follows:

$$\vec{x}_{input}^k = \left[\vec{x}_k^T - \vec{x}_f^{*T}, \vec{x}_k^T - \vec{x}_{ref,k}^T, \delta\vec{p}^T\right]^T, \tag{13}$$

where $\vec{x}_k = [R_k, V_k, \gamma_k]^T$. The mass variable is not in the input vector. The output is the current AOA α_k. Thus, the total structure is

$$Net\left(\vec{x}_{input}^k | \alpha_k\right) = [10 \times 64 \times 32 \times 16 \times 1]. \tag{14}$$

where the numbers in brackets are the number of neurons in each layer. The outline of the DNN is given in Fig. 1.

Further, the hyperbolic tangent sigmoid transfer function (*tansig*) is adopted as the activation function for the hidden layers. The learning rate is 1×10^{-3}. The loss function is mean absolute error (MAE). The update rule is scaled conjugate

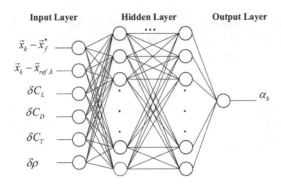

Fig. 1. The outline structure of the proposed deep network.

gradient backpropagation (*trainscg*). The input data and output are Min-Max scaling to $[-0.9, 0.9]$. Should be noted that the above hyper-parameters play an important role in the effectiveness of the DNN. However, it is not the focus of our study.

The above offline trajectory optimization method will generate 500 trajectories under uncertain parameters. The uncertain parameters have an uniformly distribution between $[-10\%, 10\%]$. The hypersonic vehicle flies 180 s in the ascent phase. Taking into the computational burden, the sampling step is selected as 100 ms. That is, each trajectory has 1,800 mesh points. We have total 900,000 state-control data. The data are randomly divided into three sets, the training set, the testing set, and the validation set according to 70%, 15%, and 15%.

3.3   The Sequential Calling Strategy and the Overall Scheme

The sequence of the trajectory optimization problems is solved by the pseudo-spectral method offline. Although the pseudo-spectral method is an efficient solver, it still has room for improvement in online applications for the generation of guidance laws under uncertain parameters. Thus, this paper trains deep neural networks to approximate the guidance laws to lessen the influence of uncertain parameters. The aim of this paper is to adopt the DNNs to approximate the relationship between the current state and the ideal controller from the vast trajectory data.

However, no matter how accurately the neural network is trained, there will always be a fitting error between the ideal controller and the DNN guidance law. The fitting error of DNNs and the calculation error will raise a gradually increasing deviation in the closed-loop guidance. When a single DNN is used, a compensation mechanism is absent to correct the fitting error and calculation error. Therefore, a strategy of multiple DNNs invocation sequentially strategy is proposed in this paper. Several DNNs are trained separately. When the hypersonic vehicle flies, the different DNNs will be called according to the remaining flight time to get the guidance law online. The aim is to let the latter DNN correct the integral truncation error induced by the former network. The integral truncation error can be viewed as the initial state error for the latter DNN. The larger the number of DNNs, the more storage space and computational time is required. To balance the efficiency and accuracy, this paper designs four DNNs according to the remaining time which is exponentially decreasing. Let $\tau = t_f - t_0$ refer to the total flight time, this paper trains four DNNs at $t_f - \tau$, $t_f - \frac{\tau}{2}$, $t_f - \frac{\tau}{4}$, and $t_f - \frac{\tau}{8}$, respectively. Accordingly, the sampling data are different. The sampling data for the first DNN are from $[t_0, t_f]$, and the sampling data for the second DNN are from $\left[t_f - \frac{\tau}{2}, t_f\right]$. The third and fourth sampling databases of the DNNs are $\left[t_f - \frac{\tau}{4}, t_f\right]$ and $\left[t_f - \frac{\tau}{8}, t_f\right]$, respectively. In the online process, the guidance controller will switch to different DNNs when the remaining time decreases.

The method consists of three parts: the offline generation of the database, the offline training of the DNNs, and the online application of the DNN guidance laws. When the DNNs are used online, the ascent dynamics will generate the real-time states, and the real-time control is generated online based on the well-trained DNNs, where the inputs are the real-time states. Figure 2 shows the overall scheme of the proposed method.

4   Numerical Simulations

This section gives the numerical simulations of the proposed method. All the numerical simulations are carried out on a PC with an *i9-11900K CPU (@3.50 GHz)*, and the software environment is *MATLAB 2020b*. The *MATLAB* neural network toolbox is adopted to train the DNNs.

4.1   The Database Establishment

The hypersonic vehicle model is from Ref. [16]. The initial states and terminal states are given in Table 1.

Table 1. The initial states and terminal states.

States	Symbol (Uni.)	Initial	Terminal
Height	$h(m)$	23164.8	48768
Velocity	$V(m/s)$	1676.4	4267.2
Flight path angle	$\gamma(\circ)$	0	0
Mass	$m(kg)$	127005.8636	NAN[a]

[a] NAN means that the terminal mass is not constrained.

The trajectories are generated under uncertain parameters with uniform distribution, which are drawn in Fig. 3. We can see that the pseudo-spectral method can get trajectories under uncertain parameters with high precision. All trajectories are converge to the desire terminal state.

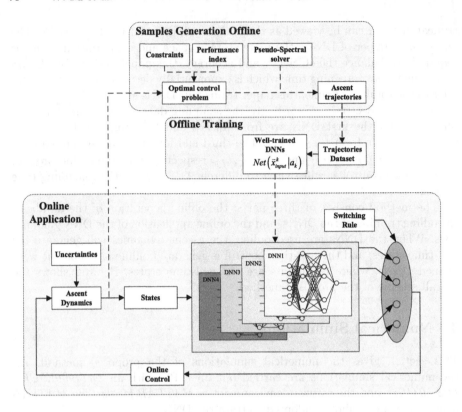

Fig. 2. The flow chart of the proposed method.

4.2　Real-Time Performance

In this section, the 1000 times Mento-Carlo simulations are carried out under uncertainties. 500 times results are from the train set, and the others are from the uniformly random. The results are given in Fig. 4. The results statistics for the robustness analysis are given in Table 2.

From Fig. 4, we can see that the terminal height error with 93% of the testing data are within $[-11.80\,\text{m}, 11.27\,\text{m}]$. The percentage of data with terminal speed error within $[-2.63\,\text{m/s}, 2.37\,\text{m/s}]$ is 94.4%. The terminal flight path angle error within $[-0.0057°, 0.0054°]$ accounts for 93% of the testing data. Our proposed method has a high-accuracy result because the terminal errors of the real-time simulations are tiny.

Figure 5 gives the trajectories under untrained uncertain parameters in real-time flight. We can see that the proposed method can guide the hypersonic vehicle to the desired terminal even in the presence of uncertain parameters which are untrained.

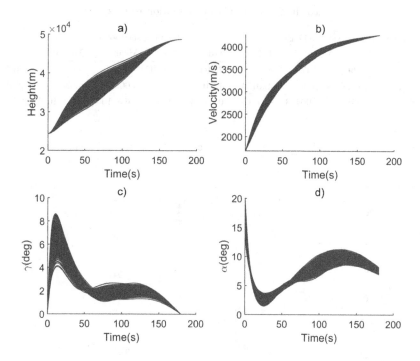

Fig. 3. The offline optimal trajectories under uncertainties: a) the height curves; b) the velocity curves; c) the flight path angle curves; d) the AOA curves.

4.3   Comparison of Different Numbers of the DNNs

In this paper, we come up with a sequential calling strategy with multiple DNNs to lessen the effect of the fitting error caused by the DNN and calculation error induced by the integration process in the closed-loop guidance. This section gives the comparison of different numbers of the DNNs.

In Fig. 6, we can see that the single DNN has the worst result. Compared with the one-layer DNN, two DNNs have a significant improvement in terminal precision. Three DNNs get an improvement over the two DNNs because the maximum error is decreasing. Because the variance is smaller, the results of four DNNs concentrate in a smaller area than three DNNs. Thus, we use four DNNs to design the guidance law in this paper.

Table 2. Monte Carlo simulation results

Parameters	Testing set (500 cases)				Training set (500 cases)			
	Max	Mean	Median	Std.	Max	Mean	Median	Std.
$\delta h_f(m)$	45.829	−0.547	−0.519	7.094	−19.308	−0.517	−0.337	4.465
$\delta v_f(m/s)$	−30.896	−0.149	−0.0154	1.916	−5.808	0.00921	0.0388	0.981
$\delta\gamma_f(deg)$	−0.116	−0.000443	-7×10^{-5}	0.00611	−0.0170	-6×10^{-5}	-3×10^{-6}	0.00264

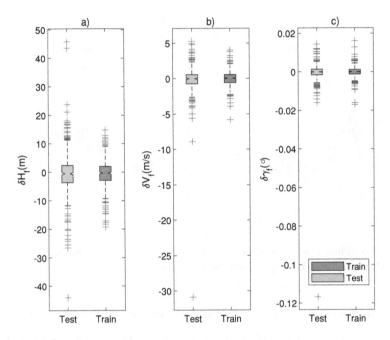

Fig. 4. The statistics of the Monte Carlo results under uncertainties: a) the terminal height; b) the terminal velocity; c) the terminal flight path angle.

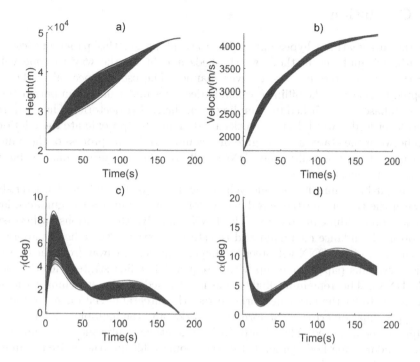

Fig. 5. The online flight trajectories under uncertainties: a) the height curves; b) the velocity curves; c) the flight path angle curves; d) the AOA curves.

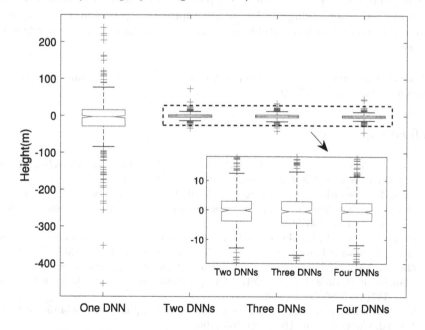

Fig. 6. The comparison of different numbers of the DNNs.

5   Conclusion

For the air-breathing hypersonic vehicle ascent phase, this paper proposes a DNN-based guidance method, which sheds a light on the way to cope with uncertain parameters in a data-driven manner. The pseudo-spectral method is adopted to generate the offline optimal trajectories under uncertain parameters. The database is established by sampling from the vast trajectories. Then, several multi-layer feedforward deep networks are designed to approximate the relationship between the states and the guidance command. In the process of the online implementation, the well-trained DNNs will give the guidance command based on the real-time states.

The air-breathing hypersonic vehicle gets a strong coupling characteristic between the thrust and the aerodynamic forces, which makes the guidance law design problem difficult to solve. The DNNs can give the relationship between the optimal guidance command and the states. However, the inherent approximation error of the DNN will steer the hypersonic vehicle away from the desired trajectory. This paper comes up with a sequential calling strategy for the multiple DNNs. The remaining flight time is designed as the switching variable. The aim is to let the latter networks correct the integral deviation generated by the approximation error of the former networks, and the high-precision terminal states can be achieved. The Mento-Carlo simulation results show that our proposed guidance method can guide the hypersonic vehicle to the desired terminal state with high precision.

However, there is still room for improvement. For example, in further work, we will study the DNN-based uncertain parameter estimation to make the proposed method more intelligent. To further improve the precision, the composite guidance combined with a suitable compensation mechanism is another field of our study.

Acknowledgments. This work is supported by the National Natural Science Foundation of China (No. 61873319).

References

1. Ding, Y., Yue, X., Chen, G., Si, J.: Review of control and guidance technology on hypersonic vehicle. Chin. J. Aeronaut. **35**(7), 1–18 (2022)
2. Chai, R., Tsourdos, A., Savvaris, A., Chai, S., Xia, Y., Chen, C.L.P.: Review of advanced guidance and control algorithms for space/aerospace vehicles. Prog. Aerosp. Sci. **122**, 100696 (2021)
3. Drake, D., Xin, M., Balakrishnan, S.N.: New nonlinear control technique for ascent phase of reusable launch vehicles. J. Guid. Control. Dyn. **27**(6), 930–937 (2004)
4. Zhai, S., Yang, J.: Piecewise analytic optimized ascent trajectory design and robust adaptive finite-time tracking control for hypersonic boost-glide vehicle. J. Franklin Inst. **357**(9), 5485–5501 (2020)
5. Lu, P.: Nonlinear trajectory tracking guidance with application to a launch vehicle. J. Guid. Control. Dyn. **19**(1), 99–106 (1996)

6. Sun, Z., Chao, T., Wang, S., Yang, M.: Ascent trajectory tracking method using time-varying quadratic adaptive dynamic programming. Proc. Inst. Mech. Eng. Part G: J. Aerosp. Eng. **233**(11), 4154–4165 (2019)
7. Nie, W., Li, H., Zhang, R.: Model-free adaptive optimal design for trajectory tracking control of rocket-powered vehicle. Chin. J. Aeronaut. **33**(6), 1703–1716 (2020)
8. Liu, L., He, Q., Wang, B., Fu, W., Cheng, Z., Wang, Y.: Ascent trajectory optimization for air-breathing hypersonic vehicles based on IGS-MPSP. Guid. Navig. Control **1**(02), 2150010 (2021)
9. Wang, X., et al.: An online generation method of ascent trajectory based on feedforward neural networks. Aerosp. Sci. Technol. **128**, 107739 (2022)
10. Diao, Y., Mu, R., Guan, Y., Cui, N.: Boost-phase trajectory planning with the nonregular reachable area constraints. Int. J. Aerosp. Eng. **2022**, 25 (2022)
11. Liu, C., Zhang, C., Xiong, F., Wang, J.: Multi-stage trajectory planning of dual-pulse missiles considering range safety based on sequential convex programming and artificial neural network. Proc. Inst. Mech. Eng. Part G J. Aerosp. Eng. 09544100221127058 (2022)
12. Wang, J., Ma, H., Li, H., Chen, H.: Real-time guidance for powered landing of reusable rockets via deep learning. Neural Comput. Appl. **35**, 6383–6404 (2022)
13. Li, W., Gong, S.: Free final-time fuel-optimal powered landing guidance algorithm combing lossless convex optimization with deep neural network predictor. Appl. Sci. **12**(7), 3383 (2022)
14. Federici, L., Benedikter, B., Zavoli, A.: Deep learning techniques for autonomous spacecraft guidance during proximity operations. J. Spacecr. Rocket. **58**(6), 1774–1785 (2021)
15. Viavattene, G., Grustan-Gutierrez, E., Ceriotti, M.: Multi-objective optimization of low-thrust propulsion systems for multi-target missions using ANNs. Adv. Space Res. **70**(8), 2287–2301 (2022)
16. Murillo Jr., O.J.: A fast ascent trajectory optimization method for hypersonic air-breathing vehicles. ProQuest dissertations and theses, Doctor of Philosophy, Iowa State University (2010)
17. Patterson, M., Rao, A.: GPOPS-II: a MATLAB software for solving multiple-phase optimal control problems using HP-adaptive Gaussian quadrature collocation methods and sparse nonlinear programming. ACM Trans. Math. Softw. (TOMS) **41**(1), 1–37 (2014)
18. Shi, Y., Wang, Z.: Onboard generation of optimal trajectories for hypersonic vehicles using deep learning. J. Spacecr. Rocket. **58**(2), 400–414 (2021)

Machine Learning and Deep Learning for Data Mining and Data-Driven Applications

Image Intelligence-Assisted Time-Series Analysis Method for Identifying "Dispersed, Disordered, and Polluting" Sites Based on Power Consumption Data

Xiao Zhang[1], Yong-Feng Zhang[1(✉)], Yi Zhang[2], and Jun Xiong[3]

[1] University of Jinan, Jinan 250022, SD, China
857489192@qq.com, cse_zhangyf@ujn.edu.cn
[2] Fuzhou University, Fuzhou 350108, FJ, China
zhangyi@fzu.edu.cn
[3] State Grid Fujian Electric Power Company, Fuzhou 350000, FJ, China

Abstract. A novel effective method for identifying "Dispersed, Disordered, and Polluting" (DDP) sites was proposed for the purpose of promoting the modernization of ecological and environmental governance capabilities and building a big data application platform for enterprises' pollution prevention. This paper aggregated data from the electricity consumption information collection system and characteristic sensing terminals, including user daily total electricity consumption records, peak and valley electricity consumption, and other information. Firstly, we used the hierarchical K-means algorithm to cluster the time series of user electricity consumption data. After ranking by cluster features, the electricity consumption time series of the selected suspicious users were encoded into Gramian angular field (GAF) images. Finally, we adopted the perceptual hash algorithm to build the model to identify "Dispersed, Disordered, and Polluting" sites. The case analysis results verified this method's feasibility, rationality, and effectiveness.

Keywords: Power consumption data · Image intelligence · Perceptual hash algorithm · Pollution prevention

1 Introduction

The greening of industrial structures is an important way to solve the problem of environmental constraints in sustainable economic development [1]. In recent years, China's investment in environmental pollution control policies and laws is unprecedented. After continuous control and management, the pollution control level of large-scale industrial enterprises has been gradually improving and the management has become more standardized. The pollutant emission intensity has been decreasing year by year. However, the pollution problem of "Dispersed, Disordered, and Polluting" enterprises has gradually become a key issue to further improve the level of environmental governance [2].

© The Author(s), under exclusive license to Springer Nature Singapore Pte Ltd. 2023
H. Zhang et al. (Eds.): NCAA 2023, CCIS 1869, pp. 79–91, 2023.
https://doi.org/10.1007/978-981-99-5844-3_6

"Dispersed, Disordered, and Polluting" enterprises [3] are those that do not comply with local industrial layout planning, industrial policies, and pollutant emission standards, or those that have incomplete procedures for land, business, quality supervision, environmental protection, etc. These enterprises are scattered, small in scale, deep in concealment, quick to move, and easy to rebound. They are located in complex areas such as urban and rural areas, administrative boundaries, remote rural areas, and other blind spots for supervision [4].

With the deployment and promotion of national power coverage [5] and online real-time carbon emission monitoring terminals, electricity, as an indispensable energy source for enterprise production activities, can timely and accurately reflect the production status and equipment usage of enterprises [6, 7]. Given the above challenges, using big data technology to help pollution prevention and ecological construction and insisting on driving environmental quality improvement through data-based innovation has become a consensus. Relying on enterprise power consumption data, we can monitor and analyze the production behavior of suspected "Dispersed, Disordered, and Polluting" enterprises, which can help improve the supervision efficiency of ecological and environmental departments.

Domestic researchers have increasingly focused on the cross-disciplinary application of big data in ecological and environmental regulation in recent years. Based on satellite remote sensing technology, [8] established a methodological system for dynamic monitoring and assessing the remediation effect of "Dispersed, Disordered, and Polluting" enterprises by using various technical means, such as online monitoring of pollution sources, air quality monitoring at ground stations, vehicle-borne navigation telemetry, etc. However, due to the problems of geometric correction and accuracy of positioning tools, the capacity to precisely locate the emission sources is relatively weak. [9] summarized the technical characteristics of electric power big data in the application of environmental protection supervision and air pollution prevention, and presented the technical route to establish the relationship model among enterprise electricity consumption, production, and pollutant emissions. In [10], the key features of "Dispersed, Disordered, and Polluting" users were screened out by constructing multi-dimensional features on power data through feature engineering, sample balancing was performed by adaptive sampling method based on BSMOTE algorithm, and suspected "Dispersed, Disordered, and Polluting" enterprises were analyzed by CatBoost algorithm.

It is in this context that this paper proposes the image intelligence-assisted time-series analysis method for identifying "Dispersed, Disordered, and Polluting" sites based on power consumption data. The clustering algorithm is used to group the time series for analysis, and then the Gramian angular field is used to image the time series. The calculation of mutual information and the perceptual hash algorithm are combined to achieve the purpose of time-series data analysis assisted by image intelligence.

2 Preliminary

In this paper, based on the usual electricity consumption data of a large number of users, we dug deeper into the electricity consumption patterns of normal users and "Dispersed, Disordered, and Polluting" users. By studying the clustering algorithm [11–13], we found a suitable clustering method for users' electricity consumption time-series data. In addition, they were converted into image data using the Gramian angular field to perform a more in-depth analysis of the filtered suspicious users.

2.1 Clustering Algorithm

As the most popular algorithm in clustering, the K-means algorithm was first used by MacQueen in 1967 [14]. Compared with other clustering algorithms, the advantage of the K-means algorithm is that it has simpler ideas and better results of clustering. The limitations are that the clustering number k in the algorithm needs to be determined in advance, the initial clustering centers are generated by random selection, and the outlier points will have an impact on the clustering results. To address these drawbacks, scholars from various fields presented different improvement methods based on different points of view [15–19].

The hierarchical K-means clustering algorithm [20, 21] is capable of adaptively obtaining the optimal or near-optimal number of clusters k, avoiding the manual setting of k value. In the beginning, as in the traditional K-means algorithm, k initial clustering centers are randomly selected for one iteration. After that, to perform finer-level clustering, the clustering measure values are calculated as follows:

$$J = \sqrt{\frac{\sum_{i=1}^{k} \sum_{j=1}^{n_i} \left(x_{ij} - c_i\right)^2}{n - 1}} \tag{1}$$

In Eq. (1), c_i is the center of the i-th cluster, x_{ij} is the j-th object of the i-th cluster, n_i is the number of objects of the i-th cluster, k is the number of classes, and n is the size of the dataset.

2.2 Gramian Angular Field

Rapidly expanding computer vision techniques inspired Zhiguang Wang and Tim Oates [22], who proposed the method of encoding time series as Gramian angular field (GAF) images to enable machines to visually recognize, classify, and learn structures and patterns.

In GAFs time increases with the position from top-left to bottom-right, so they provide a way to preserve time dependency and contain time correlations. However, the GAFs are very large since when the length of the original time series is n, the size of the Gramian matrix is $n \times n$. To downsize GAFs, we can apply the Piecewise Aggregation Approximation (PAA) [23] to smooth the time series while preserving the trends. Figure 1 gives the procedure for generating the GAF image.

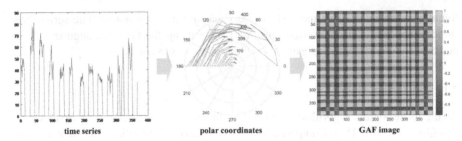

Fig. 1. Conversion of Time-Series Data to Image Data

3 Algorithm Procedure

In this paper, a three-stage strategy of "scanning, ranking, and detailed analyzing" (see Fig. 2) was used to achieve accurate capturing of "Dispersed, Disordered, and Polluting" (DDP) users: first, the electricity consumption of residential and non-residential users was scanned separately and clustering was completed using a high-performance hierarchical K-means (H-K-means) clustering technique; then, the number of members of each cluster was sorted and the cluster in which customers most likely to be DDP users is selected; finally, the suspected DDP users were analyzed in detail.

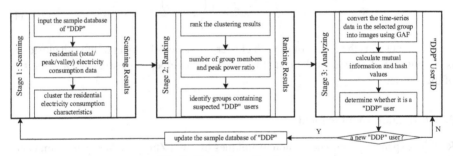

Fig. 2. Technology Roadmap

3.1 H-K-means

For a given dataset, taking the centroid of one set of patterns close to each other as a representative pattern, the whole dataset can be represented as several representative patterns with a distribution similar to that of the original dataset. The hierarchical structure of the original dataset is built up as in Fig. 3, where the number of levels L is defined by some criteria, with the first level being the original dataset and each subsequent level consisting of smaller datasets from the level above it.

Based on the H-K-means method with this multi-level structure [24], combining the data from the electricity information collection system and the data from the electricity characteristic sensing terminal, it is possible to accurately classify the massive time-series data and achieve the detection of abnormal samples.

Moreover, we studied the correlation analysis method and the principles of the evolution law of the power consumption characteristics of DDP sites to realize the analysis of the power consumption characteristics of DDP sites. After sorting out the results of the power consumption characteristic analysis, groups containing suspected DDP users were filtered by considering multiple indicators, such as the number of group members and peak-to-valley power consumption ratio.

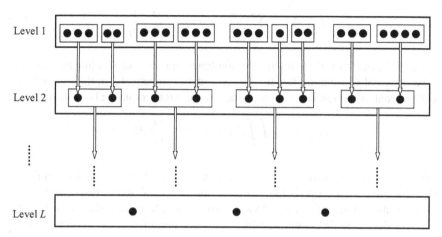

Fig. 3. Hierarchical Structure of the Original Dataset

3.2 Imaging Time Series

Considering that the composite signal after multi-source data fusion is a complex integration of various electrical equipment load signals, based on the analysis results of the electricity consumption characteristics of DDP places, a method to map the power consumption time-series data into image data [25] is employed, that is, to encode the time-series data into GAF images [22, 26].

Figure 1 has shown us the framework for encoding time series into images. We represent the time series in a polar coordinate system instead of the typical Cartesian coordinates. In the Gramian matrix [27, 28], each element is the cosine of the summation of the angles.

After obtaining the GAF images, we can build an image intelligence-based DDP sites recognition model to identify the suspicious users hidden in the residential electricity consumption group. Using mutual information and perceptual hash algorithm, we calculate the similarity between the GAF image of electricity consumption of the suspected DDP user and that of the sample in the DDP database, determine whether the suspected DDP user is a real DDP user, find its ID and locate the user. If the suspected DDP user is found to be a real DDP site after the inspection, but its GAF image is not in the sample database, it will be appended. As the sample database updating, the accuracy of the determination will gradually improve.

3.3 Mutual Information

Mutual information [29] represents the amount of information shared between two variables or among more variables. The greater the mutual information, the stronger the correlation between the variables. We can view the mutual information as an extension of the correlation coefficient in the high-dimensional nonlinear case. Formally, the mutual information of two discrete random variables X and Y can be defined as:

$$I(X;Y) = \sum_{y \in Y} \sum_{x \in X} p(x,y) log \frac{p(x,y)}{p(x)p(y)} \tag{2}$$

In Eq. (2), $p(x,y)$ is the joint distribution function of X and Y, while $p(x)$ and $p(y)$ are the marginal distribution functions of X and Y, respectively. In the case where X and Y are continuous random variables, the mutual information can be defined as:

$$I(X;Y) = \int_Y \int_X f(x,y) log \frac{f(x,y)}{f(x)f(y)} dxdy \tag{3}$$

In Eq. (3), $f(x,y)$ is the joint probability density function of X and Y, while $f(x)$ and $f(y)$ are the marginal probability density functions of X and Y, respectively.

Intuitively, mutual information [30] measures the information shared by X and Y. That is, it describes the extent to which knowing one of the two variables reduces uncertainty about the other.

In addition, the mutual information is non-negative ($I(X;Y) \geq 0$). Obviously, the worst case of obtaining information about one event from another event is 0, and the uncertainty of an event does not increase due to the knowledge of another event. Furthermore, the mutual information is symmetrical ($I(X;Y) = I(Y;X)$). That is, the amount of information extracted from Y about X is the same as the amount of information extracted from X about Y.

3.4 Perceptual Hash Algorithm

In a browser image search, a user can upload an image then the browser will display pictures on the Internet that are identical or similar to that image. The key technology to achieve this function is the perceptual hash algorithm [31] and each image will generate a fingerprint (in string format). The more similar the fingerprints of the two images are, the more similar the two original images are.

The following are the specific steps of the perceptual hash algorithm:

a. Read the image data in matrix form;
b. Convert to 256-level grayscale images;
c. Scale to 32 × 32 size and remove image details;
d. Perform a two-dimensional discrete cosine transform (it is still a 32 × 32 matrix after the transformation);
e. Intercept the 8 × 8 part of the upper left corner of the matrix (the high-frequency section is likely to gather in the upper left corner);
f. Calculate the mean of this 8 × 8 matrix;

g. Iterate through the matrix, assign the element greater than or equal to the mean value to 1, otherwise define it to 0, and generate an 8×8 hash fingerprint map;

h. Compare the hash values of the two maps, calculate the Hamming distance, and get the similarity (the hash fingerprint map represents a string of 64-bit binary hash values).

It should be noted that the larger the Hamming distance [32], the smaller the similarity. The Hamming distance of 0 means two maps are identical.

4   Practical Validation

4.1   Background

To promote the modernization of ecological and environmental governance capabilities and help the construction of ecological provinces and digital cities, a power company proposes to build a big data application platform for enterprise pollution prevention and provide a series of application services to help environmental protection departments accurately manage enterprise emissions.

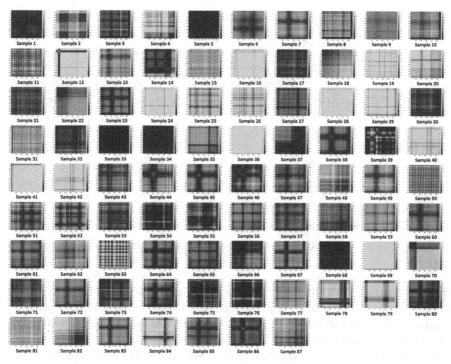

Fig. 4. DDP Database of GAF Images

Based on the available data, a sample database of identified DDP users has been created to generate GAF images of the electricity consumption of DDP users, as shown in

Fig. 4. By aggregating data from user electricity consumption files, emission enterprises, and environmental protection equipment and using big data algorithms to establish relevant analysis models, we can provide application services for environmental protection departments, such as analysis of scattered pollution users.

4.2 Implementation

In this paper, we take the electricity consumption data of a city in southeastern China as an example. The data contains electricity consumption information of 3060 households for a total of 366 days from July 2019 to July 2020, including daily total electricity consumption, peak hour electricity consumption, and low hour electricity consumption data.

Stage I: Scanning and Clustering. Using a high-performance H-K-means clustering technique, 366 days of residential electricity consumption data were scanned and clustered into 11 class groups. The clustering results and the center of these groups are shown in Fig. 5.

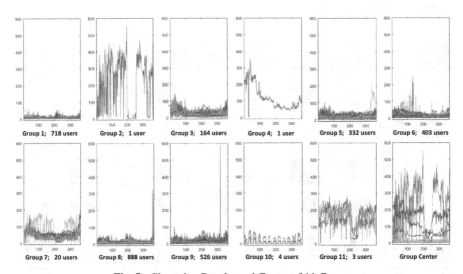

Fig. 5. Clustering Results and Center of 11 Groups

Stage II: Ranking and Filtering. The groups containing suspected DDP users were sorted according to the number of group members and filtered into the third stage for detailed analysis. A total of four groups with less than five users were selected for this case: Group 2 (1 user), Group 4 (1 user), Group 10 (4 users) and Group 11 (3 users). Users in these groups were respectively marked as User 2-1, User 4-1, User 10-1, User 10-2, User 10-3, User 10-4, User 11-1, User 11-2, User 11-3, and the time-series graphs of electricity consumption for these 9 users were shown in Fig. 6.

Stage III: Detailed Analyzing. We calculated the GAF images of the selected nine users' electricity consumption as shown in Fig. 7. Then, it is necessary to calculate

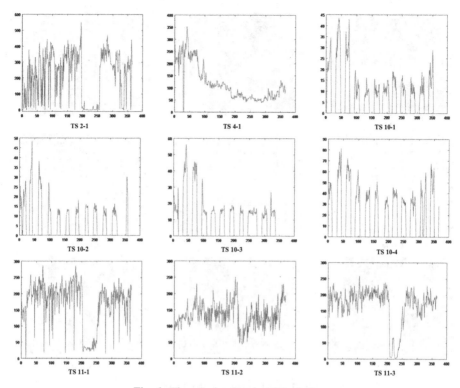

Fig. 6. Time-Series Graphs of the 9 Users

the similarity between the GAF images of the electricity consumption of these nine suspected users and the DDP database samples.

If there is a sample with high similarity to the user, we need to use the perceptual hash algorithm to compare the hash fingerprint maps between the user and this sample. A user with very high similarity to the sample can directly be identified as a DDP user.

If no sample in the DDP database with significant similarity to the user, but the electricity consumption pattern of this user is different from that of other customers and does not match the characteristics of daily residential electricity consumption, it is better to check this site in the field. After verification, if this user is confirmed to be a DDP user, the GAF image of this user needs to be added to the DDP database for updating purposes.

4.3 Results

The four suspected users in Group 10 have a noticeable pattern of electricity consumption, which may be related to the production process. By calculating the similarity with the samples in the DDP database, the GAF images of all four users in this group were found to be similar to Sample 50 in the DDP database, and the hash fingerprint maps had similarities of 96.875%, 100%, 96.875%, and 98.4375% with Sample 50, respectively.

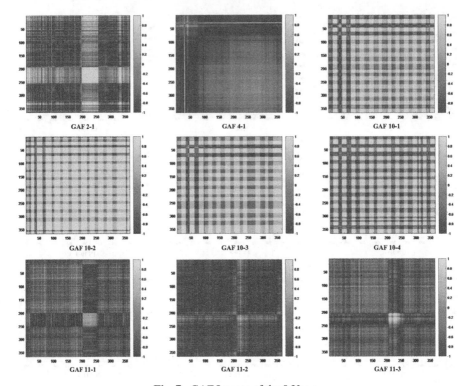

Fig. 7. GAF Images of the 9 Users

Figure 8 gives the comparison process of the hash fingerprint maps with User 10-1 as an example.

In the same way, in Group 11, User 11-1 and User 11-3 were similar to Sample 65 in the DDP database, while User 11-2 was similar to Sample 1, and all the hash fingerprint maps in Group 11 had achieved 100% similarity after comparison. Therefore, the users in Group 10 and Group 11 could be identified as DDP users.

The similarity between the electricity consumption data of User 2-1 and the samples in the DDP database was calculated, and no samples with high similarity were found. However, the electricity consumption pattern of User 2-1 is obviously different from other households and does not match the characteristics of daily residential electricity consumption. Therefore, it is suspected to be a DDP subscriber. The same situation occurs with User 4-1, and it is recommended to check them. If they are confirmed as DDP users after verification, the DDP database will be updated.

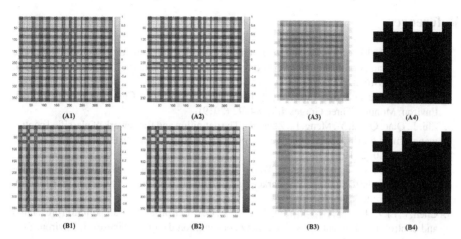

Fig. 8. This is the comparison process of the hash fingerprint maps. The letter (A) represents the sample in the DDP database (Sample 50 here) and the letter (B) represents the suspected user (User 10-1 here). The number (1) represents the original GAF image of the electricity consumption, the number (2) represents the 256-level grayscale image, the number (3) represents the image scaled to 32*32 pixels, and the number (4) represents the hash fingerprint map. For example, (B4) represents the hash fingerprint map of User 10-1.

5 Conclusion

This paper elucidates the problem of "Dispersed, Disordered, and Polluting" sites hiding among ordinary electricity consumers [33], summarizes scholars' work on the cross-disciplinary application of big data in ecological and environmental supervision, and proposes the image intelligence-assisted time-series analysis method for identifying "Dispersed, Disordered, and Polluting" sites based on electricity consumption data, which is validated in a real case. The results of the later on-site inspection also confirm the effectiveness of this strategy.

Subsequent work can revolve around integrating this method into a big data application platform for electricity and keeping the DDP database updated with the results of each check. Based on the intelligent analysis of the electricity consumption characteristics of DDP locations, combined with the electricity consumption files and network topology information of the users in their power supply areas, visualization technology can help fully display the distribution of DDP users in the area. Those high-risk users for alarm are marked to realize the intelligent diagnosis and early warning of such places. In addition, image intelligence is developing rapidly, and better computer vision algorithms can be tried for deeper data processing and mining after converting the time-series data into image data.

Acknowledgments. This research is sponsored by the Shandong Province Higher Educational Youth and Innovation Talent Introduction and Education Program.

References

1. Zhong, M.C., Li, M.J., Du, W.J.: Can environmental regulation force industrial structure adjustment: an empirical analysis based on provincial panel data (in Chinese). China Popul. Resourc. Environ. **25**(8), 107–115 (2015)

2. Cai, H.Y.: The pollution intensity of different corporations in China (in Chinese). Ecol. Environ. Monitor. Three Gorges **4**(1), 84–92 (2019)

3. Zhi, J., Qiao, Q., Li, Y., Meng, L., Zhao, R.: The definition and framework of classified control management for dispersed, violative and polluted enterprises (in Chinese). Environ. Protect. **47**(20), 46–50 (2019)

4. Sun, K., Chen, Z., Fan, M., Liu, Y., Wang, T., Shi, H.: Design of omni channel collaborative management platform for power marketing based on microservice architecture (in Chinese). Distrib. Utilizat. **38**(4), 28–36 (2021)

5. Chen, J., Li, Z., Liu, H., Gao, J., Yang, Y., Zhu, S.: Research progress of air pollution prevention and control based on enterprise electricity consumption data (in Chinese). J. Environ. Eng. Technol. 1–12 (2022)

6. Jiang, Y., Wang, Z.Q., Dai, B.: Research and analysis of residential electricity consumption behavior based on big data (in Chinese). Electric Power Inform. Commun. Technol. **13**(11), 7–11 (2015)

7. Wu, L., Zhou, Y., Chen, H., Yang, Z.: Emission characteristics of industrial air pollution by using smart-grid big data (in Chinese). Chinese J. Environ. Manage. **8**(4), 37–42 (2016)

8. Wang, Q., Li, Q., Wang, Z., Ping, F., Qiao, Q., Chen, G., et al.: Dynamic supervision technology and application of 'Dispersed, Disordered and Polluted' enterprises based on remote sensing (in Chinese). Res. of Environ. Sci. **34**(3), 511–522 (2021)

9. Li, R.: Applications and suggestions of electric power big data in the field of precise prevention and controlling of air pollution (in Chinese). Environ. Protect. **50**(10), 28–31 (2022)

10. Huang, H., Huang, R., Deng, Y., Zheng, W.: Identifying "Dispersed, Disordered and Polluted" enterprises method based on BSMOTE and CatBoost algorithms (in Chinese). Electric Power Inform. Commun. Technol. **20**(11), 105–113 (2022)

11. Rodriguez, A., Laio, A.: Clustering by fast search and find of density peaks. Science **344**(6191), 1492–1496 (2014)

12. Cheng, W., Lu, Y.: Adaptive clustering algorithm based on maximum and minimum distances, and SSE (in Chinese). J. Nanjing Univ. Posts Telecommun. (Natl. Sci. Edn.) **35**(2), 102–107 (2015)

13. Yang, J., Zhao, C.: Survey on K-means clustering algorithm (in Chinese). Comput. Eng. Appl. **55**(23), 7–14+63 (2019)

14. MacQueen, J.: Some methods for classification and analysis of multivariate observation. In: Proceedings of the 5th Berkley Symposium on Mathematical Statistics and Probability, pp. 281–297. University of California Press, Berkeley (1967)

15. Visalakshi, N. K., Suguna, J.: K-means clustering using max-min distance measure. In: 28th North American Fuzzy Information Processing Society Annual Conference, pp. 1–6. IEEE, Cincinnati (2009)

16. Lei, J., Jiang, T., Wu, K., Du, H., Zhu, G., Wang, Z.: Robust K-means algorithm with automatically splitting and merging clusters and its applications for surveillance data. Multimedia Tools Appl. **75**(19), 12043–12059 (2016)

17. Chakraborty, S., Das, S.: K-means clustering with a new divergence-based distance metric: convergence and performance analysis. Pattern Recogn. Lett. 10067–10073 (2017)

18. Jia, R., Li, Y.: K-means algorithm of clustering number and centers self-determination (in Chinese). Comput. Eng. Appl. **54**(7), 152–158 (2018)

19. Wang, J., Ma, X., Duan, G.: Improved K-means clustering k-value selection algorithm (in Chinese). Comput. Eng. Appl. **55**(8), 27–33 (2019)
20. Hu, W.: Improved hierarchical K-means clustering algorithm (in Chinese). Comput. Eng. Appl. **49**(2), 157–159 (2013)
21. Liu, Y., Li, B.: Bayesian hierarchical K-means clustering. Intell. Data Anal. **24**(5), 977–992 (2020)
22. Wang, Z., Oates, T.: Imaging time-series to improve classification and imputation (2015)
23. Keogh, E.J., Pazzani, M.J.: Scaling up dynamic time warping for datamining applications. In: Proceedings of the Sixth ACM SIGKDD International Conference on Knowledge Discovery and Data Mining, pp. 285–289. ACM, Boston (2000)
24. Xu, T.S., Chiang, H.D., Liu, G.Y., Tan, C.W.: Hierarchical K-means method for clustering large-scale advanced metering infrastructure data. IEEE Trans. Power Deliv. **32**(2), 609–616 (2017)
25. Hatami, N., Gavet, Y., Debayle, J.: Bag of recurrence patterns representation for time-series classification. Pattern Anal. Appl. **22**(3), 877–887 (2019)
26. Cui, J., Zhong, Q., Zheng, S., Peng, L., Wen, J.: A lightweight model for bearing fault diagnosis based on Gramian angular field and coordinate attention. Machines **10**(4), 282 (2022)
27. Pan, Y., Qin, C.: Identification method for distribution network topology based on two-stage feature selection and Gramian angular field (in Chinese). Autom. Electric Power Syst. **46**(16), 170–177 (2022)
28. Zhang, S., Du, L., Wang, C., Jiang, A., Xu, L.: Wind power forecasting method based on GAF and improved CNN-ResNet (in Chinese). Power Syst. Technol. 1–10 (2023)
29. Lv, Q.W., Chen, W.F.: Image segmentation based on mutual information (in Chinese). Chin. J. Comput. **2**, 296–301 (2006)
30. Gong, W.: Watershed model uncertainty analysis based on information entropy and mutual information (in Chinese) (2012)
31. Zeng, Y.: Image perceptual hashing research and application (in Chinese) (2012)
32. Ding, K., Chen, S., Meng, F.: A novel perceptual hash algorithm for multispectral image authentication. Algorithms **11**(1), 6 (2018)
33. Sun, Y., Li, G.: Discussion on investigation and classified rectification of "Dispersed, Disrupted and Polluted" enterprises (in Chinese). Chem. Eng. Manage. **655**(4), 58–62 (2023)

Non-intrusive Load Identification Based on Steady-State V-I Trajectory

Dingrui Zhi, Jiachuan Shi, and Rao Fu[✉]

Shandong Key Laboratory of Intelligent Buildings Technology, School of Information
and Electrical Engineering, Shandong Jianzhu University, Jinan 250101, China
2021080123@stu.sdjzu.edu.cn, {jcshi,furao20}@sdjzu.edu.cn

Abstract. The key step of non-intrusive load identification technology
is feature extraction, mainly indicating two classifications, steady-state
features and transient features. At the same time, due to different load
models, the current and power waveforms of the same type of electrical
appliances vary greatly during operation, but significant difference may
not be distinct in their V-I trajectory shapes. Thus, as a load feature for
identification, pixelized and continuous traditional V-I trajectory was
proposed in this article to highlight more information, taking the processed V-I trajectory map as the inputs of the CNN network for training. Aiming at the influence of hyper-parameters on recognition accuracy, Bayesian algorithm was applied to optimize the hyper-parameters.
The trained CNN network obtained was validated based on the public
dataset of PLAID. The results showed that the multi-period steady-state
VI trajectory was selected as the object of study in this paper, which can
better reflect the changing characteristics of the load during operation
compared with the traditional single-period VI trajectory map. At the
same time, a Bayesian algorithm was implemented to optimize the CNN
network's hyperparameters, which resulted in a high-accuracy recognition framework that can generalize well.

Keywords: Bayesian algorithm · Convolutional neural network ·
Hyper-parametric optimization · Non-intrusive load identification

1 Introduction

With the continuous growth of power demand and the continuous development
of smart grid technology [1], people increasingly require the ability to accurately
monitor and identify different loads. Load identification is an important component of the implementation of intelligent power systems. As a method with
strong versatility and low cost, non-intrusive load identification was first proposed by Hart in 1992 [2]. With the development of computer technology and
the popularity of smart meters, non-intrusive load identification has become a
research hot spot, and many non-intrusive load identification algorithms have
also emerged.

© The Author(s), under exclusive license to Springer Nature Singapore Pte Ltd. 2023
H. Zhang et al. (Eds.): NCAA 2023, CCIS 1869, pp. 92–105, 2023.
https://doi.org/10.1007/978-981-99-5844-3_7

Among them, Literature [3] used the DTW algorithm to calculate the similarity between test and reference templates to achieve load identification. Literature [4] extracted the steady-state value features for clustering and weighted the feature values to achieve high accuracy. The load decomposition method based on electrical characteristics and time series information in Reference [5] was effective. BP neural network algorithms were used in paper [6] to learn and identify the active power, reactive power and harmonic characteristics of various household appliances. A Hidden Markov model based on time series features was used in paper [7] to integrate the time information into the load identification model with good results. Literature [8] described in detail eight shape features such as asymmetry displayed by V-I trajectories and used hierarchical clustering methods for recognition, and literature [9] computed elliptical Fourier descriptors of V-I trajectory profiles as inputs for the classification algorithm. Literature [10] built on Literature [8], proposed a new feature "span", and compared this feature with current harmonic content, active power and reactive power, respectively, using four classification algorithms. The results showed that the V-I trajectory had a better identification capability.

Therefore, in this article, V-I trajectories were used as features for load identification. First, the high-frequency steady-state voltages and currents were plotted as V-I trajectories, which could be converted into pixel images as the input to the CNN network. Since there were a large number of hyper-parameters in the CNN network, a Bayesian optimization algorithm was proposed to optimize the hyper-parameters of the CNN algorithm to obtain the best hyper-parameters for load recognition modeling. Finally, the validation was performed based on the PLAID dataset [11], and the experimental findings indicated that the multiperiod steady-state VI trajectory was chosen as the subject of inquiry for this research since it provides more accurate representation of variations in load characteristics during operation compared to the traditional single-period VI trajectory map. At the same time, Bayesian algorithm was used to optimize the hyperparameters of the CNN network, which makes the recognition framework have higher recognition accuracy and generalization ability.

2   Construction of V-I Trajectory Pixelization

The load characteristics include steady-state characteristics such as active power, reactive power, current harmonic amplitude [12], and V-I trajectory [13], and transient characteristics such as transient energy [14], transient current and transient power. However, the time domain waveforms presented by the same type of load vary greatly due to the difference in brand and power, while the variation of V-I trajectory is small. Therefore, the V-I trajectory is chosen as the characteristic quantity for load identification.

The V-I trajectory is a graph of the voltage-current relationship of an electrical device that enters a steady state within a specified time after starting. Traditionally, V-I traces are mapped into cells and each grid was assigned as 0 or 1. Grids with V-I traces are 1, otherwise 0. The duty cycle and the amount of space continuously occupied are determined by traversal.

Traditional methods typically represented each grid space as a binary value to indicate whether the grid was covered or contained specific information. However, this method represented less information and could not provide enough detailed information. Therefore, we proposed an improved method to increase the amount of information.

Specifically, each V-I trajectory was counted on the grid it passes through, and each time a trajectory covered a grid, the count of that grid was added by 1. After traversing all the grids, we normalized these grid counts and transformed them into continuous values between 0 and 1. This normalization process could make the information of the grid through which the trajectory passes more continuous and richer, while avoiding the problem of information loss.

With this improved method, more accurate and detailed trajectory information could be obtained, thus improving the visualization and analysis of the data. The steps were as follows:

(1) Stability determination was performed to obtain a single sequence of current I and voltage V under stable operating conditions [15]. A power time series $P = p(k)_{k=1}^{\infty}$ was set and two consecutive sliding windows W_m (mean calculation window) and W_d (transient detection window) of length m and n were defined for this series data. The mean values M_m and M_d within these two windows are calculated as shown in Eqs. (1) and (2), respectively:

$$M_m = \frac{\sum_k^{k+m-1} p(k)}{m} \tag{1}$$

$$M_d = \frac{\sum_{k+m}^{k+m+n-1} p(k)}{n} \tag{2}$$

A load entered a steady state when the average values M_m and M_d are constants;

(2) Standardized the steady-state voltage and current and plotted using standardized voltage and current data to control the V-I trajectory. The standardized formula was as follows:

$$\Delta V_m = \frac{V_m}{\max |V|} \tag{3}$$

$$\Delta I_m = \frac{I_m}{\max |I|} \tag{4}$$

where $\max |V|$ is the maximum voltage in the steady-state sequence, and $\max |I|$ is the maximum current in the steady-state sequence. V_m, I_m represents the voltage and current values of the m-th sampling point in the sequence, and ΔV_m, ΔI_m represents the normalized voltage and current values of the m-th sampling point;

(3) Used normalized current and voltage sequences for V-I trajectory rendering;
(4) A $50 * 50$ grid was used to cover, and all sampling points were traversed, and if the sampling point was within the grid, the corresponding grid value was added by one;

(5) Normalized the grid values to a range of [0,1];

Figure 1 below showed the drawing of the V-I trajectory and the resulting graph of the normalized pixel map:

3   Construction and Optimization of Convolutional Neural Networks

3.1   Construction of Neural Networks

Convolutional neural networks (CNN) have powerful advantages in processing two-dimensional input data. Therefore, this article selected the AlexNet network of CNN network for load identification. The model was proposed in 2012 and was one of the major breakthroughs in depth learning in the field of image recognition. The structure of the model was shown in Fig. 2. The network included 5 layers of convolutional layers, 3 layers of pooling layers, 2 layers of Dropout layers, and 3 layers of full connectivity layers to load pixelated images (an $n \times n$ matrix, where $n = 224$) was used as model input.

Convolution layer was the key to extracting input information features in convolutional neural networks, while pooling layer was used to reduce the dimension and aggregate the features of convolutional matrix, including two methods: maximum pooling and average pooling. In the process of feature extraction, the maximum pooling method could retain more local details, which was conducive to the recognition of easily confused electrical appliances by the model. In addition, the dropout layer was a component that set thresholds and compares the weights of certain hidden layer nodes to make certain nodes not work to speed up network operations and prevent overfitting. Finally, the output layer set K nodes (where K is the number of different device types), and used the *softmax* function to set the node output value between 0 and 1. The combination of these components constituted a complete convolutional neural network that could effectively perform load identification tasks.

In order for a CNN to have recognition capability, a large amount of training data was needed for it to learn. The training samples $X = (X_1, X_2, \ldots \ldots, X_n)$ had corresponding labels $t = (t_1, t_2, \ldots \ldots, t_n)$, where X_i was the input V-I pixel map and t_i was the corresponding load type. This training network gave it a high recognition accuracy and a small loss function.

Although the classical AlexNet network performed well in image recognition, it could not be directly applied to the load recognition task and needed to be improved to meet the specific needs of the task. In addition, due to the large number of hyper-parameters of the network, including those of the convolutional layer, pooling layer and dropout layer, it needed to be optimized to achieve higher recognition accuracy. Therefore, in this article, Bayesian optimization method was chosen to optimize the hyper-parameters of the CNN network.

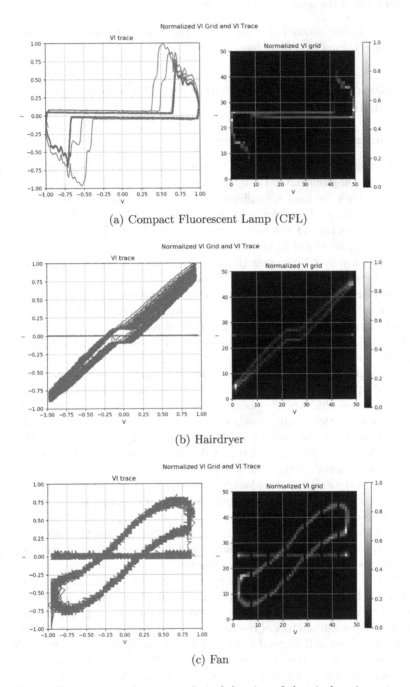

(a) Compact Fluorescent Lamp (CFL)

(b) Hairdryer

(c) Fan

Fig. 1. V-I trajectory diagram and pixel drawing of electrical equipment.

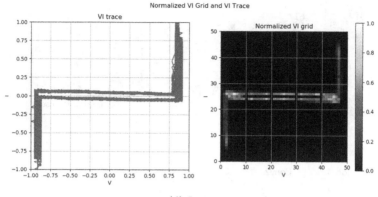

(d) Laptop

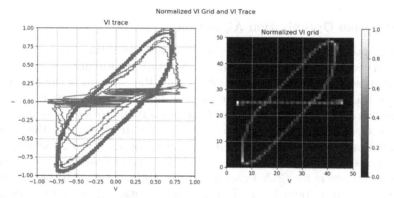

(e) Air Conditioner

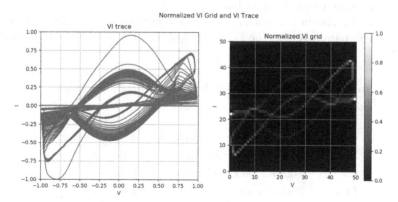

(f) Microwave Oven

Fig. 1. (*continued*)

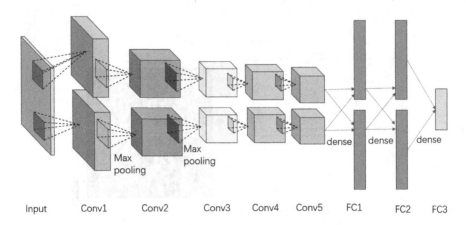

Fig. 2. AlexNet network model diagram.

3.2   Bayesian Optimization Algorithm

Bayesian optimization algorithm (BOA) is an algorithm that can estimate the maximum value of a function based on the available sampling points and is suitable for searching the optimal solution when the equation of the function is unknown [16,17]. In the field of machine learning and deep learning, Bayesian optimization is often used to solve complex optimization problems with high evaluation costs, such as hyper-parametric optimization. In this article, the number of convolutional kernels, convolutional kernel size, convolutional kernel step size, pooling kernel size, pooling step size, and dropout rate of the layer were chosen as super-parameters for CNN hyper-parameter optimization. The optimization range was shown in Table 1. The Bayesian optimization algorithm was used to initialize 10 sampling points, to find the point with the lowest executable degree as the next sampling point according to the Gaussian process, and to perform 15 iterations to obtain the optimal hyper-parameter configuration after optimization. The parameters in the Bayesian optimization process were defined in Table 2.

Table 1. Hyper-parameters to be optimized and their ranges.

Layers	Hyper-parameters	Dynamic-range
Conv	Number of convolution kernels	30–135
	Convolution kernels size	2–6
	Convolution kernels step	1–3
Pool	Pool core size	2–6
	Pool nucleation step size	1–3
Dropout	Dropout rate	0–1

Table 2. Parameter Definition in Bayesian Optimization Algorithms.

Particle parameters	Parameters to be optimized
a1, a2, a3, a4, a5	The number of convolution kernels in five convolutional layers
b1, b2, b3, b4, b5	Convolutional kernel size for five convolutional layers
c1, c2, c3, c4, c5	Convolutional kernel step size for five convolution layers
d1, d2, d3	The number of pooling kernels in the three pooling layers
e1, e2, e3	Step size of pooling kernels in three pooling layers
f1, f2	Dropout rate of the two layers

When training a CNN using the Bayesian optimization algorithm, a probabilistic proxy model and an ensemble function were selected, and then a hyperparameter was chosen during each iteration and the ensemble function was used to evaluate and optimize. As each iteration proceeded, the most promising evaluation points were added to the historical data until the termination condition was satisfied. The input of the Bayesian optimization algorithm proposed in this article included the set of parameters X to be optimized, the objective function f, the collection function S, and the Gaussian process model M. The output was the optimal CNN model. The flow of CNN Bayesian optimization was shown in Fig. 3.

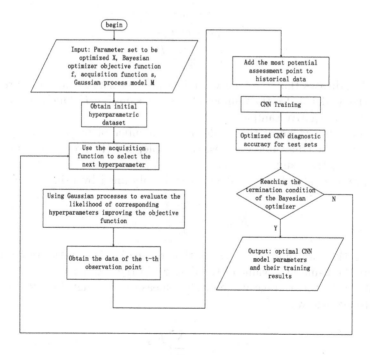

Fig. 3. Bayesian optimization CNN based recognition model training flow chart.

4   Experiment and Result Analysis

4.1   Datasets and Evaluation Criteria

This experimental dataset used the PLAID dataset from Carnegie Mellon University, which collected 30 kHz current and voltage samples from 56 homes in Pittsburgh, Pennsylvania, USA, for 11 different electrical types. Each appliance type was represented by dozens of different model instances. For each appliance, three to six measurements were collected and post-processed. The dataset contained data for transient and steady-state operation at start-up.

In the evaluation of the effectiveness of non-intrusive load recognition, the commonly used recognition accuracy (Acc), F-measure [18] score and macro-average F_{macro} were mainly used as the evaluation criteria of recognition models in this article.

The accuracy (Acc) calculation formula was shown in Eq. (5):

$$Acc = \frac{m}{n} \tag{5}$$

where m represents the correct classification number of the model, and n represents the total number of samples.

The F-measure score is obtained from the following equation:

$$precisioin_i = \frac{TP_i}{TP_i + FP_i} \tag{6}$$

$$F_i = 2 \times \frac{precision_i \times recall_i}{presicion_i + recall_i} \tag{7}$$

$$recall_i = \frac{TP_i}{TP_i + FN_i} \tag{8}$$

where TP indicates that the true value is a positive sample, and the predicted value of the model is also the number of positive samples; FP indicates that the true value is a negative sample.

The predicted value of the model is the number of positive samples; FN indicates that the true value is a positive sample, and the predicted value of the model is a negative sample number.

The average value for each device can be obtained by using the following formula:

$$F_{i\_mean} = \frac{1}{L} \sum_{g=1}^{L} F_{g\_i} \tag{9}$$

where L is the total number of times that device i exists in the test set; F_{g_i} represents the F-measure value of the device i in the g-th time.

Finally, take the average value of all F-measures through Eq. (9) to obtain the macro average value F_{macro}.

$$F_{macro} = \frac{1}{A} \sum_{i=1}^{A} F_{i\_mean} \tag{10}$$

where A is the total number of different device types.

4.2   Case Analysis

To verify the effectiveness of the method proposed, the study case used the PLAID dataset for experiments. The dataset contained 1074 sets of data, including high-frequency voltage and high-frequency current data for 11 different brands of loads. In the paper, the dataset was first converted into 1074 pixel maps of V-I trajectories and inputted into a CNN network. Next, a Bayesian optimization algorithm was used to optimize the super-parameters of the CNN network, including 23 super-parameters for 5 convolutional layers, 3 pooling layers, and 2 dropout layers. After optimization, the comparison of the super-parameters was shown in Table 3. The 33/48/1 of Conv1 layer meant the convolutional kernel size was 33, the number of convolutional kernels was 48, and the step size was 1, while the 0.5 of Dropout1 meant the dropout rate was 0.5.

Table 3. Comparison of super parameter information between CNN and BOA-CNN.

Category	CNN	BOA-CNN
Conv1	$3 \times 3/48/1$	$4 \times 4/60/1$
Pool1	$3 \times 3/2$	$2 \times 2/1$
Conv2	$5 \times 5/128/2$	$4 \times 4/121/1$
Pool2	$3 \times 3/2$	$2 \times 2/1$
Conv3	$3 \times 3/192/1$	$4 \times 4/126/1$
Conv4	$3 \times 3/192/1$	$3 \times 3/55/1$
Conv5	$3 \times 3/192/1$	$3 \times 3/125/1$
Droout1	0.5	0.4
Droout1	0.5	0.1

After comparing the optimized BOA-CNN model with the CNN model and the load recognition algorithms in the literature [9,19,20], the study presented the recognition accuracy of each algorithm in Table 4. From the table, it was observed that the BOA-CNN model outperformed the other models in terms of load recognition performance. On the PLAID dataset, the recognition accuracy of the BOA-CNN model was 93.20%. Compared with the CNN and classification algorithms in the literature [7,9,15], the recognition accuracy of the BOA-CNN model was improved by 8.4%, 13.8%, 12.5%, and 10.8%, respectively.

To test the stability of the BOA-CNN algorithm, the study conducted 12 experiments on the PLAID dataset to verify its stability. As shown in Fig. 4, in all experiments, the recognition accuracy of the BOA-CNN algorithm outperformed the ordinary CNN model with good robustness. However, in terms of stability, the BOA-CNN algorithm performed poorly and fluctuated greatly. Specifically, the difference between the maximum and minimum accuracy of the BOA-CNN algorithm was about 12% in 12 experiments, while the difference of the ordinary CNN model was about 10%. This indicated that the BOA-CNN algorithm was unstable

Table 4. Comparison of recognition accuracy of various algorithms.

Model	Accuracy
BOA-CNN	93.2%
CNN	84.8%
Literature [9]	80.7%
Literature [20]	79.4%
Literature [?]	82.4%

across experiments. Further analysis showed that, this instability could have been caused by the stochastic nature of the Bayesian optimization algorithm, and the local optimal solution. However, its k could automatically search for the optimal hyper-parameters of the convolutional, pooling, and dropout layers, reducing the time and labor cost of manual-parameter tuning.

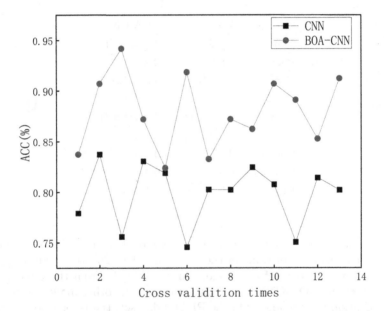

Fig. 4. Comparison of recognition accuracy between BOA-CNN and CNN.

The model confusion matrix was shown in Fig. 5, where the number in each cell represented the number of corresponding appliances, the horizontal axis represented the predicted values of the appliances, and the vertical axis represented the actual values of the appliances. It was observed that the PSO-CNN model reduced the number of false negatives (FN) and false positives (FP) for each appliance in the PLAID dataset compared to the normal CNN model, and also reduced the number of appliance types that were misclassified for each appliance. This indicated that the PSO-CNN model demonstrated excellent results in the accuracy of electrical identification and had practical value.

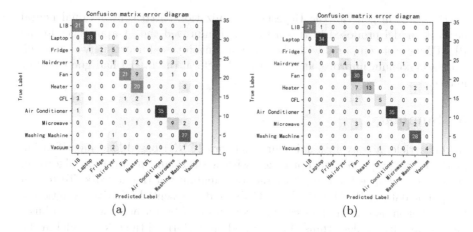

(a)   (b)

Fig. 5. Comparison of BOA-CNN and CNN confusion matrices.

For the F-measure metric and the F_{macro} metric, as shown in Fig. 6, the F_{macro} metric for the BOA-CNN was 83%, while the F_{macro} metric for the CNN was 75%. In addition, the F-measure values for blowers and heaters were lower than the macro average. This might be due to the small amount of measurement data for some devices in the dataset, which limited the learning ability of the model and results in lower F-measure values for some devices.

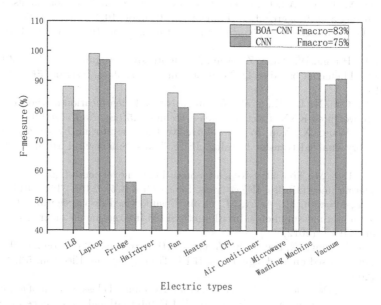

Fig. 6. Comparison between BOA-CNN and CNNF-measure indicators and F_{macro} indicators.

However, in general, the Bayesian optimization-based convolutional neural network model proposed in this article had improved the 11 appliances in the PLAID dataset compared with the original convolutional neural network model, with a maximum improvement of 30%. Therefore, the model had good performance.

5   Summary

In this article, a convolutional neural network (CNN) based load identification method was proposed and optimized with a Bayesian optimization algorithm (BOA), to enhance the feature extraction capability of the network. Tested on the public dataset PLAID, the BOA-CNN model presented effective reduction of the confusion among appliances and acquisition of better accuracy and robustness against listed algorithms. The method outperformed the CNN model in the recognition performance of various appliances, yet the model was not satisfactory due to the similarity of V-I trajectories from certain loads and the lack of samples of appliance data. Therefore, future work should take the characteristics of current distribution, power distribution, and current harmonic distribution into consideration during the operation of the load for auxiliary judgment.

References

1. Cheng, X., Li, L., Wu, H., et al.: A survey of the research on non-intrusive load monitoring and disaggregation. Power Syst. Technol. **40**(10), 3108–3117 (2016)
2. Hart, G.: Nonintrusive appliance load monitoring. Proc. IEEE **80**(12), 1870–1891 (1992)
3. Hua, L., Huang, W., Yang, Z., et al.: A nonintrusive recognition method of household load behavior based on DTW algorithm. Electr. Meas. Instrum. **56**(14), 17–22 (2019)
4. Zhu, H., Cao, N., Lu, H., et al.: Non-intrusive load identification method based on feature weighted KNN. Electron. Meas. Technol. **45**(08), 70–75 (2022)
5. Liu, Y., Wang, X., You, W.: Non-intrusive load monitoring by voltage current trajectory enabled transfer learning. IEEE Trans. Smart Grid **10**(5), 5609–5619 (2019)
6. Lin, Y., Tsai, M.: Non-intrusive load monitoring by novel neuro-fuzzy classification considering uncertainties. IEEE Trans. Smart Grid **5**(5), 2376–2384 (2014)
7. Kim, H., Marwah, M., Arlitt, M., et al.: Unsupervised disaggregation of low frequency power measurements. In: Eleventh SIAM International Conference on Data Mining. DBLP (2012)
8. Lam, H.Y., Fung, G., Lee, W.K.: A novel method to construct taxonomy electrical appliances based on load signatures. IEEE Trans. Consum. Electron. **53**(2), 653–660 (2007)
9. Bates, L.D., Develder, C., Dhaene, T., et al.: Automated classification of appliances using elliptical Fourier descriptors. In: IEEE International Conference on Smart Gird Communications (SmartGridComm). IEEE (2017)
10. Hassan, T., Javed, F., Arshad, N.: An empirical investigation of V-I trajectory based load signatures for non-intrusive load monitoring. IEEE Trans. Smart Grid **5**(2), 870–878 (2014)

11. Jiang, X., Dawson-Haggerty, S., Dutta, P., et al.: Design and implementation of a high-fidelity AC metering network. In: International Conference on Information Processing in Sensor Networks, IPSN 2009, pp. 253–264. IEEE (2009)
12. Chen, X.: Research on Non-intrusive Load Decomposition Technology and Application. South China University of Technology, Guangzhou (2018)
13. Jie, Y., Mei, F., Zheng, J., et al.: Non-intrusive load monitoring method based on V-I trajectory color coding. Autom. Electric Power Syst. **46**(10), 93–102 (2022)
14. Chang, H., Chen, K., Tsai, Y., et al.: A new measurement method for power signatures of nonintrusive demand monitoring and load identification. IEEE Trans. Ind. Appl. **48**(2), 764–771 (2012)
15. Ding, S., Wang, X., Yong, J., et al.: Research on event-based non-intrusive load monitoring method. Build. Electr. **36**(07), 57–64 (2017)
16. Zhu, H., Liu, X., Liu, Y.: Research on hyperparametric optimization based on bayesian new deep learning data communication, (2), 35–38+46 (2019)
17. Jones, D., Schonlau, M., Welch, W.: Efficient global optimization of expensive black-box functions. J. Global Optim. **13**(4), 455–492 (1998)
18. Dumoulin, V., Visin, F.: A guide to convolution arithmetic for deep learning (2016)
19. Zheng, Z., Chen, H.N., Luo, X.W.: A supervised event based non-intrusive load monitoring for non-linear appliances. Sustainability **10**(4), 1001 (2018)
20. Gao, J., Kara, E., Giri, S., et al.: A feasibility study of automated plug-load identification from high-frequency measurements. In: IEEE Global Conference on Signal and Information Processing, pp. 220–224 (2015)

Prediction of TP in Effluent at Multiple Scales Based on ESN

Yilong Liang[✉] [iD], Cuili Yang, and Yingxing Wan [iD]

Beijing University of Technology, Chaoyang District, Beijing, China
liangyilong97@outlook.com, clyang5@bjut.edu.cn, wanyx@emails.bjut.edu.cn

Abstract. Waste water treatment is related to environmental protection and human health. Aiming at the difficulty of accurate prediction of effluent TP in wastewater treatment plants (WWTPs), a multi-scale data-driven method for effluent total phosphate (TP) prediction is proposed in this paper. Firstly, the time-scale characteristics of water quality data were analyzed. Secondly, a weighted average data processing (WAP) method is designed to reconstruct the data set for multi-scale data. By matching the time scales of different variables, the effective data quantity is expanded and the prediction frequency is improved. Then, the prediction model of effluent TP based on echo state network (ESN) was constructed to achieve accurate prediction of effluent TP. Finally, the proposed method is applied to the actual effluent data set. Experimental results show that the proposed method can accurately predict effluent TP.

Keywords: Effluent total phosphate · Echo state network · Mutliscale data

1 Introduction

Water is involved in every aspect of life. However, improper operation of WWTPs will damage the water environment [1]. The total phosphate (TP) in effluent will harm health and aggravate eutrophication. Therefore, the effluent TP of WWTPs must meet increasingly stringent discharge standards [2]. In addition, due to the complex nonlinear and dynamic characteristics of wastewater treatment process [3], the stable operation and optimal control of WWTPs have become major problems in recent years. To solve this problem, a model that can accurately predict the effluent TP of WWTPs is needed. The commonly used prediction models of effluent quality are divided into two groups [4]: (1) mechanism model and (2) data-driven model.

Mechanism model [5] requires long-term observation of wastewater treatment process, formulation of appropriate model according to reaction mechanism, and

H. Zhang et al. (Eds.): NCAA 2023, CCIS 1869, pp. 106–118, 2023.
https://doi.org/10.1007/978-981-99-5844-3_8

determination of accurate values of many model parameters [6]. However, these methods are difficult to provide a completely satisfactory description of causality in WWTPs, and may show poor estimation or prediction ability [7]. At the same time, the model needs to regulate a large number of kinetic and stoichiometric parameters, which brings difficulties to the optimization of the model [8]. Therefore, the establishment of acceptable mechanism models for WWTPs is not an easy task.

Data-driven model construction methods mainly include regression analysis method, state estimation method and artificial neural network(ANN) method, among which ANN is widely used in data-driven modeling method [9]. Huang M et al. [10] effectively predicted the concentrations of effluent TP through ANN modeling. Bagheri M et al. [11] adopted a modeling method combining multi-layer perceptron and RBF neural network to predict variables such as effluent TP. Simulation results show that this method has high accuracy for effluent TP prediction. Compared with the traditional mechanism model, the data-driven method does not need a specific mathematical model, and only needs to input data to realize the online prediction of TP in effluent, which is more widely used [5]. Data-driven ANN modeling relies on data at the same scale. However, due to different detection methods, it is difficult to synchronize the collection frequency of wastewater quality data [12], which will cause the problem of multi-scale data. Prediction of effluent TP based on multi-time scale data modeling is still a difficult problem to be solved.

Above all, a multi-scale data-driven method for predicting effluent TP concentration is proposed in this paper. Firstly, a weighted average data preprocessing method is designed for multi-scale effluent water quality data, which can reconstruct the data set. Then, the prediction model of effluent TP based on ESN was established to predict effluent TP concentration. Finally, the effectiveness of the proposed method is verified by actual data.

2 Preliminary

2.1 Multiscale Time Series

Table 1 shows the time scales of some parameters in the wastewater treatment system of a WWTP. It can be seen that the concentration of pollutants such as TP is collected at a two-hour interval, while other data such as inlet flow rate, inlet pH, fan current and pump current are collected and updated at a 3-minute interval. The time interval of the original data is 3 min, that is, the data such as flow changes in real time, while the data such as TP changes once in 120 min, and about every 40 groups of data are updated. The data between two collection and other data updates are the data collected in the previous one. The difference of time scale between variables reduces the number of effective samples, which brings difficulties to modeling and prediction. At the same time, the collection and updating frequency of the actual data often do not match the corresponding time scale. There are many factors that can cause this phenomenon, such as the failure of sensors and detection instruments, network fluctuations, manual misoperation, etc., so the variables in the data are not strictly consistent with the time scale, which brings challenges for data processing.

Table 1. Time scale diagram of some variables in a WWTP.

Variable	Time scale	Variable	Time scale
TP	2 h	Old system influent PH	3 min
NH4-N	2 h	Old system fan current	3 min
COD	2 h	Old system return water	3 min
TN	2 h	Old system dosing pump	3 min
Discharge water	3 min	New system influent PH	3 min
Lift pump current	3 min	New system fan current	3 min
Old Anaerobic intake water	3 min	New system return water	3 min
New Anaerobic intake water	3 min	New system dosing pump	3 min

In order to facilitate the analysis of multi-scale time series, the data matrix can be defined as follows according to the water quality data of a WWTP.

$$X(t) = \begin{pmatrix} x_1(t_1) & \cdots\cdots & x_i(t_1) & \cdots\cdots & x_n(t_1) \\ x_1(t_1) & \cdots\cdots & x_i(t_1) & \cdots\cdots & x_n(t_2) \\ \vdots & \ddots\ddots & \vdots & \ddots\ddots & \vdots \\ x_1(t_1) & \cdots\cdots & x_i(t_{\tau_i}) & \cdots\cdots & x_n(t_{\tau_i}) \\ x_1(t_1) & \cdots\cdots & x_i(t_{\tau_i}) & \cdots\cdots & x_n(t_{\tau_i+1}) \\ \vdots & \ddots\ddots & \vdots & \ddots\ddots & \vdots \\ x_1(t_{\tau_1}) & \cdots\cdots & \cdots & \cdots\cdots & x_n(t_{\tau_1}) \\ x_1(t_{\tau_1}) & \cdots\cdots & \cdots & \cdots\cdots & x_n(t_{\tau_1+1}) \\ \vdots & \ddots\ddots & \vdots & \ddots\ddots & \vdots \end{pmatrix} \quad (1)$$

where $x_i(t_i)$ represent the ith variable in the time step $t = t_i (i = 1, 2, ...)$, and τ_i is the time scale of the ith variable, t_{τ_i} represent the collection time of the ith variable. The order of the column vectors in Eq. (1) is arranged from largest to smallest according to the time scale τ_i of the different variables, as $\tau_1 > \tau_2 > \cdots > \tau_i > \cdots > \tau_n$. Then the corresponding sampling time of each variable can be expressed as follows.

$$\begin{cases} t(x_1) = t_0 + k\tau_1 \\ t(x_2) = t_0 + k\tau_2 \\ \cdots\cdots \\ t(x_n) = t_0 + k\tau_n \end{cases} \quad (2)$$

where $t(x_i), i = 1, 2, ..., n$ denote the sampling time of the ith variable, where t_0 is the initial time, $k = 0, 1, 2, 3, ..., m$ and m is a natural number. The variable in the data matrix will update the data only when the time step reaches the acquisition moment of the variable, and the data will remain unchanged at other moments.

2.2   Echo State Network

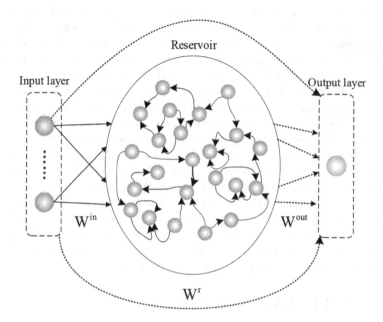

Fig. 1. The basic structure of an ESN.

As a new recursive neural network, ESN have a three-layer structure which is shown in Fig. 1. ESN is composed of input layer, reservoir, output layer, the core is randomly connected reservoir. In ESN, only output weight matrix $\boldsymbol{W}^{out} \in \mathbb{R}^{m \times (N+n)}$ needs to be trained, input weight matrix $\boldsymbol{W}^{in} \in \mathbb{R}^{N \times n}$ and the reservoir weight matrix $\boldsymbol{W}^{r} \in \mathbb{R}^{N \times N}$ are randomly generated and remain unchanged. The reservoir state $\boldsymbol{x}(t) \in \mathbb{R}^{N}$ and the output $\boldsymbol{y}(t) \in \mathbb{R}^{m}$ of ESN can be calculated as follow.

$$\boldsymbol{x}(t) = \boldsymbol{f}(\boldsymbol{W}^{r}\boldsymbol{x}(t-1) + \boldsymbol{W}^{in}\boldsymbol{u}(t)) \tag{3}$$

$$\boldsymbol{y}(t) = \boldsymbol{W}^{out}\boldsymbol{X}(t) \tag{4}$$

where $\boldsymbol{f}(\cdot) = [f_1(\cdot), \ldots, f_N(\cdot)]^T$ represents the activation function for reservoir, $\boldsymbol{u}(t) \in \mathbb{R}^{n}$ is input vector, and $\boldsymbol{X}(t) = \left[\boldsymbol{x}(t)^T, \boldsymbol{u}(t)^T\right]^T \in \mathbb{R}^{N+n}$. The output weight matrix can be solved by solving the following formula.

$$\boldsymbol{W}^{out} = \arg\min_{\boldsymbol{W}^{out}} \left\| \boldsymbol{H}\boldsymbol{W}^{out} - \boldsymbol{T} \right\|_2^2 \tag{5}$$

where reservoir state matrix $\boldsymbol{H} \in \mathbb{R}^{(N+n) \times L}$ is composed by collecting $\boldsymbol{X}(t)$, $\boldsymbol{T} \in \mathbb{R}^{L \times m}$ is the targer output vector. Then $\boldsymbol{W}^{out}$ can be calculated as follows

$$\boldsymbol{W}^{out} = \boldsymbol{H}^{\dagger}\boldsymbol{T} = (\boldsymbol{H}^T\boldsymbol{H})^{-1}\boldsymbol{H}^T\boldsymbol{T} \tag{6}$$

where $\boldsymbol{H}^{\dagger}$ is the Moore-Penrose generalized inverse of $\boldsymbol{H}$.

3   Proposed Methodology

In the previous section, the characteristics of multiscale time series were introduced. In practical scenarios, due to interference from factors such as sensor failures, environmental noise, and manual misoperation, there is often a deviation in the data scale corresponding to the collected variables, and early or delayed collection often occurs. Due to the fact that the wastewater treatment process is a complex biochemical reaction process, there are many variables within it, and the length of data is often measured in months and quarters, which can make aligning the scale of data too cumbersome. And many variables may have conflicts on time scales, increasing the difficulty of data processing.

Wastewater treatment is a complex biochemical reaction system with large time delays, and there is a strong correlation between the water quality data of wastewater treatment effluent, and there is a coupling between various variables. If we focus on the time scale of a certain water quality variable, it is difficult to effectively handle it. Therefore, based on the characteristics of wastewater effluent data, we propose a data weighted average processing(WAP) method, which can unify the data scales of different variables, thereby increasing the density of the dataset and improving the prediction frequency.

3.1   Data Weighted Average Processing

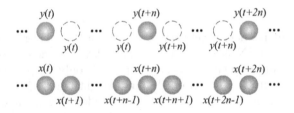

Fig. 2. Schematic diagram of multi-time scale data.

Data Review. TP data is updated once every 120 min, which is manifested in every 40 groups of data in the data set. At other times, the sensor sends the latest data, and the data of input variables such as traffic is updated every 3 min. In the data set, each group of data is the data that is collected and updated. Figure 2 is a schematic diagram of the data. The top row represents the TP variable updated at an interval of two hours, while the bottom row represents the variable updated every three minutes. It is assumed that the time scale of the output variable is $\tau = p$, that is, the value of the output variable is updated once every p time steps, and the data between each two updates is the data of the previous update and represented by dotted line ball. The input variable updates the data for each time step.

Calculate the Average Time Scale. The acquisition time of variables does not strictly abide by its corresponding time scale, so it is necessary to calculate the average time scale of output variables. The specific method is to traverse the entire data set from the initial moment and count the number of data q_i between two adjacent data acquisition moments in turn, where $i = 1, 2, ..., m$ and m represent the number of data acquisition. Then the average time scale $\bar{\tau}$ of output variables can be calculated by the following formula:

$$\bar{\tau} = \frac{1}{m} \sum_{i=1}^{m} q_i \tag{7}$$

Reconstruct the Output Variable Sequence. In order to unify the scale of input and output variables and expand the data density, the scale of output variables is subdivided. Considering that wastewater treatment is a process of continuous change, the data at one moment is related to the data at each previous moment, and this relationship gradually fades over time. Therefore, a weighted average algorithm is designed to reconstruct the output variable sequence. It is assumed that the average time scale of the output variable is $\bar{\tau}$, and the auxiliary time scale $\tilde{\tau} = 0.7\bar{\tau}$ is adopted. With $\bar{\tau}/2$ as the limit, two values $\bar{y}_1, \bar{y}_2$ are inserted between the data collected twice to assist the modeling. The specific methods are as follow.

$$\begin{cases} \bar{y}_1 = \dfrac{\sum\limits_{i=t-\frac{\bar{\tau}}{2}-\tilde{\tau}}^{t-\frac{\bar{\tau}}{2}} \frac{1}{t-\frac{\bar{\tau}}{2}-i} y(i)}{\sum\limits_{i=t-\frac{\bar{\tau}}{2}-\tilde{\tau}}^{t-\frac{\bar{\tau}}{2}} \frac{1}{t-\frac{\bar{\tau}}{2}-i}} \\[20pt] \bar{y}_2 = \dfrac{\sum\limits_{i=t-\tilde{\tau}}^{t-1} \frac{1}{t-i} y(i)}{\sum\limits_{i=t-\tilde{\tau}}^{t-1} \frac{1}{t-i}} \end{cases} \tag{8}$$

where t is the current time, $y(i)$ is the value of output variable at a certain time, $\frac{1}{t-\frac{\bar{\tau}}{2}-i}, \frac{1}{t-i}$ is the time forgetting factor. The older the data, the less impact it has on the current data. Combined with the calculated $\bar{y}_1, \bar{y}_2$ and the originally collected data $y(i)$, the data density of the output variable is expanded by three times.

However, it is worth noting that $\bar{y}_1, \bar{y}_2$ are calculated values, which cannot represent the real values. Therefore, these two variables are only used as auxiliary variables for modeling, and only the actual collected values are used to calculate the accuracy.

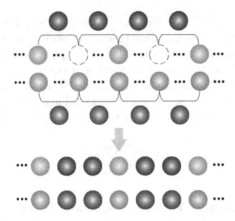

Fig. 3. Weighted processing data chart. (Color figure online)

Reconstruct the Input Variable Sequence. Similar to the output variable, the average scale $\bar{\tau}$ of the output variable is taken as the reference, and the input variable sequence is reconstructed through the auxiliary scale $\tilde{\tau}$. The scale of the input variable is expanded to the time scale of the reconstructed output variable, as follow.

$$\begin{cases} \bar{x}_1 = \dfrac{\displaystyle\sum_{i=t-\frac{\bar{\tau}}{2}-\tilde{\tau}}^{t-\frac{\bar{\tau}}{2}} \dfrac{1}{t-\frac{\bar{\tau}}{2}-i} x(i)}{\displaystyle\sum_{i=t-\frac{\bar{\tau}}{2}-\tilde{\tau}}^{t-\frac{\bar{\tau}}{2}/2} \dfrac{1}{t-\frac{\bar{\tau}}{2}-i}} \\ \\ \bar{x}_2 = \dfrac{\displaystyle\sum_{i=t-\tilde{\tau}}^{t-1} \frac{1}{t-i} x(i)}{\displaystyle\sum_{i=t-\tilde{\tau}}^{t-1} \frac{1}{t-i}} \end{cases} \tag{9}$$

where t is the current time, $x(i)$ is the value of the output variable at a certain time, and $\frac{1}{t-\frac{\bar{\tau}}{2}-i}, \frac{1}{t-i}$ is the time forgetting factor.

The schematic diagram of WAP is shown in Fig. 3. The blue dots in the figure represent output variables with larger time scales, and the green dots represent input variables with smaller time scales. The dotted lines represent the values of the output variable between two acquisition moments. After WAP, the data interval of the output variable is reduced to one-third of the original.

3.2   Prediction Method Based on WAP

The data processed by WAP can be used to train the prediction model. ESN has high prediction accuracy and strong ability to extract nonlinear information, which can be used to model TP concentration of effluent in complex wastewater

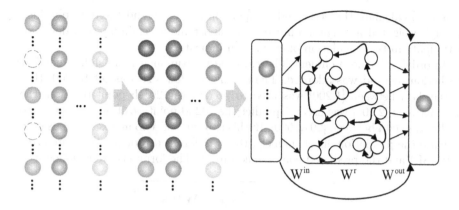

Fig. 4. Schematic diagram of the prediction model.

treatment process. So ESN is adopted as a prediction model. Figure 4 shows the schematic diagram of the prediction model. The detailed steps are as follows:

- 1. Calculate the average time scale of the output variable according to the original output variable;
- 2. The original output variable was reconstructed according to the average time scale, and the time scale of the output variable was reduced to obtain a new output variable data set;
- 3. The original input variable was reconstructed according to the average time scale of the output variable, and the time scale of the input variable was expanded to match the new time scale of the output variable, and a new input variable data set was obtained;
- 4. The ratio of training set and test set is 7:3;
- 5. Initialize the ESN, input the training set ESN, and train the prediction model;
- 6. Test the model and calculate the evaluation index.

4   Simulation

In order to verify the effectiveness of the proposed method, water quality data mentioned above are used in this section for simulation experiments. After screening, five input variables are determined, which are the current of the lifting pump of the regulating pool, the PH of the inlet water, the current of the dosing pump, the current of the fan and the flow of the return water. There are 1000 sets of original data, and 3000 sets of data can be obtained after WAP. Commonly used interpolation methods were used to compare and verify the performance of WAP, which were linear interpolation (Liner-ESN), Spline interpolation (Spline-ESN) and piecewise cubic Hermite interpolation (Pchip-ESN). Each of the three interpolation methods inserts two auxiliary points between two collections. It is worth noting that in the following experimental comparison, only the actual data is used to calculate the error and comparison accuracy, and the data obtained through weighted average is only used to assist model training, and does not participate in the calculation of evaluation indicators.

In order to achieve effluent TP real-time measurement, the sliding window prediction method was adopted. The sliding window size was set to 200, that is, the data between [1,200] was used to predict the 201th point position, the data between [2,201] was used to predict the 202nd point position, and so on, to predict the data between [201,1000]. In this experiment, two indexes were used to evaluate the prediction effect, namely, the standardized root mean square error (NRMSE) and the predicted compliance rate φ. The formulas are as follows:

$$\text{NRMSE} = \sqrt{\frac{\sum_{i=1}^{n_{max}} \|\mathbf{t}(i) - \mathbf{W}^{out}\mathbf{x}(i)\|^2}{\sum_{i=1}^{n_{max}} \|\mathbf{t}(i) - \tilde{\mathbf{t}}\|^2}} \tag{10}$$

$$\varphi = \frac{c}{s} \times 100\% \tag{11}$$

where the predicted standard fulfillment rate φ represents the percentage of the number of predicted data whose prediction accuracy meets the requirements c in the total number of predicted data s. The required prediction accuracy interval is $[0.8, 1.2]$, that is, the point position of prediction accuracy $\alpha \in [0.8, 1.2]$ is the point position of qualified prediction. The prediction accuracy can be calculated by the following formula:

$$\alpha = \frac{\tilde{y}}{y} \tag{12}$$

where $\tilde{y}, y$ represent the predicted and actual output values, respectively. ESN set the size of the reservoir $N = 300$, sparsity $D = 0.1$, and spectral radius $\rho = 0.9$.

The predicted output and test error curves of these methods are summarized in Fig. 5. All the interpolation methods can well fit the predicted output. The output curve of WAP-ESN is closer to the actual output curve and has a smaller test error curve. Therefore, compared with O-ESN and the other three interpolation methods, WAP-ESN is more suitable for predicting the TP concentration of effluent in the actual wastewater treatment process.

Figure 6 shows the box diagram of prediction error of several methods, in which the prediction error interval of O-ESN is the largest and that of WAP-ESN is the smallest. At the same time, the prediction error of WAP-ESN has fewer outliers and the smallest degree of dispersion, indicating that the prediction error distribution of WAP-ESN is dense, with small fluctuation and higher stability. Therefore, WAP-ESN is more consistent with the goal of stable operation of wastewater treatment. Figure 7 shows the comparison of prediction accuracy of O-ESN and WAP-ESN. The two black lines in the figure represent the prediction reach interval $[0.8, 1.2]$, that is, the prediction point within the two black lines is qualified prediction, while the point outside the black line is unqualified

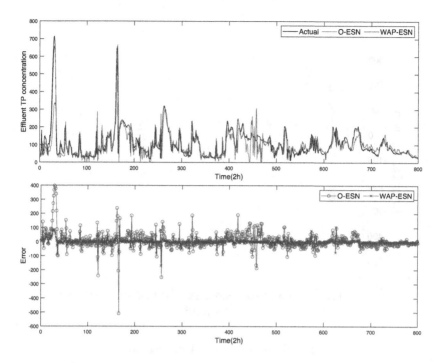

Fig. 5. Output fitting and error comparison chart.

prediction. It can be seen that most of the prediction accuracy of WAP-ESN is within the reach range, while many unqualified prediction points appear for O-ESN. The figure in the figure is the local precision map with the prediction interval between [400,500]. It can be clearly seen from the figure that WAP-ESN only has 3 unqualified prediction points, while O-ESN has 47 unqualified points, and the prediction accuracy of WAP-ESN is closer to 1, indicating that its prediction accuracy is relatively high.

The simulation results of O-ESN, interpolation method and WAP-ESN are listed in Table 2, including training time, training NRMSE value, average qualified points and predicted rate of reaching the standard. For each case, 30 simulations were carried out independently. The data in Table 2 shows that the training time of WAP-ESN is about four times that of O-ESN. This is because WAP-ESN expands the original data set from 1000 groups to 3000 groups, and more data causes the training time to become. Considering that wastewater treatment is a system with large time delay, the training time of WAP-ESN is completely acceptable. In addition, WAP-ESN has the smallest test NRMSE. In 800 slip-window predictions, the average number of qualified prediction points of WAP-ESN is 762.36, and the average prediction compliance rate is 95.35%, both of which are the minimum values. This indicates that WAP-ESN can better predict the TP of wastewater treatment effluent.

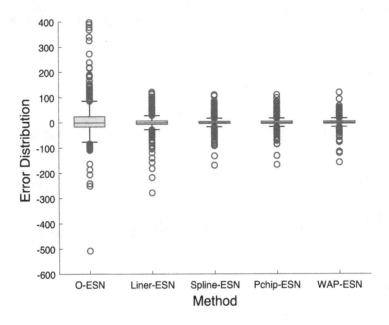

Fig. 6. Error distribution box diagram.

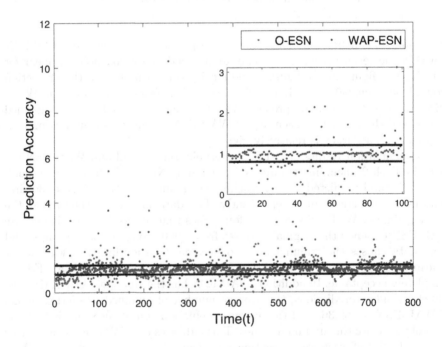

Fig. 7. Comparison Chart of Prediction Accuracy.

Table 2. Prediction Results of TP Concentration in Effluent

Method	Time(s)	Test NRMSE	Qualifying point	Pass rate
O-ESN	30.56	0.6141	383.04	0.4788
Liner-ESN	118.46	0.4288	690.42	0.8675
Spline-ESN	119.74	0.2262	738.96	0.9237
Pchip-ESN	118.78	0.1727	747.76	0.9347
WAP-ESN	119.08	0.1447	762.36	0.9535

5   Conclusion

This paper first introduces the characteristics of multi-scale time series, and then proposes a weighted average data reconstruction method according to the characteristics of multi-scale time series data. This method can increase the number of effective data, improve the prediction frequency, and thus improve the prediction accuracy. After that, the ESN prediction model is established using the reconstructed data. Finally, the performance of the method is proved by the prediction experiment of TP in wastewater treatment effluent.

Acknowledgement. This work was supported by the National Key Research and Development Project under Grant 2021ZD0112002, the National Natural Science Foundation of China under Grants 61973010, 62021003 and 61890930-5.

References

1. Hamed, M.M., Khalafallah, M.G., Hassanien, E.A.: Prediction of wastewater treatment plant performance using artificial neural networks. Environ. Modell. Softw. **19**(10), 919–928 (2004)
2. Foschi, J., Turolla, A., Antonelli, M.: Artificial neural network modeling of full-scale UV disinfection for process control aimed at wastewater reuse. J. Environ. Manage. **300**, 113790 (2021). https://doi.org/10.1016/j.jenvman.2021.113790
3. Hong, Y., Paik, B.C.: Evolutionary multivariate dynamic process model induction for a biological nutrient removal process. J. Environ. Eng. **133**(12), 1126–1135 (2007)
4. Lee, J.W., et al.: Sequential modelling of a full-scale wastewater treatment plant using an artificial neural network. Bioproc. Biosyst. Eng. **34**(8), 963–973 (2011)
5. Sorour, M.T., Bahgat, L.M.F.: Simulation analysis of ASM/Takács models in the BSM1 configuration. Environ. Technol. Lett. **27**(10), 1163–1170 (2006)
6. Vrečko, D., Hvala, N., Strazar, M.: The application of model predictive control of ammonia nitrogen in an activated sludge processs. Water Sci. Technol. **64**(5), 1115–1121 (2011)
7. Haimi, H., Mulas, M., Corona, F., et al.: Data-derived soft-sensors for biological wastewater treatment plants: an overview. Environ. Model. Softw. **47**, 88–107 (2013)
8. Wang, R., Yu, Y., Chen, Y., et al.: Model construction and application for effluent prediction in wastewater treatment plant: data processing method optimization and process parameters integration. J. Environ. Manag. **302**, 114020 (2022)

9. Cong, Q.M., Chai, T.Y., Yu, W.: Modeling wastewater treatment plant via hierarchical neural networks. Control Theory Appl. **26**(1), 8–14 (2009)

10. Huang, M., Ma, Y., Wan, J., et al.: A sensor-software based on a genetic algorithm based on neural fuzzy system for modeling and simulating a wastewater treatment process. Appl. Soft Comput. **27**, 1–10 (2015)

11. Bagheri, M., Mirbagheri, S.A., Ehteshami, M., et al.: Modeling of a sequencing batch reactor treating municipal wastewater using multi-layer perceptron and radial basis function artificial neural networks. Process Saf. Environ. Prot. **93**, 111–123 (2015)

12. Baban, A.: Biodegradability assessment and treatability of high strength complex industrial park wastewater. Clean-Soil Air Water **41**(10), 976–983 (2013)

Application of WOA-SVM Based Algorithm in Tumor Cell Detection Research

Zhiheng Luo, Yanping Li[(✉)], Dongyue Luo, Lin Wang, and Jiajun Zhu

School of Information and Electrical Engineering, Shandong Jianzhu University,
Jinan 250101, China
Liyanping0531@126.com

Abstract. Early screening of cancer is crucial for patients' survival time and quality of life. Based on this, this essay suggests a tumor cell detection technique based on a combination of whale optimization technique is proposed and support vector machine. Firstly, tumor cells are categorized using the support vector machine technique; secondly, the support vector machine's parameters have been optimized using the whale optimization technique to continue increasing the model's accuracy; eventually, a variety of approaches are dreamed up for comparison in order to confirm the success of the suggested approach. The program's findings demonstrate that the previously proposed model's accuracy is around 5% better than that of the conventional support vector machine training regimen, demonstrating the proposed model's cost effectiveness in the field of breast cancer cell proof of identity.

Keywords: Whale optimization algorithm · Support vector machine algorithm · Tumor tell detection · Parameter finding

1 Introduction

Over the past few years, the incidence of tumors has been on the rise due to changes in living environment and increased life pressures. Breast cancer is considered one of the most devastating illnesses for women, making cancer screening a crucial component of medical check-ups. As a result, accurate screening for breast tumor cells has become a major research focus. However, the current adjuvant cancer screening heavily relies on the subjective judgment of pathologists, which is both time-consuming and labor-intensive. To overcome this challenge, researchers both nationally and internationally have found that modern computer-aided diagnostic techniques can effectively reduce the impact of subjective judgment and improve the precision of diagnosing benign and malignant breast tumors in patients.

Due to the wide application of artificial intelligence algorithms, good results have also been achieved in the field of breast tumor cell detection. Existing studies have mostly used a single model to screen tumor cells, and the literature [1] used the construction of deep convolution neural net-work (deep-CNN) to achieve the task of segmentation of breast cancer cell nuclei based on the morphology of breast cancer cell nuclei. The literature [2] obtained the average volume, specific surface area, isovolume sphere radius and

© The Author(s), under exclusive license to Springer Nature Singapore Pte Ltd. 2023
H. Zhang et al. (Eds.): NCAA 2023, CCIS 1869, pp. 119–131, 2023.
https://doi.org/10.1007/978-981-99-5844-3_9

volume percentage of the nucleus, mitochondria and whole cells by 3D reconstruction of microscopic images of apoptosis-inducing breast cancer cells, and then used a BP neural network model to classify the cells. The literature [3] trained models based on U-Net networks using physicians' pre-labelled datasets, which were used to implement cell detection. The literature [4] used neural networks to simulate the brain recognition process, with the bottom layer extracting primary features and the top layer combining and abstracting the bottom features to propose an optimized deep neural network cancer cell recognition method. In the literature [5], VGG16 was improved to reduce the number of convolutional layers and convolutional kernels in the model, and a streamlined convolutional neural network model SVGG16 was proposed for the recognition of lumps in the region of interest, in response to the feature redundancy problem in the process of mammography lump recognition based on deep learning. However, the accuracy of a single model does not achieve rational results, and numerous researchers have used a hybrid model combining multiple algorithms to improve detection accuracy. A breast cancer prediction model based on the combination of the sequence forward selection algorithm (SFS) and the support vector machine algorithm (SVM) classifier was proposed in the literature [6] in the interest of enhancing the precision of computer-aided diagnostic techniques for fine needle aspiration cytopathology of breast cancer. In [7], a biogeographic optimization algorithm (BBO) was proposed to optimize the model of BP neural network for breast cancer diagnosis. 10 quantitative features of microscopic images of the nuclei of breast tumor tissue were used as the input to the network, and benign and malignant breast tumors were used as the output of the network. In [8], the original features of breast cancer were transformed using random sampling to enhance their feature representation, and the enhanced features were learned layer by layer through a cascaded random forest, and finally the classification results were output by a classifier. The literature [9] wants to introduce a made by mixing model based on support vector machines and systematic clustering (H-SVM), in which the systematic clustering algorithm was used for feature selection, and the similarity between the original tumor data and the hidden pattern was calculated by the slave function for feature reconstruction; the reconstructed dataset was used as the new feature set to train the classifier by method is proposed for support vector machines to check the classification effect. The literature [10] predicted the same dataset by using logistic regression model (LR), Gaussian kernel function support vector machine (SVM), and feedforward neural network (MLP), and concluded that among them, SVM had the shortest iteration time and feedforward neural network had the highest prediction accuracy. Based on the above analysis, a hybrid model combining machine learning algorithm and heuristic algorithm for tumor cell detection can effectively improve the detection accuracy and has good application prospects.

Therefore, this paper proposes a breast cancer tumor detection model using the Whale Optimization Algorithm (WOA) in combination with SVM. Firstly, this article employs the support vector machine algorithm to classify tumor cells; secondly, in the interest of improving the efficiency of the SVM, the WOA method is utilized to optimize the SVM parameters; finally, the method is validated through comparative experiments and is found to be effective in improving the performance of the tumor cell detection model.

2 Methodology

2.1 Support Vector Machine

SVM is a classification algorithm suggested by Vapnik et al. [11] in 1964 based on statistical theory. SVM seeks to categorize the data samples by finding the most appropriate hyperplane in the feature space. Suppose the training set $\{(x_i, y_i), i = 1, 2, ..., n\}$ of size n has two classes, if x_i belongs to class 1, y_i is 1. If x_i relates to class 2, y_i is -1. If there is a hyperplane $\omega^T x + b = 0$ that can classify the samples correctly, then the optimal hyperplane for classification is the maximizes the total amount of the shortest distances from the samples of both classes to the plane. The process of finding the optimal hyperplane is expressed in the following equation:

$$\min \frac{1}{2}\|\omega\|^2 + c \sum_{i=1}^{n} \xi_i, \xi_i 0, i = 1, 2, ..., n \tag{1}$$

$$s.t. y_i(\omega^T x + b)1 - \xi_i, \quad i = 1, 2, ..., n \tag{2}$$

where c is the punishment factor, used to regulate the severity of the penalty for incorrectly categorized samples, to achieve a compromise between misclassified samples and algorithm complexity, and is one of the parameters to be optimized. ξ_i is the slack variable, which is used to solve the minority sample classification anomaly problem. When the data set is linearly indistinguishable, the data could well be mapped into a high-dimensional space for classification using the kernel function. Since the kernel's operation simply replaces the dot product in the higher dimensional feature space, the calculation is simply done by computed $K(x_i, x_j) = \Phi(x_i)\Phi((x_j)$. After mapping placing the information in a higher dimensional space, the pairwise problem becomes:

$$\max \sum_{i=1}^{n} \alpha_i - \frac{1}{2} \sum_{i=1}^{n} \sum_{j=1}^{n} \alpha_i \alpha_j y_i y_j K(x_i, x_j) \tag{3}$$

$$s.t. \sum_{i=1}^{n} \alpha_i y_i = 0, \quad \alpha_i 0, \quad i = 1, 2, ..., n \tag{4}$$

where α_i is the Lagrange multiplier. By solving for α, using the sequential minimum optimization algorithm, ω and b are found and the optimal hyperplane is finally obtained.

Linear, polynomial, RBF (Radial Basis Function), as well as sigmoid kernel functions are different forms of traditional SVM kernel functions. In this essay, the SVM models all adopt the sigmoid kernel function with simple structure and strong generalization ability, namely:

$$K(x_i, x_j) = \tanh(gx_i^T x_j + r) \tag{5}$$

2.2 Whale Optimization Algorithm

The whale optimization algorithm [12] incorporates in the hunting process: encircling prey, random search hunting, and spiral trajectory hunting. In this case, the parameter A is used to determine whether the hunting behavior of the humpback whale is encircling the prey or randomly searching for prey.

(1) Surrounding prey phase. In the WOA algorithm, the position of the ideal whale is taken into account as the issue variable to be fully tested, and rather than defining the victim location, the whale approaches towards this location to gradually surround the food. Relatively, in the WOA algorithm, the distance between an individual and the optimal whale needs to be calculated first:

$$\vec{D} = |\vec{C} \cdot \vec{X}^*(t) - \vec{X}(t)| \tag{6}$$

where t is the number of recent iterations, X^* is the recent local optimum response. $\vec{X}(t)$ represents a whale's specific place within the tth generation. $\vec{X}^*(t)$ will continue to modify its position with each iteration. Calculating $\vec{C}$, also known as the oscillation factor, is done as follows:

$$\vec{C} = 2 \cdot \vec{r} \tag{7}$$

Using the following phrase, the location of individuals together within whale population is updated in order to conform with the optimal individual position:

$$\vec{X}(t+1) = \vec{X}^*(t) - \vec{A} \cdot \vec{D} \tag{8}$$

where $\vec{A}$ is the convergence factor and is calculated from Eq. (7) as:

$$\vec{A} = 2\vec{a} \cdot \vec{r} - \vec{a} \tag{9}$$

where $\vec{a}$ denotes a linearly decreasing vector from 2 to 0 and r is a random number between 0 and 1.

(2) Bubble attack phase. Before swimming in a spiral contraction, the process of spiral update of position appears to have started with a computation of the distance between the individual humpback whale and the current absolutely perfect whale. The mathematical model for the spiral contraction mode during the food search process is:

$$\vec{X}(t+1) = \vec{D}' \cdot e^{bl} \cdot \cos(2\pi l) + \vec{X}^*(t) \tag{10}$$

$$\vec{D}' = \left| \vec{X}^*(t) - \vec{X}(t) \right| \tag{11}$$

where: $\vec{D}$ denotes the separation vector among the individual and the currently optimal whale; l is an arbitrary number between 0 and 1; and b is a constant to constrain the spiral logarithmically shape. In order to be able to sustain contraction and go along the spiral path towards food at the exact same time, a mathematical model of this synchronous behavior is developed as follows:

$$\vec{X}(t+1) = \begin{cases} \vec{X}^*(t) - \vec{A} \cdot \vec{D} & p < 0.5 \\ \vec{D} \cdot e^{bl} \cdot \cos(2\pi l) + \vec{X}^*(t) & p \geq 0.5 \end{cases} \tag{12}$$

(3) Random search phase. The whale controls the |A| vector to swim away to capture the prey, and when |A|>1, the individual whale conducts a random search according to the relative position, by which the worldwide search of the particular whale can be facilitated, and the ideal response globally can be acquired by processing in this way, and the specific mathematical model can be referred to the following equation:

$$\vec{D} = \left| \vec{C} \cdot \vec{X}_{rand} - \vec{X} \right| \tag{13}$$

$$\vec{X}(t+1) = \vec{X}_{rand} - \vec{A} \cdot \vec{D} \tag{14}$$

where $\vec{X}_{rand}$ is the reference whale's location vector as determined at random.

3   WOA-SVM Algorithm Model

3.1   WOA-SVM Algorithm Model Construction

When confronted with a real-world problem that needs to be solved, the Whale Optimization Algorithm (WOA) can be categorized into two types: the maximum problem and the minimum problem based on their fitness objectives. In the case of a maximum problem, the fitness is predominantly evaluated using classification accuracy. Conversely, for a minimum problem, the fitness is typically assessed using the misclassification rate. In this study, the WOA algorithm iteratively optimizes two parameters of the Support Vector Machine (SVM) to continuously enhance the generalization accuracy of the SVM model. The fitness value, in terms of cross-validation, is determined by the financial support given to the misclassification rate of the SVM model.

Setting the parameters correctly is crucial to enhance the classification performance of Support Vector Machines (SVM). Tharwa [13] used the Whale Optimization Algorithm (WOA) optimized SVM in 2017 to classify unknown drug toxicity, achieving high accuracy and sensitivity. In this study, the same algorithm is employed to identify breast cancer data. The WOA-SVM starts with initializing the population and considers two key parameters - the individual search range and the population size. In every iteration of the WOA algorithm, each individual in the population is updated based on their search outcomes. Individuals beyond the search area's boundary are re-initialized, while individuals within the territorial limits undergo either local or global searches, depending on the parameter settings. The flowchart illustrating the WOA-SVM algorithm for breast cancer identification is depicted in Fig. 1.

3.2   Evaluation Criteria for the WOA-SVM Algorithm

This research paper focuses on evaluating the performance of the WOA-SVM algorithm [14] by analyzing two components: the algorithm model and the input data. The evaluation process involves two specific testing methods. Firstly, different parameters are selected and inputted into the same model to assess their impact on the algorithm's performance. Secondly, the same dataset is fed into different models to compare their evaluation metrics, allowing for a comprehensive analysis of both the algorithm's performance and the characteristics of the input data. By quantifying the evaluation parameters

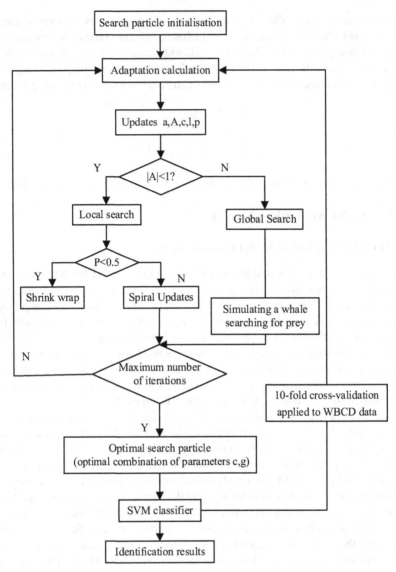

Fig. 1. Flow chart of breast cancer identification by WOA-SVM algorithm

and analyzing the corresponding algorithm data, this study aims to provide a comprehensive understanding of the WOA-SVM algorithm and its effectiveness in handling the given dataset.

In this paper, the WOA-SVM algorithm is applied to a binary classification problem. The algorithmic model first categorizes the samples into two classes: True (Positive) and False (Negative). The classification results yield four possible outcomes for each sample. Samples that are correctly predicted as true are considered True Positives (TP), while samples that are incorrectly predicted as false when they are actually true are termed

False Negatives (FN). Conversely, samples that are incorrectly predicted as true when they are actually false are referred to as False Positives (FP), and samples that are correctly predicted as false are True Negatives (TN). These categories allow for evaluating the performance of the algorithm and assessing the accuracy of the classification results.

Being the proportion of samples correctly recognized by the algorithmic model relative to all samples, accuracy is the simplest and most straightforward way to assess the performance of a model:

$$Accuracy = \frac{TP + TN}{TP + TN + FP + FN} \tag{15}$$

The percentage of samples in the true category that are both expected to be true and actually are, also known as precision or accuracy, is calculated using the method below:

$$Precision = \frac{TP}{TP + FP} \tag{16}$$

Sensitivity is the model's capacity to identify malignant patient data with accuracy. According to the following formula, the algorithm model's missed diagnosis rate decreases as the algorithm's sensitivity increases:

$$Sensitivity = \frac{TP}{TP + FN} \tag{17}$$

The model's specificity measures how well it can identify non-patients. According to the formula below, the model's misdiagnosis rate decreases as the specificity value increases:

$$Specificity = \frac{TN}{TP + FP} \tag{18}$$

4 Analysis of Arithmetic Cases Based on the WOA-SVM Algorithm

4.1 Data set Analysis

The experimental research conducted in this paper utilized Microsoft Windows 10 as the operating system, and the MATLAB R2019a software platform as the main tool. To facilitate the research work, certain functions were incorporated from the LIBSVM toolbox [15] developed by Prof. Zhiren Lin. This toolbox offers a range of application features and source codes related to SVM models, enabling users to conduct data analysis and study. The combination of the Windows 10 operating system, MATLAB software, and the LIBSVM toolbox provided a comprehensive and efficient research platform for the experiments carried out in this paper.

The data used in this paper is the Wisconsin Breast Cancer Dataset (WBCD) [16] donated by Dr. William H. Wolberg from the UCI Machine Learning Database, which was taken from the nucleus features of cells taken from fine needle aspirations of breast lumps. The dataset contains 699 samples, each with nine feature attributes and one category label, and sample data are shown in Table 1.

Table 1. Sample data from WBCD section

Attribute	value	value	value	value	value
Clump Thickness	5	5	3	6	4
Uniformity of Cell Size	1	4	1	8	1
Uniformity of Cell Shape	1	4	1	8	1
Marginal Adhesion	1	5	1	1	3
Single Epithelial Cell Size	2	7	2	3	2
Bare Nuclei	1	10	2	4	1
Bland Chromatin	3	3	3	3	3
Normal Nucleoli	1	2	1	7	1
Mitoses	1	1	1	1	1
Class	2	2	2	2	2

Table 2. WBCD dataset information

Name	Samples	Begin	Malignant	Characteristics	Category
WBCD	683	444	239	9	2

16 sample data items were largely decided to be lacking data during the sample data authentication process. These samples have now been completely eradicated, and the new data set has been managed to acquire, as shown in Table 2.

The experiments are conducted in two validation methods, the first one uses the conventional hold-out strategy, which first separates the data set D into two sets that are mutually exclusive., assuming that the two sets are S and T respectively, $S \cup U = D, S \cap T = 0$. The algorithm model is trained using the set S, where the algorithm's capacity for generalization needs to be analyzed, and this testing process needs to be realized by the set T. The only way to maintain the consistency of the data distribution is to choose the test set or training set based on two categories and randomly according to a medium ratio. For example, if 70% of the data samples are required as the training set, then 70% of the samples are randomly selected from the benign and malignant samples, respectively, for model training, and the remaining samples are used for evaluation. The following validation proposition uses 10-foldcross-Validation [17] and divides the dataset into 10 equal parts, with 1 part as the test data and 9 of them in turn as the training data. The benign and malignant samples in this study are each separated into 10 equal portions. One benign and one malignant sample are then chosen as the validation set without being repeated, and the remaining data samples are utilized as the test set and put through simulation trials. The folded cross-validation schematic is shown in Fig. 2.

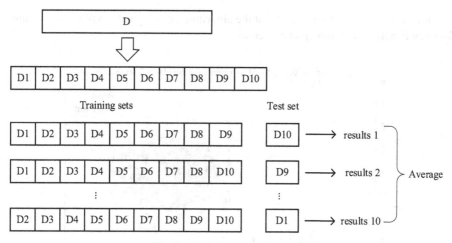

Fig. 2. 10-foldcross-Validation

4.2 Analysis of the Findings from Experiments

The WOA-SVM model proposed in this paper dichotomizes the benign and malignant cell data. As can be seen from Fig. 3, the conventional SVM model alone is not effective in classifying the cell data, and the data within the hyperplane interval cannot be accurately judged, which may lead to misdiagnosis by doctors.

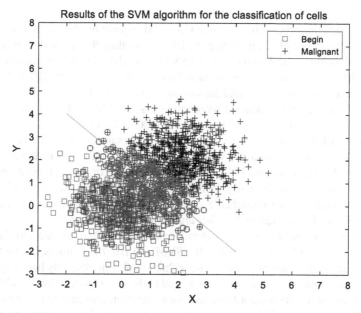

Fig. 3. SVM algorithm for the classification of benign and malignant tumor cells

However, it is clear from Fig. 4 that the algorithm does a good job of distinguishing data that is outside the hyperplane interval.

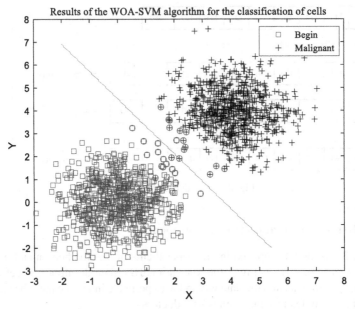

Fig. 4. WOA-SVM algorithm for the classification of benign and malignant cells

To showcase the superiority of the WOA-SVM algorithm, this study compared it with several other algorithms, including the BBO algorithm, the H-SVM algorithm, the SFS-SVM algorithm, and the conventional SVM algorithm. The WOA population size was set to 50, and the training was conducted for 500 iterations. In the BBO algorithm, the migration rate was set to 0.2, and the variance was set to 0.1. The number of clusters K in the H-SVM algorithm was set within the range of 2 to 30. For the conventional SVM model, C was set to 0.1, and g was set to 0.005. The evolutionary curves of the four optimization algorithms were compared, and the results are presented in Fig. 5.

Figure 5 demonstrates that the WOA-SVM algorithm outperforms the SFS-SVM algorithm, SVM algorithm, and H-SVM algorithm in terms of data optimization during the first 50 iterations. Additionally, as the number of iterations increases, the WOA-SVM algorithm displays a remarkable capability to seek superiority and has a faster convergence speed compared to the other algorithms at 150 iterations.

Table 3 clearly demonstrates the superior test metrics of the WOA-SVM, H-SVM, and SFS-SVM algorithms compared to the SVM and BBO algorithms for the WBCD dataset. These three optimization-based algorithms exhibit significantly higher accuracy and benign/malignant accuracy. When compared to the SVM model with default parameters, the three optimization algorithms achieve accuracy improvements of 5.41%, 3.9%, and 2.89% respectively, highlighting the substantial impact of key parameter values on SVM model performance.

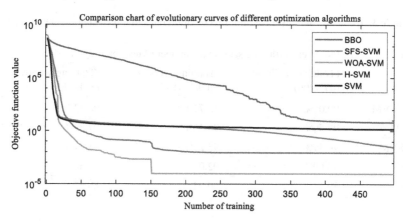

Fig. 5. Comparison of evolutionary curves for different optimization algorithms

Table 3. Ratio of tumor identification from simulation results comparison (70% of the selected training set) (477/206)

Methods	ACC	Benign Diagnosis Rate	Malignancy confirmation rate
WOA-SVM	99.01%	98.50%	99.96%
H-SVM	97.50%	98.16%	97.86%
SFS-SVM	96.49%	98.62%	96.50%
SVM	93.60%	96.11%	87.57%
BBO	89.14%	91.49%	68.18%

In terms of training set samples, with 70% of the data used for training, the WOA-SVM algorithm outperforms the H-SVM and SFS-SVM algorithms. Specifically, WOA-SVM achieves a 1.51% higher accuracy rate than H-SVM and a 2.52% higher accuracy rate than SFS-SVM. Furthermore, in terms of accuracy for malignant samples, WOA-SVM attains an impressive 99.96%, surpassing H-SVM by 2.1% and SFS-SVM by 3.46%. These results highlight the effectiveness of the WOA-SVM algorithm in achieving high accuracy and improving the classification performance for the WBCD dataset.

The number of samples in the test set was raised, whilst the total amount of samples used for training was gradually decreased to better represent the real performance of each algorithm in identifying and classifying the WBCD dataset. The decreasing trend of accuracy results can be observed for each algorithm in Table 4, which can be attributed to the reduction of learnable samples. However, even with this reduction, the WOA-SVM algorithm maintains its superiority over the other four algorithms in terms of accuracy. This further emphasizes the effectiveness and stability of the WOA-SVM algorithm in handling reduced sample sizes and highlights its ability to consistently deliver accurate classification results.

Table 4. Comparison of average tumor recognition rates in three experimental groups

Methods	(Number of training set groups/number of test set groups)		
	(Training set 70%) (477/206)	(Training set 60%) (409/274)	(Training set 50%) (341/342)
WOA-SVM	99.01%	98.71%	98.33%
H-SVM	97.50%	98.25%	97.31%
SFS-SVM	96.49%	93.21%	91.37%
SVM	93.60%	92.02%	90.56%
BBO	89.14%	90.02%	88.52%

5 Conclusion

This paper proposes a WOA-SVM-based algorithm to optimize breast cancer tumor cell accuracy continuously. The SVM is used as the primary classifier for the WBCD dataset. To overcome the fluctuation of experimental results caused by different penalty factors and kernel function parameters selection in SVM, a novel heuristic algorithm, WOA, is introduced for parameter optimization. The experimental results demonstrate that the WOA-SVM algorithm outperforms traditional SVM, H-SVM, SFS-SVM, and BBO algorithms, with 5.41%, 1.51%, 2.52%, and 9.87% accuracy improvement, respectively. The proposed algorithm can improve the ability of doctors to correctly identify breast cancer malignant tumors, which is significant for breast cancer patients to receive timely and effective treatment. The approach suggested in this research offers a fresh perspective regarding the way to investigate pathology. To further increase the precision of breast cancer cell detection, additional in-depth investigations will be carried out in the future on the data optimization method.

References

1. Krithiga, R., Geetha, P.: Deep learning based breast cancer detection and classification using fuzzy merging techniques. Mach. Vis. Appl. **31**(7–8) (2020). https://doi.org/10.1007/s00138-020-01122-0
2. Hu, H., et al.: Machine learning for identification of apoptotic breast cancer cells based on 3D morphological parameters. Sci. Technol. Innov. **20**, 117–118 (2020)
3. Han, H., Wei, B., Sui, D., Li, S.: A U-Net-Based method for detection of cancer cells in pathological sections of breast cancer. Precision Clin. Med. **33**(06), 471–473+477 (2018)
4. Yang, X., Wang, Z., Li, J.: Research cancer cell recognition system based on deep neural network. Software Guide **19**(03), 65–68 (2020)
5. Pan, A., Xu, S., Cheng, Y., She, Y.: Breast mass image recognition based on SVGG16. J. South-Central Univ. Natl. (Natl. Sci. Edn.) **40**(04), 410–416 (2021)
6. Lai, S., Liu, Q., Yu, L., Liu, W., Yang, R., Jin, H.: Construction of breast cancer prediction model based on SFS-SVM. Chin. J. Med. Phys. **36**(07), 826–829 (2019)
7. Li, H.: BBO-based optimized BP neural network for breast cancer diagnosis. Shanxi Electron. Technol. (5), 35–36, 44 (2018)

8. Li, J., He, R.: Research on breast cancer detection method based on a deep forest algorithm. New Gener. Technol. **8**, 11–16 (2021)
9. Yu, Y., Fan, C., Zhu, R., Xiong, H.: Breast cancer diagnosis based on combined H-SVM model. Intell. Comput. Appl. **10**(11), 97–100+105 (2020)
10. Wang, D., Huang, Y.: Prediction of breast cancer based on SVM-MLP. Microcomput. Appl. **38**(01), 130–133+138 (2022)
11. Vapnik, V.: The nature of statistical learning theory. Springer science&. Business media (2013)
12. Mirjalili, S., Lewis, A.: The whale optimization algorithm. Adv. Eng. Softw. **95**, 51–67 (2016)
13. Tharwat, A., Moemen, Y.S., Hassanien, A.E.: Classification of toxicity effect sof biotransformed hepatic drugs using whale optimized support vector machines. J. Biomed. Inform. **68**, 132–149 (2017)
14. Handelman, G.S., Kok, H.K., Chandra, R.V., et al.: Peering into the black box ofartificial intelligence: evaluation metrics of machine learning methods. Am. J. Roentgenol. **212**(1), 38–43 (2019)
15. Chang, C.C., Lin, C.J., LIBSVM: A library for support vector machines. ACM Trans. Intell. Syst. Technol. **2**(3), 27 (2011)
16. Oskouei, R.J., Kor, N.M., Maleki, S.A.: Data mining and medical world: breast cancers' diagnosis, treatment, prognosis and challenges. Am. J. Cancer Res. **7**(3), 610–627 (2017)
17. Liang, Z., Li, Z., Lai, J.: Application of 10-fold cross-validation in the evaluation of generalization ability of prediction models and the realization in R. Chin. J. Hospital Stat. **27**(04), 289–292 (2020)

MM-VTON: A Multi-stage Virtual Try-on Method Using Multiple Image Features

Guojian Li, Haijun Zhang[(✉)], Xiangyu Mu, and Jianghong Ma

Department of Computer Science, Harbin Institute of Technology, Shenzhen 518055, China
{hjzhang,majianghong}@hit.edu.cn,
21b951013@stu.hit.edu.cn

Abstract. Virtual try-on allows users to see how they look without actually trying the clothes on during their purchase. This technology has numerous applications in the display of clothing effects and is especially useful during the pandemic, because it enables remote try-on without physical contact. The major limitations of current virtual try-on methods, however, lie in the difficulty of addressing clothing deformation, edge synthesis, etc. In this study, we present a new three-stage virtual try-on method to reduce the reliance on clothing regions in human images. To achieve this, we design a new semantic prediction module to fully remove clothing-related information from human images. Additionally, we introduce a new try-on module to fuse the extracted features using an adversarial loss, resulting in significant improvements on the try-on image quality. Experimental results have demonstrated the effectiveness of our method, which achieves competitive results in comparison to state-of-the-art methods.

Keywords: Densepose · Image synthesis · Semantic prediction · Virtual try-on

1 Introduction

Virtual try-on refers to the technology of changing clothes from clothes image to human body image. With the development of artificial intelligence (AI) and virtual reality (VR), virtual try-on technology has been widely used by many fashion retailers to replace the clothes on models, reducing the costs of hiring models and the time spent on changing clothes for users. At the same time, the development of the Internet has also made more and more people choose online shopping. Especially with the emergence of the epidemic, almost every person who owns a smartphone has the experience of buying clothes online. Virtual try-on offers significant benefits to consumers in terms of enhancing purchase experience. By allowing users to visualize the effect of clothing on themselves without physically trying them on in stores, this technology not only reduces the

© The Author(s), under exclusive license to Springer Nature Singapore Pte Ltd. 2023
H. Zhang et al. (Eds.): NCAA 2023, CCIS 1869, pp. 132–146, 2023.
https://doi.org/10.1007/978-981-99-5844-3_10

time and efforts of customers required to complete a purchase, but also enables them to try on multiple outfits in home. Consequently, this not only minimizes the inconvenience of repeated trips to physical stores but also facilitates the sales process for merchants.

The earliest virtual try-on method is VITON [14], which used a two-stage method to generate try-on images from coarse to fine. Afterwards, most mainstream virtual try-on methods are constructed based on VITON, such as CP-VTON [27], CP-VTON+ [20], Viton-GAN [15], etc. These methods usually employ a two-stage mode, which warp the cloth in the first stage and generat the try-on images in the second stage. However, those methods still have the following issues: (1) to a certain extent, a try-on image depends on the clothing region of the original human image; (2) the expression ability of features extracted from human images is not strong enough; and (3) the effect of the generated image is not photo-realistic enough.

At present, there exist several methods to address these issues, such as CP-VTON [27] and its improved version of the clothing-irrelevant feature expression, which reduce the dependence on the clothing region of an original human image to a certain extent. Moreover, SieveNet [16] and ACGPN [29] add a semantic prediction module to generate semantic information and provide richer features for the final generator. In addition, the Viton-GAN [15] model adds an adversarial loss function based on CP-VTON to improve the quality of the generated image. Although these methods have addressed the above problems to some extent, they have not achieved particularly remarkable results.

To further deal with these problems, this paper proposes a 2D image-based multi-stage and multi-feature virtual try-on method, called MM-VTON, which introduces a new semantic prediction module, using multiple features of a human image as inputs to predict the semantic distribution of the targeted try-on image. First, our proposed semantic prediction module leverages the densepose [12] features, instead of the foreground mask used in traditional methods. This approach ensures that the try-on image remains unaffected by clothing semantics present in the original image. Second, a new generation module is proposed by using the warped cloth and the predicted semantic distribution information to generate the final try-on image. At the same time, an adversarial loss is added to make the model generated from the generator more realistic. Third, our work constructs a new three-stage try-on model. In the first stage, it takes the irrelevant clothing features and cloth image as input, and uses an FPN [18] network to extract multi-level features. Then, in line with the ClothFlow [13] and the PFAFN [8], an appearance flow module is utilized in our work to warp a cloth image. In the second stage, a semantic prediction module is developed to predict the semantic distribution of the try-on image. In the third stage, the features extracted in the first two stages are integrated to generate the final try-on image. The experimental results show that the quality of try-on images generated by our method is better than that of compared methods.

The remainder of this article is organized as follows. In Sect. 2, we briefly introduce the current technology about virtual try-on and image synthesis. Our proposed MM-VTON is presented in Sect. 3. Experimental results including both quantitative and qualitative results in comparison with state-of-the-art methods are illustrated in Sect. 4. Finally, we conclude the paper in Sect. 5.

2   Related Work

2.1   Virtual Try-on

Virtual try-on techniques are mainly divided into the methods based on 3D modeling and the methods based on neural networks. Traditional virtual try-on is largely based on 3D simulation technology, e.g., the Drape [11] and the research on 3D methods proposed by Sekine *et al.* [26], as well as later studies such as Clothcap [23], Tailornet [22] and Pix2surf [17]. The methods based on 3D simulation, however, need to address several challengeable issues. For example, the construction of a 3D dataset is extremely complex, the cost of 3D equipment is high, and the computational complexity of model calculations is large. Due to these problems, it is difficult for methods based on 3D images to achieve widespread applications.

The first virtual try-on method based on neural network is VITON [14], which first generates a coarse try-on image and warps the cloth by Thin Plate Spline (TPS) [2]),and then refines the coarse try-on image to generate the final try-on image. Many subsequent methods were improved based on VITON, such as CP-VTON [27], CP-VTON+ [20], SieveNet [16], ACGPN [29], MG-VTON [7], VITON-HD [6], etc. In terms of cloth deformation, we employ the appearance flow method, which is the same as the PFAFN [8], to warp the cloth. Moreover, we propose a new three-stage virtual try-on network, which shows superior performance on our dataset in comparison to current mainstream models.

2.2   Image Synthesis

Generative Adversarial Network (GAN) [10,30,32,33]was firstly proposed to synthesize images in many tasks. The two most important structures of the GAN are the generator and discriminator. The generator is able to generate images similar to the corresponding real samples, while the discriminator is able to distinguish the truth or false of the generated images and the real images. During this confrontation training, an image generated by the generator tends to be constantly close to its corresponding real image. In particular, DCGAN [28] uses a convolutional neural network to improve the basic structure of GAN. In order to add prior information to the network, CGAN [21] adds some conditions in the inputs of the network to control the process of image synthesis. Moreover, some variants of GANs such as improvedGAN [25], AW-GAN [4], infoGAN [5], and bigGAN [3], have been reported recently.

In addition to GANs, U-Net [24] network,based on the framework of encoder and decoder, is also widely used in the field of image synthesis. It is a classic fully convolutional network, with no full connection operation. For its special framework, it is widely used in the field of image segmentation and image synthesis, especially in the field of medical image processing. In recent years, there have emerged many variants of U-Net, such as V-UNet [19], UNet++ [35], Res-UNet [31].

2.3 Features of Our Work

In the virtual try-on domain, the training of a model is based on a number of data pairs consisting of human images and their corresponding cloth images. However, due to the lack of images of the same person wearing different cloth, it is imperative to manually extract clothing-irrelevant features. This results in two issues. Firstly, if the original cloth information present in a human image is not entirely eliminated during the training process, the model may become overly dependent on it, leading to try-on images containing elements of the original cloth information. Secondly, this may cause the try-on image to display artifacts, ultimately affecting its authenticity. Therefore, it is critical to address these challenges to ensure accurate virtual try-on applications.

To deal with the above problems, we propose an MM-VTON, which adds a semantic prediction module and takes advantage of the densepose features to completely remove the cloth information in human images as well as eliminate the cloth artifacts in their try-on images.

3   MM-VTON

3.1   Overview of MM-VTON

This paper proposes a multi-stage virtual try-on network, called MM-VTON, which utilizes multiple features of images. The overall framework of MM-VTON is shown in Fig. 1. We extract a variety of image features, including the foreground mask of a cloth image, the human parsing feature, and the densepose- and openpose-based features. The irrelevant clothing features are then combined. The proposed approach for virtual try-on consists of three stages: a clothes warping module (CWM), a semantic prediction module (SPM), and a generative module (GM). In the first stage, the CWM uses two FPN networks to extract the features of an input image to generate the cloth appearance flow information, and then utilizes an appearance flow model to warp the cloth image. In the second stage, the SPM uses densepose to remove the clothing information in the human image, and predicts semantic distribution information of the try-on image based on the Res-UNet network. Finally, the GM fuses the preserved human image regions and the effective features generated by the CWM and the SPM at previous stages, and then produces the final try-on image.

3.2   Clothes Warping Module (CWM)

TPS [1] has been widely used as a clothing transformation method in previous studies. However, its efficacy is severely limited when the degree of cloth warping

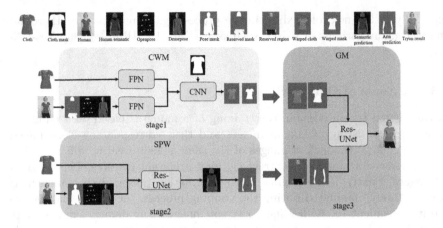

Fig. 1. The framework of MM-VTON

is too high or the cloth is blocked. In order to address this issue, we utilize the appearance flow model [34] to warp cloth, generating an offset of the pixels from a source image to the target image. The specific process is shown in Fig. 2.

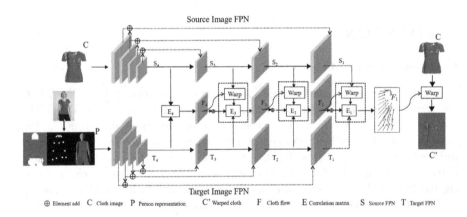

Fig. 2. The process of the CWM

As shown in Fig. 2, the CWM can be simplified into three steps: Frist, the FPN network is designed to generate multi-level features of cloth image and irrelevant clothing features. Second, for each level of the feature map, we use the current level of appearance flow information to warp clothing features, and then calculate the correlation coefficient with the target feature map. Third, we take the calculated correlation coefficient as the CNN input to calculate the next level of the characteristic map and the corresponding residual features. Subsequently, CWM executes through the second and third steps in the feature map of each layer.

The loss function in CWM includes L1 loss and VGG loss, which can be expressed as:

$$L_{warp} = \lambda_l L_1 + \lambda_p L_p,$$ (1)

where the λ_l and the λ_p represent the proportion of L1 loss and VGG loss, respectively. The L_1 and L_p in Eq. 1 are formulated as follows:

$$L_1 = \|S_I - I\| + \|S_{mask} - I_{mask}\|,$$ (2)

$$L_p = \sum_m \|\varphi_m(S_I) - \varphi(I)\|,$$ (3)

where S_I represents a warped cloth image, I represents a real image, which is the clothing region in the human image, S_{mask} represents the mask of the warped cloth, I_{mask} represents the mask of the clothing area in the model image, and φ_m represents the output features of layer m in the pre-trained VGG network.

3.3 Semantic Prediction Module (SPM)

SieveNet [16] is the first model to apply the semantic prediction method to the virtual try-on task. This model employed the cloth image, human mask, and openpose as input. However, these inputs may lead to inaccurate predictions of human body parsing due to the lack of human body pose information. Moreover, a human mask may make the model rely on the cloth region of an original human image, due to the fact that human mask features contain the cloth contour of a human image. To address these issues, we introduce the densepose feature and a special network to predict the semantic information of try-on images according to human body posture features and cloth images. The process of this stage is shown in Fig. 3.

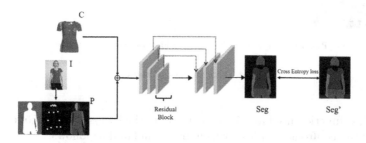

Fig. 3. The process of the SPM

The input of SPM includes densepose features, clothing irrelevant features, and cloth image, and its output conveys the semantic information of a try-on image. The SPM is constructed on the basis of the Res-UNet architecture, which consists of five layers of U-Net. Each layer's basic unit comprises a residual block composed of a three-layer convolutional neural network.

The loss function used in this stage is Cross-Entropy Loss in the form of:

$$L_{seg} = -\frac{1}{N} \sum_{i=1}^{N} \sum_{j=1}^{C} y_{ij} \log(p_{ij}), \tag{4}$$

where N indicates the number of samples, that is, the total number of pixels in the image, C indicates the number of all categories, y_{ij} indicates whether the true category of the i pixel is the j category, p_{ij} indicates that the model predicts the i pixel is the probability of class j.

3.4   Generative Module (GM)

During the GM stage, the Res-UNet architecture is extensively employed to generate the ultimate try-on image, by using the same structure as the SPM. Its structure is shown in Fig. 4. GM fuses a warped cloth image, the mask of cloth image, irrelevant clothing features, and the predicted semantic information of the try-on image corresponding to the final try-on image.

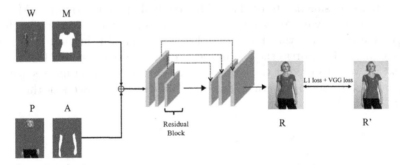

⊕ Features concat   P Preserved area   A Arm mask   W Warped cloth   M Mask of warped cloth   R Generated image   R' Real image

Fig. 4. The process of the GM

The loss function used in this stage, includes the L1 loss, the VGG perception loss, and the confrontation loss, which are formulated as follows:

$$L_{tryon} = \lambda_{l1} L_{l1} + \lambda_{tp} L_{tp}, \tag{5}$$

where λ_{l1} and λ_{tp} represent the proportion of L1 loss and VGG loss, respectively. These three losses are shown in the following equations:

$$L_{l1} = \frac{1}{CHW} \left(\|S_I - I\|_1 \right), \tag{6}$$

$$L_{tp} = \sum_j \frac{1}{C_j H_j W_j} \|\phi_j (S_I) - \phi_j (I)\|_2^2, \tag{7}$$

where ϕ_j represents the jth convolutional layer of the VGG network, S_I represents the try-on image generated by our method, and I represents the human image.

4   Experiments

Extend experiment were conducted to demonstrate the performance of MM-VTON in this section. Firstly, we presented the details of experimental setup, such as dataset, evaluation metrics, and implementation details. Afterword, the quantitative and qualitative results were reported with the state-of-art methods.

4.1   Experimental Setup

Dataset. In order to examine the performance of our model, we utilized a large number of cloth images with the resolution of 256*192 and human images with the resolution of 256*192 from the Website. We collected a total of 3,1847 image pairs, of which 27,866 image pairs were used as training set and 3,891 image pairs were used as test set. In addition to the acquisition of original images, there are four features of an image extracted before training the model: (1) cloth mask, which is the foreground area of the cloth image; (2) human parser features, which are the analytic diagram of the human image; (3) openpose-based features, which are the human posture features represented by 18 different key points; and (4) densepose features, which are the human posture features represented by 24 different planes. As a result, our virtual try-on dataset contains these four features and the pair of cloth and human images.

Evaluation Metrics. Different metrics are employed for evaluating different modules. For the semantic prediction model, we use the Mean Pixel Accuracy (MPA) to evaluate the accuracy of semantic prediction. MPA calculates the accuracy of each pixel prediction category in the image. In our network, the human image pixels are divided into 20 categories according to the Lool Into Person (LIP) dataset [9]. For the try-on image, Structure Similarity Index Measure (SSIM) and Fréchet Inception Distance (FID) are selected as the evaluation indicators to compare with the other virtual try-on models. SSIM calculates the similarity of images from three aspects, which contain luminance, contrast, and structure. FID calculates the connection between the generated image and the real image through the inception model. In addition to these two indicators, we also select some images for visual comparison.

Compared Methods and Implementation Details. We compared our SPM with SieveNet-seg, which is the semantic prediction module in SieveNet [16].

Our SPM leverages Res-UNet as the primary network, while the origin U-Net was employed in SieveNet-seg. To verify the semantic prediction ability of our adopted Res-UNet, we compare SPM with the semantic prediction module of SieveNet adopting original UNet and Res-UNet. Moreover, five state-of-art methods for the virtual try-on task were compared, including CP-VTON [27], Viton-GAN [15], SieveNet [16], ACGPN [29], and PFAFN [8].

MM-VTON is implemented in Python using PyTorch. We employed Adam to optimize each module in our model, and the learning rate was set to 0.00005. Our network was training with batch size of 4 on an NVIDIA 3060 GPU. We ended the training of our network at 100 epochs.

4.2   Quantitative Results

To evaluate semantic prediction models, Table 1 compares the proposed SPM with SieveNet-seg in terms of MPA. Our findings indicate that replacing the origin U-Net with Res-UNet in SieveNet led to a significant improvement in MPA. It is observed that the performance of SPM, when using densepose as the new feature input, is comparable to that of the SieveNet-seg model with Res-UNet, which does not utilize the densepose feature. However, our SPM outperforms the latter model in terms of actual semantic prediction, as depicted in Fig. 5 and discussed in Subsect. 4.3.

Table 1. Comparison of SPM and SieveNet-seg

Model	Densepose	MPA
SieveNet-seg with U-Net	No	85.62%
SieveNet-seg with Res-UNet	No	**89.38%**
SPM	Yes	**88.24%**

We compared MM-VTON with five baseline methods to evaluate the try-on results in Table 2. In general, we note that our method produces better or comparable performance than other methods. Specifically, our method has almost the same SSIM and FID as PBAFN. In addition, MM-VTON has obvious advantages over other four methods. For example, our method achieves 67% increase in SSIM and 90% reduction in FID over Viton-GAN. The main reason of exceptional performance was ascribed to the fact that the densepose feature we adopted can represent the pose of human model in a image accurately.

4.3   Qualitative Results

We present a visual analysis of the performance of our SPM in comparison to state-of-the-art baseline methods, as depicted in Fig. 5. The visual depictions in the second row, specifically in the first and third columns, effectively demonstrate

Table 2. Comparison of MM-VTON and five baseline methods

Model	SSIM	FID
CP-VTON	0.70	58.3
Viton-GAN	0.52	57.74
SieveNet	0.78	67.37
ACGPN	0.80	26.39
PBAFN	0.86	5.41
MM-VTON	**0.87**	**5.34**

the compromised semantic predictions in the arm region when a short-sleeved vest is replaced with a long-sleeved T-shirt on a fashion model for SieveNet-seg with orign U-Net and SieveNet-seg with Res-UNet. In contrast, our SPM achieves better performance. Furthermore, After analyzing images in the second, fourth, and sixth columns, it becomes evident that SieveNet-seg is not successful in making precise predictions for human hands. In contrast, our proposed method exhibits superior performance in the same task.

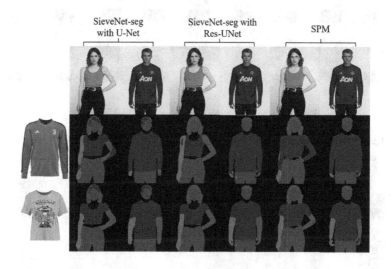

Fig. 5. The generation result of the three modules

Our study addresses the challenging task of changing short to long sleeves in virtual try-on. The effectiveness of our proposed method is demonstrated in Fig. 6, where it is evident that the model's arm is fully covered by the garment, and the hand area is accurately preserved in the first, fourth, sixth, and eighth columns. To further evaluate the performance of our approach, we conduct visual comparisons with other competing models, as illustrated in Fig. 7. From the presented examples, it is evident that our proposed MM-VTON approach generates the most realistic and compelling try-on results, particularly in terms of hand and on-shirt graphics synthesis. Specifically, CP-VTON and SieveNet suffer from compromised color richness in their generated images, and the body

Fig. 6. The generation results of MM-VTON

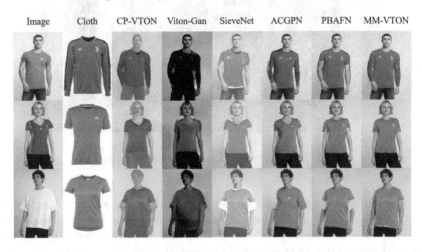

Fig. 7. The results of try-on images generated by MM-VTON and five baseline methods

parts in these images appear blurred. Viton-GAN generates images with dark colors, which blur details such as texture on the clothing. FBAFN exhibits limitations in retaining the original human body image, resulting in compromised performance in trying on different types of clothing.

5 Conclusion

In this paper, we introduce a multi-stage virtual try-on method to change clothes in a human body. The proposed network extracts and warps the target clothes using multiple image features. Specifically, we develop an MM-VTON, which consists of three modules for clothing warping, human body semantic prediction, and human body and clothing synthesis. To evaluate the performance of our proposed network, we performed comparative experiments on our dataset collected from websites with SSIM and FID. Experimental results demonstrate that our proposed model achieves promising results in terms of image quality in comparison with state-of-the-art methods.

Acknowledgement. This work was supported in part by the National Natural Science Foundation of China under Grant no. 61972112 and no. 61832004, the Guangdong Basic and Applied Basic Research Foundation under Grant no. 2021B1515020088, the Shenzhen Science and Technology Program under Grant no. JCYJ20210324131203009, and the HITSZ-J&A Joint Laboratory of Digital Design and Intelligent Fabrication under Grant no. HITSZ-J&A-2021A01.

References

1. Belongie, S.J., Malik, J., Puzicha, J.: Shape matching and object recognition using shape contexts. IEEE Trans. Pattern Anal. Mach. Intell. **24**(4), 509–522 (2002). https://doi.org/10.1109/34.993558
2. Bookstein, F.L.: Principal warps: thin-plate splines and the decomposition of deformations. IEEE Trans. Pattern Anal. Mach. Intell. **11**(6), 567–585 (1989). https://doi.org/10.1109/34.24792
3. Brock, A., Donahue, J., Simonyan, K.: Large scale GAN training for high fidelity natural image synthesis. In: 7th International Conference on Learning Representations, ICLR 2019, New Orleans, LA, USA, 6–9 May 2019. OpenReview.net (2019)
4. Chandaliya, P.K., Nain, N.: AW-GAN: face aging and rejuvenation using attention with wavelet GAN. Neural Comput. Appl. **35**(3), 2811–2825 (2023)
5. Chen, X., Duan, Y., Houthooft, R., Schulman, J., Sutskever, I., Abbeel, P.: Infogan: interpretable representation learning by information maximizing generative adversarial nets. In: Lee, D.D., Sugiyama, M., von Luxburg, U., Guyon, I., Garnett, R. (eds.) Advances in Neural Information Processing Systems 29: Annual Conference on Neural Information Processing Systems 2016, 5–10 December 2016, Barcelona, Spain, pp. 2172–2180 (2016)
6. Choi, S., Park, S., Lee, M., Choo, J.: VITON-HD: high-resolution virtual try-on via misalignment-aware normalization. In: IEEE Conference on Computer Vision and Pattern Recognition, CVPR 2021, virtual, 19–25 June 2021, pp. 14131–14140. Computer Vision Foundation/IEEE (2021). https://doi.org/10.1109/CVPR46437.2021.01391

7. Dong, H., et al.: Towards multi-pose guided virtual try-on network. In: 2019 IEEE/CVF International Conference on Computer Vision, ICCV 2019, Seoul, Korea (South), 27 October–2 November 2019, pp. 9025–9034. IEEE (2019).https://doi.org/10.1109/ICCV.2019.00912

8. Ge, Y., Song, Y., Zhang, R., Ge, C., Liu, W., Luo, P.: Parser-free virtual try-on via distilling appearance flows. In: IEEE Conference on Computer Vision and Pattern Recognition, CVPR 2021, virtual, 19–25 June 2021, pp. 8485–8493. Computer Vision Foundation/IEEE (2021). https://doi.org/10.1109/CVPR46437.2021.00838

9. Gong, K., Liang, X., Zhang, D., Shen, X., Lin, L.: Look into person: self-supervised structure-sensitive learning and a new benchmark for human parsing. In: 2017 IEEE Conference on Computer Vision and Pattern Recognition, CVPR 2017, Honolulu, HI, USA, 21–26 July 2017, pp. 6757–6765. IEEE Computer Society (2017). https://doi.org/10.1109/CVPR.2017.715

10. Goodfellow, I.J., et al.: Generative adversarial nets. In: Ghahramani, Z., Welling, M., Cortes, C., Lawrence, N.D., Weinberger, K.Q. (eds.) Advances in Neural Information Processing Systems 27: Annual Conference on Neural Information Processing Systems 2014, 8–13 December 2014, Montreal, Quebec, Canada, pp. 2672–2680 (2014)

11. Guan, P., Reiss, L., Hirshberg, D.A., Weiss, A., Black, M.J.: DRAPE: dressing any person. ACM Trans. Graph. 31(4), 35:1–35:10 (2012). https://doi.org/10.1145/2185520.2185531

12. Güler, R.A., Neverova, N., Kokkinos, I.: Densepose: dense human pose estimation in the wild. In: 2018 IEEE Conference on Computer Vision and Pattern Recognition, CVPR 2018, Salt Lake City, UT, USA, 18–22 June 2018, pp. 7297–7306. Computer Vision Foundation/IEEE Computer Society (2018). https://doi.org/10.1109/CVPR.2018.00762

13. Han, X., Huang, W., Hu, X., Scott, M.R.: Clothflow: a flow-based model for clothed person generation. In: 2019 IEEE/CVF International Conference on Computer Vision, ICCV 2019, Seoul, Korea (South), 27 October–2 November 2019, pp. 10470–10479. IEEE (2019). https://doi.org/10.1109/ICCV.2019.01057

14. Han, X., Wu, Z., Wu, Z., Yu, R., Davis, L.S.: VITON: an image-based virtual try-on network. In: 2018 IEEE Conference on Computer Vision and Pattern Recognition, CVPR 2018, Salt Lake City, UT, USA, 18–22 June 2018, pp. 7543–7552. Computer Vision Foundation/IEEE Computer Society (2018). https://doi.org/10.1109/CVPR.2018.00787

15. Honda, S.: VITON-GAN: virtual try-on image generator trained with adversarial loss. In: Fusiello, A., Bimber, O. (eds.) 40th Annual Conference of the European Association for Computer Graphics, Eurographics 2019 - Posters, Genoa, Italy, 6–10 May 2019, pp. 9–10. Eurographics Association (2019). https://doi.org/10.2312/egp.20191043

16. Jandial, S., Chopra, A., Ayush, K., Hemani, M., Kumar, A., Krishnamurthy, B.: Sievenet: a unified framework for robust image-based virtual try-on. In: IEEE Winter Conference on Applications of Computer Vision, WACV 2020, Snowmass Village, CO, USA, 1–5 March 2020, pp. 2171–2179. IEEE (2020). https://doi.org/10.1109/WACV45572.2020.9093458

17. Lei, J., Sridhar, S., Guerrero, P., Sung, M., Mitra, N., Guibas, L.J.: Pix2Surf: learning parametric 3D surface models of objects from images. In: Vedaldi, A., Bischof, H., Brox, T., Frahm, J.-M. (eds.) ECCV 2020. LNCS, vol. 12363, pp. 121–138. Springer, Cham (2020). https://doi.org/10.1007/978-3-030-58523-5_8

18. Lin, T., Dollár, P., Girshick, R.B., He, K., Hariharan, B., Belongie, S.J.: Feature pyramid networks for object detection. In: 2017 IEEE Conference on Computer Vision and Pattern Recognition, CVPR 2017, Honolulu, HI, USA, 21–26 July 2017, pp. 936–944. IEEE Computer Society (2017). https://doi.org/10.1109/CVPR.2017.106
19. Milletari, F., Navab, N., Ahmadi, S.: V-net: fully convolutional neural networks for volumetric medical image segmentation. In: Fourth International Conference on 3D Vision, 3DV 2016, Stanford, CA, USA, 25–28 October 2016, pp. 565–571. IEEE Computer Society (2016). https://doi.org/10.1109/3DV.2016.79
20. Minar, M.R., Tuan, T.T., Ahn, H., Rosin, P., Lai, Y.K.: CP-VTON+: clothing shape and texture preserving image-based virtual try-on. In: The IEEE/CVF Conference on Computer Vision and Pattern Recognition (CVPR) Workshops (2020)
21. Mirza, M., Osindero, S.: Conditional generative adversarial nets. CoRR abs/1411.1784 (2014). https://arxiv.org/abs/1411.1784
22. Patel, C., Liao, Z., Pons-Moll, G.: Tailornet: predicting clothing in 3D as a function of human pose, shape and garment style. In: 2020 IEEE/CVF Conference on Computer Vision and Pattern Recognition, CVPR 2020, Seattle, WA, USA, 13–19 June 2020, pp. 7363–7373. Computer Vision Foundation/IEEE (2020). https://doi.org/10.1109/CVPR42600.2020.00739
23. Pons-Moll, G., Pujades, S., Hu, S., Black, M.J.: Clothcap: seamless 4D clothing capture and retargeting. ACM Trans. Graph. **36**(4), 73:1–73:15 (2017). https://doi.org/10.1145/3072959.3073711
24. Ronneberger, O., Fischer, P., Brox, T.: U-Net: convolutional networks for biomedical image segmentation. In: Navab, N., Hornegger, J., Wells, W.M., Frangi, A.F. (eds.) MICCAI 2015. LNCS, vol. 9351, pp. 234–241. Springer, Cham (2015). https://doi.org/10.1007/978-3-319-24574-4_28
25. Salimans, T., Goodfellow, I.J., Zaremba, W., Cheung, V., Radford, A., Chen, X.: Improved techniques for training GANs. In: Lee, D.D., Sugiyama, M., von Luxburg, U., Guyon, I., Garnett, R. (eds.) Advances in Neural Information Processing Systems 29: Annual Conference on Neural Information Processing Systems 2016, 5–10 December 2016, Barcelona, Spain, pp. 2226–2234 (2016)
26. Sekine, M., Sugita, K., Perbet, F., Stenger, B., Nishiyama, M.: Virtual fitting by single-shot body shape estimation. In: International Conference on 3D Body Scanning Technologies, pp. 406–413. Citeseer (2014)
27. Wang, B., Zheng, H., Liang, X., Chen, Y., Lin, L., Yang, M.: Toward characteristic-preserving image-based virtual try-on network. In: Ferrari, V., Hebert, M., Sminchisescu, C., Weiss, Y. (eds.) ECCV 2018. LNCS, vol. 11217, pp. 607–623. Springer, Cham (2018). https://doi.org/10.1007/978-3-030-01261-8_36
28. Wu, Q., Chen, Y., Meng, J.: DCGAN-based data augmentation for tomato leaf disease identification. IEEE Access **8**, 98716–98728 (2020). https://doi.org/10.1109/ACCESS.2020.2997001
29. Yang, H., Zhang, R., Guo, X., Liu, W., Zuo, W., Luo, P.: Towards photo-realistic virtual try-on by adaptively generating↔preserving image content. In: 2020 IEEE/CVF Conference on Computer Vision and Pattern Recognition, CVPR 2020, Seattle, WA, USA, 13–19 June 2020, pp. 7847–7856. Computer Vision Foundation/IEEE (2020). https://doi.org/10.1109/CVPR42600.2020.00787
30. Zhang, H., Sun, Y., Liu, L., Wang, X., Li, L., Liu, W.: Clothingout: a category-supervised GAN model for clothing segmentation and retrieval. Neural Comput. Appl. **32**, 4519–4530 (2020)

31. Zhang, Z., Liu, Q., Wang, Y.: Road extraction by deep residual U-net. IEEE Geosci. Remote Sens. Lett. **15**(5), 749–753 (2018). https://doi.org/10.1109/LGRS.2018. 2802944

32. Zhou, D., et al.: Learning to synthesize compatible fashion items using semantic alignment and collocation classification: an outfit generation framework. IEEE Trans. Neural Netw. Learn. Syst. (2022)

33. Zhou, D., Zhang, H., Li, Q., Ma, J., Xu, X.: Coutfitgan: learning to synthesize compatible outfits supervised by silhouette masks and fashion styles. IEEE Trans. Multimedia (2022)

34. Zhou, T., Tulsiani, S., Sun, W., Malik, J., Efros, A.A.: View synthesis by appearance flow. In: Leibe, B., Matas, J., Sebe, N., Welling, M. (eds.) ECCV 2016. LNCS, vol. 9908, pp. 286–301. Springer, Cham (2016). https://doi.org/10.1007/978-3-319-46493-0_18

35. Zhou, Z., Rahman Siddiquee, M.M., Tajbakhsh, N., Liang, J.: UNet++: a nested U-net architecture for medical image segmentation. In: Stoyanov, D., et al. (eds.) DLMIA/ML-CDS -2018. LNCS, vol. 11045, pp. 3–11. Springer, Cham (2018). https://doi.org/10.1007/978-3-030-00889-5_1

Self-Attention-Based Reconstruction
for Planetary Magnetic Field

Ziqian Yan, Zhao Kang$^{(\boxtimes)}$, and Ling Tian

University of Electronic Science and Technology of China, Chengdu 611731, China
2020270903008@std.uestc.edu.cn, {zkang,lingtian}@uestc.edu.cn

Abstract. In space exploration missions, missing data is a common issue in planetary magnetic field measurements due to various reasons. As a feasibility validation, this study introduces, for the first time, models based on self-attention mechanisms into the task of planetary magnetic field data reconstruction and selects the SAITS (Self-Attention-based Imputation for Time Series) model for validation due to its outstanding performance. Experimental results on real magnetic field datasets demonstrate that the self-attention-based temporal model can effectively reconstruct missing values in planetary magnetic field data associated with multiple variables.

Keywords: Planetary Magnetic Field · Missing Value Reconstruction · Self-Attention Mechanism

1 Introduction

A planet's magnetic field is generated by its internal physical and chemical processes, and studying these magnetic fields is an important task of space exploration [1–3]. For example, studying the geomagnetic field can provide key information for mineral resource development, seismology research [4, 5], and more. However, accurately measuring the planetary magnetic field can be challenging in space exploration missions due to the complex and unknown nature of the space environment, resulting in missing values or low quality [6]. To comprehensively understand planets and facilitate further research, it is essential to develop methods for missing value reconstruction.

In recent decades, several methods for reconstructing missing values in geomagnetic field data have been proposed. For instance, [7] proposed a data assimilation-based method to predict geomagnetic field changes, which shows a good linear regression relationship between the predicted values and the actual values. [8] compared the performance of artificial neural networks, support vector regression (SVR), generalized regression neural networks (GRNN), and the K-nearest neighbor algorithm (KNN) in missing value reconstruction on the MAGDAS-9 ground electromagnetism dataset and found that SVR had good reconstruction performance. [9] compared the performance of support vector machine, random forest, gradient boosting, long short-term memory (LSTM) in time series and spatial magnetic field data and demonstrated that LSTM had good performance in both types of geomagnetic data. [10] proposed a coupled matrix

© The Author(s), under exclusive license to Springer Nature Singapore Pte Ltd. 2023
H. Zhang et al. (Eds.): NCAA 2023, CCIS 1869, pp. 147–159, 2023.
https://doi.org/10.1007/978-981-99-5844-3_11

factorization method, which can reconstruct incomplete magnetic data from multiple sites and model their spatial correlation, and its reconstruction accuracy is higher than that of previous methods. However, there is no research specifically focused on reconstructing missing values in planetary magnetic field data beyond Earth, which could be more challenging in practice.

With the introduction of the transformer proposed by [11], self-attention-based models for missing value reconstruction have gained popularity, such as [12–14]. Numerous experiments have shown that self-attention-based models can more effectively learn the correlations in data, and their reconstruction performance is superior to that of traditional machine learning models. However, there have been no records of developing and applying such models to magnetic field data reconstruction.

As a feasibility study of using the self-attention mechanism in magnetic field data reconstruction, this paper introduces the use of the Self-Attention-based Imputation for Time Series (SAITS) model [15] for magnetic field data reconstruction and compares its reconstruction results with traditional RNN models (M-RNN [16] and BRITS [17]) and the classic transformer model. Experiments show that the self-attention-based SAITS model can achieve good performance in magnetic field data reconstruction for missing values.

2   SAITS Model Introduction

The SAITS model utilizes the self-attention mechanism and improves it to better model time series. At the same time, it uses a joint optimization training approach, which can effectively improve the training performance of the model.

2.1   Joint-Optimization Training Approach

This joint optimization training approach includes two tasks: Masked Imputation Task (MIT) and Observed Reconstruction Task (ORT) (Fig. 1).

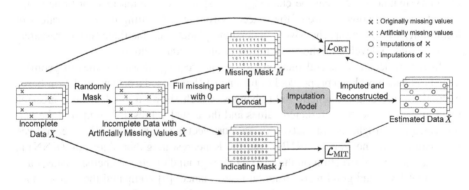

Fig. 1. A schematic illustration of the joint training method in [15]

Masked Imputation Task (MIT). The masked imputation task is a task of reconstructing values with artificially masked values. In this method, some values in the input of the model that are not missing will be randomly masked by humans, which means that these values are also missing for the model. After random masking, the actual input time series is denoted as $\hat{X}$, its corresponding missing mask is $\hat{M}$, and the model output is denoted as $\tilde{X}$.

To distinguish between missing values caused by human masking and missing values in the data itself, an indicator mask I is introduced. The definitions of $\hat{M}$ and I are shown as follows, where d represents the dimension of the time series, and t represents the time step:

$$\hat{M}_t^d = \begin{cases} 1 \text{ if } X_t^d \text{ is observed} \\ 0 \text{ if } X_t^d \text{ is missing} \end{cases} \quad I_t^d \begin{cases} 1 \text{ if } X_t^d \text{ is artifically masked} \\ 0 \text{ otherwise} \end{cases} \tag{1}$$

After completing the masked imputation task, the model calculates the loss between the artificially masked values and their corresponding reconstructed values, which is computed using the following equation:

$$L_{\text{MIT}} = l_{\text{MAE}}\left(\tilde{X}, X, I\right) \tag{2}$$

Here, the calculation of mean absolute error (MAE) is defined by Eq. (3), where X_e, X_t, and M represent the reconstructed values, actual values, and mask of the time series, respectively:

$$l_{\text{MAE}}(X_e, X_t, M) = \frac{\sum_{d}^{D} \sum_{t}^{T} |(X_e - X_t) \odot M|_t^d}{\sum_{d}^{D} \sum_{t}^{T} M_t^d} \tag{3}$$

Observed Reconstruction Task (ORT). The observed reconstruction task is a reconstruction task based on the observed (i.e., non-missing) values. The loss in observed reconstruction task is calculated as follows:

$$L_{\text{ORT}} = l_{\text{MAE}}\left(\tilde{X}, X, \hat{M}\right) \tag{4}$$

In the SAITS model, the masked imputation task and observed reconstruction task are inseparable. The masked imputation task allows the model to predict missing values as accurately as possible, and the observed reconstruction task ensures that the model converges to the distribution of non-missing data.

2.2 SAITS Model

SAITS consists of two diagonally masked self-attention (DMSA) blocks and a weighted combination block (Fig. 2).

Diagonally-Masked Self-Attention. In the traditional transformer model [9], its self-attention mechanism can be described by Eq. (5), where $\underline{d_k}$ is the dimensionality of matrices Q and K, d_v is the dimensionality of matrix V, and the product of Q and K is the attention scores.

$$SA(Q, K, V) = \text{softmax}\left(\frac{QK^T}{\sqrt{d_k}}\right)V \tag{5}$$

SAITS improves the self-attention mechanism of the transformer by introducing diagonal masking. Diagonal masking sets the values on the diagonal of the attention scores to $-\infty$, making the attention weights of the corresponding time steps close to 0 after the function. After masking, the value at time step t in the input time series cannot "see" itself, making the model only be able to reconstruct the value at that position based on other time steps during training. The self-attention mechanism with diagonal masking is called diagonally masked self-attention (DMSA). The feed-forward network in the model can be described by Eq. (6), diagonal masking and DMSA can be described by Eqs. (7) and (8), the positional encoding can be described by Eq. (9), and the illustration of diagonal masking is shown in Fig. 3. The diagonally-masked multi-head attention (DiagMaskedMHA) can be described by Eqs. (10) and (11).

$$\text{FFN}(x) = \text{ReLU}(xW_1 + b) \tag{6}$$

$$\text{DiagMask}(x_{i,j}) = \begin{cases} x_{i,j}, & i \neq j \\ -\infty, & i = j \end{cases} \tag{7}$$

$$\text{DMSA}(Q, K, V) = \text{softmax}(\text{DiagMask}(\frac{QK^T}{\sqrt{d_k}}))V \tag{8}$$

$$\text{PosEnc}(pod, 2i) = \sin(\frac{pos}{10000^{\frac{2i}{d_{\text{model}}}}}) \quad \text{PosEnc}(pod, 2i+1) = \cos(\frac{pos}{10000^{\frac{2i}{d_{\text{model}}}}}) \tag{9}$$

$$\text{DiagMaskedMHA}(x) = \text{concat}(h_1, h_2, ..., h_h)W^o \tag{10}$$

$$h_i = \text{DMSA}(xW_i^Q, xW_i^K, xW_i^V) \tag{11}$$

Diagonally-Masked Self-Attention Block (DMSA Block). In the first DMSA block, the input is the concatenation of the observed sequence $\hat{X}$ and its missing mask $\hat{M}$. The output of this block, denoted as $\tilde{X}_1$ (Learned Representation 1), is used to fill in the missing values in $\tilde{X}_1$ to obtain the imputed sequence $\hat{X}'$. The computation in the first DMSA block can be described by Eqs. (12), (13), (14) and (15), where p in Eq. (12) represents the positional encoding and N in Eq. (13) denotes the number of stacked layers.

$$e = \left[\text{concat}(\hat{X}, \hat{M})W_e + b_e\right] + p \tag{12}$$

$$z = \{\text{FFN}(\text{DiagMaskedMHA}(e))\}^N \tag{13}$$

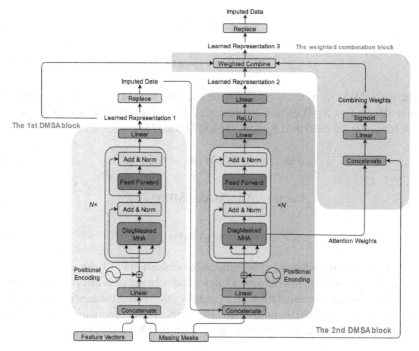

Fig. 2. Schematic illustration of the SAITS model architecture in [15]

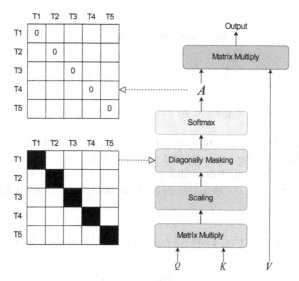

Fig. 3. Schematic illustration of the diagonal masking in [15]

$$\tilde{X}_1 = zW_z + b_z \tag{14}$$

$$\hat{X}' = \hat{M} \odot \hat{X} + (1 - \hat{M}) \odot \tilde{X}_1 \tag{15}$$

The second DMSA block is similar to the first one, with input $\hat{X}'$, and the computed result is denoted as $\tilde{X}_2$ (Learned Representation 2). The computation in the second DMSA block can be described by Eqs. (16), (17) and (18).

$$a = \left[\text{concat}(\hat{X}', \hat{M})W_a + b_a \right] + p \tag{16}$$

$$\beta = \{\text{FFN}(\text{DiagMaskedMHA}(a))\}^N \tag{17}$$

$$\tilde{X}_2 = \text{ReLU}(\beta W_\beta + b_\beta)W_\gamma + b_\gamma \tag{18}$$

Weighted Combination Block. The weighted combination block dynamically weights $\tilde{X}_1$ and $\tilde{X}_2$ based on the temporal dependencies and missing data information to obtain the final $\tilde{X}_3$ (Learned Representation 3). $\tilde{X}_3$ replaces the missing values in the input $\hat{X}$ to obtain $\hat{X}_c$, which is the final reconstructed time series. The computation in the weighted combination block can be described by Eqs. (18), (19), (20) and (21).

$$\hat{A} = \frac{1}{h} \sum_i^h A_i \tag{19}$$

$$\eta = \text{sigmoid}(\text{concat}(\hat{A}, \hat{M})W_\eta + b_\eta) \tag{20}$$

$$\tilde{X}_3 = (1 - \eta) \odot \tilde{X}_1 + \eta \odot \tilde{X}_2 \tag{21}$$

$$\tilde{X}_c = \hat{M} \odot \tilde{X} + (1 - \hat{M}) \odot \tilde{X}_3 \tag{22}$$

Loss Function. The total loss function of the model is a weighted sum of the mean squared errors between the reconstructed values and the observed values for both the mask reconstruction task and the value reconstruction task, where the weight is an adjustable parameter. The computation of the loss function is shown in Eqs. (23), (24), and (25).

$$L_{\text{ORT}} = \frac{1}{3}(l_{\text{MAE}}(\tilde{X}_1, X, \hat{M}) + l_{\text{MAE}}(\tilde{X}_2, X, \hat{M}) + l_{\text{MAE}}(\tilde{X}_3, X, \hat{M})) \tag{23}$$

$$L_{\text{MIT}} = l_{\text{MAE}}(\hat{X}_c, X, I) \tag{24}$$

$$L = L_{\text{ORT}} + \lambda L_{\text{MIT}} \tag{25}$$

3 Experiments

3.1 Description of Experimental Data

The dataset used in the experiments consists of Jupiter's magnetic field data obtained by the Juno spacecraft on December 5–6, 2020 [18] and the Beijing Ming Tombs station geomagnetic dataset from 1991 to 2001 [19].

Jupiter Magnetic Field Data from Juno Spacecraft. The magnetic field data includes 3 dimensions: magnetic field intensity in the X, Y, and Z directions, with a sampling frequency of 1Hz. The experiment selects the data from December 6, 2020 as the test set, with a total of 86,400 time steps, while sequence length is set as 1,000. After preprocessing, data missing was randomly added to the test set to simulate the missing

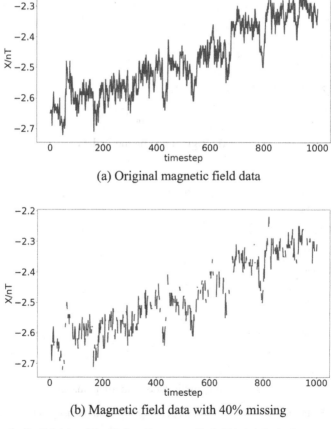

(a) Original magnetic field data

(b) Magnetic field data with 40% missing

Fig. 4. Partial data of the X-direction magnetic field intensity in the test set.

magnetic field data in actual applications, with a total missing rate of 40%. Figure 4b shows some of the data in test set.

Beijing Ming Tombs Station Geomagnetic Dataset. The data is in IAGA-2002 format and includes the horizontal magnetic field intensity (H), magnetic declination (D), and vertical magnetic field intensity (Z), with a sampling frequency of 1 per minute. The experiment selects data from January to July 2001, with a ratio of 7:1:1 for the training set, validation set, and test set. The test set contains 40,000 time steps, while sequence length is set as 1,000. After preprocessing, missing data were randomly added to the test set to simulate missing magnetic field data in practical applications, with a total missing rate of 40%. Figure 5b shows some of the data in test set.

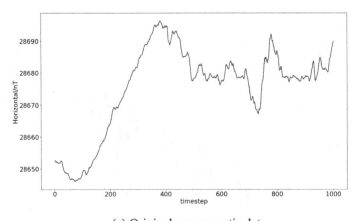

(a) Original geomagnetic data

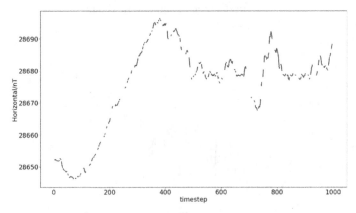

(b) Geomagnetic data with 40% missing

Fig. 5. Partial data of the horizontal geomagnetic field intensity in the test set

3.2 Reconstruction Results

To verify the reconstruction performance of SAITS on geomagnetic data, M-RNN, BRITS, and the classic transformer model were selected for comparison, and the mean absolute error (MAE), root mean square error (RMSE), and R-squared were used as metrics. The MAE and RMSE reflect the error of the reconstruction result relative to the actual value, while the R-squared reflects the correlation between the reconstruction result and the actual value. Higher R-squared indicates stronger correlation. Tables 1 and 2 show the evaluation results of different reconstruction methods on the test sets of the two datasets in different dimensions.

Table 1. Evaluation results of different reconstruction methods on Jupiter magnetic field data

Dimension	Evaluation metrics	loss			
		M-RNN	BRITS	transformer	SAITS
X	MAE	0.9129	0.2339	0.0307	**0.0121**
	RMSE	1.5453	0.4543	0.0576	**0.0241**
	R-squared	0.4883	0.9558	0.9993	**0.9999**
Y	MAE	0.8893	0.1611	0.0287	**0.0117**
	RMSE	1.4575	0.3369	0.0546	**0.0241**
	R-squared	0.5545	0.9762	0.9994	**0.9999**
Z	MAE	0.0911	0.0215	0.0134	**0.0080**
	RMSE	0.1658	0.0430	0.0266	**0.0160**
	R-squared	0.5883	0.9723	0.9894	**0.9961**

Tables 1 and 2 record the results of the four reconstruction models on different dimensions of the test sets of two datasets, with the best results highlighted in bold. From Tables 1 and 2, it can be seen that the M-RNN model exhibits significant errors in the reconstruction results for both datasets, while the BRITS model performs relatively well, although most of its metrics are lower than those of the self-attention based model. The reconstruction results shown in Figs. 6 and 7 further demonstrate the presence of noticeable glitches in the reconstructions produced by the M-RNN and BRITS models, making them less unsuitable for practical applications.

In contrast, the transformer model and the SAITS model, which based on self-attention mechanisms, achieve good reconstruction accuracy on both magnetic field datasets, highlighting the advantage of self-attention mechanisms in capturing the temporal dependences of magnetic fields. Additionally, due to the improvements in self-attention mechanism, the SAITS model outperforms the transformer model and other RNN-based models in reconstructing time-series magnetic field data.

Table 2. Evaluation results of different reconstruction methods on geomagnetic data

Dimension	Evaluation metrics	loss			
		M-RNN	BRITS	transformer	SAITS
H	MAE	6.0399	1.7105	0.3599	**0.1014**
	RMSE	10.8435	3.4152	0.6927	**0.2427**
	R-squared	0.3048	0.9310	0.9972	**0.9997**
D	MAE	0.5110	0.0656	0.0247	**0.0086**
	RMSE	0.9419	0.1615	0.0505	**0.0212**
	R-squared	0.3202	0.9800	0.9980	**0.9997**
Z	MAE	1.7404	0.3930	0.0679	**0.0262**
	RMSE	3.6235	0.9558	0.1351	**0.0517**
	R-squared	0.6210	0.9736	0.9995	**0.9999**

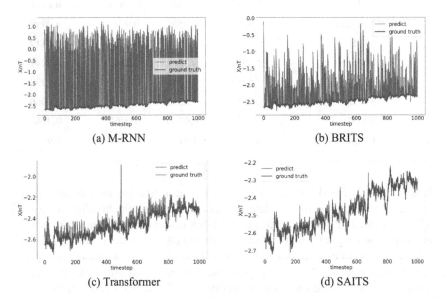

(a) M-RNN

(b) BRITS

(c) Transformer

(d) SAITS

Fig. 6. Reconstruction results of different reconstruction methods on Jupiter magnetic field data

The effectiveness and superiority of the SAITS model in reconstructing planetary magnetic field data demonstrate the potential of self-attention mechanisms in this field. However, it is noted that there are still some glitches in the reconstruction results of the Transformer and SAITS models on the Jupiter magnetic field dataset, indicating the complexity of the deep-space environment and the difficulty of data reconstruction.

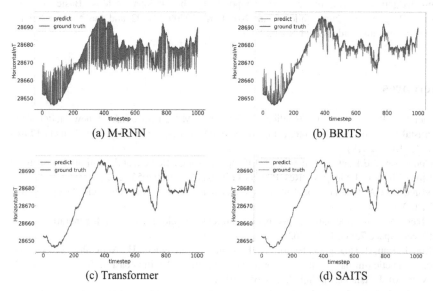

Fig. 7. Reconstruction results of different reconstruction methods on geomagnetic data

Therefore, although the SAITS model achieves promising results in terms of reconstruction accuracy, further research and improvements are necessary to better adapt to the complex task of reconstructing planetary magnetic field data.

4   Conclusion and Future Work

This feasibility study introduces the self-attention mechanism for reconstructing missing values in planetary magnetic field data. Experimental results on real datasets demonstrate that the transformer and SAITS models, based on self-attention mechanism, achieve good reconstruction performance, showcasing the potential of attention mechanism in planetary magnetic field reconstruction.

However, in space missions, the unique characteristics of planetary magnetic fields and the complexity of the space environment make data reconstruction challenging. Thus, it is necessary to developing a model specifically tailored for reconstructing planetary magnetic field data, comparing to general time series imputation models. Furthermore, leveraging the spatio-temporal correlations among data for improved reconstruction accuracy, as observed in the fusion of different measurements in the Earth's magnetic field, is not applicable to planetary magnetic fields.

In future work, we will investigate reconstruction models that capture the unique characteristics of planetary magnetic field data compared to other types of sequential data. Additionally, we will explore the feasibility of utilizing measurement data from scientific payloads other than magnetometers to assist in reconstructing magnetic field data, aiming for better reconstruction performance.

Acknowledgements. This research was supported by the National Defense Basic Scientific Research Program of China under Grant No. JCKY2020903B002 and the Natural Science Foundation of China under Grant No. 62276053.

References

1. Langlais, B., Thébault, E., Houliez, A., Purucker, M.E., Lillis, R.J.: A new model of the crustal magnetic field of Mars using MGS and MAVEN. J. Geophys. Res. Planets **124**(6), 1542–1569 (2019)
2. Connerney, J.E.P., et al.: A new model of Jupiter's magnetic field at the completion of Juno's Prime Mission. J. Geophys. Res. Planets **127**(2), e2021JE007055 (2022).
3. Stanley, S.: A dynamo model for axisymmetrizing Saturn's magnetic field. Geophys. Res. Lett. **37**(5) (2010)
4. Alken, P., et al.: International geomagnetic reference field: the thirteenth generation. Earth, Planets Space **73**(1), 1–25 (2021)
5. Xu, G., Han, P., Huang, Q., Hattori, K., Febriani, F., Yamaguchi, H.: Anomalous behaviors of geomagnetic diurnal variations prior to the 2011 off the Pacific coast of Tohoku earthquake (Mw9. 0). J. Asian Earth Sci. **77**, 59–65 (2013)
6. Liu, L., Tian, L., Kang, Z., Wan, T.: Spacecraft anomaly detection with attention temporal convolution networks. Neural Comput. Appl. **35**, 9753–9761 (2023)
7. Lhuillier, F., Aubert, J., Hulot, G.: Earth's dynamo limit of predictability controlled by magnetic dissipation. Geophys. J. Int. **186**(2), 492–508 (2011)
8. Muhammad Asraf, H., Nur Dalila, K.A., Abd Latiff, Z.I., Jusoh, M.H., Akimasa, Y.: Missing data imputation of MAGDAS-9's ground electromagnetism with supervised machine learning and conventional statistical analysis models. Alexandria Eng. J. **61**(1), 937–947 (2022)
9. Liu, H., et al.: A nonlinear regression application via machine learning techniques for geomagnetic data reconstruction processing. IEEE Trans. Geosci. Remote Sens. **57**(1), 128–140 (2018)
10. Liu, H., et al.: SGCast: a new forecasting framework for multilocation geomagnetic data with missing traces based on matrix factorization. IEEE Trans. Instrumen. Measure. **70**, 1–11 (2021)
11. Vaswani, A., Shazeer, N., Parmar, N., Uszkoreit, J., Jones, L., Gomez, A.N., et al.: Attention is all you need. Adv. Neural Inform. Process. Syst. **30** (2017)
12. Ma, J., Shou, Z., Zareian, A., Mansour, H., Vetro, A., Chang, S.F.: CDSA: cross-dimensional self-attention for multivariate, geo-tagged time series imputation. arXiv preprint arXiv:1905. 09904 (2019)
13. Shan, S., Li, Y., Oliva, J.B.: Nrtsi: Non-recurrent time series imputation. arXiv preprint arXiv: 2102.03340 (2021)
14. He, K., Chen, X., Xie, S., Li, Y., Dollár, P., Girshick, R.: Masked autoencoders are scalable vision learners. In: Proceedings of the IEEE/CVF Conference on Computer Vision and Pattern Recognition, pp. 16000–16009 (2022)
15. Du, W., Côté, D., Liu, Y.: Saits: Self-attention-based imputation for time series. arXiv preprint arXiv:2202.08516 (2022)
16. Yoon, J., Zame, W.R., van der Schaar, M.: Estimating missing data in temporal data streams using multi-directional recurrent neural networks. IEEE Trans. Biomed. Eng. **66**(5), 1477–1490 (2019)

17. Cao, W., Wang, D., Li, J., et al.: BRITS: Bidirectional recurrent imputation for time series. Adv. Neural Inform. Process. Syst. 31 (2018)
18. Juno magnetometer Jupiter archive. https://pds-ppi.igpp.ucla.edu/search/view?f=yes&id= pds://PPI/JNO-J-3-FGM-CAL-V1.0. Accessed 23 Jan 2023
19. The geomagnetic dataset of Beijing Ming Tombs station. http://www.csdata.org/p/35/. Accessed 23 Jan 2023

Semi-supervised Multi-class Classification Methods Based on Laplacian Vector Projection

Yangtao Xue and Li Zhang[✉]

School of Computer Science and Technology,
Soochow University, Suzhou 215006, China
20184027008@stu.suda.edu.cn, zhangliml@suda.edu.cn

Abstract. Laplacian pair-weight vector projection (LapPVP) is a binary classifier for semi-supervised learning, which seeks a pair of projection vectors only for two-class data. This paper extends LapPVP to semi-supervised multi-class classification tasks and proposes two novel semi-supervised multi-class methods, named one-versus-one LapPVP (OVO-LapPVP) and one-versus-rest LapPVP (OVR-LapPVP). By using the strategy of "one-versus-one", OVO-LapPVP decomposes a semi-supervised multi-class classification task into multiple binary problems that can be directly solved by multiple LapPVPs. Considering the concept of "one-versus-rest", OVR-LapPVP is designed for generating multiple hyperplanes for multiple classes, one for each class. The above proposed semi-supervised multi-class classification methods both consider the discriminative information of labeled data and graph structure of unlabeled data. Experiments are conducted on nine UCI datasets to display the classification performance for multi-class data. Compared with other popular semi-supervised multi-class classification methods based on manifold regularization, the proposed semi-supervised multi-class classification methods hold the advantage of LaPVP and have better performance.

Keywords: Multi-class classification · Semi-supervised learning · Manifold regularization · Vector Projection

1 Introduction

Twin support vector machine (TSVM) has become one of the popular binary classifiers because it can greatly reduce the computational complexity of support vector machine (SVM) [5]. TSVM obtains two nonparallel hyperplanes by two smaller-sized and related SVM-type problems and is a supervised binary classifier that require a large amount of labeled data. However, it is difficult for TSVM to deal with multi-classification problems. In order to inherit the advantages of TSVM, some multi-class versions of TSVM have been proposed for wider applications [9,10].

Inspired by the manifold regularization, Laplacian TSVM (LapTSVM) was proposed for semi-supervised learning and become a useful extension of TSVM [8]. Similar to LapTSVM, Laplacian twin parametric-margin SVM

H. Zhang et al. (Eds.): NCAA 2023, CCIS 1869, pp. 160–174, 2023.
https://doi.org/10.1007/978-981-99-5844-3_12

(LapTPMSVM) [12] and Laplacian least squares TSVM (LapLSTSVM) [1] are the semi-supervised versions of twin parametric-margin SVM [7] and least squares TSVM [6], respectively, which are variants of TSVM. The above three semi-supervised methods all aim to generate two nonparallel hyperplanes so that each hyperplane is closer to its own class as far as possible from the other. Laplacian pair-weight vector projection (LapPVP) was also motivated by the idea of nonparallel hyperplanes [11]. LapPVP is an excellent binary classifier compared with other nonparallel hyperplanes methods. Furthermore, LapPVP integrates the between-class scatter, the within-class scatter, and the Laplacian regularization together for semi-supervised binary classification. Each class data provides the class-specific information by computing the between-class and within-class scatters, which enhances the power of discriminative representation. The Laplacian regularization provides the graph structure of labeled and unlabeled data, which is the intrinsic geometrical structure of data.

Although these semi-supervised binary classifiers are promising, their multi-class versions have been rarely explored. To solve the semi-supervised multi-class classification problems, we extend the formulation of LapPVP to its multi-class versions and propose two novel semi-supervised multi-class methods, named one-versus-one LapPVP (OVO-LapPVP) and one-versus-rest LapPVP (OVR-LapPVP). Researchers have developed the ideas of "one-versus-one" (OVO) and "one-versus-rest" (OVR) to decompose a multi-class problem into multiple binary sub-problems [3,9]. Originally, LapPVP obtains a pair of projection vectors for two-class data. Both OVO-LapPVP and OVR-LapPVP solve a multi-class classification task using multiple LapPVPs, but differ in the number of LapPVPs. In specific, the multi-class versions of LapPVP have the following characteristics:

(1) OVO-LapPVP obtains $C(C-1)/2$ pairs of projection vectors for C-class data, whereas OVR-LapPVP achieves C nonparallel hyperplanes for C-class data, one for each class.
(2) The proposed multi-class classification methods are all solved by eigenvalue decomposition, which avoids finding solutions to complex quadratic programmings.
(3) Experimental results on UCI datasets demonstrate the effectiveness of OVO-LapPVP and OVR-LapPVP.

2   Semi-supervised Multi-class LapPVPs

In this section, we describe the proposed OVO-LapPVP and OVR-LapPVP detailly. Now, we consider a semi-supervised multi-class classification task. Let $\mathbf{X}_l = [\mathbf{x}_1, \cdots, \mathbf{x}_l]^T \in \mathbb{R}^{l \times m}$ and $\mathbf{X}_u = [\mathbf{x}_{l+1}, \cdots, \mathbf{x}_n]^T \in \mathbb{R}^{(n-l) \times m}$ be the labeled and unlabeled sample matrices, respectively, $\mathbf{x}_i \in \mathbb{R}^m$ is an m-dimensional sample, l and n are the numbers of labeled and total samples, respectively. Let y_i be the label of $\mathbf{x}_i$, where $y_i \in [1, 2, ..., C]$, and C is the number of classes. Let $\mathbf{X}_{l,c}$ be the labeled sample matrix belonging to the cth class. The total sample matrix can be denoted as $\mathbf{X} = [\mathbf{X}_{l,1}; \cdots; \mathbf{X}_{l,C}; \mathbf{X}_u] \in \mathbb{R}^{n \times m}$.

In LapPVP, we need to construct an adjacency graph using the K nearest neighbor method. The ith row and jth column element of the adjacency matrix $\mathbf{S}$ induced by the adjacency graph can be represented as

$$S_{ij} = \begin{cases} 1, & if \ \mathbf{x}_i \in N_K(\mathbf{x}_j) \ \ or \ \ \mathbf{x}_j \in N_K(\mathbf{x}_i) \\ 0, & otherwise \end{cases}, \tag{1}$$

where $N_K(\mathbf{x}_i)$ is the set of K nearest neighbors of $\mathbf{x}_i$. Then we get the classical Laplacian matrix $\mathbf{L} = \mathbf{D} - \mathbf{S}$, where $\mathbf{D}$ is the diagonal matrix with $D_{ii} = \sum_j S_{ij}$.

2.1 OVO-LapPVP

The one-versus-one strategy is a popular technique that can easily extend binary classifiers to multi-class ones. By using the one-versus-one strategy, OVO-LapPVP is presented, which can generate $C(C-1)/2$ binary classifiers for C-class data. Given a C-class dataset, we obtain the set of class pairs $Z = \{z = (c_1, c_2) | c_1, c_2 = 1, 2, \cdots, C, c_1 \neq c_2\}$. To construct a binary classifier with class pair z, suppose class c_1 is the positive class, and class c_2 as the negative one. The corresponding LapPVP classifier is trained with labeled samples in both classes c_1 and c_2 and all unlabeled samples.

Let $\mathbf{X}_z = [\mathbf{X}_{l,c_1}; \mathbf{X}_{l,c_2}; \mathbf{X}_u] \in \mathbb{R}^{n_z \times m}$ with $z = (c_1, c_2)$, where n_z is the total number of samples in both classes c_1 and c_2 and without labels. The objective functions of OVO-LapPVP with respect to z can be expressed as follows:

$$\max_{\mathbf{v}_{c_1}} \mathbf{v}_{c_1}^T \mathbf{H}_{c_1} \mathbf{v}_{c_1} - \rho_{c_1} \mathbf{v}_{c_1}^T \mathbf{X}_z^T \mathbf{L}_z \mathbf{X}_z \mathbf{v}_{c_1}$$
$$s.t. \quad \mathbf{v}_{c_1}^T \mathbf{v}_{c_1} = 1 \tag{2}$$

and

$$\max_{\mathbf{v}_{c_2}} \mathbf{v}_{c_2}^T \mathbf{H}_{c_2} \mathbf{v}_{c_2} - \rho_{c_2} \mathbf{v}_{c_2}^T \mathbf{X}_z^T \mathbf{L}_z \mathbf{X}_z \mathbf{v}_{c_2}$$
$$s.t. \quad \mathbf{v}_{c_2}^T \mathbf{v}_{c_2} = 1 \tag{3}$$

where $\mathbf{v}_{c_1} \in \mathbb{R}^m$ and $\mathbf{v}_{c_2} \in \mathbb{R}^m$ are projection vectors, ρ_{c_1} and ρ_{c_2} are regularization parameters that are greater than 0, the Laplacian matrix $\mathbf{L}_z \in \mathbb{R}^{n_z \times n_z}$ is computed based on $\mathbf{X}_z$, and the discriminative matrices $\mathbf{H}_{c_1} \in \mathbb{R}^{m \times m}$ and $\mathbf{H}_{c_2} \in \mathbb{R}^{m \times m}$ are defined as

$$\mathbf{H}_{c_1} = \alpha_{c_1} \left(\mathbf{X}_z - \mathbf{e}_z \mathbf{u}_{c_1}^T \right)^T \left(\mathbf{X}_z - \mathbf{e}_z \mathbf{u}_{c_1}^T \right)$$
$$- (1 - \alpha_{c_1})(\mathbf{X}_{c_1} - \mathbf{e}_{c_1} \mathbf{u}_{c_1}^T)^T (\mathbf{X}_{c_1} - \mathbf{e}_{c_1} \mathbf{u}_{c_1}^T) \tag{4}$$

and

$$\mathbf{H}_{c_2} = \alpha_{c_2} \left(\mathbf{X}_z - \mathbf{e}_z \mathbf{u}_{c_2}^T \right)^T \left(\mathbf{X}_z - \mathbf{e}_z \mathbf{u}_{c_2}^T \right)$$
$$- (1 - \alpha_{c_2})(\mathbf{X}_{c_2} - \mathbf{e}_{c_2} \mathbf{u}_{c_2}^T)^T (\mathbf{X}_{c_2} - \mathbf{e}_{c_2} \mathbf{u}_{c_2}^T), \tag{5}$$

where $\alpha_{c_1} \in [0,1]$ and $\alpha_{c_2} \in [0,1]$ are parameters to balance the between-class scatter matrix and the within-class one, class centers $\mathbf{u}_{c_j} = \frac{1}{|X_{l,c_j}|} \sum_{\mathbf{x}_i \in X_{l,c_j}} \mathbf{x}_i$

$(j = 1, 2)$, X_{l,c_j} is the set of labeled samples in the c_jth class, $\mathbf{e}_z \in \mathbb{R}^{n_z}$ and $\mathbf{e}_{c_j} \in \mathbb{R}^{|X_{l,c_j}|}$ $(j = 1, 2)$ are vectors of all ones.

The following theorems describe the solutions to the optimization problems (2) and (3).

Theorem 1. *The optimal solution* $\mathbf{v}_{c_1}$ *of the optimization problem (2) is the eigenvector corresponding to the maximum eigenvalue of the matrix* $(\mathbf{H}_{c_1} - \mathbf{X}_z^T \mathbf{L}_z \mathbf{X}_z)$.

Proof. To find the solution to (2), we first generate the corresponding Lagrangian function with positive multipliers λ_{c_1}. That is

$$L(\mathbf{v}_{c_1}, \lambda_{c_1}) = \mathbf{v}_{c_1}^T \mathbf{H}_{c_1} \mathbf{v}_{c_1} - \rho_{c_1} \mathbf{v}_{c_1}^T \mathbf{X}_z^T \mathbf{L}_z \mathbf{X}_z \mathbf{v}_{c_1} - \lambda_{c_1} (\mathbf{v}_{c_1}^T \mathbf{v}_{c_1} - 1). \qquad (6)$$

Next, we derive the partial derivative of $L(\mathbf{v}_{c_1}, \lambda_{c_1})$ with respect to the primal variable $\mathbf{v}_{c_1}$, and then make it equal zero, which results in

$$\begin{aligned} \frac{\partial L(\mathbf{v}_{c_1}, \lambda_{c_1})}{\partial \mathbf{v}_{c_1}} &= \left(\mathbf{H}_{c_1} - \rho_{c_1} \mathbf{X}_z^T \mathbf{L}_z \mathbf{X}_z \right) \mathbf{v}_{c_1} - \lambda_{c_1} \mathbf{v}_{c_1} = 0 \\ &\Rightarrow \left(\mathbf{H}_{c_1} - \rho_{c_1} \mathbf{X}_z^T \mathbf{L}_z \mathbf{X}_z \right) \mathbf{v}_{c_1} = \lambda_{c_1} \mathbf{v}_{c_1}. \end{aligned} \qquad (7)$$

By introducing the Tikhonov regularization term in Eq. (7), the solution can be obtained by solving a classical eigenvalue problem. Finally, the solution $\mathbf{v}_{c_1}$ can computed as the eigenvector corresponding to the maximum eigenvalue of $\left(\mathbf{H}_{c_1} - \rho_{c_1} \mathbf{X}_z^T \mathbf{L}_z \mathbf{X}_z \right)$. $\qquad \square$

Theorem 2. *The optimal solution* $\mathbf{v}_{c_2}$ *of the optimization problem (3) is the eigenvector corresponding to the maximum eigenvalue of the matrix* $(\mathbf{H}_{c_2} - \mathbf{X}_z^T \mathbf{L}_z \mathbf{X}_z)$.

The proof of Theorem 2 is similar to that of Theorem 1; thus, we omit the proof of Theorem 2. Theorems 1 and 2 provide the optimal projection vectors $\mathbf{v}_{c_1}$ and $\mathbf{v}_{c_2}$ by performing eigen-decomposition, respectively. After obtaining $\mathbf{v}_{c_1}$ and $\mathbf{v}_{c_2}$ for class pair $z = (c_1, c_2)$, we can respectively compute the distances between an unknown test point $\mathbf{x}$ and class centers $\mathbf{u}_{c_j}$ $(j = 1, 2)$ by

$$d(\mathbf{x}, \mathbf{u}_{c_1}) = \|\mathbf{v}_{c_1}^T \mathbf{x} - \mathbf{v}_{c_1}^T \mathbf{u}_{c_1}\| \qquad (8)$$

and

$$d(\mathbf{x}, \mathbf{u}_{c_2}) = \|\mathbf{v}_{c_2}^T \mathbf{x} - \mathbf{v}_{c_2}^T \mathbf{u}_{c_2}\| \quad , \qquad (9)$$

where $\|\cdot\|$ is the 2-norm of a vector. If $d(\mathbf{x}, \mathbf{u}_{c_1}) \leq d(\mathbf{x}, \mathbf{u}_{c_2})$ for class pair z, then $\mathbf{x}$ is more like a sample in class c_1.

Algorithm 1. OVO-LapPVP

Training Phase

Input: Sample matrices $\mathbf{X}_{l,c}$, $c = 1, \cdots, C$, and $\mathbf{X}_u$;

1: Obtain the set of class pairs $Z = \{z = (c_1, c_2) | c_1, c_2 = 1, 2, \cdots, C, c_1 < c_2, c_1 \neq c_2\}$.

2: **for** $\forall z \in Z$ **do**

3:     Construct an LapPVP classifier for class pair $z = (c_1, c_2)$;

4:     Obtain the training sample matrix $\mathbf{X}_z = [\mathbf{X}_{l,c_1}; \mathbf{X}_{l,c_2}; \mathbf{X}_u] \in \mathbb{R}^{n_z \times m}$ for the LapPVP with class pair (c_1, c_2), where samples in class c_1 are positive, and ones in class c_2 negative;

5:     Compute the matrices $\mathbf{H}_{c_1}$ and $\mathbf{H}_{c_2}$ by Eqs. (4) and (5), respectively;

6:     Generate the adjacency matrix $\mathbf{S}_z$ by Eq. (1) based on the training sample matrix $\mathbf{X}_z$;

7:     Compute the Laplacian matrix $\mathbf{L}_z = \mathbf{D}_z - \mathbf{S}_z$, where $\mathbf{D}_z$ is the diagonal matrix with $D_{z_{ii}} = \sum_j S_{z_{ij}}$;

8:     Obtain the projection vectors $\mathbf{v}_{c_1}$ and $\mathbf{v}_{c_2}$ according to Theorems 1 and 2, respectively;

9: **end for**

Output: Projection vector pairs $(\mathbf{v}_{c_1}, \mathbf{v}_{c_2})$, $(c_1, c_2) \in Z$.

Testing Phase

Input: Unknown test sample $\mathbf{x}$, sample matrices $\mathbf{X}_{l,c}$, $c = 1, \cdots, C$, $\mathbf{X}_u$, and projection vector pairs $(\mathbf{v}_{c_1}, \mathbf{v}_{c_2})$, $(c_1, c_2) \in Z$;

1: Initialize $\hat{\mathbf{y}} = [0, \cdots, 0]^T \in \mathbb{R}^C$;

2: **for** $\forall z \in Z$ **do**

3:     Compute the distances between $\mathbf{x}$ and class centers by Eq. (9) with optimal vectors $\mathbf{v}_{c_1}$ and $\mathbf{v}_{c_2}$;

4:     Update the c_1th element $\hat{y}_{c_1}$ or the c_2th element $\hat{y}_{c_2}$ in $\hat{\mathbf{y}}$ by Eq. (10);

5: **end for**

6: Assign the class label for $\mathbf{x}$ with Eq. (11);

Output: Estimated label for $\mathbf{x}$.

To assign a class label to $\mathbf{x}$, we construct a vote vector $\hat{\mathbf{y}} = [\hat{y}_1, \cdots, \hat{y}_C]^T$, where $\hat{y}_c$ is the vote for the cth class. Generally, the initial value of $\hat{y}_c$ is 0 for all $c = 1, \cdots, C$. For each class pair $z = (c_1, c_2)$, we update the vote for only class c_1 or class c_2. That is

$$\begin{cases} \hat{y}_{c_1} \leftarrow \hat{y}_{c_1} + 1, \; if \;\; d(\mathbf{x}, \mathbf{u}_{c_1}) \leq d(\mathbf{x}, \mathbf{u}_{c_2}) \\ \hat{y}_{c_2} \leftarrow \hat{y}_{c_2} + 1, \; otherwise \end{cases}. \qquad (10)$$

The update procedure is performed on the whole set Z. Thus, the strategy of classification rule for $\mathbf{x}$ is defined as

$$c^* = \arg \max_{c=1,\cdots,C} \hat{y}_c, \qquad (11)$$

where c^* is the estimated class label for $\mathbf{x}$. We summarize the specific procedure of OVO-LapPVP in Algorithm 1.

2.2   OVR-LapPVP

Originally, LapPVP obtains a projection vector for each class that keeps data points in the same class as close to one another, meanwhile as far from points in the other class. In LapPVP, the within-class scatter requires only one-class training data, and the Laplacian regularization requires all data to construct the adjacent graph. However, the between-scatter captures the information of different classes. Hence, it is easy to generate multiple projection vectors for multi-class data by using the one-versus-rest strategy. This section proposes OVR-LapPVP by constructing C projection vectors for C-class training data. In other words, each class has its own projection vector.

To construct a binary classifier for the cth class, we suppose samples in class c are positive, and samples in other rest $(C-1)$ classes are negative. Let $\mathbf{X}_{l,c}$ be the labeled sample matrix for the cth class, and l_c is the number of samples in the cth class. Then the total labeled sample matrix is $\mathbf{X}_l = [\mathbf{X}_{l,1}; \cdots ; \mathbf{X}_{l,C}]$, and the training sample matrix is $\mathbf{X} = [\mathbf{X}_l; \mathbf{X}_u]$, where $\mathbf{X}_u$ is the unlabeled sample matrix.

Thus, the cth optimization formulation of OVR-LapPVP is given by

$$\max_{\mathbf{v}_c} \quad \mathbf{v}_c^T \mathbf{H}_c \mathbf{v}_c - \rho_c \mathbf{v}_c^T \mathbf{X}^T \mathbf{L} \mathbf{X} \mathbf{v}_c,$$
$$s.t. \quad \mathbf{v}_c^T \mathbf{v}_c = 1, \tag{12}$$

where the discriminative matrix $\mathbf{H}_c$ is defined as

$$\mathbf{H}_c = \alpha_c \mathbf{B}_c - (1 - \alpha_c) \mathbf{W}_c, \tag{13}$$

where $\alpha_c \in [0,1]$ is the parameter to balance the between-class scatter matrix $\mathbf{B}_c$ and within-class scatter matrix $\mathbf{W}_c$ of the cth class, $\mathbf{B}_c$ and $\mathbf{W}_c$ respectively have the forms

$$\mathbf{B}_c = \left(\mathbf{X} - \mathbf{e}\mathbf{u}_c^T\right)^T \left(\mathbf{X} - \mathbf{e}\mathbf{u}_c^T\right) \tag{14}$$

and

$$\mathbf{W}_c = (\mathbf{X}_c - \mathbf{e}_c\mathbf{u}_c^T)^T(\mathbf{X}_c - \mathbf{e}_c\mathbf{u}_c^T), \tag{15}$$

where $\mathbf{e} \in \mathbf{R}^n$ and $\mathbf{e}_c \in \mathbb{R}^{l_c}$ are vectors of ones. To find the solution of the optimization problem (12), we have the following theorem.

Theorem 3. *The optimal solution $\mathbf{v}_c$ of the optimization problem (12) is the eigenvector corresponding to the maximum eigenvalue of the matrix $(\mathbf{H}_c - \mathbf{X}^T \mathbf{L} \mathbf{X})$.*

Proof. The Lagrangian function of Eq. (12) can be written as

$$L(\mathbf{v}_c, \lambda_c) = \mathbf{v}_c^T \mathbf{H}_c \mathbf{v}_c - \rho_c \mathbf{v}_c^T \mathbf{X}^T \mathbf{L} \mathbf{X} \mathbf{v}_c - \lambda_c(\mathbf{v}_c^T \mathbf{v}_c - 1). \tag{16}$$

Then, we derive the partial derivative of $L(\mathbf{v}_c, \lambda_c)$ with respect to $\mathbf{v}_c$ and make it equal zero and have

$$\frac{\partial L(\mathbf{v}_c, \lambda_c)}{\partial \mathbf{v}_c} = \left(\mathbf{H}_c - \rho_c \mathbf{X}^T \mathbf{L} \mathbf{X}\right) \mathbf{v}_c - \lambda_c \mathbf{v}_c = 0$$
$$\Rightarrow \left(\mathbf{H}_c - \rho_c \mathbf{X}^T \mathbf{L} \mathbf{X}\right) \mathbf{v}_c = \lambda_c \mathbf{v}_c. \tag{17}$$

Finally, the cth optimization problem of OVR-LapPVP is also transferred to an eigenvalue decomposition problem. The solution $\mathbf{v}_c$ is obtained as the eigenvectors corresponding to the maximum eigenvalue of the matrix $\left(\mathbf{H}_c - \rho_c \mathbf{X}^T \mathbf{L} \mathbf{X}\right)$.   □

After obtaining C projection vectors, one for each class, the unknown test data is predicted by the projected distances between it and all class centers. The decision function is defined as:

$$c^* = \arg \min_{c=1,2,\ldots,C} \|\mathbf{v}_c^T \mathbf{x} - \mathbf{v}_c^T \mathbf{u}_c\| \tag{18}$$

where c^* is the estimated class label of $\mathbf{x}$. We summarize the specific procedure of OVR-LapPVP in Algorithm 2.

Algorithm 2. OVR-LapPVP

Training Phase
Input: Sample matrices $\mathbf{X}_{l,c}$, $c = 1, \cdots, C$, and $\mathbf{X}_u$;
  1: Let $\mathbf{X} = [\mathbf{X}_{l,1}; \cdots; \mathbf{X}_{l,C}; \mathbf{X}_u] \in \mathbb{R}^{n \times m}$ be the training samples;
  2: Obtain the adjacency matrix $\mathbf{S}$ by Eq. (1) based on $\mathbf{X}$;
  3: Compute the Laplacian matrix $\mathbf{L} = \mathbf{D} - \mathbf{S}$, where $\mathbf{D}$ is the diagonal matrix with
      $D_{ii} = \sum_j S_{ij}$;
  4: **for** $c = 1$ to C **do**
  5:     Compute the discriminative matrix $\mathbf{H}_c$ by Eq. (13) for the cth class data points;
  6:     Obtain the vectors $\mathbf{v}_c$ by solving eigenvalue decomposition problem Eq. (17);
  7: **end for**
Output: Projection vectors $\mathbf{v}_c$, $c = 1, \cdots, C$.
Testing Phase
Input: Unknown test sample $\mathbf{x}$, sample matrices $\mathbf{X}_{l,c}$, $c = 1, \cdots, C$, $\mathbf{X}_u$, and projection vectors $\mathbf{v}_c$, $c = 1, \cdots, C$;
  1: **for** $c = 1$ to C **do**
  2:     Calculate the distance from $\mathbf{x}$ to the cth class;
  3: **end for**
  4: Assign the class label to $\mathbf{x}$ by Eq. (18).
Output: Estimated label for $\mathbf{x}$.

2.3   Comparison of Multi-class LapPVP

We make a comparison for the proposed two multi-class methods here. OVO-LapPVP constructs $C(C-1)/2$ binary classifiers by solving $C(C-1)/2$ pairs of eigenvalue decomposition problems. The computational complexity of OVO-LapPVP is approximately $O(C^2 m^3)$. OVR-LapPVP generates C projection vectors by solving C eigenvalue decomposition problems. The computational complexity of OVR-LapPVP is approximately $O(Cm^3)$.

Obviously, OVO-LapPVP requires a large amount of training time as compared to OVR-LapPVP. Moreover, the strategy of classification in OVO-LapPVP is more complex than that of OVR-LapPVP because OVO-LapPVP handles more binary classifiers. Of course, OVR-LapPVP has its own disadvantage of having a greater space complexity because its optimization problems are performed on all training data.

3 Experiments

3.1 Experimental Setup

To confirm the feasibility and effectiveness of the proposed multi-class methods, we need to compare them with other multi-class ones. However, few paper discusses the semi-supervised multi-class tasks. Thus, algorithms compared here are extended by binary semi-supervised classifiers (LapTSVM, LapTPMSVM, and LapLSTSVM) using strategies of both OVO and OVR, where LapTSVM, LapTPSVM and LapLSTSVM are recently popular nonparallel classifiers for semi-supervised learning as well as LapPVP. Finally, we get six multi-class algorithms: OVO-LapTSVM, OVR-LapTSVM, OVO-LapTPMSVM, OVR-LapTPMSVM, OVO-LapLSTSVM and OVR-LapLSTSVM.

Experiments are conducted on benchmark datasets that are from the University of California at Irvine (UCI) Machine Learning Repository [2]. The details of datasets are presented in Table 1. For each dataset, we randomly select 70% of samples from each class for training, and the remaining 30% for test. The features of data are all normalized to the interval $[0, 1]$. To reduce the running time of parameter selection, hyper-parameters are set to the same value for all binary classifiers in one multi-class method. In addition, the grid search method [4] is used for selecting the optimal hyper-parameters. In OVO-LapPVP and OVR-LapPVP, α takes its value from the set $\{2^{-10}, 2^{-9}, \ldots, 2^{0}\}$, and ρ from $\{2^{-5}, 2^{-4}, \ldots, 2^{5}\}$. Moreover, the number of nearest neighbors varies in the set $\{1, 3, 5, 7, 9\}$.

Table 1. Information of benchmark datasets

Dataset	#Sample	#Feature	#Class
Balance	625	4	3
Dnatest	1186	180	3
Glass	214	9	6
Iris	150	4	3
Lungcancer	32	56	3
Waveform	5000	21	3
Wine	178	13	3
X8D5K	1000	8	5
Zoo	101	16	7

All Matlab scripts of the involved algorithms are written by ourselves. For a fair comparison, we exploit the Matlab toolbox of quadratic programming to solve quadratic programmings in the relevant methods. All methods are implemented in MATLAB R2015b on a personal computer, whose system configuration is Intel Core i5 (3.6 GHz) and 8 GB random access memory.

Parameters Analysis. It is well known, the parameters of a classifier may have a great influence on its classification performance. Here, we conduct the grid search method to analyze the impact of parameters on proposed methods through the Glass dataset. As mentioned before, the pair of parameters take the same values. In OVO-LapPVP and OVR-LapPVP, α takes its value from the set $\{2^{-10}, 2^{-9}, \ldots, 2^0\}$, and ρ from $\{2^{-5}, 2^{-4}, \ldots, 2^5\}$. Here, K is selected from the set $\{1, 3, 5, 7, 9\}$.

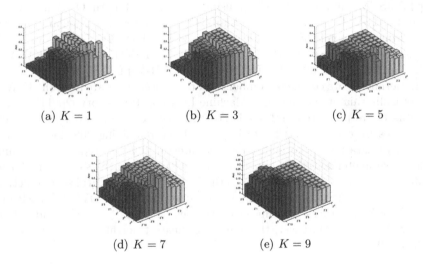

(a) $K = 1$ (b) $K = 3$ (c) $K = 5$

(d) $K = 7$ (e) $K = 9$

Fig. 1. Accuracy vs. parameters (α, ρ) of OVO-LapPVP on Glass dataset under different K

Figures 1 and 2 separately show the influence of parameters of OVO-LapPVP and OVR-LapPVP with 20% of labeled data and 50% of unlabeled data on the Glass dataset. It is obviously observed that the pair of parameters (α, ρ) can also greatly affect the performance of OVO-LapPVP and OVR-LapPVP. Additionally, when varying the number of nearest neighbors, the optimal pair of parameters (α, ρ) is totally different. Thus, the pair of optimal parameters cannot be fixed for all datasets. Therefore, parameter selection may be an issue for our method. However, the grid search method can help us to find appropriate parameters for OVO-LapPVP and OVR-LapPVP in experiments.

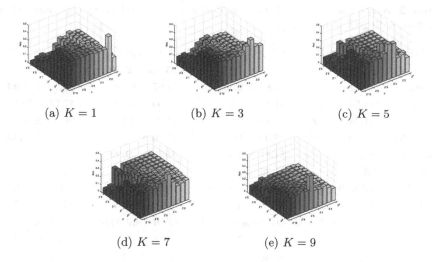

(a) $K = 1$ (b) $K = 3$ (c) $K = 5$

(d) $K = 7$ (e) $K = 9$

Fig. 2. Accuracy vs. parameters (α, ρ) of OVR-LapPVP on Glass dataset under different K

3.2   Results and Discussion

Experiments are conducted on nine UCI datasets to compare the classification performance of semi-supervised multi-class methods. We run 10 trials on each dataset and report the average accuracy.

Table 2 lists the results of eight semi-supervised multi-class classification methods on nine datasets with 10% of labeled data and 50% of unlabeled data. It is clear that the proposed methods including OVO-LapPVP and OVR-LapPVP can perform better. On Dnatest, Iris, Lungcancer, Wine, X85DK and Zoo datasets, OVO-LapPVP has the highest accuracy, followed by OVR-LapPVP. On the Glass dataset, OVR-LapPVP has the highest accuracy, followed by OVO-LapPVP.

Tables 3 and 4 separately summary the results of eight multi-class methods obtained with 20% and 30% of labeled data, respectively. Both OVO-LapPVP and OVR-LapPVP outperform other methods on all datasets except Balance and Waveform. The classification performance on the X8D5K dataset shows that the proposed semi-supervised classification methods are effective for multi-class classification tasks.

Observation on Tables 2, 3 and 4 indicates that OVO-LapPVP achieves the highest performance on 15 cases and the second highest on 6 cases. Note that there are 27 cases totally. At the same time, we can see that OVR-LapPVP is the best on 9 cases and the second best on 12 cases. To be simply, our methods are superior to other methods in 24 out of 27 cases. Findings suggests that OVO-LapPVP has the best performance for multi-class classification tasks among compared methods, followed by OVR-LapPVP.

Table 2. Average accuracy and standard deviation (%) obtained by eight multi-class methods with 10% of labeled data

Dataset	OVO-LapTSVM	OVO-LapTPMSVM	OVO-LapLSTSVM	OVO-LapPPV
Balance	84.13 ± 2.41	73.17 ± 4.67	**85.93 ± 2.70**	84.13 ± 1.28
Dnatest	79.52 ± 3.13	69.86 ± 2.50	52.67 ± 0.79	**83.17 ± 1.94**
Glass	44.55 ± 8.69	41.82 ± 5.98	45.00 ± 6.70	47.12 ± 7.01
Iris	91.33 ± 4.12	92.89 ± 3.60	94.89 ± 2.78	**96.44 ± 1.55**
Lungcancer	45.00 ± 10.80	46.00 ± 10.75	37.00 ± 12.52	**64.00 ± 8.43**
Waveform	84.38 ± 0.77	81.81 ± 0.55	**85.37 ± 0.93**	79.95 ± 1.46
Wine	95.27 ± 2.74	94.36 ± 2.90	93.64 ± 3.76	**95.45 ± 2.14**
X8D5K	**100.00 ± 0.00**	**100.00 ± 0.00**	**100.00 ± 0.00**	100.00 ± 0.00
Zoo	88.79 ± 4.56	89.70 ± 4.09	90.00 ± 5.16	**91.70 ± 3.79**
Dataset	OVR-LapTSVM	OVR-LapTPMSVM	OVR-LapLSTSVM	OVR-LapPPV
Balance	81.06 ± 5.24	68.41 ± 5.04	85.56 ± 2.60	83.02 ± 2.83
Dnatest	66.80 ± 4.10	73.46 ± 2.16	60.98 ± 2.17	82.92 ± 1.87
Glass	42.73 ± 6.05	41.97 ± 4.95	46.21 ± 4.24	**48.79 ± 5.19**
Iris	84.22 ± 8.41	86.44 ± 6.58	82.44 ± 4.85	95.56 ± 3.47
Lungcancer	39.00 ± 11.97	43.00 ± 12.52	37.00 ± 13.37	60.00 ± 12.47
Waveform	85.29 ± 0.87	78.75 ± 1.87	83.42 ± 1.31	78.30 ± 2.54
Wine	92.55 ± 5.10	93.27 ± 2.28	92.91 ± 5.17	95.64 ± 5.09
X8D5K	99.97 ± 0.11	99.77 ± 0.39	99.97 ± 0.11	**100.00 ± 0.00**
Zoo	89.09 ± 4.56	86.67 ± 3.83	90.61 ± 4.15	88.79 ± 3.21

The running time of eight methods on nine datasets with 20% of labeled and 50% of unlabeled training data is recorded in Fig. 3. The first, third, fifth, and seventh columns in every sub-figure represent the running time obtained by methods using the OVO strategy, and the rest columns represent those using OVR. Obviously, the OVR-based methods spend less time than OVO-based ones because the OVR-based methods need to solve less optimization problems. Four methods, OVO-LapLSTSVM, OVR-LapLSTSVM, OVO-LapPVP and OVR-LapPVP, that avoid the complex quadratic programmings can save much time. Therefore, OVO-LapPVP and OVR-LapPVP are promising when applied to multi-class classification tasks.

Additionally, we give the rank of individual methods in Table 5. A method that has the highest accuracy is ranked the first, that has the second highest accuracy is ranked the second, and so on. Digits in the row of "10%" mean the average rank obtained from Table 2, "20%" and "30%" are from Tables 3 and 4, respectively. The row of "Average rank" provides the average rank over 27 cases, and that of "Friedman test" shows the rank difference between methods

Table 3. Average accuracy and standard deviation (%) obtained by eight multi-class methods with 20% of labeled data

Dataset	OVO-LapTSVM	OVO-LapTPMSVM	OVO-LapLSTSVM	OVO-LapPPV
Balance	86.72 ± 2.37	73.81 ± 2.58	**87.30 ± 1.75**	80.32 ± 2.61
Dnatest	82.81 ± 1.52	77.50 ± 1.88	56.97 ± 0.92	86.07 ± 1.61
Glass	48.03 ± 5.05	43.03 ± 6.11	47.88 ± 5.59	**54.85 ± 7.03**
Iris	95.11 ± 2.73	94.89 ± 1.83	94.67 ± 2.81	96.00 ± 4.03
Lungcancer	40.00 ± 14.91	41.00 ± 15.95	40.00 ± 16.33	**68.00 ± 9.19**
Waveform	84.95 ± 1.02	81.72 ± 1.06	**86.59 ± 0.94**	81.09 ± 1.09
Wine	95.64 ± 1.76	95.09 ± 2.28	94.55 ± 4.62	**96.00 ± 1.43**
X8D5K	**100.00 ± 0.00**	**100.00 ± 0.00**	**100.00 ± 0.00**	100.00 ± 0.00
Zoo	92.42 ± 2.58	90.91 ± 3.50	90.30 ± 5.68	**93.33 ± 2.39**
Dataset	OVR-LapTSVM	OVR-LapTPMSVM	OVR-LapLSTSVM	OVR-LapPPV
Balance	81.85 ± 4.11	68.84 ± 3.33	86.83 ± 1.68	83.86 ± 4.54
Dnatest	75.51 ± 2.20	78.51 ± 1.97	70.34 ± 2.72	**86.26 ± 1.70**
Glass	47.88 ± 5.63	39.55 ± 7.71	48.03 ± 5.67	49.73 ± 9.37
Iris	87.33 ± 6.63	88.22 ± 8.38	83.56 ± 4.34	**96.44 ± 3.51**
Lungcancer	40.00 ± 17.64	40.00 ± 9.43	39.00 ± 17.92	62.00 ± 9.19
Waveform	85.69 ± 0.83	77.95 ± 1.62	85.77 ± 0.93	79.13 ± 0.99
Wine	94.00 ± 3.84	93.64 ± 4.39	94.00 ± 3.54	96.73 ± 4.91
X8D5K	**100.00 ± 0.00**	99.73 ± 0.34	99.97 ± 0.11	**100.00 ± 0.00**
Zoo	90.61 ± 3.90	90.61 ± 3.33	89.39 ± 3.57	92.73 ± 2.93

and reference method, where OVR-LapPVP is taken as the reference method. The smaller the value of average rank and Friedman test is, the better performance the corresponding method has. It is obvious that OVO-LapPVP is slightly better than OVR-LapPVP since the value of OVO-LapPVP is less than 0 in the Friedman test. According to Table 5, we can conclude OVO-LapPVP has a greater superiority, and OVR-LapPVP ranks the second.

Table 4. Average accuracy and standard deviation (%) obtained by eight multi-class methods with 30% of labeled data

Dataset	OVO-LapTSVM	OVO-LapTPMSVM	OVO-LapLSTSVM	OVO-LapPPV
Balance	**88.89 ± 1.60**	74.50 ± 3.47	87.46 ± 1.22	84.13 ± 2.51
Dnatest	84.38 ± 1.82	81.97 ± 2.13	60.65 ± 1.75	86.60 ± 0.96
Glass	49.70 ± 5.50	43.48 ± 8.89	51.36 ± 5.73	**51.52 ± 3.71**
Iris	94.89 ± 2.78	93.11 ± 3.70	94.89 ± 2.58	95.78 ± 3.22
Lungcancer	58.00 ± 12.29	43.00 ± 14.94	53.00 ± 16.36	76.00 ± 9.66
Waveform	85.19 ± 1.01	81.80 ± 0.79	**86.86 ± 1.00**	81.46 ± 0.84
Wine	96.55 ± 1.81	96.36 ± 1.21	95.09 ± 3.64	**97.45 ± 1.53**
X8D5K	**100.00 ± 0.00**	**100.00 ± 0.00**	**100.00 ± 0.00**	100.00 ± 0.00
Zoo	92.42 ± 1.28	91.82 ± 2.49	90.30 ± 1.92	**93.33 ± 2.58**
Dataset	OVR-LapTSVM	OVR-LapTPMSVM	OVR-LapLSTSVM	OVR-LapPPV
Balance	82.28 ± 3.04	67.99 ± 6.32	87.72 ± 1.43	84.42 ± 5.07
Dnatest	80.93 ± 2.52	81.66 ± 2.29	76.83 ± 2.91	**87.25 ± 0.65**
Glass	49.39 ± 5.06	40.61 ± 6.69	47.27 ± 5.57	50.61 ± 6.45
Iris	90.00 ± 8.13	88.67 ± 6.83	81.33 ± 3.66	**96.00 ± 3.44**
Lungcancer	45.00 ± 15.81	46.00 ± 16.47	45.00 ± 8.50	**77.00 ± 6.75**
Waveform	85.99 ± 0.87	78.02 ± 1.36	86.42 ± 0.79	78.22 ± 0.76
Wine	95.45 ± 3.46	96.00 ± 2.24	96.55 ± 2.18	97.64 ± 1.50
X8D5K	**100.00 ± 0.00**	99.80 ± 0.28	**100.00 ± 0.00**	**100.00 ± 0.00**
Zoo	91.21 ± 3.01	91.21 ± 2.65	90.00 ± 4.75	91.82 ± 3.21

Table 5. Rank of eight methods on UCI datasets

Classifier	10%	20%	30%	Average rank	Friedman test
OVO-LapTSVM	3.67	3.00	2.67	3.11	0.52
OVR-LapTSVM	5.22	4.56	4.67	4.81	2.22
OVO-LapTPMSVM	4.44	4.22	4.67	4.44	1.85
OVR-LapTPMSVM	5.78	5.67	5.67	5.70	3.11
OVO-LapLSTSVM	3.67	3.89	3.78	3.78	1.19
OVR-LapLSTSVM	4.67	4.67	4.44	4.59	2.00
OVO-LapPPV	2.00	2.44	2.44	2.30	−0.30
OVR-LapPPV	3.00	2.33	2.44	2.59	0.00

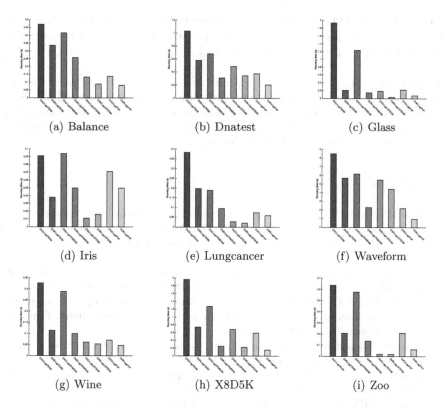

(a) Balance   (b) Dnatest   (c) Glass

(d) Iris   (e) Lungcancer   (f) Waveform

(g) Wine   (h) X8D5K   (i) Zoo

Fig. 3. Running time of eight methods on UCI datasets

4   Conclusion

In this paper, we propose OVO-LapPVP and OVR-LapPVP for multi-class classification tasks. As the extension of LapPVP, the proposed methods both can take advantage of the discriminative information of the labeled data and the graph structure of unlabeled data to obtain the optimal projection vectors. OVO-LapPVP adopts the OVO strategy, and OVR-LapPVP adopts the OVR strategy. That is to say, OVO-LapPVP obtains $C(C-1)/2$ pairs of projection vectors, and OVR-LapPVP obtains C ones for the C-class data. Since OVO-LapPVP needs more optimization problems to solve than OVR-LapPVP, which results that OVO-LapPVP costs more time than OVR-LapPVP on the same dataset.

Experiments conducted on the UCI datasets have demonstrated that the proposed methods have better classification performance compared with other popular semi-supervised methods based on manifold regularization and twin support vector machine. OVO-LapPVP outperforms OVR-LapPVP in classification performance, but is worse than OVR-LapPVP in running efficiency. Therefore, OVO-LapPVP and OVR-LapPVP can be applied to different scenes.

OVO-LapPVP and OVR-LapPVP are both excellent on solving multi-class classification problems for linear cases, in the following work, we can consider the methods with kernel ticks for nonlinear cases.

Acknowledgment. This work was supported in part by the Natural Science Foundation of the Jiangsu Higher Education Institutions of China under Grant Nos. 19KJA550002 and 19KJA610002, by the Priority Academic Program Development of Jiangsu Higher Education Institutions, and by the Collaborative Innovation Center of Novel Software Technology and Industrialization.

References

1. Chen, W., Shao, Y., Deng, N., Feng, Z.: Laplacian least squares twin support vector machine for semi-supervised classification. Neurocomputing **145**, 465–476 (2014)
2. Dua, D., Graff, C.: UCI machine learning repository (2017). https://archive.ics.uci.edu/ml
3. Goyal, N., Gupta, K., Kumar, N.: Multiclass twin support vector machine for plant species identification. Multimedia Tools Appl. **78**(19), 27785–27808 (2019)
4. Iwan, S., Adam, P., Gary, W.: SVM parameter optimization using grid search and genetic algorithm to improve classification performance. Telecommun. Comput. Electron. Control **12**, 1502–1509 (2016)
5. Jayadeva, Khemchandani, R., Chandra, S.: Twin support vector machines for pattern classification. IEEE Trans. Pattern Anal. Mach. Intell. **29**(5), 905–910 (2007)
6. Kumar, M.A., Gopal, M.: Least squares twin support vector machines for pattern classification. Expert Syst. Appl. **36**(4), 7535–7543 (2009)
7. Peng, X.: TPMSVM: a novel twin parametric-margin support vector machine for pattern recognition. Pattern Recogn. **44**(10–11), 2678–2692 (2011)
8. Qi, Z., Tian, Y., Shi, Y.: Laplacian twin support vector machine for semi-supervised classification. Neural Netw. **35**, 46–53 (2012)
9. Tomar, D., Agarwal, S.: A comparison on multi-class classification methods based on least squares twin support vector machine. Knowl. Based Syst. **81**, 131–147 (2015)
10. Xie, J., Hone, K.S., Xie, W., Gao, X., Shi, Y., Liu, X.: Extending twin support vector machine classifier for multi-category classification problems. Intell. Data Anal. **17**(4), 649–664 (2013)
11. Xue, Y., Zhang, L.: Laplacian pair-weight vector projection for semi-supervised learning. Inf. Sci. **573**, 1–19 (2021)
12. Yang, Z., Xu, Y.: Laplacian twin parametric-margin support vector machine for semi-supervised classification. Neurocomputing **171**, 325–334 (2016)

Integrating EMD, LMD and TCN Methods for COVID-19 Forecasting

Lulu Sun[1(✉)], Zhouming Liu[1], Peilin Yang[1], Choujun Zhan[2], and Kim-Fung Tsang[1]

[1] School of Electrical and Computer Engineering, Nanfang College Guangzhou, Guangzhou 510970, China
`sunll@nfu.edu.cn`, {`liuzhm190043,yangpl210051`}`@stu.nfu.edu.cn`, `ee330015@cityu.edu.hk`
[2] School of Computer Science, South China Normal University, Guangzhou 510641, China

Abstract. The coronavirus disease 2019, abbreviated as COVID-19 pandemic, has caused a global turmoil encompassing not only public health but also the economic sphere. Therefore, timely and reliable forecasting of the disease becomes critical to enhance the efficiency of government decision-making, effectively manage resources in the supply chain, and facilitate challenging political choices . Recently, deep learning-based network models have shown great success in lots of time series forecasting applications. In this study, we propose a hybrid empirical mode decomposition (EMD) and local mean decomposition (LMD) based neural network framework for COVID-19 predicting. EMD, as well as LMD, can efficiently decompose the original data into more stationary signals with different frequencies. What is more, a temporal convolutional network (TCN) with two TCN blocks was used to learn a precise mapping from input coronavirus series to the output ones. Because of the unique construction of our network, it can solve time series foresting problems with different nonlinear tendencies and oscillation. We showcase the capabilities of our model through two notable applications in COVID-19 prediction, including weekly new confirmed cases and deaths forecasting. The experimental results demonstrate that our model achieves significantly better performance compared to numerous cutting-edge forecasting methods when evaluated based on root mean squared log error, mean absolute percentage error, pearson correlation, and coefficient of determination. For instance, when predicting weekly new confirmed cases and deaths, our proposed method achieves an impressive R^2 score of 0.9744 and 0.9886, respectively, with a forecasting horizon of 1. This outperforms the second-best method by a significant margin, ranging from 0.0131 to 0.1099, across various scenarios.

Keywords: COVID-19 predicting · Temporal convolutional network · Empirical mode decomposition · Local mean decomposition

H. Zhang et al. (Eds.): NCAA 2023, CCIS 1869, pp. 175–190, 2023.
https://doi.org/10.1007/978-981-99-5844-3_13

1   Introduction

COVID-19, the full name is coronavirus disease 2019, broke out in full swing in early 2020. In just a few months, the virus has spread to more than 200 countries around the world. As of October 18, 2022, 6,548,492 deaths and 622,389,418 cases of COVID-19 were confirmed worldwide[1]. Besides the unprecedented global health crisis, the economic impact is equally astonishing. Thus, early and accurate prediction of COVID-19 is of paramount importance for governments and policymakers to adopt appropriate measures to hold back the spread of COVID-19. Given the severity of the pandemic and the demand for accurate forecasting of COVID-19 spread, many works have been carried out to estimate the progression of COVID-19 since its global outbreak.COVID-19 forecasting, like epidemic forecasting in general, has long been recognized as a challenging task, for which many classic methods have been developed . These methods can be broadly classified into three categories:

Compartmental Models. Traditional susceptible-infected-recovered model (SIR) [13], susceptible-exposed-infected-recovered model (SEIR) [3] and susceptible-infected-dead-recovered model (SIDR) [4] belong to this class,these models have been extensively utilized for infectious disease modeling since the early 20th century. These models categorize the population into distinct groups based on their infection and recovery status, and simulate the transmission process using differential equations. . The major limitation of compartmental models is that they rely on lots of parameters, such as the rate of treatment and quarantine, which is hard to determine as they are influenced by many uncontrollable and dynamically changing factors. Therefore, there is not much chance for these models to achieve very accurate forecasting results.

Statistical Models. Auto regressive integrated moving average (ARIMA) [8] is a classic method, which combines autoregressive, moving average, and differencing components to forecast future short-term values referring to the historical series. In fact, ARIMA has been employed to predict the increasing cases and deaths of COVID-19 in various countries [2]. One limitation of statistical models is their inability to learn nonlinear modes, leading to unsatisfactory predictive performance.

Machine Learning Models. The machine learning (ML) models, like XGBoost Classifier [17], support vector machine (SVM) [20], and Deep learning (DL) models [10,19,23,24] have been used widely in the COVID-19 forecasting problem as well. ML models aim to mimic human intelligence by iteratively learning from previous computations. It enables the generation of reliable and repeatable decisions and results. While ML models exhibit improved capacity and flexibility in learning from covariates and forecasting epidemic progression, they often struggle to perform well in long-term forecasting tasks. Deep learning (DL) techniques refer to a class of complex artificial neural network architectures that

---

[1] https://covid19.who.int.

consist of multiple layers of information processing stages in hierarchical structures. These architectures are trained using the backpropagation algorithm. For example, multiple layer perceptron [18], recurrent neural network [12], gated recurrent unit [11], long short-term memory [14] and transformer [22]. DL models have achieved remarkable success across natural language processing, computer vision, entertainment, biology, biomedical research, and financial applications. In recent times, a variety of DL methods [10,19,23,24] have been applied to forecast the epidemiological trends of the COVID-19 pandemic. However, it is not without its challenges, which can make its application difficult in certain fields. For instance, it requires a significant amount of high-quality data and may suffer from overfitting when the dataset size is small. Additionally, deep neural networks often encounter training issues such as vanishing and exploding gradients.

Predicting daily new confirmed cases and deaths accurately for COVID-19 is a challenging task, primarily because COVID19 time series is typically regarded as a non-linear and non-stationary temporal data that is influenced by numerous interacting factors. It is worth mentioning that local mean decomposition (LMD), as well as empirical mode decomposition (EMD) has been extensively utilized for analyzing nonlinear and nonstationary signals since the early 21st century. Through the process of decomposition, a complex series can be decomposed into some intrinsic mode functions (IMFs) and product functions (PFs), which exhibit simpler spectral components and stronger correlations. Taking into account the findings from the temporal convolutional network (TCN) [5], which suggest that a simple convolutional architecture outperforms traditional recurrent networks across different tasks. we hope to employ such a TCN architecture for further investigation.

Inspired by the above observations, this paper introduces a novel approach for COVID-19 forecasting by integrating the EMD and LMD techniques within a neural network. The main contributions of this study can be summarized as follows:

(1) We propose an EMD-LMD-TCN framework for COVID-19 forecasting. Based on our current knowledge, it signifies the primary occurrence to introduce EMD or LMD for employing neural networks within the area of COVID-19 pandemic. This time series decomposition method generates simplified frequency components, which is advantageous for achieving accurate forecasting results and mitigating the problem of overfitting during neural network training.

(2) Our model can handle many COVID-19 forecasting problems, such as weekly new confirmed cases and deaths at the same time. Moreover, we could deliver precise predictions across both short-term and long-term time frames simultaneously, which is a huge advantage compared to other methods.

(3) In our analysis, we conduct a comprehensive evaluation of our proposed method, comparing its performance with other published results in terms of forecasting weekly new confirmed cases and deaths. Experimental results demonstrate that our network surpasses other machine learning models, showcasing its superior forecasting capabilities.

2 Related Works

2.1 Empirical Mode Decomposition (EMD)

The empirical mode decomposition (EMD) technique, initially introduced by [15], is specifically designed for nonlinear and nonstationary signals. Its core principle involves decomposing complex datasets into a limited number of intrinsic mode functions (IMFs)In the EMD, an IMF must meet two requirements: (1) the number of extrema and zero crossings in the entire dataset must either be equal or differ by at most one, and (2) at any point, the mean value of the envelope defined by the local maxima and the envelope defined by the local minima is zero.

2.2 Local Mean Decomposition (LMD)

Local mean decomposition (LMD) is another modulated signal analysis method, proposed by [21]. Its is to decompose nonlinear signals into a limited number of product functions (PFs) and one residue. A PF is formed by multiplying an envelope signal in terms of amplitude with a frequency-modulated (FM) signal. Essentially, the LMD scheme is designed to iteratively separate an FM signal from an amplitude envelope signal.

2.3 Temporal Convolutional Network (TCN)

The popular temporal convolutional network architecture (TCN) [5] is designed for handling sequential learning tasks, where earlier studies are mostly based on recurrent neural networks (RNNs). TCN has two distinguishing characteristics: (1) the length of the input should be preserved in the output of the network; (2) the network is causal, meaning it can solely leverage the information from preceding time steps. Dilated convolutions are commonly adopted in TCN architectures to rapidly expand the receptive field. This allows for the construction of networks that excel in sequential tasks, surpassing the efficiency and prediction performance of Recurrent Neural Networks (RNNs).

Given a sequence of input $x_0, ..., x_T$, the goal is to forecast the corresponding outputs $y_0, ..., y_T$ at each time. In formal terms, a sequence modeling network can be defined as a nonlinear mapping relationship $f : X \rightarrow Y$, where X represents the input sequence and Y represents the forecasted sequence. This mapping function is obtained through supervised learning, which involves training the network using loss functions that valuate the divergence between the ground truth sequence and the forecasted sequence. A crucial constraint in sequence modeling is that when predicting the output y_t at a specific time step t, the model can purely leverage the inputs that have been previously observed, namely $x_0, ..., x_t$.

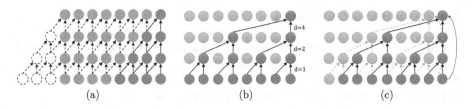

Fig. 1. (a) Casual convolution (filter size k = 2) (b) Dilated casual convolution (dilated factors d = 1, 2, 4 and filter size k = 2) (c) An example of residual connection in a TCN. The red lines are identity mappings. (Color figure online)

Causal Convolutions. To fulfill the first principle as mentioned above, we use a 1-D fully-convolutional architecture with zero padding in the TCNs, ensuring that the network can capture and model temporal dependencies in the input sequence efficiently. To accomplish the second principle, we use causal convolutions, because the output at each time step t only depends on the input up to that time step and not on any future time step t, as shown in Fig. 1(a) .

Dilated Convolutions. The use of causal convolutions in the network limits the receptive field size, which can be a challenge in capturing long-range dependencies in the input sequence. One approach to overcome this limitation is to stack multiple layers, but this can lead to increased computational complexity. To address such a problem, dilated convolutions are applied to facilitate an exponentially expanding receptive field within a restricted number of layers. The dilated convolution operation, denoted as F, applied to element s of a 1-D sequence $\mathbf{x} \in R^n$ for a filter $f : \{0, ..., k-1\} \rightarrow R$:

$$F(s) = (\mathbf{x} *_d f)(s) = \sum_{i=0}^{k-1} f(i) \cdot \mathbf{x}_{s-d\cdot i}, \tag{1}$$

where k is the filter size, d is the dilation factor, $*$ is the convolution operator, and $s - d \cdot i$ takes into account the temporal order of preceding elements.

Residual Connections. In a generic TCN model, there is a 1D fully-convolutional network component along with numerous residual blocks. One such block consists of a branch that includes a set of modifications represented by function F. The outputs of these transformations are then combined with the input $\mathbf{x}$ of the block,

$$o = \text{Activation}(\mathbf{x} + \mathcal{F}(\mathbf{x})). \tag{2}$$

The incorporation of these residual modules contributes to the stabilization of training deep and larger TCNs. Besides, in order to match the original channel size, a supplementary 1×1 convolutional layer is employed.

3   Problem Formulation

COVID-19 forecasting primarily involves forecasting future values of temporal date, It aims to develop a model that accurately captures the patterns and relationships within the data, By analyzing historical data, allowing for the prediction of future values. Given a n driving series, $i.e.,$

$$\mathbf{X} = \left(\mathbf{x}^1, \mathbf{x}^2, \cdots, \mathbf{x}^n\right)^\top = (\mathbf{x}_1, \mathbf{x}_2, \cdots, \mathbf{x}_T) \in \mathbb{R}^{n \times T}, \tag{3}$$

where T is the length of window size, $\mathbf{x}^k = \left(x_1^k, x_2^k, \cdot, x_T^k\right)^\top \in \mathbb{R}^T$ denotes a sequential series of length T, $\mathbf{x}_t = \left(x_t^1, x_t^2, \cdots, x_t^n\right)^\top \in \mathbb{R}^n$ denotes a vector of n exogenous input series at time t. Usually, given the previous values of the target series, $i.e.,$

$$\mathbf{y} = (y_1, \ldots, y_T) \in \mathbb{R}^T, \tag{4}$$

as well as the current and previous values of n additional series $(\mathbf{x}_1, \mathbf{x}_2, \cdots, \mathbf{x}_T) \in \mathbb{R}^{n \times T}$, the task is to forecast the upcoming values $y_{T+H} \in \mathbb{R}$ at a time point $T + H$, where H represents the future time period of the prediction. We denote $\hat{y}_{T+H}$ as the forecast of y_{T+H}. Our model is designed to construct a non-linear relationship function F that captures the complex relationship between previous observations and future outcomes. It aims to understand and predict the future values based on the information available from the past:

$$\hat{y}_{T+H} = F\left(y_1, \cdots, y_T, \mathbf{x}_1, \mathbf{x}_2, \cdots, \mathbf{x}_T\right) \tag{5}$$

4   Methodology

In this paper, a novel approach integrating EMD, LMD, and TCN methods (called EMD-LMD-TCN model), which feeds the original time series, IMF components, and PF components into TCN, is proposed for COVID-19 prediction. The algorithm is outlined below, and the flowchart is depicted in Fig. 2. The EMD-LMD-TCN approach comprises three stages, and the methods utilized will be briefly introduced in the subsequent sections.

Stage 1: By applying the EMD and LMD techniques, we can split the original COVID-19 datasets into multiple sub-sequences. These sub-series consist of simpler frequency components and exhibit stronger correlations. This decomposition process simplifies the data and makes it easier to model and analyze.

Stage 2: Concatenate these IMFs, PFs, and residues decomposed by EMD and LMD incorporating the initial COVID-19 dataset as the input.

Stage 3: Construct the unique TCN model for input series, and apply the trained model to obtain the prediction values of given COVID-19 datasets.

4.1   Time Series Decomposition

The first stage (including EMD and LMD process) decomposes the original COVID-19 series into a number of IMFs, PFs, and residues. The extracted IMF

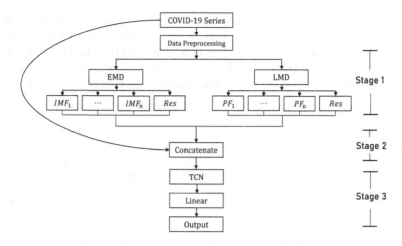

Fig. 2. Flowchart of the EMD-LMD-TCN model

components in the EMD represent a broad spectrum of frequencies, capturing a range of periodic fluctuations of COVID-19. Each IMF component captures a specific local characteristic time scale. However, there can be mode mixing, where IMFs may contain overlapping frequencies. To mitigate this, the LMD method is employed in this study, which reduces the effects of mode mixing and provides a more balanced time-varying spectral analysis. Additionally, the LMD method is employed to extract latent characteristics from the non-stationary and nonlinear COVID-19 signals. Phase functions (PFs) are also used, as they offer a better understanding of modulated characteristics and provide a more precise depiction of the given modulated series. Considering all this, the employment of these IMFs, PFs, and residues is beneficial to explore the variability and formation of COVID-19 from a new perspective. More detailed information on signal decomposition can be found in Sect. 5.2.

4.2 Proposed EMD-LMD-TCN Model

Following the aforementioned decomposition procedure, the extracted hidden information from the prediction variables and the original series is utilized as input to build the TCN forecasting model. This approach allows for the incorporation of valuable information into the model for more accurate predictions. As shown in Fig. 3, since two TCN blocks are sufficient to model any complex system with desired accuracy, the designed EMD-LMD-TCN model in this study will have only two TCN blocks. One block consists of a 1×2 convolutional layer, one weight norm layer, and one Relu layer. Simultaneously, incorporate a 1×1 convolution alongside one TCN block when there is a difference in dimensions between the input and output. The next is one Relu activation layer, which is then followed by a skip connection layer starting point at the original input. Afterward, another Relu activation is utilized to extract useful features. In the final step, a linear layer is employed to reduce the dimensionality of the high-dimensional features to one-dimensional vectors. These vectors encompass the predicted information, including the combined provided COVID-19 data along with the associated EMD and LMD decompositions. By utilizing the one-step look-ahead method, we can generate final predictions for the future range.

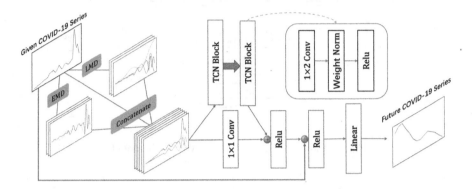

Fig. 3. Scheme of the EMD-LMD-TCN model

5   Experiment Setup

5.1   Data Description

COVID-19 surveillance data is collected and accessed through the COVID-19 monitoring dashboard[2], containing daily reported case and death count at the resolution of the world, which includes the USA, India, Brazil, Germany, France, UK, Russian Federation, and the Republic of Korean, etc. When considering multiple similar COVID-19 forecasting scenarios [1,23], the daily case counts (Fig. 4(a)) and death counts (Fig. 4(b)) are further aggregated on a weekly basis, ending on Saturdays, starting from the week ending February 8, 2020, and ending on the week ending April 16, 2022 (115 weeks). The dataset initially consists of a total of $(115 - T)$ data points after subtracting the window size T. These data points are then divided into a training set, which accounts for 85% of the total sample points, and a testing set, which consists of the remaining 15% of points.

5.2   IMF and PF Components Extraction

Previous prediction studies primarily focused on minimizing the discrepancy between fitting values and actual values. However, the chaotic and inherently complex nature of the original COVID-19 datasets makes it challenging to

---

[2] https://ourworldindata.org/coronavirus.

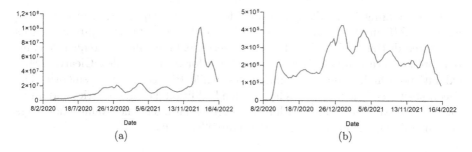

Fig. 4. (a) Daily new confirmed cases (b) Daily new confirmed deaths

directly capture the movement patterns of COVID-19 using conventional prediction models. To improve the prediction accuracy, in this study, EMD and LMD are applied to split the original COVID-19 datasets into different components.

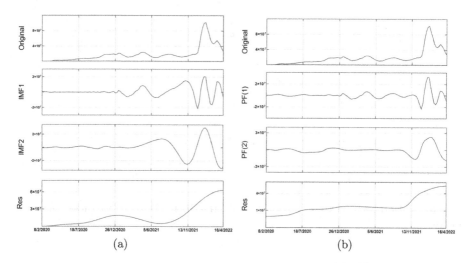

Fig. 5. (a) EMD for weekly new confirmed cases (b) LMD for weekly new confirmed cases

Figure 5 presents the individual EMD and LMD decomposition processes applied to newly reported cases data. The application of the EMD method on the new confirmed cases results in two Intrinsic Mode Functions (IMFs) and one residue component. These components exhibit a stable and regular variation pattern, similar to the decomposition results obtained through the LMD method. Likewise, we apply the EMD and LMD techniques to the weekly new confirmed deaths. Due to the trend of death curve seeming to be shakier, more IMFs and PFs can be taken out from the original signal.

By examining Fig. 6, it becomes evident that the application of either the EMD or LMD method allows us to decompose the weekly new confirmed deaths into three distinct sub-series. These sub-series exhibit simpler frequency components and are comparatively more straightforward to model. Clearly, IMFs exhibit more high-frequency components, while PFs seem to be smooth. The decomposition of the time series has revealed additional characteristics and information that were embedded within the original data. This newfound knowledge aids in training the forecasting model more effectively.

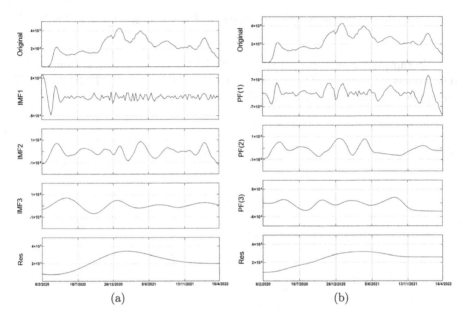

Fig. 6. (a) EMD for weekly new confirmed deaths (b) LMD for weekly new confirmed deaths

5.3   Training Details

Training window size T is set to 21 for the proposed EMD-LMD-TCN model, and forecasting horizon is made to 1, 2, and 3 weeks (representing the short, medium, and long term respectively.) ahead at each time point. As shown in Fig. 3, our model consists of 2 TCN blocks with 512 hidden units and 1 linear layer with 512 hidden units, etc. In addition, we set batch size as 16, epoch number as 1000, mean squared error (MSE) as a loss function, learning rate as 0.3, Adam optimizer as an optimizer, and all models are trained using stochastic initialization, where the initial values of the model parameters are randomly sampled from a distribution with a mean of 0 and a variance of 1.

6 Experiment Results

In this section, we conduct a comparative analysis of the proposed EMD-LMD-TCN model with various state-of-the-art time series forecasting methods as well as recent popular deep learning models, including ARIMA [8], XGBoost [9], MLP [18], RNN [12], LSTM [14], GRU [11], and TCN [5].

6.1 Evaluation Metrics

The following metrics: root mean squared log error (RMSLE) [24], mean absolute percentage error (MAPE) [16], Pearson correlation (PCORR) [7], coefficient of determination (R^2) [6] are widely employed to assess forecasting performance. To simplify the notation, let us define the following terms: $\hat{y}$ represents the forecasted values, $\overline{\hat{y}}$ represents the mean values of $\hat{y}$, y represents the real values, and $\bar{y}$ represents the mean values of y.

$$\text{RMSLE} = \sqrt{\frac{1}{n} \sum_{i=1}^{n} \left(\log\left(\hat{y}_i + 1\right) - \log\left(y_i + 1\right) \right)^2} \tag{6}$$

$$\text{MAPE} = \frac{100}{n} \sum_{i=1}^{n} \frac{|y_i - \hat{y}_i|}{|y_i|} \tag{7}$$

$$\text{PCORR} = \frac{\sum_{i=1}^{n} \left(\hat{y}_i - \overline{\hat{y}}\right)\left(y_i - \bar{y}\right)}{\sqrt{\sum_{i=1}^{n} \left(\hat{y}_i - \overline{\hat{y}}\right)^2} \sqrt{\sum_{i=1}^{n} \left(y_i - \bar{y}\right)^2}} \tag{8}$$

$$R^2 = 1 - \frac{\sum_{i=1}^{n} \left(y_i - \hat{y}_i\right)^2}{\sum_{i=1}^{n} \left(y_i - \bar{y}\right)^2} \tag{9}$$

6.2 Forecasting Performance

Our EMD-LMD-TCN model is versatile and capable of addressing various time series forecasting tasks, including the prediction of weekly new reported **cases** and **deaths** in the context of COVID-19. During the training process of each model, we utilize only the past values of the target series, along with their associated EMD and LMD components, as inputs.

We record RMSLE, MAPE, PCORR, and R^2 comparisons to other state-of-the-art algorithms on weekly new reported **cases** datasets with horizon $H = 1$, 2, and 3 in Table 1. Overall, our EMD-LMD-TCN model establishes overwhelming superiority over the other methods in all evaluation assessments, without considering the prediction window. For instance, when considering the RMSLE with a H of 1, our method achieves the top ranking, surpassing TCN, GRU, LSTM, RNN, ARIMA, MLP, and XGBoost. When considering the R^2 indicator, which measures the correlation between the predicted and original COVID-19 series, our model achieves a score of 0.9744 for a horizon of 1. This outperforms

Table 1. RMSLE, MAPE, PCORR and R^2 results of various techniques on weekly new confirmed **cases** datasets with a horizon $H = 1$, 2, and 3. The most favorable results for each row is **emphasized**. The second most favorable results for each row is underlined.

Metric	ARIMA	XGBoost	MLP	RNN	LSTM	GRU	TCN	**Ours**
$H = 1$								
Method								
RMSLE	0.1001	0.2125	0.1049	0.0984	0.0949	0.0857	<u>0.0807</u>	**0.0661**
MAPE	9.0312	20.3873	8.8266	8.5399	8.3417	7.5616	<u>6.8626</u>	**5.5153**
PCORR	<u>0.9828</u>	0.8705	0.9722	0.9752	0.9763	0.9813	0.9805	**0.9888**
R^2	0.9158	0.7217	0.9424	0.9495	0.9520	<u>0.9613</u>	0.9603	**0.9744**
$H = 2$								
Method								
RMSLE	0.2605	0.3220	0.2914	0.2199	<u>0.1614</u>	0.2005	0.1896	**0.1149**
MAPE	26.0922	31.8797	21.8906	17.7330	<u>13.6745</u>	16.1576	15.3756	**9.0088**
PCORR	0.9319	0.7186	0.8631	0.8868	<u>0.9462</u>	0.8970	0.9137	**0.9657**
R^2	0.4272	0.2540	0.7051	0.7765	<u>0.8775</u>	0.7944	0.8242	**0.9306**
$H = 3$								
Method								
RMSLE	0.4658	0.3888	0.4653	0.2568	0.2766	<u>0.2487</u>	0.2772	**0.2460**
MAPE	46.7617	39.3972	33.4711	<u>19.2096</u>	22.1574	20.2606	21.8623	**18.0647**
PCORR	<u>0.8328</u>	0.5723	0.7015	0.8093	0.8234	0.8166	0.7573	**0.9129**
R^2	−1.3783	−0.9831	0.2261	0.6181	0.6309	<u>0.6552</u>	0.5335	**0.7451**

the second best result of 0.9613 obtained by the GRU model. This can also be noticed in Fig. 7, which figure illustrates the overall trends of the original and predicted weekly new confirmed cases. Due to limited space, in Fig. 7 we only show three results by representative methods. Obviously, our predicting curve fits the ground truth best, which supports the numerical findings in Table 1.

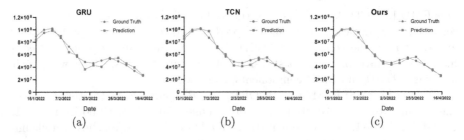

Fig. 7. Forecasted results for weekly new confirmed **cases** with state-of-the-art approaches ($H = 1$)

In a similar way, RMSLE, MAPE, PCORR, and R^2 comparisons to other advanced algorithms on weekly new reported **deaths** datasets with horizon H

= 1, 2, and 3 can be found in Table 2. Similarly, there is no question that our model significantly outperforms other competing methods across all forecasting horizons on the testing datasets. To be specific, under the metric of R^2 on $H = 1$, our model reaches 0.9886, exceeding the second-best method of 0.0324, which is a vast transcendence relative to other methods. Moreover, we utilize visual curve comparisons as a visual representation to highlight the strong efficacy of our model, as shown in Fig. 8. Clearly, among these methods, our model could draw the curve, which fits the original series the most perfectly.

Table 2. RMSLE, MAPE, PCORR and R^2 results of various techniques on weekly new confirmed **deaths** datasets with a horizon $H = 1$, 2, and 3. The most favorable results for each row is **emphasized**. The second most favorable results for each row is underlined.

$H = 1$								
Method								
Metric	ARIMA	XGBoost	MLP	RNN	LSTM	GRU	TCN	Ours
RMSLE	0.1020	0.1791	0.0918	0.1039	0.0902	<u>0.0831</u>	0.0874	**0.0588**
MAPE	8.5908	17.6928	7.3365	8.4579	7.4020	<u>6.8182</u>	7.3714	**4.0156**
PCORR	0.9658	0.9329	0.9722	0.9740	0.9765	<u>0.9780</u>	0.9722	**0.9967**
R^2	0.9291	0.7818	0.9449	0.9455	0.9530	<u>0.9562</u>	0.9449	**0.9886**
$H = 2$								
Method								
Metric	ARIMA	XGBoost	MLP	RNN	LSTM	GRU	TCN	Ours
RMSLE	0.1863	0.3074	0.1869	0.2286	0.2003	0.1795	<u>0.1627</u>	**0.1573**
MAPE	16.4714	33.2331	16.5905	19.6354	18.4190	15.8999	<u>13.1374</u>	**12.4858**
PCORR	0.8662	0.7591	0.9072	0.8793	0.9130	<u>0.9163</u>	0.9091	**0.9815**
R^2	0.7172	0.3712	0.8129	0.7605	0.8003	<u>0.8285</u>	0.8254	**0.8897**
$H = 3$								
Method								
Metric	ARIMA	XGBoost	MLP	RNN	LSTM	GRU	TCN	Ours
RMSLE	0.3135	0.4342	0.2817	0.3434	0.2923	0.2767	<u>0.2647</u>	**0.2606**
MAPE	30.0468	49.5606	25.3509	33.2951	29.7262	26.0744	**21.9652**	<u>22.7387</u>
PCORR	0.6608	0.5122	0.7611	0.7442	<u>0.7947</u>	0.7761	0.7431	**0.9809**
R^2	0.2916	−0.2275	<u>0.5747</u>	0.4587	0.5268	0.5704	0.5458	**0.6846**

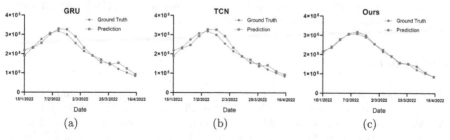

Fig. 8. Forecasted results for weekly new confirmed **deaths** with state-of-the-art approaches ($H = 1$)

In conclusion, the incorporation of EMD and LMD techniques in TCN enables it to accurately estimate the mapping function from the provided COVID-19 series to future ones. This results in the generation of fitting curves that are visually more appealing and aligned with the data. That is, our prediction framework can effectively boost the performance of deep learning methods when applied to datasets of limited size. Actually, we have applied the same decomposition techniques to the other four deep learning methods, including MLP, RNN, LSTM, and GRU. However, TCN model shows arrogant superiority over the other methods, and that is the reason we choose it as the forecasting network. The findings suggest that when combined with time decomposition techniques, a straightforward temporal convolutional architecture outperforms recurrent architectures in the task of COVID-19 forecasting.

7 Conclusion

In this study, we propose a hybrid framework combining EMD and LMD with a TCN for COVID-19 forecasting. Our approach involves dividing the original COVID-19 datasets into multiple component series using EMD and LMD techniques. These sub-series are then concatenated with the initial COVID-19 series to form input sequences. The TCN model is trained to effectively model the relationships between the given COVID-19 series and future ones. Notably, the EMD and LMD decomposition techniques result in simpler frequency components and stronger correlations, which contribute to more accurate forecasting and help address the issue of overfitting.

In future studies, our focus will be on conducting in-depth investigations into various time series decomposition methods. Through the analysis and comparison of diverse decomposition methods, our aim is to uncover additional rules and strategies that can enhance the precision of predictions. By thoroughly examining the similarities and distinctions among these techniques, we anticipate discovering valuable insights.

Acknowledgement. This work was supported by, the Natural Science Foundation of Guangdong Province, China (2023A1515011618), Key Scientific Research Platform and Project of Guangdong Provincial Education Department (2022ZDZX1040) and Educational Science Planning Topic of Guangdong Province (2022GXJK382), Research Project of Nanfang College Guangzhou (2022XK25).

References

1. Al-Qaness, M.A., Ewees, A.A., Fan, H., Abd El Aziz, M.: Optimization method for forecasting confirmed cases of COVID-19 in China. J. Clin. Med. **9**(3), 674 (2020)
2. Alabdulrazzaq, H., Alenezi, M.N., Rawajfih, Y., Alghannam, B.A., Al-Hassan, A.A., Al-Anzi, F.S.: On the accuracy of ARIMA based prediction of COVID-19 spread. Results Phys. **27**, 104509 (2021)

3. Almeida, R.: Analysis of a fractional SEIR model with treatment. Appl. Math. Lett. **84**, 56–62 (2018)
4. Anastassopoulou, C., Russo, L., Tsakris, A., Siettos, C.: Data-based analysis, modelling and forecasting of the COVID-19 outbreak. PLoS ONE **15**(3), e0230405 (2020)
5. Bai, S., Kolter, J.Z., Koltun, V.: An empirical evaluation of generic convolutional and recurrent networks for sequence modeling. arXiv preprint arXiv:1803.01271 (2018)
6. Behnood, A., Golafshani, E.M., Hosseini, S.M.: Determinants of the infection rate of the COVID-19 in the us using ANFIS and virus optimization algorithm (VOA). Chaos Solit. Fractals **139**, 110051 (2020)
7. Benesty, J., Chen, J., Huang, Y., Cohen, I.: Pearson correlation coefficient. In: Noise Reduction in Speech Processing, pp. 1–4. Springer, Heidelberg (2009). https://doi.org/10.1007/978-3-642-00296-0_5
8. Box, G.E., Jenkins, G.M., Reinsel, G.C., Ljung, G.M.: Time Series Analysis: Forecasting and Control. Wiley, Hoboken (2015)
9. Chen, T., Guestrin, C.: XGBoost: a scalable tree boosting system. In: Proceedings of the 22nd ACM SIGKDD International Conference on Knowledge Discovery and Data Mining, pp. 785–794 (2016)
10. Chimmula, V.K.R., Zhang, L.: Time series forecasting of COVID-19 transmission in Canada using LSTM networks. Chaos Solit. Fractals **135**, 109864 (2020)
11. Cho, K., Van Merriënboer, B., Bahdanau, D., Bengio, Y.: On the properties of neural machine translation: Encoder-decoder approaches. arXiv preprint arXiv:1409.1259 (2014)
12. Elman, J.L.: Finding structure in time. Cogn. Sci. **14**(2), 179–211 (1990)
13. Harko, T., Lobo, F.S., Mak, M.: Exact analytical solutions of the susceptible-infected-recovered (SIR) epidemic model and of the sir model with equal death and birth rates. Appl. Math. Comput. **236**, 184–194 (2014)
14. Hochreiter, S., Schmidhuber, J.: Long short-term memory. Neural Comput. **9**(8), 1735–1780 (1997)
15. Huang, N.E., et al.: The empirical mode decomposition and the Hilbert spectrum for nonlinear and non-stationary time series analysis. Proc. Royal Soc. Lond. Ser. A Math. Phys. Eng. Sci. **454**(1971), 903–995 (1998)
16. Makridakis, S., Hibon, M.: The M3-competition: results, conclusions and implications. Int. J. Forecast. **16**(4), 451–476 (2000)
17. Rahman, M.S., Chowdhury, A.H., Amrin, M.: Accuracy comparison of ARIMA and XGBoost forecasting models in predicting the incidence of COVID-19 in Bangladesh. PLOS Global Public Health **2**(5), e0000495 (2022)
18. Rosenblatt, F.: The perceptron, a perceiving and recognizing automaton Project Para. Cornell Aeronautical Laboratory (1957)
19. Shoeibi, A., et al.: Automated detection and forecasting of COVID-19 using deep learning techniques: a review. arXiv preprint arXiv:2007.10785 (2020)
20. Singh, V., et al.: Prediction of COVID-19 corona virus pandemic based on time series data using support vector machine. J. Discrete Math. Sci. Cryptogr. **23**(8), 1583–1597 (2020)
21. Smith, J.S.: The local mean decomposition and its application to EEG perception data. J. R. Soc. Interface **2**(5), 443–454 (2005)
22. Vaswani, A., et al.: Attention is all you need. In: Advances in Neural Information Processing Systems, vol. 30 (2017)

23. Wang, L., Adiga, A., Venkatramanan, S., Chen, J., Lewis, B., Marathe, M.: Examining deep learning models with multiple data sources for COVID-19 forecasting. In: 2020 IEEE International Conference on Big Data (Big Data), pp. 3846–3855. IEEE (2020)
24. Zeroual, A., Harrou, F., Dairi, A., Sun, Y.: Deep learning methods for forecasting COVID-19 time-series data: a comparative study. Chaos Solit. Fractals **140**, 110121 (2020)

Computational Intelligence, Nature-Inspired Optimizers, and Their Engineering Applications

LOS Guidance Law for Unmanned Surface Vehicle Path Following with Unknown Time-Varying Sideslip Compensation

Zijie Dai, Qiuxia Zhang, Jinyao Cheng, and Yongpeng Weng[✉]

School of Marine Electrical Engineering, Dalian Maritime University, Dalian 116024, Liaoning, China
wengyongpengneu@163.com, zhangqx@dlmu.edu.cn

Abstract. This paper proposes a new approach to solving the path-following problem for underdriven unmanned surface vehicles (USVs) with unknown sideslip angles. The method employs a sideslip observer to estimate and compensate for time-varying sideslip angles caused by external perturbations or curved paths, and further incorporates the line-of-sight (LOS) guidance law to deliver a path-following guidance scheme. The methodology rests on the theory of cascade stability and ensures the global asymptotic stability of the closed-loop system. Experimental results have verified the validity of the approach.

Keywords: path following · unmanned surface vehicles · line-of-sight guidance

1 Introduction

In recent decades, USVs have been extensively used in practical situations [1–6], and hence, have garnered significant attention from the control community. In motion control, the guidance law plays a critical role, acting as the key factor that enables USVs to achieve tracking motion, including path following and target tracking. Several approaches exist for path tracking, such as LOS [7,8], vision-based [9], and Lyapunov-based [10] methods. Of these, the LOS method has been widely adopted for its simplicity, fast response time, and easy adjustment of parameters, especially in practical scenarios.

Under complex environmental conditions, USVs are vulnerable to exposure to disturbances and are thus plagued by sideslip angles. Although the angle may be small, it can lead to an increase in tracking error, ultimately resulting in poor performance for the LOS guidance law. This is primarily because underactuated USVs lack the ability to sway and offset the sideslip angle, with only surge and yaw being able to control the vehicle's direction. Therefore, accounting for the sideslip angle to enhance path tracking performance can present a significant challenge [11]. To address the issue of sideslip angle, researchers have conducted extensive research in this area, the most straightforward approach being to utilize high-precision sensors that can measure it in real time [12]. However, due to the

© The Author(s), under exclusive license to Springer Nature Singapore Pte Ltd. 2023
H. Zhang et al. (Eds.): NCAA 2023, CCIS 1869, pp. 193–205, 2023.
https://doi.org/10.1007/978-981-99-5844-3_14

high cost of such sensors, it is not practically feasible to use them. Therefore, several alternative methods have been proposed to address this issue. Researchers have developed an integral LOS (ILOS) method based on the LOS guidance law, which compensates for the sideslip angle by incorporating the integral term [11,14–16]. For instance, the ILOS guidance law proposed by Caharija et al. [14] compensates for the drift effect induced by environmental disturbances, allowing the vessel with sideslip angles to move along a straight path. Moreover, to track curved paths, the side slip angle depends on the path curvature and the USV's speed. Fossen et al. [15] achieved curved path tracking based on the ILOS method that provides the required heading angle. On the other hand, Lekkas et al. [11] used ILOS to offset constant external disturbances for side slip angle elimination. Additionally, adaptive LOS (ALOS) guidance methods were developed. Fossen et al. [17] extended the LOS to Dubins path for path following and adaptively estimated the unknown side slip angle as a parameter. Liu et al. [18] proposed an advanced LOS (ALOS) guidance law that enables accurate path-following, even when the path has a significant degree of curvature. Furthermore, Mu et al. [19] proposed a fuzzy optimization algorithm-based ALOS guidance law that enables vehicles to track both straight and curved paths. However, the side slip angle in practical applications cannot be assumed to be constant, which presents a challenge for both ALOS and ILOS methods.

Based on the above proposed ILOS and ALOS compared to the proposed sideslip observer based LOS (SLOS) guidance law, the key elements of the proposed SLOS guidance law in this paper are summarized as follows:

1) In comparison to conventional ALOS and ILOS methods, the proposed SLOS method is capable of addressing time-varying sideslip angles and is deemed more practical for applications.
2) The integration term used in both ALOS and ILOS guidance laws presents a challenge in eliminating the steady-state error, which may affect system stability. In contrast, the proposed SLOS guidance law offers a solution to eliminate this problem altogether.
3) The stability of the closed-loop system is rigorously analyzed and demonstrated in the paper, showcasing the robustness of the proposed approach. Furthermore, simulation results reveal the superior effectiveness of the scheme in achieving asymptotic tracking control of underactuated USVs.

The rest of this paper is organized as: Sect. 2 gives some preliminaries. Section 3 presents the design and stability analysis of SLOS guidance law. Section 4 and Sect. 5 provide a simulation study and some conclusions about SLOS, respectively.

2   Problem Formulation

As shown in Fig. 1, this paper utilizes both the earth-fixed inertial frame and the body-fixed frame to represent the motion of the USV and the accompanying sideslip angle.

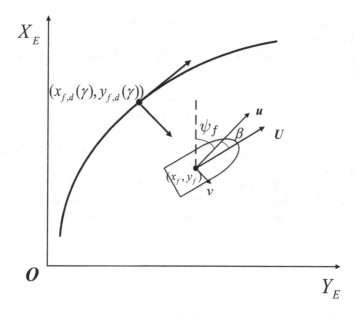

Fig. 1. Schematic diagram of β.

According to [20], the kinetics of USV can be expressed as

$$
\begin{aligned}
\dot{x}_f &= u \cos \psi - v \sin \psi \\
\dot{y}_f &= u \sin \psi + v \cos \psi \\
\dot{\psi}_f &= r
\end{aligned}
\tag{1}
$$

where (x_f, y_f) represents the position of the vehicle in the earth-fixed frame, $\psi_f \in (-\pi, \pi]$ represents the heading angle and (u, v, r) refer to the surge, sway and yaw velocities in the body-fixed frame, respectively.

One possible approach is to introduce a path parameter γ to represent the inertial position of the reference path as $(x_{f,d}(\gamma), y_{f,d}(\gamma))$. The path-tangential angle $\alpha(\gamma)$ is denoted as $\alpha(\gamma) = \text{atan2}(y'_{f,d}(\gamma), x'_{f,d}(\gamma))$, where $x'_{f,d}(\gamma) = \partial x_{f,d}(\gamma)/\partial \gamma$ and $y'_{f,d}(\gamma) = \partial y_{f,d}(\gamma)/\partial \gamma$.

Then, letting $x_{f,e}$, $y_{f,e}$ denote the errors in the path-tangential reference frame, corresponding to the along-track and cross-track errors, respectively, we have

$$
\begin{bmatrix} x_{f,e} \\ y_{f,e} \end{bmatrix} = \begin{bmatrix} \cos \alpha(\gamma) & -\sin \alpha(\gamma) \\ \sin \alpha(\gamma) & \cos \alpha(\gamma) \end{bmatrix}^T \begin{bmatrix} x_f - x_{f,d}(\gamma) \\ y_f - y_{f,d}(\gamma) \end{bmatrix}
\tag{2}
$$

Derive the time derivatives of $x_{f,e}$ and $y_{f,e}$ to yield

$$
\begin{cases} \dot{x}_{f,e} = U \cos(\psi_f - \alpha(\gamma) + \beta) + \dot{\alpha}(\gamma) y_{f,e} - \dot{\gamma}\sqrt{x'_{f,d}{}^2(\gamma) + y'_{f,d}{}^2(\gamma)} \\ \dot{y}_{f,e} = U \sin(\psi_f - \alpha(\gamma) + \beta) - \dot{\alpha}(\gamma) x_{f,e} \end{cases}
\tag{3}
$$

where $U = \sqrt{u^2 + v^2}$ represents the course speed of USV and $\beta = \text{atan2}(v, u)$ denotes the sideslip angle.

Then, on the basis of the first-order Euler transformation technique, Eq. (3) can be further equivalent as

$$\begin{cases} x_{f,e,t+1} = x_{f,e,t} + T_s \left\{ U_t \cos(\psi_{f,t} - \alpha(\gamma_t) + \beta_t) + \dot{\alpha}(\gamma_t) y_{f,e,t} - u_{p,t} \right\} \\ y_{f,e,t} = y_{f,e,t} + T_s \left\{ U_t \sin(\psi_{f,t} - \alpha(\gamma_t) + \beta_t) - \dot{\alpha}(\gamma_t) x_{f,e,t} \right\} \end{cases} \quad (4)$$

where $u_{p,t} = \Delta\gamma_t \sqrt{x'_{f,d}{}^2(\gamma) + y'_{f,d}{}^2(\gamma)}$ denotes the speed of the virtual reference point and $\Delta\gamma_t = \gamma_{t+1} - \gamma_t$.

Control objective: To effectively overcome the unknown time-varying sideslip angle of the USV, this paper will design an observer-based guidance law that allows the USV to start from any position and still converge to the reference path $(x_f(\gamma_t), y_f(\gamma_t))$, i.e.:

$$\begin{cases} \lim_{t\to\infty} x_{f,e,t} \leq \delta_1 \\ \lim_{t\to\infty} y_{f,e,t} \leq \delta_2 \end{cases} \quad (5)$$

where δ_1 and δ_2 are some small constants.

3   SLOS Guidance Design and Analysis

3.1   Sideslip Angle Identification

In this section, an observer to evaluate the time-varying sideslip angle is presented. Since the magnitude of the sidelobe angle is small in practice, there exists a positive constant β^* such that for all t, $|\beta_t| \leq \beta^*$.

From Eq. (4), we can further obtain that

$$\begin{cases} x_{f,e,t+1} = x_{f,e,t} + T_s\{U_t \cos(\psi_{f,t} - \alpha(\gamma_t))\cos\beta_t \\ \qquad - U_t \sin(\psi_{f,t} - \alpha(\gamma_t))\sin\beta_t + \dot{\alpha}(\gamma_t)y_{f,e,t} - u_{p,t}\} \\ y_{f,e,t+1} = y_{f,e,t} + T_s\{U_t \sin(\psi_{f,t} - \alpha(\gamma_t))\cos\beta_t \\ \qquad + U_t \cos(\psi_{f,t} - \alpha(\gamma_t))\sin\beta_t - \dot{\alpha}(\gamma_t)x_{f,e,t}\} \end{cases} \quad (6)$$

It is noted that the value of β_t is typically quite small, generally less than $5°$ [17]. Consequently, it can be deduced that $\cos\beta_t \approx 1$ and $\sin\beta_t \approx \beta_t$, so Eq. (6) is equivalent to

$$\begin{cases} x_{f,e,t+1} = x_{f,e,t} + T_s\{U_t \cos(\psi_{f,t} - \alpha(\gamma_t)) \\ \qquad - U_t \beta_t \sin(\psi_{f,t} - \alpha(\gamma_t)) + \dot{\alpha}(\gamma_t)y_{e,t} - u_{p,t}\} \\ y_{f,e,t+1} = y_{f,e,t} + T_s\{U_t \sin(\psi_{f,t} - \alpha(\gamma_t)) \\ \qquad + U_t \cos(\psi_{f,t} - \alpha(\gamma_t))\beta_t - \dot{\alpha}(\gamma_t)x_{f,e,t}\} \end{cases} \quad (7)$$

To estimate the sideslip angle online, we design the observer as follows

$$
\begin{cases}
\hat{x}_{f,e,t+1} = \hat{x}_{f,e,t} + T_s\{U_t\cos(\psi_{f,t} - \alpha(\gamma_t)) - U_t sin(\psi_{f,t} \\
\qquad -\alpha(\gamma_t))\hat{\beta}_t + \dot{\alpha}(\gamma_t)y_{f,e,t} - u_{p,t}\} - c_1\tilde{x}_{f,e,t} \\
\hat{y}_{f,e,t+1} = \hat{y}_{f,e,t} + T_s\{U_t\sin(\psi_{f,t} - \alpha(\gamma_t)) \\
\qquad +U_t\cos(\psi_t - \alpha(\gamma_t))\hat{\beta}_t - \dot{\alpha}(\gamma_t)x_{f,e,t}\} - c_2\tilde{y}_{f,e,t} \\
\hat{\beta}_{t+1} = k_1k_2\hat{\beta}_t - T_s\{U_t sin(\psi_t - \alpha(\gamma_t))\tilde{x}_{e,t} \\
\qquad -U_t\cos(\psi_{f,t} - \alpha(\gamma_t))\tilde{y}_{f,e,t}\}
\end{cases}
\tag{8}
$$

with $\tilde{x}_{f,e,t} = \hat{x}_{f,e,t} - x_{f,e,t}$, $\tilde{y}_{f,e,t} = \hat{y}_{f,e,t} - y_{f,e,t}$ denoting the estimation error, c_1 and c_2 being the positive constants, $k_1 = c_1 - 1$, $k_2 = c_2 - 1$, and $\hat{\beta}_t$ being the estimate of β_t.

Thus, the error dynamics can be expressed as $\tilde{x}_{f,e,t}$, $\tilde{y}_{f,e,t}$ and $\tilde{\beta}_t$ below:

$$
\begin{cases}
\tilde{x}_{f,e,t+1} = -T_sU_t sin(\psi_{f,t} - \alpha(\gamma_t))\tilde{\beta}_t - k_1\tilde{x}_{f,e,t} \\
\tilde{y}_{f,e,t+1} = T_sU_t\cos(\psi_{f,t} - \alpha(\gamma_t))\tilde{\beta}_t - k_2\tilde{y}_{f,e,t} \\
\tilde{\beta}_{t+1} = k_1k_2\tilde{\beta}_t - (1 - k_1k_2)\beta_t - T_s\{\frac{U_t sin(\psi_{f,t}-\alpha(\gamma_t))}{k_2}\tilde{x}_{f,e,t} - \frac{U_t\cos(\psi_{f,t}-\alpha(\gamma_t))}{k_1}\tilde{y}_{f,e,t}\}
\end{cases}
\tag{9}
$$

Simplifying the above equation yields

$$
\begin{cases}
\tilde{x}_{f,e,t+1} = -a_{1,t}\tilde{\beta}_t - k_1\tilde{x}_{f,e,t} \\
\tilde{y}_{f,e,t+1} = a_{2,t}\tilde{\beta}_t - k_2\tilde{y}_{f,e,t} \\
\tilde{\beta}_{t+1} = k_1k_2\tilde{\beta}_t - (1 - k_1k_2)\beta - \frac{a_{1,t}}{k_2}\tilde{x}_{f,e,t} + \frac{a_{2,t}}{k_1}\tilde{y}_{f,e,t}
\end{cases}
\tag{10}
$$

where $a_{1,t} = T_sU_t sin(\psi_{f,t} - \alpha(\gamma_t))$, $a_{2,t} = T_sU_t\cos(\psi_{f,t} - \alpha(\gamma_t))$.

3.2 Guidance Law Design

To ensure simultaneous convergence of $x_{f,e,t}$ and $y_{f,e,t}$, we design the guidance law in the following manner:

$$
\psi_{f,d,t} = \alpha(\gamma_t) + arctan(-\frac{y_{f,e,t}}{\Delta} - \hat{\beta}_t)
\tag{11}
$$

where Δ denotes the look-ahead distance.

To stabilize $x_{f,e,t}$, $u_{p,t}$ is constructed to be

$$
\begin{aligned}
u_{p,t} = U_t\cos(\psi_{f,t} - \alpha(\gamma_t)) - U_t\sin(\psi_{f,t}) \\
-\alpha(\gamma_t))\hat{\beta}_t + \dot{\alpha}(\gamma_t)y_{f,e,t} + \kappa\hat{x}_{f,e,t}
\end{aligned}
\tag{12}
$$

in which κ is a constant parameter.

Further, γ_t is the path parameter variable designed for guidance purposes, and the updating law for γ_t can be obtained from Eq. (12), which is given by

$$
\begin{aligned}
\gamma_{t+1} = \gamma_t + \frac{U_t\cos(\psi_{f,t}-\alpha(\gamma_t))}{\sqrt{x'^2_f(\gamma_t)+y'^2_f(\gamma_t)}} \\
+\frac{-U_t sin(\psi_{f,t}-\alpha(\gamma_t))\hat{\beta}_t+\dot{\alpha}(\gamma_t)y_{f,e,t}+\kappa\hat{x}_{f,e,t}}{\sqrt{x'^2_f(\gamma_t)+y'^2_f(\gamma_t)}}
\end{aligned}
\tag{13}
$$

Therefore, $x_{f,e,t}$ in Eq. (7) can be transformed into

$$\hat{x}_{f,e,t+1} = (1 - T_s \kappa) \, \hat{x}_{f,e,t} - c_1 \tilde{x}_{f,e,t} \tag{14}$$

Considering

$$\begin{cases} \sin\left(\arctan\left(-\frac{y_{f,e,t}}{\Delta} - \hat{\beta}_t\right)\right) = -\frac{y_{f,e,t} + \Delta\hat{\beta}_t}{\sqrt{\Delta^2 + \left(y_{f,e,t} + \Delta\hat{\beta}_t\right)^2}} \\ \cos\left(\arctan\left(-\frac{y_{f,e,t}}{\Delta} - \hat{\beta}_t\right)\right) = \frac{\Delta}{\sqrt{\Delta^2 + \left(y_{f,e,t} + \Delta\hat{\beta}_t\right)^2}} \end{cases} \tag{15}$$

Therefore, $y_{f,e,t}$ in Eq. (7) can be expressed as

$$\begin{aligned} \hat{y}_{f,e,t+1} = & \left(1 - \frac{T_s U_t}{\sqrt{\Delta^2 + \left(y_{f,e,t} + \Delta\hat{\beta}_t\right)^2}}\right) \hat{y}_{f,e,t} - T_s \dot{\alpha}(\gamma_t) \hat{x}_{f,e,t} \\ & - T_s \dot{\alpha}(\gamma_t) \tilde{x}_{f,e,t} - c_2 \tilde{y}_{f,e,t} \end{aligned} \tag{16}$$

4   Stability Analysis

Define following Lyapunov function $V_{1,t} = \tilde{x}_{f,e,t}^2 + \tilde{y}_{f,e,t}^2 + \tilde{\beta}_t^2$, then

$$\begin{aligned} \Delta V_{1,t} = & \, \tilde{x}_{f,e,t+1}^2 + \tilde{y}_{f,e,t+1}^2 + \tilde{\beta}_{t+1}^2 - \tilde{x}_{f,e,t}^2 - \tilde{y}_{f,e,t}^2 - \tilde{\beta}_t^2 \\ \leq & - \left[1 - k_1^2 - \frac{3a_{1,t}^2}{k_2}\right] \tilde{x}_{f,e,t}^2 - \left[1 - k_2^2 - \frac{3a_{2,t}^2}{k_1}\right] \tilde{y}_{f,e,t}^2 \\ & - \left[1 - a_{1,t}^2 - a_{2,t}^2 - 2k_1^2 k_2^2\right] \tilde{\beta}_t^2 + 3(1 - k_1 k_2)^2 \beta_t^2 \end{aligned} \tag{17}$$

Let $h = \min\left\{\left[1 - k_1^2 - \frac{3a_{1,t}^2}{k_2}\right], \left[1 - k_2^2 - \frac{3a_{2,t}^2}{k_1}\right], \left[1 - a_{1,t}^2 - a_{2,t}^2 - 2k_1^2 k_2^2\right]\right\}$, then we can obtain

$$\Delta V_{1,t} \leq -h\left(1 - \theta\right) \|E_t\|^2 - h\theta \|E_t\|^2 + 3(1 - k_1 k_2)^2 \beta_t^2 \tag{18}$$

with $E_t = [\tilde{x}_{f,e,t}, \tilde{y}_{f,e,t}, \tilde{\beta}_t]^T$.

Since when $\|E\| \geq \frac{\sqrt{3}(1 - k_1 k_2)\beta_t}{h\theta}$, the above equation yields

$$\Delta V_{1,t} \leq -h\left(1 - \theta\right) \|E_t\| \tag{19}$$

Therefore, $\tilde{x}_{f,e,t}$, $\tilde{y}_{f,e,t}$ and $\tilde{\beta}_t$ will converge to a region with finite boundaries. Next, Consider Lyapunov function $V_{2,t} = |\hat{y}_{f,e,t}| + |\hat{x}_{f,e,t}|$, then

$$\begin{aligned} \Delta V_{2,t} = & \, |\hat{y}_{f,e,t+1}| + |\hat{x}_{f,e,t+1}| - |\hat{y}_{f,e,t}| - |\hat{x}_{f,e,t}| \\ \leq & - \frac{T_s U_t}{\sqrt{\Delta^2 + \left(y_{f,e,t} + \Delta\hat{\beta}_t\right)^2}} |\hat{y}_{f,e,t}| - (T_s \kappa - T_s \dot{\alpha}(\gamma_t)) |\hat{x}_{f,e,t}| \\ & + c_1 |\tilde{x}_{f,e,t}| + T_s |\dot{\alpha}(\gamma_t)\tilde{x}_{f,e,t}| + c_2 |\tilde{y}_{f,e,t}| \end{aligned} \tag{20}$$

Let $h = \min\left\{\dfrac{T_sU_t}{\sqrt{\Delta^2+(y_{f,e,t}+\Delta\hat{\beta}_t)^2}}, (T_s\kappa - T_s\dot{\alpha}(\gamma_t))\right\}$, then from the above equation we get

$$\begin{aligned}\Delta V_{2,t} \leq &-h\,(1-\theta)\,\|E_t\| - h\theta_t\,\|E_t\| + c_1\,|\tilde{x}_{f,e,t}| \\ &+T_s\,|\dot{\alpha}(\gamma_t)\tilde{x}_{f,e,t}| +c_2\,|\tilde{y}_{f,e,t}|\end{aligned} \tag{21}$$

with $E_t = [\hat{x}_{f,e,t}, \hat{y}_{f,e,t}]^T$.

When $\|E\| \geq \dfrac{(c_1|\tilde{x}_{f,e,t}|+T_s|\dot{\alpha}(\gamma_t)\tilde{x}_{f,e,t}|+c_2|\tilde{y}_{f,e,t}|)}{h\theta}$, then

$$\Delta V_{2,t} \leq -h\,(1-\theta)\,\|E_t\| \tag{22}$$

So, $\hat{x}_{f,e,t}$, $\hat{y}_{f,e,t}$ eventually converges to a bounded region.

At this point, we can express the entire closed-loop system as a series expansion, given by:

$$\Sigma_1 : \begin{cases} \hat{x}_{e,t+1} = (1 - T_s\kappa)\,\hat{x}_{e,t} - c_1\tilde{x}_{e,t} \\ \hat{y}_{e,t+1} = \left(1 - \dfrac{T_sU_t}{\sqrt{\Delta^2+(y_{e,t}+\Delta\hat{\beta}_t)^2}}\right)\hat{y}_{e,t} - T_s\dot{\alpha}(\omega_t)\hat{x}_{e,t} \\ \qquad\quad - T_s\dot{\alpha}(\omega(k))\hat{x}_e(k) - T_s\dot{\alpha}(\omega_t)\tilde{x}_{e,t} - c_2\tilde{y}_{e,t} \end{cases} \tag{23}$$

$$\Sigma_2 : \begin{cases} \tilde{x}_{f,e,t+1} = -T_sU_t\sin(\psi_{f,t} - \alpha(\gamma_t))\tilde{\beta}_t - k_1\tilde{x}_{f,e,t} \\ \tilde{y}_{f,e,t+1} = T_sU_t\cos(\psi_{f,t} - \alpha(\gamma_t))\tilde{\beta}_t - k_2\tilde{y}_{f,e,t} \end{cases} \tag{24}$$

Therefore, we can infer that stability is preserved throughout the entire closed-loop cascade control system.

5   Simulation Experiments

The simulation and experimental results presented in this section provide evidence of the efficacy of the proposed SLOS guidance law. The simulation model employed in this study is constructed using the CyberShip II [21] as a basis for modeling the dynamics of the USV:

$$M\dot{v} = \tau - C(v)v - D(v)v - g(\eta_f) + d$$

with $\eta_f = [x_f, y_f, \psi_f]^T$, $v = [u, v, r]^T$ representing, respectively, the surge, sway velocities and yaw rate, $\tau = [\tau_u, 0, \tau_r]^T$ representing the control input, and $d = [d_u, d_v, d_r]^T$ referring to total interferences. Moreover, M, C, D and g denote nonlinearities that are completely considered as unknown because they are difficult to obtain in practice.

According to CyberShip II we can obtain the interference vector as

$$\begin{bmatrix} d_u \\ d_v \\ d_r \end{bmatrix} = \begin{bmatrix} 1.2 + 0.1|u_t|^3 \\ 0.1 + 0.1|u_t|^2 + randn(1) \\ -0.1r_t^3 + 0.2\sin(v_t) \end{bmatrix}. \tag{25}$$

At the control level, we use a classical proportional-integral-differential (PID) strategy to track the desired heading angle.

$$\tau_{u,t} = -K_{up}(u_t - u_{d,t}) - K_{ui}\sum_{j=1}^{t}(u_j - u_{d,j})$$
$$- K_{ud}(u_t - u_{d,t} - (u_{t-1} - u_{d,t-1})) \tag{26}$$

$$\tau_{r,t} = -K_{rp}(\psi_{f,t} - \psi_{f,r,t}) - K_{ri}\sum_{j=1}^{t}(\psi_{f,j} - \psi_{f,r,j})$$
$$- K_{rd}(\psi_{f,t} - \psi_{f,r,t} - (\psi_{f,t-1} - \psi_{f,r,t-1})) \tag{27}$$

with K_{up}, K_{ui}, K_{ud}, K_{rp}, K_{ri} and K_{rd} being positive parameters.
The desired parameter path is set to

$$\begin{cases} x_d(\gamma_t) = -10cos(0.01\gamma_t) + 10 \\ y_d(\gamma_t) = 10sin(0.01\gamma_t) \end{cases} \tag{28}$$

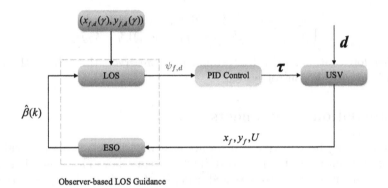

Observer-based LOS Guidance

Fig. 2. Path following architecture.

The proposed observer-based LOS guidance law consists of two basic elements, namely the observer and the command heading angle, as depicted in Fig. 2. The guidance system operates independently of the control system and is compatible with all kinds of heading autopilots. To provide a more comprehensive understanding of the SLOS guidance law proposed in this paper, we have listed the step-by-step details in Table 1.

Figures 3, 4, 5, 6, 7 and 8 present the simulation results. Specifically, Fig. 3 illustrates the path following performance of the SLOS method. Figure 4 displays the update of the path parameter. Figure 5 and Fig. 6 show the tracking of surge speed and heading angle. Figure 7 portrays the path following error

Table 1. The SLOS guidance law step

Method: A novel LOS guidance law for USV with sideslip compensation
1: **Initialization**:
2:    Set initial states $x_0, y_0, \psi_0, u_0, v_0, r_0$ and the parameterized paths γ_0.
3:    Set the controller parameters as the Table 2.
4: **Loop for USV**:
5:    Generate parameter path $x_{f,d}(\gamma_t)$, $y_{f,d}(\gamma_t)$ and the path-tangential angle $\alpha(\gamma_t)$.
6:    Caculate the path following error $x_{f,e,t}$ and $y_{f,e,t}$.
7:    Caculate the guidancelaw $\psi_{f,d,t+1}$, $u_{f,d,t+1}$.
8:    Caculate control signals $\tau_{u,t}$, $\tau_{r,t}$.
9:    Set the observer $\hat{x}_{f,e,t+1}$, $\hat{y}_{f,e,t+1}$ and caculate $\hat{\beta}_{t+1}$.
10:    Update the speed of the virtual reference point $u_{p,t}$ and path parameter γ_{t+1}.
11:    Caculate the output of USV.
12: **End loop**

Table 2. The simulation parameters.

Parameter	Value
Initial location (x_f, y_f)	$(-2, 2)$
Initial speed (u, v, r)	$(0, 0, 0)$
c_1	1.5
c_2	1.8
Δ	5
K_{up}	3
K_{ui}	0.15
K_{ud}	150
K_{rp}	0.8
K_{ri}	0.08
K_{rd}	350

and its observation effect. Figure 8 shows the sideslip angle together with the observations. The tracking error and observation error were found to be minimal throughout the simulation. The small error indicates the effectiveness of the SLOS method in accurately tracking the desired path, as well as the reliability of the observer in providing precise measurements of system performance. Figure 9 highlights the superiority of the proposed SLOS guidance law by demonstrating the impact of sideslip observation under the ALOS guidance law. The results unequivocally demonstrate that the SLOS method is better suited to handle the effects caused by the sideslip angle.

Figures 10 and 11 provide additional evidence of the effectiveness of the sideslip angle observer. By setting $\hat{\beta}_t = 0$ in Eq. (11), we can observe the impact of

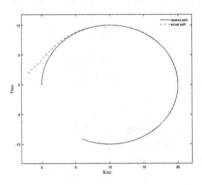

Fig. 3. Path following performance

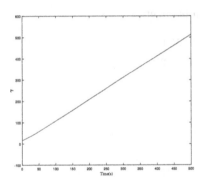

Fig. 4. Path parameter γ.

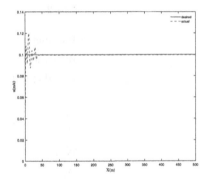

Fig. 5. Surge speed u.

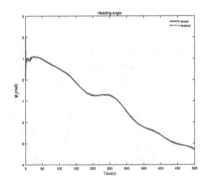

Fig. 6. Heading angle ψ_f.

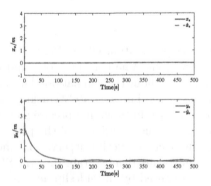

Fig. 7. Path following performance.

the sideslip angle on path-tracking performance. The results demonstrate that even a small sideslip angle can have a substantial effect on path-tracking.

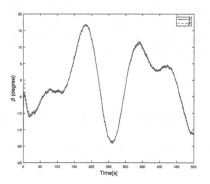

Fig. 8. Sideslip β.

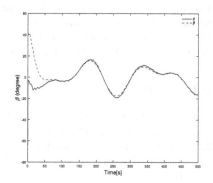

Fig. 9. Sideslip $\hat{\beta}$ under ALOS guidance law.

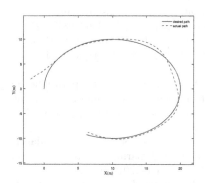

Fig. 10. Path following performance with $\hat{\beta} = 0$.

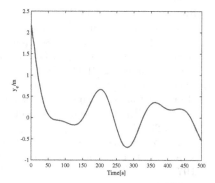

Fig. 11. y_e under with $\hat{\beta} = 0$.

6   Conclusions

The proposed research involves the development of a novel guidance law method based on the application of online estimation and compensation techniques, which are capable of effectively compensating for the USV sideslip angle resulting from various external factors such as ocean currents, waves, and other disturbances. The advanced nature of the proposed methodology has been demonstrated through a rigorous theoretical analysis that establishes the stability of the entire closed-loop system. It is pertinent to note that the effectiveness of the designed guidance method has been verified by in-depth simulation studies.

Acknowledgement. This work was supported by the National Natural Science Foundation of China of P. R. China (under Grants 51009017, 51379002 and 61803063), the Fund for Dalian Distinguished Young Scholars (under Grant 2016RJ10), the Fund for Liaoning Innovative Talents in Colleges and Universities (under Grant LR2017024), the Fundamental Research Funds for the Central Universities (under Grants 3132019344, 3132019108 and 3132018126), and the Stable Supporting Fund of Science and Technology on Underwater Vehicle Laboratory (SXJQR2018WDKT03).

References

1. Breivik, M., Hovstein V.E., Fossen I.: Straight-line target tracking for unmanned surface vehicles, pp. 131–149 (2008)
2. Wang, N., Ahn, C.K.: Coordinated trajectory-tracking control of a marine aerial-surface heterogeneous system. IEEE ASME Trans. Mechatron. **26**(6), 3198–3210 (2021)
3. Hong, Z., Yu, H., Guo, C.: Finite time PAILOS based path following control of underactuated marine surface vessel with input saturation. ISA Trans. (2022). https://doi.org/10.1016/j.isatra.2022.09.027
4. Wang, N., Ahn, C.K.: Hyperbolic-tangent LOS guidance-based finite-time path following of underactuated marine vehicles. IEEE Trans. Ind. Electron. **67**(10), 8566–8575 (2019)
5. Rout, R., Cui, R., Yan, W.: Sideslip-compensated guidance-based adaptive neural control of marine surface vessels. IEEE Trans. Cybern. **52**(5), 2860–2871 (2020)
6. Liu, Z.: Improved ELOS based path following control for underactuated surface vessels with roll constraint. Ocean Eng. **245**, 110348 (2022). https://doi.org/10.1016/j.oceaneng.2021.110348
7. Borhaug, E., Pavlov, A., Pettersen, K.Y.: Integral LOS control for path following of underactuated marine surface vessels in the presence of constant ocean currents. In: 47th IEEE Conference on Decision Control, Cancun, Mexico, pp. 4984–4991. IEEE (2008)
8. Weng, Y., Wang N., Carlos G.S.: Data-driven sideslip observer-based adaptive sliding-mode path-following control of underactuated marine vessels. Ocean Eng. **197**, 106910 (2020). https://doi.org/10.1016/j.oceaneng.2019.106910
9. Wang, K., Liu, Y., Li, L.: Vision-based tracking control of underactuated water surface robots without direct position measurement. IEEE Trans. Control Syst. Technol. **23**(6), 2391–2399 (2015)
10. Bibuli, M., Caccia, M., Lapierre, L., Bruzzone, G.: Guidance of unmanned surface vehicles: experiments in vehicle following. IEEE Trans. Robot. **19**(3), 92–102 (2012)
11. Lekkas, A.M., Fossen, T.I.: Integral LOS path following for curved paths based on a monotone cubic Hermite spline parametrization. IEEE Trans. Control Syst. Technol. **22**(6), 2287–2301 (2014)
12. Bevly, D.M., Sheridan, R., Pettersen, K.Y.: Integrating INS sensors with GPS velocity measurements for continuous estimation of vehicle sideslip and tire cornering stiffness. In: American Automatic Control Conference, AACC, Washington, USA, 25–27 June, pp. 483–493 (2001)
13. Fossen, T.I., Breivik, M., Skjetne, R.: Line-of-sight path following of underactuated marine craft. IFAC Proc. Vol. **36**(21), 211–216 (2003)
14. Caharija, W., et al.: Integral line-of-sight guidance and control of underactuated marine vehicles: theory, simulations, and experiments. IEEE Trans. Control Syst. Technol. **24**(5), 1623–1642 (2016)

15. Fossen, T.I., Lekkas, A.M.: Direct and indirect adaptive integral line-of-sight path-following controllers for marine craft exposed to ocean currents. Int. J. Adapt. Control Signal Process. **31**(4), 445–463 (2017)
16. Li, M., Guo, C., Yu, H.: Extended state observer-based integral line-of-sight guidance law for path following of underactuated unmanned surface vehicles with uncertainties and ocean currents. Int. J. Intell. Syst. **18**(3), 17298814211011035 (2021). https://doi.org/10.1177/17298814211011035
17. Fossen, T.I., Pettersen, K.Y., Galeazzi, R.: Line-of-sight path following for Dubins paths with adaptive sideslip compensation of drift forces. IEEE Trans. Control Syst. Technol. **23**(2), 820–827 (2014)
18. Liu, C., Negenborn, R.R., Chu, X., Zheng, H.: Predictive path following based on adaptive line-of-sight for underactuated autonomous surface vessels. J. Mar. Sci. Technol. **23**, 483–494 (2018)
19. Mu, D., Wang, G., Fan, Y., Bai, Y., Zhao, Y.: Fuzzy-based optimal adaptive line-of-sight path following for underactuated unmanned surface vehicle with uncertainties and time-varying disturbances. Math. Probl. Eng. **2018**, 1–12 (2018)
20. Wang, N., Xie, G., Pan, X.: Full-state regulation control of asymmetric underactuated surface vehicles. IEEE Trans. Ind. Electron. **6611**, 8741–8750 (2019)
21. Skjetne, R., Fossen, T.I., Kokotović, P.V.: Adaptive maneuvering, with experiments, for a model ship in a marine control laboratory. Automatica **41**(2), 289–298 (2005)

Application of Bat Algorithm to Reduce Power Loss in Electrical Power Systems

Samson Ademola Adegoke[1], Yanxia Sun[1(✉)], and Zenghui Wang[2]

[1] Department of Electrical and Electronic Engineering Science, University of Johannesburg, Johannesburg, South Africa
ysun@uj.ac.za

[2] Department of Electrical Engineering, University of South Africa, Pretoria, South Africa

Abstract. The bat algorithm (BA) is a population-based optimization that mimics the echolocation of microbats when looking for prey and avoiding obstacles. BA is a unique algorithm that gives fast convergence and optimum solutions to a problem. Reactive power dispatch (RPD) plays a vital role in the control and operation of the power system and is part of the optimal power flow (OPF) problem. It is formulated as continuous and discrete (i.e., mixed integer nonlinear programming) model. This research involves the application of BA to RPD to reduce power loss (PL) in transmission systems and applied constraints handling (penalty function) to keep it within the operating limits. Also, the optimum control settings are voltage of the generator, reactive power sources, and transformer taps were obtained without violating the limits. The method's performance was demonstrated on standard IEEE New England 39 and 57 node systems. BA can reduce the test systems' losses to 37.109 MW and 22.304 MW from the initial case of 43.6 MW and 28.46 MW, respectively. The percentage savings of the test systems are 14.94% and 21.63%, respectively, which shows that BA can reduce PL in electrical power systems. The simulation results of BA are compared with the other techniques, and BA outperformed them.

Keywords: Transmission power loss · Bat algorithm · Reactive power dispatch

1 Introduction

As a result of the lack of reactive support, which renders the generators incapable of meeting load demand, the increased demand for electricity by end users has forced power system generators to run close to their operating limits. Reactive power dispatch (RPD) significantly impacts electrical power systems' security, reliability, and operation (EPS). Because of its effect on getting the best control variables for the generator, transformer tap, and shunt compensation, it has drawn more attention from researchers over the decay. RPD is an important aspect of the optimal power flow (OPF) issue, which requires mixed integer nonlinear programming and mixes continuous (generator node voltage (VG)), discrete (reactive compensator (RC)), and continuous (transformer taps (TP)) variables [1–6]. To reduce active power loss (PL) and voltage deviation (VD) and improve the

© The Author(s), under exclusive license to Springer Nature Singapore Pte Ltd. 2023
H. Zhang et al. (Eds.): NCAA 2023, CCIS 1869, pp. 206–219, 2023.
https://doi.org/10.1007/978-981-99-5844-3_15

voltage stability index (VSI) while upholding equality and inequality constraints, RPD issues have been applied. The VG, shunt capacitors, and TP settings serve as the control variable in determining the objective function.

Additionally, lowering active PL in power systems is crucial to increase the voltage profile because doing so will increase system quality and enable better service provision [2, 7]. Numerous researchers have attempted to solve RPD using classical optimization methods such as interior point, nonlinear, linear programming methods, gradient-based, and quadratic programming. However, these methods are inaccurate to give the perfect solution to the problem [8, 9].

Recently, meta-heuristic techniques have been successfully used to solve RPD problems, such as improved pathfinder algorithm (IPFA) [8], tight-and-cheap conic relaxation approach (TCCR) [10], modified pathfinder algorithm (mPFA) [11], hybrid PA (HPFA) [12], Hybrid (HPSO-PFA) [9], ant lion optimizer (ALO) [13], improved (IALO) [14], tree seed algorithm (TSA) [15], improved social spider algorithm (ISSA) [16], whale optimization algorithm (WOA) [17], moth–flame optimizer (MFO) [2], chaotic krill herd algorithm (CKHA) [18], gravitational search algorithm based conditional selection strategies (GSA-CSS) [19], "imperialist competitive algorithm (ICA)" [3], "Artificial Bee Colony (ABC)" [20], nonlinear programming method (NLP) [21].

The improved Bat algorithm has been used successfully to solve RPD problems due to fast convergence and its optimum solution. Sylvere Mugemanyi et al. presented chaotic BA to solve optimum RPD considering three objective functions active PL, VD, and L-index [4]. Directional BA was presented to minimize active PL while satisfying the constraints [22]. Also, BA has been applied to solve various engineering problems like economic dispatch [23], the application of microelectronics, scheduling problems, image processing, engineering optimization, and RPD, etc. [4].

Therefore, this work presents the application of the BA to reduce the active transmission PL in the power system for effective operation and improvement of the power system voltage profile. BA is chosen due to its good solution and faster convergence to a problem. BA is applied to reduce the active PL in the transmission systems. It used the penalty function to keep the constraints within limits and avoid violation. Also, each case's results were compared with existing algorithms, revealing that BA gives more accurate results than others in the literature.

The remaining part of the paper is structured as follows: Sect. 2 presents the problem formulation, Sect. 3 discusses the BA and the steps involved in solving the RPD problem, Sect. 4 discusses the result and discussion, and the last Sect. 5 is the conclusion and future work.

2 Problem Formulation

This work aims to reduce the transmission PL while maintaining the system constraints. RPD problem can be formulated as the reduction function (x, u) as follows

$$f = P_{loss} = f(\text{x, u}) = \sum_{K=1}^{N_L} G_k \left(v_i^2 + v_j^2 - 2V_i V_j \cos\theta_{ij} \right) \quad (1)$$

while maintaining

$$\begin{cases} g(x, u) = 0 \\ h(x, u) \leq 0 \end{cases} \qquad (2)$$

$$u^{min} \leq u \leq u^{max} \qquad (3)$$

$$x^{min} \leq x \leq x^{max} \qquad (4)$$

Here, P_{loss} is the total PL, $f(x, u)$ is the objective function to be optimized, x is the vector of the dependent variables, u is the vector of the control variables, G_k is the branch, N_L is the total number of transmission lines, K is the between bus i and j, θ_{ij} "is the voltage angle between bus i and j, V_i is the voltage at the ith bus, V_j is the voltage at the jth bus", $g(x, u)$ is the equality constraints, $h(x, u)$ is the inequality constraints, and min and max are the minima and maximum limit values.

2.1 Equality Constraints

The equality constraints are the power system network's active and reactive load flow (LF) equation.

$$P_{gi} - P_{di} - V_i \sum_{j=1}^{N} V_j(g_{ij}\cos\theta_{ij} + b_{ij}\sin\theta_{ij}) = 0 \qquad (5)$$

$$Q_{gi} - Q_{di} - V_i \sum_{j=1}^{N} V_j(g_{ij}\sin\theta_{ij} - b_{ij}\cos\theta_{ij}) = 0 \qquad (6)$$

Here, g_{ij} and b_{ij} are conductance and susceptance, and θ_{ij} is the phase voltage difference.

2.2 Inequality Constraints

Inequality constraints are operational variables that must be kept within acceptable limits.
Generator constraints

$$V_{gi}^{min} \leq V_{gi} \leq V_{gi}^{max} \quad i = 1\ldots, N_g \qquad (7)$$

$$Q_{gi}^{min} \leq Q_{gi} \leq Q_{gi}^{max} \quad i = 1\ldots, N_g \qquad (8)$$

$$P_{gi}^{min} \leq P_{gi} \leq P_{gi}^{max} \quad i = 1\ldots, N_g \qquad (9)$$

Reactive compensation constraints

$$Q_{ci}^{min} \leq Q_{ci} \leq Q_{ci}^{max} \quad i = 1\ldots, N_C \qquad (10)$$

Transformer tap ratio constraints

$$T_k^{min} \leq T_k \leq T_k^{max} \quad i = 1\ldots, N_T \qquad (11)$$

Security constraints are the constraints of voltage at load buses and loading of transformer line

$$V_{ki}^{min} \leq V_{ki} \leq V_{ki}^{max} \quad i = 1 \ldots, N_B \tag{12}$$

$$S_k \leq S_k^{max} \quad i = 1 \ldots, N_K \tag{13}$$

Here, V_{gi}^{min} and V_{gi}^{max} are voltage magnitude limits, Q_{gi}^{min} and Q_{gi}^{max} are the generation limits of reactive power, V_{ki}^{min} and V_{ki}^{max} are the minimum and maximum limits of load voltage, V_{ki} is the voltage magnitude at load ith bus, S_k is the apparent LF at the branch, S_k^{max} is the maximum value of apparent LF at the ith bus, P_{gi}^{min} and P_{gi}^{max} are the limits of active power, Q_{ci}^{min} and Q_{ci}^{max} are the reactive compensation limits, and T_k^{min} and T_k^{max} are the transformer tap limits,

2.3 Handling of Constraints

The handling constraints help to keep objective function to the desired limits to avoid undesired solutions to the problem by adding dependent variables together with the penalty factor. Therefore, the objective function in Eq. (1) is expressed together with the penalty factor, as shown in Eq. (14).

$$f_T = f + \beta_V \sum_{K=1}^{N_B}(V_i - V_i^{lim})^2 + \beta_g \sum_{K=1}^{N_B}(Q_{gi} - Q_{gi}^{lim})^2 + \beta_s \sum_{K=1}^{N_B}(S_i - S_i^{lim})^2 \tag{14}$$

where;

$$\beta_V, \beta_g, and \beta_s \text{ are the penalty values.} \tag{15}$$

$$V_i^{lim} = \begin{cases} V_i^{lim}, if \ V_i < V_i^{min} \\ V_i^{lim}, if \ V_i > V_i^{max} \end{cases} \tag{16}$$

$$Q_{gi}^{lim} = \begin{cases} Q_{gi}^{lim}, if \ Q_{gi} < Q_{gi}^{min} \\ Q_{gi}^{lim}, if \ Q_{gi} > Q_{gi}^{max} \end{cases} \tag{17}$$

$$S_i^{lim} = \begin{cases} S_i^{lim}, if \ S_i < S_i^{min} \\ S_i^{lim}, if \ S_i > S_i^{max} \end{cases} \tag{18}$$

3 Bat Algorithm

When looking for prey or food and avoiding obstacles, microbats use echolocation, which is inspired by BA [24]. The frequency of up to 25–150 kHz, or 2–14 mm in the air, is equivalent to 8–10 ms of hundreds of seconds in echolocation. Microbats move randomly and release a brief pulse; when food is nearby, the pulse emission rate and frequency are tuned up. The echolocation wavelength will be reduced as a result of pulse emission and frequency tuning, improving the accuracy of detection. The BA was developed using the analogy of tuning the echolocation frequency of microbats [25]. The idealization of the echolocation trait of microbats follows three guidelines. Section 3.1 contains the pseudo-code for BA [26].

a. All bats employ echolocation, which is remarkable considering how it allows them to distinguish between their prey and background obstacles.
b. Bats fly randomly to find food at a velocity of v_i, a position of x_i, with a frequency of f^{min}, a wavelength λ, and a loudness A_0. Depending on how close their target is, the wavelength or frequency of their produced pulse can be automatically changed, and the pulse emission rate (r) can change from 0 to 1.
c. The A_0 is varied in various ways, from the huge value of A_0 (positive) to the fixed value of A_{min}.

Each bat has a position X^i, velocity V^i, and frequency f^i in the d-dimensional space, which are updated at each iteration in the direction of the optimal position. Equation (22) is utilized for the local search, and after one of the best current solutions is chosen, a random walk is used to create a new solution for all the bats.

$$f^i = f^{min} + r1\left(f^{max} - f^{min}\right) \tag{19}$$

$$V^i_{(t+1)} = V^i_{(t)} + f^i(X^i_{(t)} - X^{best}_{(t)}) \tag{20}$$

$$X^i_{(t+1)} = X^i_{(t)} + V^i_{(t+1)} \tag{21}$$

$$X_{new} = X_{old} + rA^i_{(t)} \tag{22}$$

Here, f^{min} and f^{max} are the minimum and maximum frequencies, f^i is the ith bats. f^{min} and f^{max} are set to 0 and 100, respectively [4, 24]. t is the current iteration, $r1$ is a random vector between (0,1), r is a random number ranging from $(-1,1)$.

The loudness decreases once the bats find food, and the r keeps increasing. The r^0 and loudness A^i are updated as follows.

$$A^{t+1}_i = \alpha A^i_{(t)} \tag{23}$$

$$r^{t+1}_i = r^0_i[1 - \exp(-\gamma t)] \tag{24}$$

Here, $A^i_{(t)}$ is the average A^i of all the bats (1,2), $X^{best}_{(t)}$ is the current global best solution, γ, and α are 0.9 and r^0_i is the emission pulse rate (0,1)

3.1 The Pseudo Code of BA

Objective function f(x), x= (x₁...,xₐ)ᵍ

Initialize the bat population x_i= (i =1,2.......n) and v_i

Define pulse frequency f_i at x_i

Define the loudness A_i and pulse rate r_i

while (t < maximum iteration number)

    Adjusting the frequency and generate new solution as give in equation [19]

    Adjusting the velocities and position equations [20-21]

      if (rand > r_i)

          Extract the best solution

          Generate a local solution among the selected best solution

      end if

      Generate a new solution randomly

      if (rand A_i and f (x_i) < f (x ᵍᵇᵉˢᵗ))

          Select the new solution

          Increase r_i and reduce A_i

      end if

      Rank the bats and select the current best x ᵍᵇᵉˢᵗ

    end

3.2 Implementation of BA to Reactive Power Dispatch

The various steps of applying BA to the RPD problem are given below.

1. Initialization of bat position X^i and velocity V^i are randomly generated.
2. Evaluate the bat objective function fitness using Eq. (1) following the result of the Newton Raphson LF method [4].
3. Adjust the frequency f^i of each bat according to Eq. (19).
4. Update the position and velocity of each bat using Eqs. (20–21), respectively.
5. Generated a new solution based on a random walk using Eq. (22).
6. Extract and store the best value solution between the old and new solution.
7. Update the value of loudness and pulse emission rate using Eq. (23–24).
8. Check if the stopping criteria are met; if not, go to step 2
9. Extract the minimum value and stop.

4 Result and Discussion

The New England 39 and IEEE 57 bus systems were used to implement BA to solve the RPD problem effectively. MATPOWER 5.1 was installed in MATLAB 2018b to run the simulation on the 8GB core i5 HP computer. The parameters for the algorithm is given in Table 1. The line and bus data for the two test systems are taken from [4].

Table 1. Parameters setting

Variables	Values
Population number	50
Maximum number of iterations	200
Amin	1
Amax	2
Alpha (α) = beta (γ)	0.9

5 IEEE 39 Bus System

This test system consists of 21 control variables, out of which 5 are transformer tap changing connected to lines 12–11, 10–32, 22–35, 2–30, and 19–20; 10 generators located at nodes 30, 31, 32, 33, 34, 35, 36, 37, 38 and 39; and 6 reactive shunt compensators at nodes 1, 5, 10, 14, 22 and 27. The control variable is given in Table 2. The convergence of power loss for BA is given in Fig. 1, which shows a fast convergence. The PL of BA reduces to 37.0852 MW from the initial case of 43.6 MW, and the % of savings of BA is 14.94%, which is also more than PSO; this shows a significant performance of BA in lower loss reduction and % save. Table 3 gives the results of the PL reduction of BA and is compared with other algorithms. From Table 3, BA gives the lowest PL result. This shows the influential significance of BA in PL reduction compared with PSO [27], which offers 41.8892 MW. The best control variables obtained by BA are compared with PSO, as illustrated in Table 4.

Table 2. IEEE 39 bus system control variable limits [4].

Vmax	V_{min}	T_{max}	T_{min}	Qcmax	Qcmin
1.1	0.9	1.07	0.93	0.25	0

5.1 IEEE 57 Bus System

This test system comprises 25 control variables, 7 generators at nodes 1, 2, 3, 6, 8, 9, and 12. 15 tap transformers and 3 reactive compensators. The control variable limits

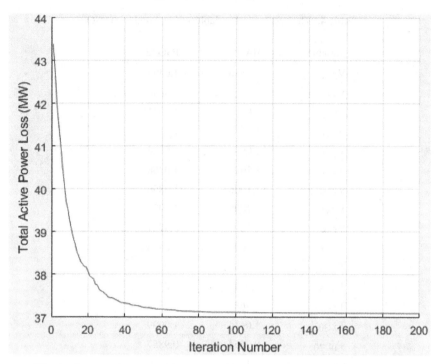

Fig. 1. Convergence curve of active PL for IEEE 39 New England bus System.

Table 3. Comparison of active PL with PSO

Algorithms	Best	Worst	Mean	STD	% save
PSO [27]	41.8892	–	–	–	3.92
BA	**37.0852**	37.109	37.097	0.0170	14.94

are given in Table 5. The "active power demand, reactive power demand, active power generation, reactive power generation, and active PL are given below" [4].

$$P_{load} = 1250.8 MW$$

$$Q_{load} = 336.4 MVAR$$

$$P_G = 1279.26 MW$$

$$Q_G = 345.45 MVAR$$

$$P_{loss} = 28.46 MW$$

Table 4. The best control variable as compared with PSO

Variable	BA	PSO [27]
Vg_{30}	1.0886	1.0496
Vg_{31}	1.1	1.0888
Vg_{32}	1.0678	1.0931
Vg_{33}	1.0967	1.0431
Vg_{34}	1.0997	1.0944
Vg_{35}	1.1008	1.0268
Vg_{36}	1.1006	1.0856
Vg_{37}	1.1008	1.0925
Vg_{38}	1.1011	0.9767
Vg_{39}	1.1003	1.0435
T_{12-11}	0. 9652	0.9917
T_{10-32}	1.0347	1.0394
T_{22-35}	0.9681	1.0304
T_{2-30}	1.0134	1.0039
T_{19-20}	0.9654	0.9857
Q_{C1}	0.1570	0.1710
Q_{C5}	0.003	0.1428
Q_{C11}	0.1449	0.1248
Q_{C14}	0.2036	0.1094
Q_{C22}	0.0617	0.0841
Q_{C27}	0.0108	0.1449

Table 5. Control variable limits [4]

IEEE 57 test system	Variables	Upper limits	Lower limits
	V	1.1	0.9
	T	1.1	0.9
	Qc	0.2	0

The simulation result of BA for loss reduction, best, worst, mean, STD, and % of saving is given in Table 6. It can be observed from Table 6 that BA gives the smallest PL reduction of 22.304 MW than other algorithms like, MOF [2], ABC [20], SOA [21], etc. BA performs better than other algorithms in obtaining the best solution. Also, the % of saving obtained by BA was 21.63% which is higher than all the compared algorithms; this demonstrates the method's effectiveness. The convergence curve of active PL is given in Fig. 2. The control variables obtained by BA are shown in Table 7 and compared with other algorithms.

Table 6. Comparison of active PL with other algorithms

Algorithms	Best MW	Worst MW	Mean MW	STD	% save
BA	**22.3035**	22.3885	22.346	0.0602	21.63
MOF [2]	24.25293	–	–	–	12.96
NLP [21]	25.9023	30.8544	27.8584	1.1677×10^{-2}	8.9934
ALC-PSO [28]	23.39	23.44	23.41	82×10^{-5}	17.82
CTFWO [29]	23.3235	–	23.6395	–	–
SR-DE [30]	23.3550	24.1656	23.4392	0.1458	17.94
ABC [20]	24.1025	–	–	–	–
PSO [19]	24.22017	26.05227		1.204	–
GSA [19]	22.76482	27.13467		0.415	–
GSA-CSS [19]	22.69912	25.56537		0.665	–
PSO-ICA [3]	25.5856	–	–	–	–
PSO [3]	27.5543	–	–	–	–
ICA [3]	26.9997	–	–	–	–

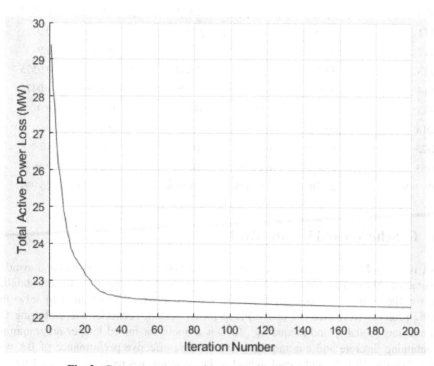

Fig. 2. Convergence curve of active PL of IEEE 57 test system

Table 7. Comparison of control variables with other techniques.

Control variable	BA	ABC [20]	PSO-ICA [3]	PSO [3]	ICA [3]
Vg1	1.1043	1.0532	1.0395	1.0284	1.06
Vg2	1.0988	1.0587	1.0259	1.0044	1.0388
Vg3	1.0858	1.0587	1.0077	0.9844	1.0078
Vg6	1.0816	1.0358	0.9982	0.9872	0.9688
Vg8	1.0965	1.0518	1.0158	1.0262	0.9715
Vg9	1.0822	1.0436	0.985	0.9834	0.9556
Vg12	1.0684	1.0455	0.9966	0.9844	0.9891
T1	1.0150	1.02	0.9265	0.9743	0.9584
T2	0.9947	1.03	0.9532	0.9716	0.9309
T3	0.9947	0.95	1.0165	1.0286	1.0269
T4	0.9157	1.03	1.0071	1.0183	1.0085
T5	1.0386	0.98	0.9414	0.9401	0.9
T6	0.9230	1.05	0.9555	0.94	0.9872
T7	0.9373	0.95	0.9032	0.9761	0.9097
T8	0.9615	0.98	0.9356	0.9211	0.9377
T9	0.9007	0.96	0.9172	0.9165	0.9166
T10	0.9794	0.99	0.9337	0.9044	0.9057
T11	1.0037	1.04	0.9	0.9118	0.9
T12	1.0037	1.08	0.9206	0.92	0.9
T13	1.0167	0.99	1.0042	0.9891	0.9575
T14	0.9546	0.97	1.0297	0.9430	1.0476
T15	1.0635	1.02	0.9294	0.9998	0.9
Q18	0.0272	0.0785	9.9846	9	0
Q25	0.0555	0.05656	10	7.0185	10
Q53	0.1073	0.04953	10	5.0387	9.5956
PL MW	**22.3035**	24.1025	25.5856	27.5543	26.9997

6 Conclusion and Future Work

BA is a population algorithm that inspires the echolocation of microbats and avoided obstacles. Active PL affects the power system networks leading to voltage instability across the countries, and many blackouts have occurred. BA was used to solve the RPD problem to reduce the active PL in power system networks while obeying the constraints (equality and inequality). RPD is a nonlinear mixed integer programming containing discrete and continuous functions. The effective performance of BA was tested on the IEEE 57 and New England 39 bus systems. BA reduced the New England

IEEE 39 bus system's loss to 37.1092 MW from the initial loss of 43.6 MW, and the % save was 14.94%. The losses and % save compared with PSO and BA gave a more significant performance. This shows the effectiveness of BA in loss reduction. However, the losses on the IEEE 57 test system were 22.3035 MW from the base case of 28.46MW, and the loss obtained was compared with other algorithms like PSO, GSA, GSA-CSS, ICA, ABC, MFO, etc. BA gives less reduction than all mentioned techniques and offers 21.63% savings higher than other methods. The simulation results showed that BA effectively reduces the system's active PL, providing better results for all two cases than the other algorithms reported in the literature. BA will be used to solve the RPD problem by incorporating flexible alternating current transmission system (FACTS) devices at the load bus for better loss reduction and voltage profile improvement in the future. Also, it will be used to solve voltage deviation and frequency control in the future.

Acknowledgment. This work was supported in part by the South African National Research Foundation under Grants 141951 and 132797.

References

1. Rajan, A., Malakar, T.: Exchange market algorithm based optimum reactive power dispatch. Appl. Soft Comput. J. **43**, 320–336 (2016). https://doi.org/10.1016/j.asoc.2016.02.041
2. Mei, R.N.S., Sulaiman, M.H., Mustaffa, Z., Daniyal, H.: Optimal reactive power dispatch solution by loss minimization using moth-flame optimization technique. Appl. Soft Comput. **59**, 210–222 (2017). https://doi.org/10.1016/j.asoc.2017.05.057
3. Mehdinejad, M., Mohammadi-Ivatloo, B., Dadashzadeh-Bonab, R., Zare, K.: Solution of optimal reactive power dispatch of power systems using hybrid particle swarm optimization and imperialist competitive algorithms. Int. J. Electr. Power Energy Syst. **83**, 104–116 (2016). https://doi.org/10.1016/j.ijepes.2016.03.039
4. Mugemanyi, S., Qu, Z., Rugema, F.X., Dong, Y., Bananeza, C., Wang, L.: Optimal reactive power dispatch using chaotic bat algorithm. IEEE Access **8**, 65830–65867 (2020). https://doi.org/10.1109/ACCESS.2020.2982988
5. Zha, Z., Wang, B., Fan, H., Liu, L.: An improved reinforcement learning for security-constrained economic dispatch of battery energy storage in microgrids. In: Zhang, H., Yang, Z., Zhang, Z., Wu, Z., and Hao, T. (eds.) Neural Computing for Advanced Applications, CCIS, vol. 1449, pp. 303–318. Springer Singapore, Singapore (2021). https://doi.org/10.1007/978-981-16-5188-5_22
6. Mu, B., Zhang, X., Mao, X., Li, Z.: An optimization method to boost the resilience of power networks with high penetration of renewable energies. In: Zhang, H., Yang, Z., Zhang, Z., Wu, Z., and Hao, T. (eds.) Neural Computing for Advanced Applications, CCIS, vol. 1449, pp. 3–16. Springer Singapore, Singapore (2021). https://doi.org/10.1007/978-981-16-5188-5_1
7. Adegoke, S.A., Sun, Y.: Power system optimization approach to mitigate voltage instability issues: a review. 1–40 (2023). https://doi.org/10.1080/23311916.2022.2153416
8. Adegoke, S.A., Sun, Y.: Diminishing Active Power Loss and Improving Voltage Profile Using an Improved Pathfinder Algorithm Based on Inertia Weight (2023)
9. Adegoke, S.A., Sun, Y.: Optimum Reactive Power Dispatch Solution using Hybrid Particle Swarm Optimization and Pathfinder Algorithm, pp. 403–410 (2022) https://doi.org/10.47839/ijc.21.4.2775

10. Bingane, C., Anjos, M.F., Le Digabel, S.: Tight-and-cheap conic relaxation for the optimal reactive power dispatch problem. IEEE Trans. Power Syst. **34**, 4684–4693 (2019). https://doi.org/10.1109/TPWRS.2019.2912889

11. Yapici, H.: Solution of optimal reactive power dispatch problem using pathfinder algorithm. Eng. Optim. **53**, 1946–1963 (2021). https://doi.org/10.1080/0305215X.2020.1839443

12. Suresh, V., Senthil Kumar, S.: Research on hybrid modified pathfinder algorithm for optimal reactive power dispatch. Bull. Polish Acad. Sci. Tech. Sci. **69**, 1–8 (2021). https://doi.org/10.24425/bpasts.2021.137733

13. Mouassa, S., Bouktir, T., Salhi, A.: Ant lion optimizer for solving optimal reactive power dispatch problem in power systems. Eng. Sci. Technol. an Int. J. **20**, 885–895 (2017). https://doi.org/10.1016/j.jestch.2017.03.006

14. Li, Z., Cao, Y., Van Dai, L., Yang, X., Nguyen, T.T.: Finding solutions for optimal reactive power dispatch problem by a novel improved antlion optimization algorithm. Energies **12**, (2019). https://doi.org/10.3390/en12152968

15. Üney, M.Ş., Çetinkaya, N.: New metaheuristic algorithms for reactive power optimization. Teh. Vjesn. **26**, 1427–1433 (2019). https://doi.org/10.17559/TV-20181205153116

16. Nguyen, T.T., Vo, D.N.: Improved social spider optimization algorithm for optimal reactive power dispatch problem with different objectives. Springer London (2020). https://doi.org/10.1007/s00521-019-04073-4

17. Medani, K. ben oualid, Sayah, S., Bekrar, A.: Whale optimization algorithm based optimal reactive power dispatch: a case study of the Algerian power system. Electr. Power Syst. Res. **163**, 696–705 (2018). https://doi.org/10.1016/j.epsr.2017.09.001

18. Mukherjee, A., Mukherjee, V.: Chaotic krill herd algorithm for optimal reactive power dispatch considering FACTS devices. Appl. Soft Comput. J. **44**, 163–190 (2016). https://doi.org/10.1016/j.asoc.2016.03.008

19. Chen, G., Liu, L., Zhang, Z., Huang, S.: Optimal reactive power dispatch by improved GSA-based algorithm with the novel strategies to handle constraints. Appl. Soft Comput. **50**, 58–70 (2017). https://doi.org/10.1016/j.asoc.2016.11.008

20. Ettappan, M., Vimala, V., Ramesh, S., Kesavan, V.T.: Optimal reactive power dispatch for real power loss minimization and voltage stability enhancement using Artificial Bee Colony Algorithm. Microprocess. Microsyst. **76**, 103085 (2020). https://doi.org/10.1016/j.micpro.2020.103085

21. Dai, C., Chen, W., Zhu, Y., Zhang, X.: Seeker optimization algorithm for optimal reactive power dispatch. IEEE Trans. Power Syst. **24**, 1218–1231 (2009). https://doi.org/10.1109/TPWRS.2009.2021226

22. Sravanthi, C.H., Karthikaikannan, D.: Optimal reactive power dispatch using directional bat algorithm. In: Arun Bhaskar, M., Dash, S.S., Das, S., Panigrahi, B.K. (eds.) International Conference on Intelligent Computing and Applications: Proceedings of ICICA 2018, pp. 311–320. Springer Singapore, Singapore (2019). https://doi.org/10.1007/978-981-13-2182-5_30

23. Adarsh, B.R., Raghunathan, T., Jayabarathi, T., Yang, X.S.: Economic dispatch using chaotic bat algorithm. Energy **96**, 666–675 (2016). https://doi.org/10.1016/j.energy.2015.12.096

24. Yang, X.S., Gandomi, A.H.: Bat algorithm: a novel approach for global engineering optimization. Eng. Comput. (Swansea, Wales). **29**, 464–483 (2012). https://doi.org/10.1108/02644401211235834

25. Gandomi, A.H., Yang, X.S.: Chaotic bat algorithm. J. Comput. Sci. **5**, 224–232 (2014). https://doi.org/10.1016/j.jocs.2013.10.002

26. Agbugba, E.E., Technologiae, M.: Hybridization of Particle Swarm Optimization with Bat Algorithm for Optimal Reactive Power Dispatch (2017)

27. Pandya, S., Roy, R.: Particle swarm optimization based optimal reactive power dispatch. In: Proceedings of the 2015 IEEE International Conference on Electrical, Computer and Communication Technologies ICECCT 2015 (2015). https://doi.org/10.1109/ICECCT.2015. 7225981
28. Singh, R.P., Mukherjee, V., Ghoshal, S.P.: Optimal reactive power dispatch by particle swarm optimization with an aging leader and challengers. Appl. Soft Comput. J. **29**, 298–309 (2015). https://doi.org/10.1016/j.asoc.2015.01.006
29. Abd-El Wahab, A.M., Kamel, S., Hassan, M.H., Mosaad, M.I., Abdulfattah, T.A.: Optimal reactive power dispatch using a chaotic turbulent flow of water-based optimization algorithm. Mathematics. **10** (2022). https://doi.org/10.3390/math10030346
30. Mallipeddi, R., Jeyadevi, S., Suganthan, P.N., Baskar, S.: Efficient constraint handling for optimal reactive power dispatch problems. Swarm Evol. Comput. **5**, 28–36 (2012). https:// doi.org/10.1016/j.swevo.2012.03.001

An Enhanced Subregion Dominance Relation for Evolutionary Many-Objective Optimization

Shuai Wang[1], Hui Wang[1(✉)], Zichen Wei[1], Futao Liao[1], and Feng Wang[2]

[1] School of Information Engineering, Nanchang Institute of Technology, Nanchang 330099, China
huiwang@whu.edu.cn
[2] School of Computer Science, Wuhan University, Wuhan 430072, China
fengwang@whu.edu.cn

Abstract. Pareto dominance-based approach is a classical method for solving multi-objective optimization problems (MOPs). However, as the number of objectives increases, the selection pressure drops sharply. Solutions with good convergence and diversity are hardly obtained. To tackle these issues, this paper proposes an enhanced subregion dominance (called ESD-dominance) relation for evolutionary many-objective optimization. In ESD-dominance, individuals in the population are associated with a set of uniform reference vectors according to the Euclidean distance. Individuals associated with the same reference vector constitute a subregion. To enhance the convergence, each subregion is re-layered based on a new convergence metric. To maintain the diversity, the density in different subregions is considered. In order to validate the performance of ESD-dominance, a modified NSGA-II (called ESD-NSGA-II) algorithm is constructed based on the proposed dominance relation. In the experiments, a set of WFG benchmark problems with 3, 5, 8, and 15 objectives are tested. Computational results show the competitiveness of ESD-NSGA-II when compared with eight other state-of-the-art algorithms.

Keywords: evolutionary many-objective optimization · Pareto dominance · subregions · enhanced subregion dominance

1 Introduction

In real world, many engineering applications involve multi-objective optimization problems (MOPs) [1–3]. All objectives are conflicting and required to be optimized simultaneously. In general, a MOP can be described as follows:

$$\begin{cases} \min F(\mathbf{x}) = (f_1(\mathbf{x}), \dots, f_M(\mathbf{x})) \\ subject\ to\ \mathbf{x} \in \Omega \end{cases} \tag{1}$$

where $\mathbf{x} = (x_1, x_2, \dots, x_D)$ is the vector of decision variables, Ω is the decision space of the problem, D is the number of decision variables, $f_i(\mathbf{x})$ is the i-th

H. Zhang et al. (Eds.): NCAA 2023, CCIS 1869, pp. 220–234, 2023.
https://doi.org/10.1007/978-981-99-5844-3_16

objective function, and M is the number of objectives. Unlike single-objective optimization problems, it is not easy to determine which solution is the best one for MOPs. When the number of objectives is larger than 3 ($M > 3$), the MOPs are called many-objective optimization problems (MaOPs).

For ordinary MOPs ($M \leq 3$), multi-objective evolutionary algorithms (MOEAs) have shown excellent performance. With the growth of objectives, the optimization performance of most MOEAs is rapidly deteriorated. The main reason is that the ratio of non-dominated individuals in the population increases quickly. It significantly reduce the selection pressure and slow down the convergence. To solve MaOPs, different improved versions of many-objective evolutionary algorithms (MaOEAs) have been proposed based on existing MOEAs. According to the type of improvements, those MaOEAs can be categorized four classes: Pareto-based, decomposition-based, indicator-based, and others.

In this paper, an enhanced subregion dominance (called ESD-dominance) relation for MaOPs. The ESD-dominance splits the objective space into several small subregions by reference vectors. Each subregion is re-layered based on a new convergence metric. To maintain diversity, dominance relations are established in different subregions. Finally, the proposed ESD-dominance is embedded into NSGA-II [2], and a modified NSGA-II (namely ESD-NSGA-II) algorithm is constructed. To validate the performance of ESD-NSGA-II, a set of WFG benchmark problems with 3, 5, 8, and 15 objectives are tested. Results of ESD-NSGA-II are compared with eight other state-of-the-art MaOEAs.

The rest of the paper is organized as follows. Section 2 introduces the background work of the proposed dominance relation. Section 3 presents the proposed ESD-NSGA-II algorithm. Sections 4 gives experimental results and discussions. Finally, this paper is concluded in Sect. 5.

2   Related Work

Pareto dominance relation is successfully used to deal with various MOPs. There are many Pareto dominance based MOEAs, and NSGA-II is one of the most popular ones [2]. As the number of objectives increases, the crowding-based selection of NSGA-II is no longer suitable. To tackle this issue, an enhanced version of NSAGA-II (NSGA-III) was proposed by introducing reference points [4]. The traditional Pareto dominance relation easily results in lacking selection pressure for MaOPs. The probability of two solutions that do not dominate each other increases gradually. Then, the dominance resistance appears [5].

To overcome the above difficulty, some scholars tried to modify the original dominance relation to distinguish non-dominated solutions. Sato et al. [6] used the controlled dominant area of solutions (CDAS) to expand the dominance region and increase the selection pressure. In addition, generalized Pareto optimality (GPO) [7] and α-dominance [8] also expanded the dominance region by modifying the dominance relation. However, they need to choose a proper parameter to stabilize the algorithm.

The dominance relation can be improved by gridding the objective space. Yang et al. [9] proposed a grid dominance, which determines the dominance

---

Algorithm 1: *Association*(P_t, **W**)

---

1   Normalize P_t;
2   **for** *each* $q \in P_t$ **do**
3 | Calculate the Euclidean distance between individual q and all weight vectors;
4 | The weight vector with the shortest distance is associated with the solution q;
5 | The size of the corresponding subregion is increased by 1 (*Density* $+ +$);
6   **end**

---

relation with grid coordinates instead of objective values. There are some other objective space grid dominance relations, such as ϵ-dominance [10], paϵ-dominance [11], and cone ϵ-dominance [12].

Defining dominance relations by fuzzy logic is also a common method, such as (1-k)-dominance [13], and L-dominance [14]. In these dominance relations, the number of objectives where one candidate solution is smaller or greater than another is usually considered as the criterion to determine the dominance relation [5].

Besides the above improvements, the idea of decomposition-based MOEAs is used to modify the dominance relations, which is defined by pre-defined uniform weight vectors, such as RP-dominance [15] and RPS-dominance [16]. The uniform weight vectors can ensure the distribution of solutions. The improved dominance relation can guide the candidate solutions to converge along the direction of the weight vector.

3   Proposed Approach

As mentioned before, the Pareto dominance relationships can result in insufficient selection pressure with increasing number of objectives. MOEAs based on Pareto dominance also cannot balance convergence and diversity well. In order to overcome these shortcomings, an enhanced subregion dominance (ESD-dominance) relation based on decomposition techniques is proposed for evolutionary many-objective optimization. Firstly, the two-layer method of DAS and Dennis [17] is introduced in the original dominance relationship to generate reference vectors and decompose the objective space. Then, a convergence metric *Rank* is defined to enhance the convergence. Furthermore, the dominance of different subregions is established by the defined subregion density *Density* to strengthen the diversity.

3.1   Population Association and Subregion

Assume that P_t is the current population at generation t. A set of uniformly distributed weight vectors **W** $= \{W_1, W_2, \ldots, W_N\}$ are generated as reference points. Firstly, the objective function values of each solution in P_t is normalized. For each solution **x** $\in P_t$, the Euclidean distances ($dist$(**x**, **W**)) between **x** and all weight vectors are calculated. Then, the weight vector with the shortest distance is associated with the solution **x**. Each weight vector may associate with multiple

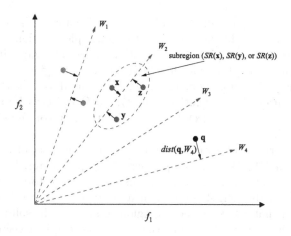

Fig. 1. Population association and subregion

solutions, which constitute a subregion. The number of solutions in a subregion are called *Density*. For the solution **x**, let $SR(\mathbf{x})$ be its associated subregion and $\mathbf{x} \in SR(\mathbf{x})$.

The main framework of population association ($Association(P_t, \mathbf{W})$) is described in Algorithm 1. According to Algorithm 1, each solution in the population is associated with a reference point (weight vector), and a set of subregions are obtained. To illustrate the above operation, Fig. 1 presents a simple example. As seen, all solutions complete the association with reference points. W_1, W_2, and W_4 associates with 2, 3, and 1 solutions, respectively. There are no solutions associated with W_3. **x**, **y**, and **z** are associated with W_2. Those three solutions constitute the subregion, which is called $SR(\mathbf{x})$, $SR(\mathbf{y})$, or $SR(\mathbf{z})$.

3.2   Enhancing Convergence and Diversity

In [15], to compare the convergence degree of individuals in the same subregion, the distance between the projective point of the reference vector nearest to the individual and the origin was chosen as the evaluation indicator. However, this metric is easily influenced by the shape of the PF surface. Therefore, a new convergence metric *Rank* is designed. For each solution **x** in the population, its ranking value is calculated for each objective. Then, $Rank(\mathbf{x})$ is defined by

$$Rank(\mathbf{x}) = \sum_{j=1}^{M} rank_j(\mathbf{x}) \tag{2}$$

where $j = 1, 2, \ldots, M$, $rank_j(\mathbf{x})$ indicates the ranking value of X for the j-th objective. As seen, $Rank(\mathbf{x})$ is the sum of the ranking values of the individual **x**. It can measure the convergence degree of an individual in the population. Generally, a smaller *Rank* means a better convergence.

In Sect. 3.1, the population P_t is divided into N subregions according to the weight vectors. Each subregion may contain multiple solutions. Then, we use *Density* to represent the number of solutions in a subregion. It can be seen that *Density* can reflect the crowding degree of the subregion. A larger *Density* means a more crowded subregion. Thus, *Density* is used to measure the diversity during the evolution. For a solution $\mathbf{x}$ in P_t, let $Density(\mathbf{x})$ be the number of solutions in the subregion $SR(\mathbf{x})$. By measuring the value of $Density(\mathbf{x})$, we can judge the crowding degree of $\mathbf{x}$.

The neighbourhood of a weight vector is defined as a set of its $N * \xi$ closest weight vectors in $\{W_1, W_2, \ldots, W_N\}$, where ξ is a parameter. Each weight vector determines a subregion. The $N * \xi$ closest weight vectors constitute $N * \xi$ subregions. Then, the neighbourhood of a subregion can be described as follows. Supposing that $SR(\mathbf{x})$ is the subregion containing the solution $\mathbf{x}$, and the associated weight vector is W. Thus, the $N * \xi$ closest weight vectors to W constitute $N * \xi$ subregions, which are the neighbourhood of the subregion $SR(\mathbf{x})$. The parameter ξ affects the neighbourhood size, which leads to the range of dominance relations in different subregions. If ξ is too large, it may result in strong dominance relations and deteriorate the performance of the algorithm. The sensitivity analysis of ξ is conducted in Sect. 4.4.

3.3   ESD-Dominance Relation

In Sect. 3.1, each solution in P_t is associated with a reference point (weight vector). Solutions associated with the same weight vector form a subregion. In Sect. 3.2, a new convergence degree (*Rank*) and a diversity metric (*Density*) are designed to ensure both convergence and diversity. Then, the definition of ESD-dominance is given as follows.

Let $\mathbf{x}$ and $\mathbf{y}$ be two solutions in P_t. $\mathbf{x}$ and $\mathbf{y}$ are located in the subregions $SR(\mathbf{x})$ and $SR(\mathbf{y})$, respectively. A solution $\mathbf{x}$ is said to dominate another solution $\mathbf{y}$ in ESD (denoted $\mathbf{x} \prec_{ESD} \mathbf{y}$), if and only if one of the following conditions is satisfied.

1. $\mathbf{x}$ Pareto dominates $\mathbf{y}$ (denoted $\mathbf{x} \prec \mathbf{y}$).
2. $\mathbf{x}$ and $\mathbf{y}$ are Pareto equivalent:
   1) $SR(\mathbf{x}) = SR(\mathbf{y})$ and $Rank(\mathbf{x}) < Rank(\mathbf{y})$;
   2) $SR(\mathbf{x}) \neq SR(\mathbf{y})$, $Rank(\mathbf{x}) < Rank(\mathbf{y})$, $Density(\mathbf{y}) > Density(\mathbf{x}) > 1$, and the two subregions ($SR(\mathbf{x})$ and $SR(\mathbf{y})$) are adjacent.

For MaOPs, there are many non-dominated individuals in the population. When $\mathbf{x}$ and $\mathbf{y}$ are Pareto equivalent, the individual with a lower *Rank* is selected if they are associated with the same reference vector. When two individuals are in two different subregions, two metrics (*Rank* and *Density*) and the neighbor relationship of the two subregions are considered. It is worth noting that ESD-dominance guarantees a strict Pareto order and emphasizes both convergence and diversity.

To clearly illustrate the second case in ESD-dominance relation, Fig. 2 presents a specific example. In Fig. 2(a), two individuals $\mathbf{x}$ and $\mathbf{y}$ are Pareto

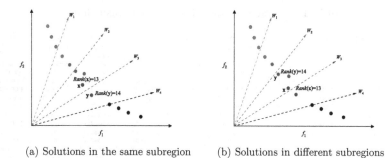

(a) Solutions in the same subregion   (b) Solutions in different subregions

Fig. 2. An example for the ESD-dominance relation.

Algorithm 2: Framework of ESD-NSGA-II

---

1 Initialize the population P_t and reference points $\mathbf{W}$;
2 **while** *not terminate* **do**
3   Generation of mating pools P' using binary tournaments;
4   Genetic operation on P' to produce offspring Q_t;
5      $R_t \leftarrow P_t \cup Q_t$;
6      $Density \leftarrow Association(R_t, \mathbf{W})$;
7   Calculate $Rank$ according to Eq. (2);
8      $(F_1, F_2, \ldots) \leftarrow$ Non-ESD-dominated-sort(R_t);
9      $i \leftarrow 1, P_{t+1} \leftarrow \emptyset$;
10   **while** $|P_{t+1}| + |F_i| \leq N$ **do**
11         $P_{t+1} \leftarrow P_{t+1} \cup F_i$;
12         $i \leftarrow i + 1$;
13   **end**
14   **for** $j \leftarrow 1 : N - |P_{t+1}|$ **do**
15   Rank individuals based on the distance of their association weights and find the smallest individual $R_t(j)$;
16         $P_{t+1} \leftarrow P_{t+1} \cup R_t(j)$;
17   **end**
18 **end**

---

equivalent, and they are associated with the same reference point W_3. Then, $\mathbf{x}$ and $\mathbf{y}$ are in the same subregion. $Rank(\mathbf{x}) < Rank(\mathbf{y})$ indicates that $\mathbf{x}$ has better convergence than $\mathbf{y}$. Therefore, $\mathbf{x}$ is said to dominate $\mathbf{y}$ in ESD (denoted $\mathbf{x} \prec_{ESD} \mathbf{y}$).

In Fig. 2(b), two individuals $\mathbf{x}$ and $\mathbf{y}$ are Pareto equivalent, but they are associated with different reference points W_3 and W_2, respectively. Then, $\mathbf{x}$ and $\mathbf{y}$ are in different subregions. $Rank(\mathbf{x}) < Rank(\mathbf{y})$ and $Density(\mathbf{y}) > Density(\mathbf{x}) > 1$ are satisfied. In addition, the two subregions $(SR(\mathbf{x})$ and $SR(\mathbf{y}))$ are neighbors. It means that $\mathbf{x}$ is better $\mathbf{y}$ in terms of both convergence and diversity metrics. Therefore, $\mathbf{x}$ is said to dominate $\mathbf{y}$ in ESD (denoted $\mathbf{x} \prec_{ESD} \mathbf{y}$).

3.4 Framework of ESD-NSGA-II

To validate the performance of the proposed ESD-dominance, a modified NSGA-II (called ESD-NSGA-II) is designed by replacing the Pareto dominance with ESD-dominance. The detailed descriptions of ESD-NSGA-II are shown in Algorithm 2. First, the population P_t is initialized, and a set of uniform reference

Table 1. HV values obtained by different algorithms.

Problem	M	RPD-NSGA-II	MOEA/D	NSGA-III	RVEA	tDEA	Two_Arch2	NSGA-II/SDR	KnEA	ESD-NSGA-II
WFG1	3	8.9042e-1 (1.03e-1) -	8.3505e-1 (3.15e-2) -	9.0568e-1 (2.49e-2) -	8.4900e-1 (4.74e-2) -	9.2889e-1 (7.22e-3) -	**9.4027e-1** (**2.36e-3**) +	9.1424e-1 (1.15e-2) -	9.2282e-1 (8.58e-3) -	9.3626e-1 (5.95e-3)
	5	9.6981e-1 (7.99e-2) -	9.1263e-1 (3.49e-2) -	9.6842e-1 (2.34e-2) -	9.7875e-1 (2.90e-2) -	9.8883e-1 (1.19e-2) -	9.9298e-1 (1.07e-3) -	9.6505e-1 (2.72e-2) -	9.9009e-1 (2.43e-3) -	**9.9811e-1** (**2.03e-4**)
	8	9.9775e-1 (1.51e-3) =	9.1807e-1 (3.26e-2) -	9.9627e-1 (2.24e-3) -	9.7576e-1 (4.17e-2) -	9.9540e-1 (1.10e-3) -	9.9620e-1 (8.75e-4) -	9.8537e-1 (1.20e-2) -	9.9586e-1 (1.68e-3) -	**9.9807e-1** (**3.08e-3**)
	15	**9.9986e-1** (**1.41e-4**) =	9.2435e-1 (4.63e-2) -	9.9910e-1 (4.28e-4) -	9.8201e-1 (4.02e-2) -	9.9750e-1 (8.59e-4) -	9.9822e-1 (8.62e-4) -	9.8960e-1 (8.75e-3) -	9.9801e-1 (1.05e-3) -	9.9984e-1 (2.17e-4)
WFG2	3	9.3101e-1 (1.31e-3) -	8.9657e-1 (1.34e-2) -	9.2769e-1 (8.76e-4) -	9.1916e-1 (2.78e-3) -	9.3181e-1 (1.40e-3) -	9.3168e-1 (2.28e-3) -	9.0435e-1 (8.90e-3) -	9.2408e-1 (3.12e-3) -	**9.3487e-1** (**8.60e-4**)
	5	9.9176e-1 (1.23e-3) -	9.2775e-1 (2.36e-2) -	9.9027e-1 (1.56e-3) -	9.8388e-1 (2.35e-3) -	9.8884e-1 (1.53e-3) -	9.9096e-1 (1.47e-3) -	9.5784e-1 (9.02e-3) -	9.8812e-1 (1.26e-3) -	**9.9694e-1** (**4.03e-4**)
	8	9.8749e-1 (2.52e-3) -	9.2514e-1 (8.34e-3) -	9.9362e-1 (2.82e-3) -	9.7923e-1 (4.92e-3) -	9.8751e-1 (6.88e-3) -	9.9459e-1 (1.50e-3) -	9.7357e-1 (6.59e-3) -	9.9210e-1 (1.33e-3) -	**9.9906e-1** (**5.24e-4**)
	15	9.9020e-1 (3.92e-3) -	9.2738e-1 (2.35e-2) -	9.9318e-1 (1.94e-3) -	9.7341e-1 (7.81e-3) -	9.0781e-1 (6.89e-2) -	9.9729e-1 (8.26e-4) -	9.8993e-1 (2.44e-3) -	9.9264e-1 (2.29e-3) -	**9.9910e-1** (**5.40e-4**)
WFG3	3	3.6653e-1 (5.34e-2) -	3.4485e-1 (1.46e-2) -	3.7641e-1 (4.93e-2) -	3.3111e-1 (9.77e-3) -	3.8097e-1 (4.72e-2) -	**3.9046e-1** (**2.53e-3**) +	3.8537e-1 (5.82e-3) -	3.8775e-1 (2.17e-2) -	3.8462e-1 (4.16e-2)
	5	1.7033e-1 (7.63e-3) +	3.2477e-2 (3.22e-2) -	1.2288e-1 (1.43e-2) -	1.0090e-1 (3.05e-2) -	1.8425e-1 (1.62e-2) +	1.7223e-1 (1.70e-2) +	**1.9436e-1** (**1.64e-2**) +	1.1672e-1 (2.17e-2) -	1.5295e-1 (1.93e-2)
	8	3.2116e-2 (2.43e-2) -	0.0000e+0 (0.00e+0) -	3.6256e-3 (7.92e-3) -	0.0000e+0 (0.00e+0) -	5.2027e-2 (1.82e-2) -	1.6833e-2 (1.10e-2) -	3.8264e-2 (2.03e-2) -	9.4480e-3 (1.16e-2) -	**7.5507e-2** (**1.49e-2**)
	15	0.0000e+0 (0.00e+0) =	0.0000e+0 (0.00e+0) =	0.0000e+0 (0.00e+0) =	0.0000e+0 (0.00e+0) =	0.0000e+0 (0.00e+0) =	0.0000e+0 (0.00e+0) =	0.0000e+0 (0.00e+0) =	0.0000e+0 (0.00e+0) =	0.0000e+0 (0.00e+0)
WFG4	3	5.5240e-1 (1.26e-3) =	5.2830e-1 (2.72e-3) -	5.5205e-1 (1.31e-3) -	5.4250e-1 (2.43e-3) -	**5.5249e-1** (**1.46e-3**) =	5.5036e-1 (1.97e-3) -	5.4231e-1 (2.82e-3) -	5.3620e-1 (3.26e-3) -	5.5236e-1 (2.60e-3)
	5	7.8246e-1 (2.98e-3) -	6.5282e-1 (2.53e-2) -	7.7951e-1 (1.61e-3) -	7.7846e-1 (3.56e-3) -	7.8002e-1 (2.12e-3) -	7.2169e-1 (4.71e-3) -	7.6349e-1 (3.50e-3) -	7.6439e-1 (2.57e-3) -	**7.9602e-1** (**3.38e-3**)
	8	9.1539e-1 (2.32e-3) -	4.1760e-1 (5.28e-2) -	9.0281e-1 (4.87e-3) -	8.9625e-1 (5.86e-3) -	9.0574e-1 (3.88e-3) -	7.9031e-1 (9.14e-3) -	9.0493e-1 (3.49e-3) -	9.0545e-1 (2.25e-3) -	**9.3281e-1** (**6.05e-3**)
	15	9.6814e-1 (3.41e-3) -	2.7651e-1 (3.57e-2) -	9.4253e-1 (8.06e-3) -	9.5306e-1 (9.01e-3) -	9.5089e-1 (5.79e-3) -	7.7990e-1 (1.90e-2) -	9.8525e-1 (1.60e-3) -	9.7666e-1 (2.80e-3) -	**9.9301e-1** (**5.23e-4**)
WFG5	3	5.1553e-1 (1.33e-3) +	4.9635e-1 (1.53e-3) -	**5.1781e-1** (**2.50e-4**) +	5.1413e-1 (1.33e-3) +	5.1780e-1 (3.31e-4) +	5.1045e-1 (3.68e-3) =	5.1130e-1 (1.27e-3) =	5.0046e-1 (3.66e-3) -	5.1110e-1 (1.74e-3)
	5	7.3954e-1 (1.56e-3) +	6.3814e-1 (1.91e-2) -	7.4132e-1 (5.15e-4) +	7.4022e-1 (1.90e-3) +	**7.4174e-1** (**4.83e-4**) +	6.8241e-1 (4.50e-3) -	7.2479e-1 (3.70e-3) -	7.2292e-1 (4.73e-3) -	7.3703e-1 (5.12e-3)
	8	8.6040e-1 (1.20e-3) -	4.9537e-1 (2.04e-2) -	8.5884e-1 (1.05e-3) -	8.5598e-1 (1.65e-3) -	8.5859e-1 (8.66e-4) -	7.3514e-1 (8.45e-3) -	8.5237e-1 (2.16e-3) -	8.4461e-1 (3.44e-3) -	**8.6994e-1** (**3.69e-3**)
	15	8.7540e-1 (9.41e-3) -	3.1440e-1 (3.44e-2) -	8.9086e-1 (5.13e-3) -	9.0807e-1 (1.50e-3) -	8.8913e-1 (5.15e-3) -	6.9661e-1 (1.55e-2) -	9.1637e-1 (1.40e-3) -	9.1655e-1 (4.27e-4) -	**9.1912e-1** (**4.76e-4**)
WFG6	3	4.9763e-1 (1.03e-2) =	4.7618e-1 (1.81e-2) -	**5.0083e-1** (**1.11e-2**) =	4.9645e-1 (1.24e-2) =	5.0044e-1 (1.40e-2) =	4.9817e-1 (1.41e-2) =	5.0055e-1 (1.18e-2) =	4.8073e-1 (1.44e-2) -	4.9910e-1 (1.69e-2)
	5	7.2223e-1 (1.40e-2) =	5.0760e-1 (4.38e-2) -	7.1913e-1 (1.99e-2) =	7.1320e-1 (2.22e-2) =	7.1302e-1 (1.40e-2) =	6.6027e-1 (1.01e-2) -	7.1231e-1 (1.37e-2) =	6.8733e-1 (1.55e-2) -	**7.2609e-1** (**1.52e-2**)
	8	8.3310e-1 (1.61e-2) =	2.7737e-1 (2.25e-2) -	8.2043e-1 (2.93e-2) -	8.1463e-1 (4.73e-2) -	8.2976e-1 (1.85e-2) -	7.1740e-1 (1.51e-2) -	8.3512e-1 (1.72e-2) -	8.2090e-1 (1.74e-2) -	**8.6106e-1** (**1.94e-2**)
	15	8.9161e-1 (1.95e-2) =	1.3338e-1 (4.80e-2) -	8.7076e-1 (2.07e-2) -	7.4618e-1 (7.85e-2) -	8.7842e-1 (1.72e-2) -	7.0698e-1 (2.61e-2) -	8.9537e-1 (2.61e-2) =	8.9365e-1 (2.67e-2) =	**8.9643e-1** (**1.74e-2**)
WFG7	3	5.5242e-1 (8.37e-4) +	5.0741e-1 (1.46e-2) -	5.5312e-1 (8.53e-4) +	5.4606e-1 (1.43e-3) -	**5.5412e-1** (**8.37e-4**) +	5.5402e-1 (1.18e-3) +	5.5053e-1 (2.31e-3) -	5.4288e-1 (3.08e-3) -	5.5093e-1 (2.45e-3)
	5	7.8700e-1 (1.51e-3) -	5.6681e-1 (3.69e-2) -	7.8433e-1 (1.74e-3) -	7.8339e-1 (1.19e-3) -	7.8615e-1 (9.62e-4) -	7.3896e-1 (4.49e-3) -	7.7331e-1 (4.67e-3) -	7.7489e-1 (4.15e-3) -	**7.9145e-1** (**3.94e-3**)
	8	9.1884e-1 (1.58e-3) -	3.5191e-1 (7.62e-3) -	9.1160e-1 (1.77e-3) -	8.9176e-1 (4.51e-3) -	9.1474e-1 (1.18e-3) -	8.0522e-1 (9.41e-3) -	9.0787e-1 (4.36e-3) -	8.9456e-1 (6.84e-3) -	**9.3525e-1** (**2.93e-3**)
	15	9.6422e-1 (3.41e-3) -	1.7177e-1 (2.59e-2) -	9.6972e-1 (3.60e-3) -	9.5417e-1 (9.47e-3) -	9.5591e-1 (5.22e-3) -	7.7192e-1 (2.01e-2) -	9.5657e-1 (7.06e-2) -	9.8278e-1 (3.03e-3) -	**9.9316e-1** (**1.03e-3**)
WFG8	3	4.5706e-1 (2.20e-3) -	4.4965e-1 (4.59e-3) -	4.6454e-1 (1.85e-3) +	4.5664e-1 (4.14e-3) -	4.6461e-1 (2.42e-3) +	**4.6704e-1** (**2.43e-3**) +	4.6158e-1 (2.14e-3) =	4.4305e-1 (4.72e-3) -	4.6009e-1 (2.40e-3)
	5	6.6939e-1 (1.98e-3) +	4.8264e-1 (9.96e-2) -	6.6883e-1 (3.11e-3) +	6.5929e-1 (4.57e-3) =	**6.7000e-1** (**2.55e-3**) +	6.0538e-1 (7.15e-3) -	6.5100e-1 (4.62e-3) -	6.3176e-1 (4.20e-3) -	6.5652e-1 (4.94e-3)
	8	7.8979e-1 (3.06e-3) -	6.7759e-2 (4.16e-2) -	7.6707e-1 (7.17e-3) -	6.9192e-1 (5.25e-2) -	7.8345e-1 (1.99e-2) -	6.0281e-1 (1.11e-2) -	7.9342e-1 (1.06e-2) -	7.6710e-1 (2.46e-2) -	**8.0924e-1** (**1.28e-2**)
	15	9.0147e-1 (7.37e-3) -	2.3759e-2 (2.06e-2) -	8.3928e-1 (5.18e-2) -	6.9395e-1 (1.02e-1) -	8.7473e-1 (1.85e-2) -	5.1088e-1 (3.95e-2) -	8.2176e-1 (1.48e-1) -	9.2629e-1 (2.88e-2) =	**9.4173e-1** (**4.15e-3**)
WFG9	3	4.9399e-1 (7.30e-2) -	4.7155e-1 (2.22e-2) -	5.1874e-1 (2.99e-2) -	5.2263e-1 (5.73e-3) -	5.2888e-1 (3.40e-3) =	5.2908e-1 (3.20e-3) +	**5.3044e-1** (**3.95e-3**) +	5.2544e-1 (4.33e-3) -	5.2818e-1 (3.32e-2)
	5	6.8990e-1 (5.08e-2) -	5.7291e-1 (4.10e-2) -	7.2996e-1 (4.89e-3) -	7.2695e-1 (6.20e-3) -	7.3512e-1 (6.75e-3) -	6.9287e-1 (1.06e-2) -	7.3574e-1 (5.06e-3) -	7.3786e-1 (5.73e-3) -	**7.5685e-1** (**1.29e-2**)
	8	7.4280e-1 (1.37e-1) -	3.5712e-1 (1.11e-1) -	8.1275e-1 (5.96e-2) -	8.2369e-1 (1.24e-2) -	8.4150e-1 (1.17e-2) -	7.1919e-1 (1.09e-2) -	8.5859e-1 (4.79e-3) -	8.5553e-1 (5.31e-3) -	**8.6231e-1** (**6.84e-2**)
	15	8.3179e-1 (3.00e-2) -	1.7976e-1 (8.75e-2) -	8.8068e-1 (2.38e-2) -	8.1797e-1 (4.74e-2) -	8.4571e-1 (2.18e-2) -	6.6661e-1 (3.64e-2) -	9.3087e-1 (4.23e-3) -	8.9401e-1 (4.75e-2) -	**9.5354e-1** (**4.41e-3**)
+/-/=		5/24/7	0/35/1	5/27/4	2/30/4	7/25/4	6/26/4	2/27/7	0/32/4	

points (weight vectors) **W** are generated. The number of reference points is equal to the population size N (Line 1). Then, the parents are selected based on binary tournaments. Genetic operations (polynomial variation and simulated binary crossover) are conducted to generate offspring Q_t. In addition, the

Table 2. IGD+ values obtained by different algorithms.

Problem	M	RPD-NSGA-II	MOEA/D	NSGA-III	RVEA	tDEA	Two_Arch2	NSGA-II/SDR	KnEA	ESD-NSGA-II
WFG1	3	1.8522e-1 (2.88e-1) -	1.6194e-1 (5.15e-2) -	1.2265e-1 (4.08e-2) -	2.6152e-1 (8.80e-2) -	8.3549e-2 (1.30e-2) -	**6.5534e-2 (5.05e-3)** +	9.1071e-2 (1.52e-2) -	8.7368e-2 (1.38e-2) -	7.0595e-2 (7.26e-3)
	5	2.7153e-1 (3.42e-1) =	1.6801e-1 (5.51e-2) +	2.6241e-1 (3.93e-2) -	2.1988e-1 (7.01e-2) =	**1.5008e-1 (2.46e-2)** +	1.6250e-1 (1.11e-2) +	1.9818e-1 (4.51e-2) =	1.7271e-1 (1.06e-2) +	1.8604e-1 (1.49e-2)
	8	2.4500e-1 (5.19e-2) +	**1.8216e-1 (5.92e-2)** +	3.1399e-1 (4.59e-2) +	1.8816e-1 (7.83e-2) +	2.5387e-1 (4.26e-2) +	2.6866e-1 (1.63e-2) +	2.6453e-1 (3.81e-2) +	2.9357e-1 (2.21e-2) +	7.2886e-1 (1.08e-1)
	15	2.7945e-1 (2.12e-2) +	2.0127e-1 (1.03e-1) +	3.7728e-1 (5.52e-2) +	**1.7066e-1 (9.00e-2)** +	3.7216e-1 (3.17e-2) +	4.0898e-1 (4.34e-2) +	2.2027e-1 (1.80e-2) +	4.5832e-1 (7.12e-2) +	1.2654e+0 (2.49e-1)
WFG2	3	4.9285e-2 (2.50e-3) -	6.7807e-2 (7.89e-3) -	5.6257e-2 (1.43e-3) -	7.9501e-2 (6.84e-3) -	4.7579e-2 (1.77e-3) -	5.2038e-2 (4.31e-3) -	5.8327e-2 (5.33e-3) -	5.5871e-2 (3.50e-3) -	**4.4515e-2 (2.26e-3)**
	5	1.3809e-1 (3.36e-2) +	1.5603e-1 (3.40e-2) -	1.8227e-1 (4.24e-2) -	1.5099e-1 (6.28e-3) =	**1.3266e-1 (2.58e-3)** +	1.6256e-1 (6.83e-3) -	1.5219e-1 (1.65e-2) =	1.7469e-1 (1.64e-2) -	1.5045e-1 (6.90e-3)
	8	1.8916e-1 (4.34e-1) +	1.7629e-1 (1.44e-2) +	2.6354e-1 (3.60e-2) +	**1.4709e-1 (6.01e-3)** +	2.9222e-1 (1.53e-1) +	2.6715e-1 (1.10e-2) +	2.1022e-1 (1.43e-2) +	2.9779e-1 (2.30e-2) +	3.7861e-1 (3.33e-2)
	15	2.6975e-1 (1.80e-2) +	1.7294e-1 (4.77e-2) +	4.2623e-1 (4.14e-2) +	**1.6570e-1 (3.11e-2)** +	3.3184e+0 (1.35e+0) +	4.4156e-1 (3.49e-2) +	2.1821e-1 (1.21e-2) +	4.5812e-1 (6.03e-2) +	7.5210e-1 (1.06e-1)
WFG3	3	1.1494e-1 (1.15e-2) -	1.1163e-1 (2.07e-2) -	1.0009e-1 (1.21e-2) -	1.8323e-1 (2.15e-2) -	9.0913e-2 (1.20e-2) -	**5.7750e-2 (4.64e-3)** +	6.0173e-2 (5.50e-3) +	5.8064e-2 (6.20e-3) +	7.0672e-2 (6.68e-3)
	5	3.1778e-1 (1.38e-2) +	5.6395e-1 (1.01e-1) -	4.1482e-1 (4.04e-2) -	4.4382e-1 (5.36e-2) -	3.1246e-1 (3.11e-2) +	2.7107e-1 (2.43e-2) +	**2.4990e-1 (3.30e-2)** +	3.0279e-1 (4.61e-2) +	3.5241e-1 (5.25e-2)
	8	8.9623e-1 (4.21e-2) +	2.2070e+0 (1.00e-1) -	1.2617e+0 (3.14e-1) -	1.9229e+0 (2.96e-1) -	8.9200e-1 (2.28e-1) +	6.6333e-1 (7.13e-2) +	7.2594e-1 (1.95e-1) +	**5.5959e-1 (6.73e-2)** +	1.0816e+0 (1.47e-1)
	15	2.5551e+0 (2.01e-1) +	3.8339e+0 (1.69e-2) -	1.5128e+0 (4.34e-1) +	3.6914e+0 (1.41e-1) -	1.4683e+0 (2.57e-1) +	**1.3624e+0 (1.42e-1)** +	2.2156e+0 (3.09e-1) +	1.6942e+0 (5.22e-1) +	2.6787e+0 (2.13e-1)
WFG4	3	8.6144e-2 (1.75e-3) -	1.2448e-1 (3.71e-3) -	8.6611e-2 (1.83e-3) -	1.0145e-1 (3.49e-3) -	8.5852e-2 (1.91e-3) -	9.3491e-2 (2.78e-3) -	1.0580e-1 (4.42e-3) -	1.0981e-1 (3.60e-3) -	**8.5757e-2 (3.94e-3)**
	5	3.4608e-1 (4.66e-3) -	5.5831e-1 (2.89e-2) -	3.4923e-1 (2.12e-3) -	3.4979e-1 (4.05e-3) -	3.4819e-1 (2.87e-3) -	4.8544e-1 (1.06e-2) -	4.0229e-1 (7.30e-3) -	3.8700e-1 (5.49e-3) -	**3.2104e-1 (7.62e-3)**
	8	7.2682e-1 (3.42e-1) -	1.2095e+0 (3.04e-2) -	7.4727e-1 (7.06e-3) -	7.3102e-1 (4.48e-3) -	7.4267e-1 (5.35e-3) -	1.1611e+0 (2.09e-2) -	8.2727e-1 (1.61e-2) -	7.6880e-1 (1.09e-2) -	**7.0989e-1 (2.85e-2)**
	15	1.3425e+0 (2.23e-2) -	1.5817e+0 (9.16e-3) -	1.5062e+0 (4.10e-1) -	**1.2809e+0 (9.50e-3)** +	1.3594e+0 (1.50e-2) -	2.2261e+0 (4.42e-1) -	1.4095e+0 (2.40e-2) -	1.6589e+0 (2.45e-1) -	1.3428e+0 (2.17e-2)
WFG5	3	1.3635e-1 (1.71e-3) +	1.5612e-1 (1.35e-3) -	1.3345e-1 (3.40e-4) +	1.3898e-1 (1.39e-3) +	**1.3336e-1 (3.39e-4)** +	1.4375e-1 (2.91e-3) =	1.4574e-1 (2.17e-3) -	1.5710e-1 (4.23e-3) -	1.4279e-1 (2.53e-3)
	5	3.9653e-1 (2.33e-3) +	5.6523e-1 (1.99e-2) -	3.9340e-1 (6.93e-4) +	3.9312e-1 (8.78e-4) +	**3.9281e-1 (8.31e-4)** +	5.2128e-1 (8.41e-3) -	4.4192e-1 (8.29e-3) -	4.3260e-1 (1.01e-2) -	4.0414e-1 (1.13e-2)
	8	7.9363e-1 (5.38e-3) -	1.1984e+0 (8.32e-3) -	7.9208e-1 (3.20e-3) =	**7.8162e-1 (4.21e-3)** =	7.9172e-1 (2.04e-3) =	1.2799e+0 (4.35e-2) -	8.7490e-1 (1.77e-2) -	8.3949e-1 (1.21e-2) -	7.8867e-1 (1.86e-2)
	15	2.2250e+0 (2.94e-1) -	1.6163e+0 (8.11e-3) -	1.5341e+0 (4.62e-2) -	**1.3222e+0 (6.49e-3)** +	1.5350e+0 (5.10e-2) -	2.7348e+0 (3.02e-1) -	1.4430e+0 (2.62e-2) -	1.5135e+0 (3.51e-2) -	1.4254e+0 (3.27e-2)
WFG6	3	1.6353e-1 (1.56e-2) -	1.9968e-1 (2.62e-2) -	**1.5959e-1 (1.66e-2)** =	1.6764e-1 (1.84e-2) -	1.6025e-1 (2.12e-2) -	1.6930e-1 (2.10e-2) =	1.6541e-1 (1.76e-2) -	1.8655e-1 (2.05e-2) -	1.6178e-1 (2.50e-2)
	5	4.2228e-1 (1.92e-2) -	7.1582e-1 (5.40e-2) -	4.2653e-1 (2.80e-2) =	4.3485e-1 (3.06e-2) =	4.3521e-1 (1.98e-2) -	5.7015e-1 (1.59e-2) -	4.5900e-1 (2.03e-2) -	4.9366e-1 (2.00e-2) -	**4.1766e-1 (2.34e-2)**
	8	8.2516e-1 (2.30e-2) -	1.3498e+0 (4.14e-2) -	8.5398e-1 (6.44e-2) -	8.2238e-1 (3.81e-2) =	8.3272e-1 (2.59e-2) =	1.2594e+0 (2.58e-2) -	8.8247e-1 (3.00e-2) -	8.8796e-1 (2.73e-2) -	**7.9974e-1 (2.99e-2)**
	15	1.4602e+0 (8.30e-2) -	1.7063e+0 (3.67e-2) -	1.4375e+0 (5.44e-2) =	1.4075e+0 (3.12e-2) =	**1.4009e+0 (2.42e-2)** +	2.4897e+0 (2.89e-1) -	1.4630e+0 (2.76e-2) -	1.4933e+0 (5.28e-2) -	1.4346e+0 (4.80e-2)
WFG7	3	8.6463e-2 (1.16e-3) -	1.5733e-1 (2.45e-2) -	8.5814e-2 (1.19e-3) +	9.7211e-2 (2.23e-3) -	**8.4259e-2 (1.23e-3)** +	8.8375e-2 (1.74e-3) -	9.3765e-2 (3.73e-3) -	9.9991e-2 (3.72e-3) -	8.8234e-2 (3.76e-3)
	5	3.4154e-1 (2.06e-3) -	6.3538e-1 (3.57e-2) -	3.4516e-1 (2.68e-3) -	3.4704e-1 (2.03e-3) -	3.4199e-1 (1.76e-3) -	4.5742e-1 (8.81e-3) -	3.8574e-1 (1.08e-2) -	3.6661e-1 (8.42e-3) -	**3.3238e-1 (8.96e-3)**
	8	7.3048e-1 (7.50e-3) -	1.2401e+0 (3.01e-3) -	7.4649e-1 (3.39e-3) -	7.3505e-1 (4.85e-3) -	7.3830e-1 (3.43e-3) -	1.1591e+0 (3.31e-2) -	8.2671e-1 (2.49e-2) -	8.0231e-1 (1.44e-2) -	**7.0570e-1 (1.41e-2)**
	15	2.9918e+0 (1.70e-1) -	1.6151e+0 (1.04e-2) -	2.9256e+0 (1.76e-1) -	**1.4293e+0 (3.50e-1)** +	2.6545e+0 (2.86e-1) -	3.7307e+0 (3.13e-1) -	1.6050e+0 (3.75e-1) =	2.3772e+0 (3.91e-1) -	1.4940e+0 (1.80e-1)
WFG8	3	2.3448e-1 (3.59e-3) -	2.4558e-1 (6.75e-3) -	2.2499e-1 (3.24e-3) +	2.3866e-1 (5.84e-3) -	**2.2476e-1 (3.99e-3)** +	2.2591e-1 (4.53e-3) +	2.3479e-1 (3.92e-3) -	2.5217e-1 (6.29e-3) -	2.2948e-1 (3.36e-3)
	5	6.4972e-1 (6.93e-3) +	8.6277e-1 (1.17e-1) -	6.1985e-1 (8.18e-3) +	6.3239e-1 (3.73e-3) +	**6.1832e-1 (6.38e-3)** +	7.7606e-1 (1.11e-2) -	6.7797e-1 (9.02e-3) -	6.9402e-1 (7.19e-3) -	6.5700e-1 (7.95e-3)
	8	1.5373e+0 (2.20e-2) +	2.0214e+0 (1.91e-1) -	1.7309e+0 (4.24e-1) -	**1.2351e+0 (1.89e-1)** +	1.3918e+0 (1.57e-1) =	1.9827e+0 (9.99e-2) -	1.3843e+0 (1.34e-1) =	1.3800e+0 (2.21e-1) =	1.4142e+0 (1.63e-1)
	15	3.0667e+0 (2.01e-1) -	2.5215e+0 (2.07e-1) -	3.6360e+0 (1.70e+0) -	**1.7182e+0 (1.85e-1)** =	3.5663e+0 (1.21e+0) -	3.6845e+0 (2.80e-1) -	2.4494e+0 (1.04e+0) =	2.5449e+0 (6.23e-1) -	1.8874e+0 (2.65e-1)
WFG9	3	1.6056e-1 (1.01e-1) -	1.9370e-1 (3.46e-2) -	1.2767e-1 (4.64e-2) -	1.2211e-1 (8.17e-3) -	**1.1198e-1 (4.23e-3)** +	1.1422e-1 (4.35e-3) +	1.1483e-1 (3.89e-3) +	1.1804e-1 (5.10e-3) -	1.1652e-1 (5.18e-2)
	5	4.8706e-1 (7.15e-2) -	6.4817e-1 (5.73e-2) -	4.1461e-1 (9.44e-3) -	4.0311e-1 (1.02e-2) -	4.0618e-1 (1.18e-2) -	5.1420e-1 (2.22e-2) -	4.2047e-1 (8.67e-3) -	4.0845e-1 (7.43e-3) -	**3.7128e-1 (2.29e-2)**
	8	1.2966e+0 (2.86e-1) -	1.4132e+0 (2.47e-1) -	9.4831e-1 (1.04e-1) -	**8.2935e-1 (1.90e-2)** +	9.0302e-1 (2.77e-2) -	1.4763e+0 (9.95e-2) -	9.1778e-1 (1.70e-2) -	8.9373e-1 (2.25e-2) -	8.5458e-1 (9.48e-2)
	15	3.1532e+0 (2.83e-1) -	1.9877e+0 (3.93e-1) -	3.3083e+0 (5.51e-1) -	**1.4086e+0 (7.60e-2)** +	2.1612e+0 (1.70e-1) -	4.2791e+0 (7.09e-1) -	1.6901e+0 (7.69e-2) +	2.3213e+0 (2.43e-1) =	2.1832e+0 (2.57e-1)
+/-/=		11/17/8	6/29/1	10/21/5	12/15/9	15/16/5	12/21/3	10/19/7	9/25/2	

offspring Q_t and P_t are merged into the population R_t (Lines 3–5). Subsequently, each individual is associated with a weight vector, and the subregion

Table 3. HV values obtained by ESD-NSGA-II with different neighborhood sizes.

Problem	M	2%	3%	4%	5%	10%	1%
WFG1	3	9.3797e-1 =	9.3294e-1 -	9.3594e-1 =	9.3294e-1 -	9.3194e-1 -	**9.3866e-01**
	5	9.9797e-1 -	9.9799e-1 -	9.9738e-1 =	9.9569e-1 -	9.9656e-1 -	**9.9814e-01**
	8	9.9748e-1 =	9.9302e-1 -	9.9764e-1 =	**9.9820e-1 =**	9.9716e-1 -	9.9786e-01
	15	9.9922e-1 -	9.9940e-1 =	9.9951e-1 =	9.9720e-1 -	9.9846e-1 -	**9.9986e-01**
WFG2	3	9.3389e-1 -	9.3142e-1 -	**9.3517e-1 =**	9.2906e-1 -	9.2179e-1 -	9.3492e-01
	5	9.9690e-1 =	9.9642e-1 =	9.9663e-1 =	9.9587e-1 -	9.9536e-1 -	**9.9698e-01**
	8	9.9864e-1 =	9.9838e-1 =	9.9891e-1 =	9.9901e-1 =	9.9860e-1 =	**9.9904e-01**
	15	**9.9931e-1 =**	9.9903e-1 =	9.9864e-1 =	9.9906e-1 =	9.9878e-1 =	9.9901e-01
WFG3	3	3.7953e-1 =	**3.8424e-1 =**	3.8344e-1 =	3.7689e-1 -	3.6784e-1 -	3.8413e-01
	5	1.5428e-1 =	1.6614e-1 +	**1.8443e-1 +**	1.4915e-1 =	1.5312e-1 =	1.4711e-01
	8	6.5866e-2 =	7.6094e-2 =	7.4872e-2 =	**7.8222e-2 =**	7.8072e-2 =	7.1482e-02
	15	**0.0000e+0 =**	0.0000e+0 =	0.0000e+0 =	0.0000e+0 =	0.0000e+0 =	0.0000e+00
WFG4	3	**5.5571e-1 +**	5.5321e-1 =	5.5166e-1 =	5.4860e-1 -	5.4421e-1 -	5.5286e-01
	5	7.9275e-1 =	7.9004e-1 -	7.8389e-1 -	7.8986e-1 -	7.8362e-1 -	**7.9609e-01**
	8	9.3246e-1 =	9.3169e-1 -	9.2429e-1 -	9.2686e-1 -	9.2246e-1 -	**9.3292e-01**
	15	9.9278e-1 =	9.9228e-1 -	9.8936e-1 -	9.9153e-1 -	9.8992e-1 -	**9.9315e-01**
WFG5	3	**5.1460e-1 +**	5.1302e-1 +	5.1157e-1 =	5.1088e-1 =	5.0628e-1 -	5.1089e-01
	5	**7.3511e-1 =**	7.3294e-1 =	7.3253e-1 =	7.3188e-1 =	7.2845e-1 =	7.3467e-01
	8	8.6878e-1 =	8.6774e-1 =	8.6676e-1 =	8.6645e-1 =	8.6063e-1 -	**8.6878e-01**
	15	9.1921e-1 =	9.1901e-1 =	9.1873e-1 -	9.1911e-1 -	9.1637e-1 -	**9.1935e-01**
WFG6	3	5.0061e-1 =	**5.0139e-1 =**	4.8837e-1 =	4.8874e-1 =	4.9165e-1 =	4.9852e-01
	5	7.2435e-1 =	7.0777e-1 -	7.1898e-1 =	7.2206e-1 =	7.0830e-1 -	**7.2880e-01**
	8	8.6162e-1 =	8.5276e-1 =	8.4374e-1 -	8.5028e-1 -	8.3753e-1 -	**8.6359e-01**
	15	8.9628e-1 =	9.0120e-1 =	**9.0140e-1 =**	8.9543e-1 =	8.9546e-1 =	8.9851e-01
WFG7	3	**5.5133e-1 =**	5.5089e-1 =	5.4811e-1 -	5.4403e-1 -	5.4002e-1 -	5.5073e-01
	5	7.8905e-1 =	7.8725e-1 -	7.8578e-1 -	7.8376e-1 -	7.7354e-1 -	**7.9172e-01**
	8	9.3376e-1 -	9.3328e-1 -	9.2970e-1 -	9.2920e-1 -	9.2161e-1 -	**9.3511e-01**
	15	9.9280e-1 =	9.9293e-1 =	9.9279e-1 =	9.9273e-1 =	9.9226e-1 =	**9.9350e-01**
WFG8	3	4.5228e-1 -	4.4968e-1 -	4.4492e-1 -	4.3984e-1 -	4.4309e-1 -	**4.5990e-01**
	5	6.4447e-1 -	6.3255e-1 -	6.3137e-1 -	6.2963e-1 -	6.3179e-1 -	**6.5649e-01**
	8	8.0186e-1 =	7.8735e-1 -	7.7704e-1 -	7.6828e-1 -	7.8462e-1 -	**8.0560e-01**
	15	**9.4677e-1 =**	9.3659e-1 =	9.0549e-1 =	8.9352e-1 -	9.2126e-1 =	9.4495e-01
WFG9	3	5.3526e-1 =	5.2611e-1 =	5.3447e-1 =	5.2982e-1 =	5.3527e-1 =	**5.3536e-01**
	5	7.4661e-1 =	7.5416e-1 =	7.3669e-1 =	7.5669e-1 =	7.5179e-1 =	**7.5721e-01**
	8	**8.8760e-1 =**	8.7554e-1 =	8.4962e-1 =	8.8400e-1 =	8.6676e-1 =	8.7571e-01
	15	9.5301e-1 =	9.5274e-1 =	9.3500e-1 =	9.0697e-1 -	9.2958e-1 -	**9.5370e-01**
+/-/=		2/6/28	2/13/21	1/11/24	0/21/15	0/24/12	

density is calculated (Line 6). The *Rank* of each individual is calculated, and the population R_t is sorted by ESD-dominance (Lines 7–8). Finally, the optimal frontier is selected based on ESD-dominance to form the next generation P_{t+1}, where the last considered frontier is selected by the distance from the individual to the association weight (Lines 10–17).

Table 4. IGD+ values obtained by ESD-NSGA-II with different neighborhood sizes.

Problem	M	2%	3%	4%	5%	10%	1%
WFG1	3	7.3099e-2 -	8.4430e-2 -	7.1488e-2 =	8.7078e-2 -	9.6664e-2 -	**6.7784e-02**
	5	1.9067e-1 =	2.0106e-1 -	2.1868e-1 -	2.5808e-1 -	2.6309e-1 -	**1.8299e-01**
	8	**6.8350e-1** =	8.1839e-1 =	6.8796e-1 =	6.9483e-1 =	8.1518e-1 =	7.2211e-01
	15	1.3650e+0 =	1.3955e+0 =	1.2982e+0 =	1.4302e+0 =	**1.1899e+0** =	1.2373e+00
WFG2	3	4.5510e-2 =	5.0789e-2 -	**4.3759e-2** =	5.4899e-2 -	6.3228e-2 -	4.4338e-02
	5	**1.4627e-1** =	1.5848e-1 -	1.7077e-1 -	1.7498e-1 -	2.0197e-1 -	1.5026e-01
	8	4.0478e-1 =	3.9858e-1 =	4.0036e-1 =	**3.7741e-1** =	4.1365e-1 =	3.9479e-01
	15	7.9683e-1 =	7.8442e-1 =	**7.4444e-1** =	8.2675e-1 =	8.4167e-1 =	7.5016e-01
WFG3	3	8.3294e-2 -	7.3479e-2 =	7.5841e-2 =	8.8851e-2 -	9.7816e-2 -	**7.1560e-02**
	5	3.4624e-1 =	3.5141e-1 =	**3.0066e-1** +	3.7246e-1 =	3.7699e-1 =	3.6881e-01
	8	1.3363e+0 -	1.1878e+0 =	**9.9009e-1** =	1.0820e+0 =	1.5108e+0 -	1.1199e+00
	15	2.7041e+0 =	2.8871e+0 =	**2.0754e+0** +	2.8462e+0 =	2.5768e+0 =	2.6926e+00
WFG4	3	**8.1058e-2** +	8.5526e-2 =	8.6812e-2 =	9.3233e-2 -	1.0048e-1 -	8.4882e-02
	5	3.2577e-1 =	3.3530e-1 -	3.4972e-1 -	3.4003e-1 -	3.5293e-1 -	**3.1992e-01**
	8	7.1167e-1 =	7.1123e-1 =	7.4683e-1 -	7.2919e-1 -	7.6216e-1 -	**7.0705e-01**
	15	1.3479e+0 =	**1.3417e+0** =	1.4022e+0 -	1.3485e+0 =	1.3752e+0 =	1.3423e+00
WFG5	3	**1.3808e-1** +	1.4098e-1 =	1.4216e-1 =	1.4461e-1 =	1.5277e-1 -	1.4294e-01
	5	**4.0271e-1** =	4.1237e-1 =	4.1702e-1 =	4.1726e-1 =	4.2671e-1 =	4.0863e-01
	8	**7.9006e-1** =	7.9478e-1 =	8.0454e-1 =	8.0744e-1 =	8.4525e-1 =	7.9285e-01
	15	1.3980e+0 +	1.4118e+0 =	**1.3956e+0** =	1.4211e+0 =	1.4707e+0 -	1.4306e+00
WFG6	3	1.5917e-1 =	**1.5861e-1** =	1.7829e-1 -	1.7850e-1 -	1.7599e-1 =	1.6225e-01
	5	4.2263e-1 =	4.4920e-1 -	4.3801e-1 -	4.3570e-1 -	4.6398e-1 -	**4.1445e-01**
	8	8.0746e-1 =	8.2273e-1 -	8.3580e-1 -	8.3210e-1 -	8.8442e-1 -	**7.9648e-01**
	15	**1.4171e+0** =	1.4386e+0 =	1.4243e+0 =	1.4316e+0 =	1.4686e+0 =	1.4446e+00
WFG7	3	**8.8266e-2** =	8.9123e-2 =	9.4075e-2 -	9.9996e-2 -	1.0729e-1 -	8.8597e-02
	5	3.3538e-1 =	3.4262e-1 -	3.4904e-1 -	3.5432e-1 -	3.7608e-1 -	**3.3187e-01**
	8	7.0872e-1 =	7.0680e-1 -	7.2805e-1 -	7.2493e-1 -	7.6596e-1 -	**7.0610e-01**
	15	1.5463e+0 =	1.4518e+0 =	1.4092e+0 =	**1.3576e+0** +	1.4052e+0 =	1.4546e+00
WFG8	3	2.4167e-1 -	2.4547e-1 -	2.5328e-1 -	2.5931e-1 -	2.5550e-1 -	**2.2962e-01**
	5	6.7663e-1 -	6.9960e-1 -	7.0550e-1 -	7.1033e-1 -	7.0258e-1 -	**6.5823e-01**
	8	**1.3689e+0** =	1.4416e+0 =	1.4439e+0 =	1.5494e+0 =	1.4050e+0 =	1.4627e+00
	15	1.8096e+0 =	**1.5987e+0** +	1.7079e+0 =	1.8281e+0 =	1.6006e+0 +	1.8379e+00
WFG9	3	1.0604e-1 =	1.2146e-1 =	1.0800e-1 -	1.1427e-1 -	1.0690e-1 =	**1.0463e-01**
	5	3.8443e-1 =	3.7152e-1 =	3.9977e-1 =	3.7271e-1 =	3.7446e-1 =	**3.7139e-01**
	8	**8.0639e-1** =	8.2181e-1 =	8.8486e-1 =	8.2729e-1 =	8.3209e-1 =	8.3945e-01
	15	2.0318e+0 =	2.0242e+0 =	2.0653e+0 =	2.0580e+0 =	**1.9611e+0** =	2.1781e+00
+/-/=		3/5/28	1/9/26	2/13/21	1/16/19	1/21/14	

4 Experimental Study

4.1 Benchmark Problems and Parameter Settings

In this section, the classical WFG benchmark set is used in the following experiments [18]. For each problem, the number of objectives M is set to 3, 5, 8, and 15, respectively. To validate the performance of ESD-NSGA-II, eight other MaOEAs are selected for comparisons. The involved algorithms are listed as follows.

- RPD-NSGA-II [15].
- MOEA/D [19].
- NSGA-III [4].
- RVEA [20].
- tDEA [21].
- Two_Arch2 [22].
- NSGA-II/SDR [23].

- KnEA [24].
- Proposed ESD-NSGA-II.

For the above nine algorithms, the parameter settings are described as follows. For $M = 3$, 5, 8, and 15, the corresponding population sizes N is set to 91, 126, 156, and 240, respectively. The maximum number of function evaluations $(MaxFEs)$ is set to 30000, 50000, 80000, and 1200000, respectively. All comparison algorithms performed crossover and variation using SBX [25] and polynomial variation, respectively. The probabilities of SBX and variance are 1.0 and $1/D$, respectively. The distribution indices are all set to 20. For other parameters, we use the default settings of PlatEMO [26]. All algorithms are independently executed 20 times. To assess the variability of the algorithm effects, the Wilcoxon rank sum test was used, where the significance level α is 0.05.

4.2   Performance Metrics

According to the suggestion of [27], the hypervolume indicator (HV) [28] and IGD+ were chosen to evaluate the performance of the algorithm. IGD+ is a modified inverted generational distance (IGD) [29]. To compute IGD+, we first generate a set of uniformly distributed weight vectors. Then, we find the intersection of the weight vectors with the true PF, which can be used as reference points. A smaller IGD+ value implies a higher quality solution. HV is a popular evaluation metric because it does not require the true PF. HV is calculated as the volume of the objective space surrounded by a set of approximate solutions and a reference point. A larger HV means a better performance of the algorithm (Fig. 3).

4.3   Comparison Results

The results of all comparison algorithms IGD+ and HV for the WFG test problem are shown in Table 1 and Table 2, respectively. where "$+/ - / =$" in the final row of the table indicates the overall comparison results. The symbols "+", "−", and "=" indicate that the proposed approach RPD-NSGA-II is worse, better, and similar than to the compared algorithm, respectively. As shown in Table 1, ESD-NSGA-II performs the best among all the comparison algorithms. ESD-NSGA-II is better than RPD-NSGA-II on 24 test cases, and for MOEA/D, ESD-NSGA-II is better than MOEA/D on 35 test cases. Compared to NSGA-III, ESD-NSGA-II obtained better results in 27 test cases. For RVEA and tDEA, ESD-NSGA-II is better on 30,25 test cases, respectively. Compared with Two_Arch2, ESD-NSGA-II obtained better results on 26 test cases. NSGA-II/SDR and KnEA are better than ESD-NSGA-II on 2 and 0 test cases, respectively. For a more visual representation, Fig. 3 shows WFG9 plotting the parallel coordinates of the non-dominated solution set at 8 objectives. In the figure, each dash line represents a solution. It can be seen from the figure that only Two_Arch2 and ESD-NSGA-II obtain better result plots. However, in terms of diversity, ESD-NSGA-II performs better.

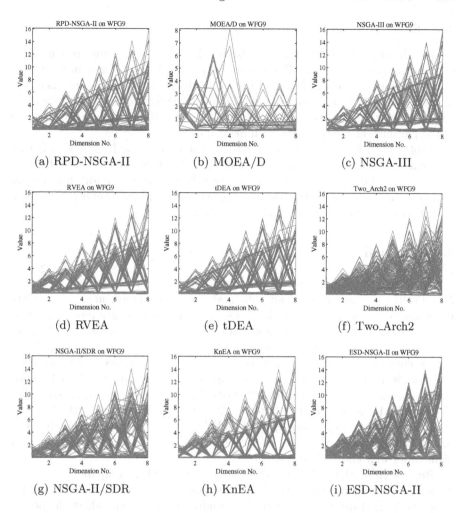

Fig. 3. Distribution of solutions obtained by ESD-NSGA-II and eight other MaOEAs on WFG9 with 8 objectives.

Table 2 shows the IGD+ values of the comparison algorithms for the test problem. From the overall view of Table 2, ESD-NSGA-II is better than RPD-NSGA-II in 17 test cases. For MOEA/D, ESD-NSGA-II outperforms ESD-NSGA-II in 29 test cases. In 10 test cases, NSGA-III is better than ESDSGA-II. ESD-NSGA-II is better than RVEA on 15 test cases. For tDEA, ESD-NSGA-II obtained 16 better results. Compared with Two_Arch2, ESD-NSGA-II has better performance on 21 test cases. For NSGA-II/SDR and KnEA, ESDSGA-II is worse on 10 and 9 test instances, respectively. The results show that ESD-NSGA-II is competitive in handling both regular and irregular PF surfaces.

4.4   Effect the Neighborhood Size

As mentioned in Sect. 3.2, the neighborhood size $(N * \xi)$ leads to a range of dominance relationships in different subregions. If it is too large, it may result in strong dominance relationship, which reduces the performance of the algorithm. To verify the effect of neighborhood size on the performance of ESD-NSGA-II, the parameter ξ is set to 1%, 2%, 3%, 4%, 5%, and 10%, respectively. The rest of the parameters are the same with Sect. 4.3.

Tables 3 and 4 present HV and IGD+ values obtained by ESD-NSGA-II with different ξ, respectively. From the results, $\xi = 1\%$ helps ESD-NSGA-II obtain the relatively best performance among six different cases. In addition, we can conclude that establishing the dominant relation within a range of neighborhoods can improve the performance of the algorithm.

5   Conclusion

To overcome the deficiencies of the traditional Pareto dominance in solving MaOPs, this paper proposes an enhanced subregion dominance (called ESD-dominance) relation. Firstly, individuals in the population are associated with a set of uniform reference vectors according to the Euclidean distance. Individuals associated with the same reference vector constitute a subregion. Then, a new convergence metric is designed to compare the convergence degree of individuals in the same subregion. Subsequently, the density in different subregions is considered to maintain the diversity. Finally, a modified NSGA-II (called ESD-NSGA-II) is proposed by replacing the Pareto dominance with the ESD-dominance. In the experiments, a set of WFG benchmark problems with 3, 5, 8, and 15 objectives are tested. Performance of ESD-NSGA-II is compared with eight other well-known MaOEAs.

Results show that ESD-NSGA-II outperforms other compared algorithms regarding convergence and diversity. It demonstrates the competitiveness of the proposed ESD-dominance relation. The neighborhood size can affect the performance of our approach, and the relatively best parameter is chosen. To test the performance of ESD-NSGA-II, unconstrained benchmark problems are utilized. In future work, the performance of ESD-dominance on some complex problems including constrained and real-world problems will be investigated.

Acknowledgements. This work was supported by the National Natural Science Foundation of China (No. 62166027), and Jiangxi Provincial Natural Science Foundation (No. 20212ACB212004).

References

1. Deb, K.: Multi-objective optimisation using evolutionary algorithms: an introduction. In: Wang, L., Ng, A., Deb, K. (eds.) Multi-objective Evolutionary Optimisation for Product Design and Manufacturing, pp. 3–34. Springer, London (2011). https://doi.org/10.1007/978-0-85729-652-8_1

2. Deb, K., Pratap, A., Agarwal, S., Meyarivan, T.: A fast and elitist multiobjective genetic algorithm: NSGA-II. IEEE Trans. Evol. Comput. **6**(2), 182–197 (2002)
3. Coello, C.A.C., Pulido, G.T., Lechuga, M.S.: Handling multiple objectives with particle swarm optimization. IEEE Trans. Evol. Comput. **8**(3), 256–279 (2004)
4. Deb, K., Jain, H.: An evolutionary many-objective optimization algorithm using reference-point-based nondominated sorting approach, part i: solving problems with box constraints. IEEE Trans. Evol. Comput. **18**(4), 577–601 (2013)
5. Tian, Y., Wang, H., Zhang, X., Jin, Y.: Effectiveness and efficiency of nondominated sorting for evolutionary multi-and many-objective optimization. Complex Intell. Syst. **3**(4), 247–263 (2017)
6. Sato, H., Aguirre, H.E., Tanaka, K.: Controlling dominance area of solutions and its impact on the performance of MOEAs. In: Obayashi, S., Deb, K., Poloni, C., Hiroyasu, T., Murata, T. (eds.) EMO 2007. LNCS, vol. 4403, pp. 5–20. Springer, Heidelberg (2007). https://doi.org/10.1007/978-3-540-70928-2_5
7. Zhu, C., Xu, L., Goodman, E.D.: Generalization of pareto-optimality for many-objective evolutionary optimization. IEEE Trans. Evol. Comput. **20**(2), 299–315 (2015)
8. Ikeda, K., Kita, H., Kobayashi, S.: Failure of pareto-based MOEAs: does nondominated really mean near to optimal? In: Proceedings of the 2001 Congress on Evolutionary Computation (IEEE Cat. No. 01TH8546), vol. 2, pp. 957–962. IEEE (2001)
9. Yang, S., Li, M., Liu, X., Zheng, J.: A grid-based evolutionary algorithm for many-objective optimization. IEEE Trans. Evol. Comput. **17**(5), 721–736 (2013)
10. Laumanns, M., Thiele, L., Deb, K., Zitzler, E.: Combining convergence and diversity in evolutionary multiobjective optimization. Evol. Comput. **10**(3), 263–282 (2002)
11. Hernández-Díaz, A.G., Santana-Quintero, L.V., Coello, C.A.C., Molina, J.: Pareto-adaptive ϵ-dominance. Evol. Comput. **15**(4), 493–517 (2007)
12. Batista, L.S., Campelo, F., Guimarães, F.G., Ramírez, J.A.: Pareto cone ε-dominance: improving convergence and diversity in multiobjective evolutionary algorithms. In: Takahashi, R.H.C., Deb, K., Wanner, E.F., Greco, S. (eds.) EMO 2011. LNCS, vol. 6576, pp. 76–90. Springer, Heidelberg (2011). https://doi.org/10.1007/978-3-642-19893-9_6
13. Farina, M., Amato, P.: A fuzzy definition of "optimality" for many-criteria optimization problems. IEEE Trans. Syst. Man Cybern.-Part A Syst. Hum. **34**(3), 315–326 (2004)
14. Zou, X., Chen, Y., Liu, M., Kang, L.: A new evolutionary algorithm for solving many-objective optimization problems. IEEE Trans. Syst. Man Cybern. Part B (Cybern.) **38**(5), 1402–1412 (2008)
15. Elarbi, M., Bechikh, S., Gupta, A., Said, L.B., Ong, Y.S.: A new decomposition-based NSGA-II for many-objective optimization. IEEE Trans. Syst. Man Cybern. Syst. **48**(7), 1191–1210 (2017)
16. Gu, Q., Chen, H., Chen, L., Li, X., Xiong, N.N.: A many-objective evolutionary algorithm with reference points-based strengthened dominance relation. Inf. Sci. **554**, 236–255 (2021)
17. Das, I., Dennis, J.E.: Normal-boundary intersection: a new method for generating the pareto surface in nonlinear multicriteria optimization problems. SIAM J. Optim. **8**(3), 631–657 (1998)
18. Huband, S., Hingston, P., Barone, L., While, L.: A review of multiobjective test problems and a scalable test problem toolkit. IEEE Trans. Evol. Comput. **10**(5), 477–506 (2006)

19. Zhang, Q., Li, H.: MOEA/D: a multiobjective evolutionary algorithm based on decomposition. IEEE Trans. Evol. Comput. **11**(6), 712–731 (2007)
20. Cheng, R., Jin, Y., Olhofer, M., Sendhoff, B.: A reference vector guided evolutionary algorithm for many-objective optimization. IEEE Trans. Evol. Comput. **20**(5), 773–791 (2016)
21. Yuan, Y., Xu, H., Wang, B., Yao, X.: A new dominance relation-based evolutionary algorithm for many-objective optimization. IEEE Trans. Evol. Comput. **20**(1), 16–37 (2015)
22. Wang, H., Jiao, L., Yao, X.: Two_arch2: an improved two-archive algorithm for many-objective optimization. IEEE Trans. Evol. Comput. **19**(4), 524–541 (2014)
23. Tian, Y., Cheng, R., Zhang, X., Su, Y., Jin, Y.: A strengthened dominance relation considering convergence and diversity for evolutionary many-objective optimization. IEEE Trans. Evol. Comput. **23**(2), 331–345 (2018)
24. Zhang, X., Tian, Y., Jin, Y.: A knee point-driven evolutionary algorithm for many-objective optimization. IEEE Trans. Evol. Comput. **19**(6), 761–776 (2014)
25. Deb, K., Agrawal, R.B., et al.: Simulated binary crossover for continuous search space. Complex Syst. **9**(2), 115–148 (1995)
26. Tian, Y., Cheng, R., Zhang, X., Jin, Y.: Platemo: a matlab platform for evolutionary multi-objective optimization [educational forum]. IEEE Comput. Intell. Mag. **12**(4), 73–87 (2017)
27. Ishibuchi, H., Masuda, H., Tanigaki, Y., Nojima, Y.: Difficulties in specifying reference points to calculate the inverted generational distance for many-objective optimization problems. In: 2014 IEEE Symposium on Computational Intelligence in Multi-Criteria Decision-Making (MCDM), pp. 170–177. IEEE (2014)
28. Zitzler, E., Thiele, L.: Multiobjective evolutionary algorithms: a comparative case study and the strength pareto approach. IEEE Trans. Evol. Comput. **3**(4), 257–271 (1999)
29. Zitzler, E., Künzli, S.: Indicator-based selection in multiobjective search. In: Yao, X., et al. (eds.) PPSN 2004. LNCS, vol. 3242, pp. 832–842. Springer, Heidelberg (2004). https://doi.org/10.1007/978-3-540-30217-9_84

A Stacked Autoencoder Based Meta-Learning Model for Global Optimization

Yue Ma[1(✉)], Yongsheng Pang[1], Shuxiang Li[1], Yuanju Qu[1], Yangpeng Wang[1], and Xianghua Chu[1,2]

[1] College of Management, Shenzhen University, Shenzhen, China
yue_ma_620@163.com, lishuxiang2020@email.szu.edu.cn,
{wangyangpeng,x.chu}@szu.edu.cn
[2] Institute of Big Data Intelligent Management and Decision, Shenzhen University, Shenzhen, China

Abstract. As optimization problems continue to become more complex, previous studies have demonstrated that algorithm performance varies depending on the specific problem being addressed. Thus, this study proposes an adaptive data-driven recommendation model based on the stacked autoencoder. This approach involves the use of a meta-learning autoencoder that is stacked with multiple supervised autoencoders, generating deep meta-features. Then the proper algorithms are identified to address the new problems. To verify the feasibility of this proposed model, experiments are conducted using benchmark functions. Experimental results indicate that both instance-based and model-based meta-learners are well suited to the advanced model, and the performance is promising.

Keywords: Recommendation model · Meta-learning · Algorithm selection · Stacked Autoencoder

1 Introduction

In the real world, the diversity of people's needs is constantly increasing, while the overall demand for solutions is also on the rise. This trend has led to the emergence of increasingly complex optimization problems. Researchers across a wide range of fields have devoted considerable effort to gaining a deeper understanding of algorithm performance. Nevertheless, as demonstrated by previous studies, the same algorithm can exhibit markedly different performance levels when applied to different problems [1]. The 'No free lunch' theorem emphasizes the fact that no single algorithm can perform optimally on all problems [2]. Instead, an algorithm that performs well on one problem may be at a disadvantage when applied to another. This phenomenon is known as performance complementarity [3]. While selecting an algorithm for a problem with prior knowledge may not pose a significant challenge, selecting a suitable algorithm for a problem without prior knowledge is a difficult task. Currently, traditional algorithm selection methods are not entirely satisfactory. Trial-and-error is computationally expensive, expert knowledge may be difficult to obtain, and objectivity cannot be guaranteed through this approach. Consequently, it is imperative to identify an appropriate and efficient method for algorithm selection.

In recent years, meta-learning has gained increasing attention among researchers and has been widely applied in various domains. The fundamental thinking of meta-learning is a process of 'learning to learn' by exploring the internal learning mechanisms inherent in the data, in order to achieve better outcomes in subsequent applications [4]. In the context of addressing the algorithm selection problem, meta-learning is a brilliant approach. However, in the existing study on algorithm selection, meta-features used to train the base-learner, which is consists of simple classifiers or regressors, are primarily obtained through manual computation. This study introduces a modification by first extracting meta-features using a meta-learner, followed by training the base-learners with the generated meta-features. By establishing the relationship between algorithms and problems, meta-learning aims to provide guidance in selecting the most appropriate algorithm for a given problem [5]. Chu et al. [6] have proposed a flexible adaptive recommendation model with robust generalization ability to identify algorithms. This model is capable of recommending algorithms with superior performance for specific optimization problems.

Meta-features are sets of data that can represent the internal information of a problem instance [7]. However, many current meta-learning-based approaches still require prior knowledge to support the extraction of meta-features. The selection of meta-features is a fundamental step in the meta-learning process. Commonly used meta-features include statistical features, geometric features [8], landmarking features [6], etc. Peng et al. [9] have proposed that features of decision trees, such as nodes, leaves, branches, and tree level, can be leveraged as meta-features to describe the dataset. In addressing multi-objective optimization problems, Chu et al. [10] unprecedentedly proposed the Pareto front-based meta-features.

In order to enhance the effectiveness and accuracy of algorithm selection for optimization problems, this paper proposes an adaptive data-driven recommendation model (ADRM) based on meta-learning autoencoder (MLAE). MLAE can output a general characterization of algorithm, it consists of multiple supervised autoencoders, which is widely utilized in the field of deep learning. In ADRM, MLAE replaces the traditional meta-learner to extract deep meta-features that are related to both the given problem and the algorithm performance. By doing so, the recommended algorithm can better match new problems while effectively utilizing the original data. To evaluate the performance of ADRM, the study selected 7 Benchmark Functions from the CEC' 2008 Competition on Large-Scale Global Optimization. The experimental results indicate that ADRM is capable of recommending the algorithm with the best practical results for most problems, and is highly adaptable.

The three main contributions of this work are summarized as follows:

1) Replacing the traditional meta-learner with MLAE to extract deep meta-features can simultaneously characterize the specific problem and the algorithm performance to train the base-learner.

2) Applying autoencoder to work on meta-learning-based algorithm selection.

3) No specialized knowledge is required to support the use of ADRM, which makes it universally applicable.

The rest of the paper is organized as follows: Sect. 2 describes the background and related work on algorithm selection problem and stacked autoencoders. Section 3 presents the details of the ADRM proposed in this paper. Section 4 discussed the feasibility of ADRM through experimental results. Section 5 drew conclusions and discussed future research directions.

2 Related Work

2.1 Algorithm Selection

Research on algorithm selection started in 1976. Rice [11] suggests that the algorithm for solving a specific problem should be chosen according to the its features. Rice abstracts the algorithm selection problem into a model as shown in Fig. 1, which is very flexible and has contributed significant ideas for subsequent researchers studying algorithm selection. There are four spaces in this model. The problem space P contains numerous problem datasets p. The feature of the problem, $f(p)$, can be obtained by feature extraction, which can characterize the problem in a simpler form. The feature space F is assembled by $f(p)$, which is implemented as a mapping from P to F. There is an infinite number of algorithms a in the algorithm space A. $f(p)$ is needed to select one a from A that maximizes the mapping F to Y. In this case, the chosen algorithm should perform the best in solving this problem.

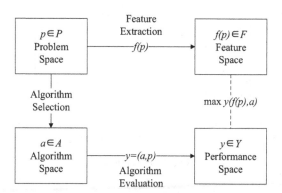

Fig. 1. Rice's algorithm selection model

Algorithm selection problem is widely applied in real life. Typical applications of algorithm selection problems include classification [12], regression [13], time series prediction [14], and constraint satisfaction [15], just to name a few. Not only above, with the divergent thoughts of researchers from different fields, the algorithm selection problem can also accomplish interdisciplinary and cross-vergence work. For example, Takahashi et al. [16] applied the framework of algorithm selection to the quality of service-based server selection problem. For Artificial Intelligence, Degroote [17] considers online algorithm selection problems, making the selection model possible to change over time while solving the given problem. Misir [18] extends the algorithm selection framework and uses it to solve Team Orienteering Problem. It helps to solve real-world problems effectively.

2.2 Stacked Autoencoder

Stacked autoencoder(SAE) is a deep neural network model, composed of multiple unsupervised neural networks Autoencoder(AE) [19], the structure of AE is shown in Fig. 2. SAE has been widely used in various fields in recent years. AE is a popular tool for extracting nonlinear features. The training process is essentially a reconstruction of the input data and minimizes the error between the reconstructed data and the input data as much as possible. The basic AE is a Multilayer Perceptron (MLP), which contains an input layer, a hidden layer, and an output layer. There are two stages involved in the training process of a basic AE, which are encoding and decoding. Encoding occurs from the input layer to the hidden layer first. The second stage, decoding is from the hidden layer to the output layer.

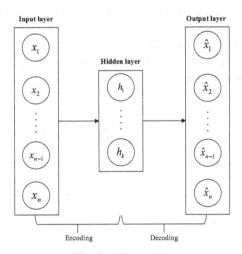

Fig. 2. AE structure

There is a similar structure between SAE and AE. Both of them have only one input layer and one output layer, the difference lies in the number of hidden layers, SAE contains several hidden layers but AE only has one. SAE is stacked with multiple AEs. The hidden layers of SAE are stacked with multiple hidden layers of trained AEs and fine-tuned by a supervised method [20].

In general, the principles of SAE and AE are quite the same, they both aim to approximate the output data to the input data infinitely. Figure 3 shows the structure of SAE. Usually, the number of nodes in the hidden layer is decreasing layer by layer. The original data in AE cannot be characterized ideally since there is only one hidden layer. Due to the hidden layers, SAE excels in reducing redundancy of data features. The output data of reconstruction characterize original data adequately because there are multiple hidden layers in SAE.

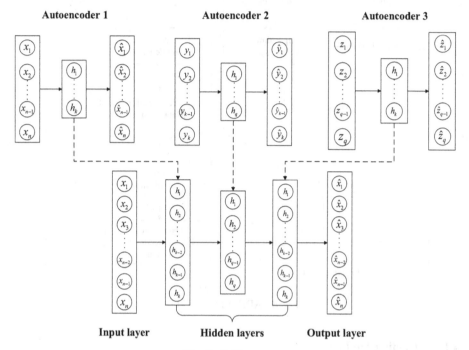

Fig. 3. SAE structure

SAE is not only popular in machine learning field, but also has remarkable practical significance. SAE can be found in many fields such as medicine, optics, and engineering. For example, predicting Drug–Target Interactions [21], classification of Hyperspectral [22] and bearing fault diagnosis [23], etc. In this study, we use SAE to extract deep meta-features for characterizing the original data. In this way, better accuracy of the algorithm selection process can be achieved.

3 Adaptive Data-Driven Recommendation Model

Based on the work that Rice [11] did on the algorithm selection problem, some improvements were made in this study. We propose an adaptive data-driven recommendation model (ADRM) based on meta-learning to recommend appropriate optimization algorithms for a given problem. ADRM consists of three modules, including extracting module, learning module, and recommending module.

Figure 4 shows the structure of the ADRM, and the details of each module will be described in the following subsections.

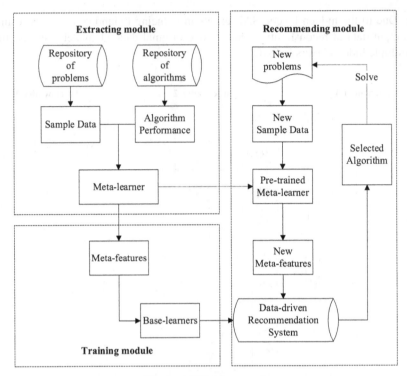

Fig. 4. ADRM framework

3.1 Extracting Module

First, making the problem identifiable is an important step in meta-level induction learning. Sample the data in the problem repository, and then compute the features of the sampled data, to obtain preliminary meta-features that represent the problem. By choosing the appropriate meta-features, the mapping relationship between meta-features and algorithm performance will be more accurate.

Next, evaluate algorithm performance, it is a crucial step for algorithm selection problems. In order to provide precise recommendation for the particular optimal problem, a thorough evaluation and characterization of the performance of the algorithms available in the algorithm repository is essential.

Upon completion of the data sampling process and algorithm performance evaluation, the acquired features and algorithm performance measures are fed into the meta-learner, MLAE. As a general rule, meta-learner will perform better when more data is used to train it, because it is a machine learning model.

MLAE improves the utilization of raw data by reconstructing it. This allows us to train base-learner with more data even if the problem instance number holds constant. Given this outcome, we suggest using MLAE as the meta-learner, with the aim of processing the preliminary meta-features and algorithm performance that have been previously obtained into deep meta-features.

MLAE is composed of two components, an objective part and an algorithm performance part. Each of them is stacked by several supervised autoencoders. Figure 5 displays the structure of MLAE. As a result of MLAE processing, deep meta-features reflect not only the given problem but also the algorithms performance. This differs from traditional algorithm recommendation frameworks which compute meta-features following data sampling.

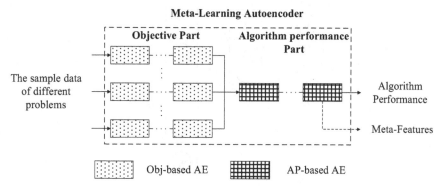

Fig. 5. MLAE structure

Figure 6 shows the structure of supervised AE. When the supervised component is the objective function value, it is an obj-based AE. If the supervised component is algorithm performance, it will be an AP-based AE. After the preliminary meta-features and the performance of algorithms are input to MLAE, they are first processed by the objective part, which is stacked by several Obj-based supervised AE. The first Obj-based supervised AE takes the input as raw data for training, with the target function values as the output. Then the hidden layer from the first Obj-based supervised AE will be input to the next Obj-based supervised AE for following training, and so on. Follow this procedure to train all Obj-based supervised AEs to complete the task of the objective part. The final obj-based features that will be input to the algorithm performance part are the final outputs obtained from the objective part, and it is undoubtedly relevant to the given problem. The operation principle of the AP part is the same as the objective part. The final AP-based features are relevant to specific problem and algorithm performance, so it's considered as deep meta-features necessary for ADRM to recommend an algorithm.

3.2 Training Module

The object of training module is the base-learner which is used for algorithm recommendation. Common base-learners are usually classified into instance-based meta-learner and model-based meta-learner [24].

The fundamental concept behind the operation of instance-based meta-learner is that the same algorithm behaves comparably when processing similar problems, and the similarity between different problems is determined by a distance metric calculation. K-nearest neighbors (KNN) is one of the most frequently used instance-based meta-learner.

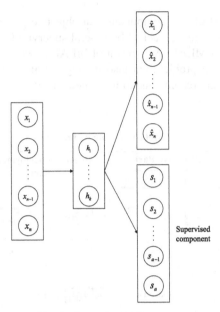

Fig. 6. Supervised AE structure

KNN is a machine learning method known for its simplicity [25], and ability to provide accurate recommendations in a short amount of time. When presented with a new problem, KNN identifies its category by measuring its distance to the k nearest neighbors in the feature space. This category information is used to predict the performance of each algorithm when solving a new problem.

The model-based base-learner learns the experience of mapping meta-features to algorithm performance when it predicts algorithm ranking. In the current research, Artificial Neural Network (ANN) often appears as a model-based base-learner. The structure of ANN is abstracted from the neuronal network of the human brain. There are a huge number of nodes inside ANN, and the connection between every two nodes from adjacent layers is called weight. ANN is not only an excellent choice for modeling nonlinear functions, but it also has impressive robustness and scalability in machine learning field.

3.3 Recommending Module

As the last module of ADRM, recommending module focused on addressing new issues. To select an appropriate optimization algorithm for a new problem, we start by sampling its data. Next, input the sampled data into the meta-learner which is pre-trained in extracting module, and output with the meta-features that characterize the new problem. Finally, the meta-features are input to the data-driven recommendation system, which will recommend a suitable optimization algorithm for the new problem from all the candidate algorithms. Following the completion of these procedures, the new problem is successfully resolved.

4 Experiment Validation

The proposed ADRM should be able to be applied in the real world. To verify such feasibility, experiments were conducted in this paper. The details will be illustrated in the following content.

4.1 Experiment Setup

To test if ADRM can perform well on optimization problems, we selected 7 Benchmark Functions from CEC'2008 Competition on Large-Scale Global Optimization and constructed a repository of problems to test ADRM [26]. A total of 1000 problem samples were extracted by Latin Hypercube Sampling for each benchmark function. The details and properties of Benchmark Functions are shown in Table 1. The last 5 columns of Table 1 are some common properties of functions. Multimodal refers to a function with several extremes. Shifted indicates whether a function can perform shifted operations. Separable and scalable are as the name implies. When a function has a certain property, it is denoted by Y in the corresponding column, or by N if it does not have the property in the corresponding column.

Table 1. Problem List

	Functions	Search Range	Multimodal	Shifted	Separable	Scalable
f_1	Shifted Sphere Function	$[-100,100]^D$	N	Y	Y	Y
f_2	Shifted Schwefel's Problem 2.21		N	Y	N	Y
f_3	Shifted Rosenbrock's Function		Y	Y	N	Y
f_4	Shifted Rastrigin's Function	$[-5,5]^D$	Y	Y	Y	Y
f_5	Shifted Griewank's Function	$[-600,600]^D$	Y	Y	N	Y
f_6	Shifted Griewank's Function	$[-32,32]^D$	Y	Y	Y	Y
f_7	FastFractal "DoubleDip" Function	$[-1,1]^D$	Y	N	N	Y

The experimental repository of algorithms contains 8 algorithms, which are CMAES [27], CSO [28], FEP [29], GA [30], GPSO [31], IMODE [32], OFA [33], SHADE [34]. We conduct experiments on platEMO [35]. The parameters of the experiment are set as follows. The dimension of all problems is 100. Population Size is set to 100, and the maximum number of Fes (MAX_FES) is set to 100000 in 5 independent runs. The default parameters are not displayed here.

MLAE is a tool to generate meta-features. The MLAE in the experiment is stacked by two obj-based AEs and two AP-based AEs, so it contains four hidden layers. Table 2 shows the specific parameter settings of MLAE. Every single AE is trained with optimizer Adam, whose maximum training epoch is fixed at 100. Early stopping is used to prevent overfitting by setting patience at every 5 epochs and stopping the model in time when training results worsen. Additionally, KNN is built by calling the Scikit-learn [36] package, setting the K value of KNN to 2, and assigning the weight of each instance according to the distance. All of the neural network models in this paper were constructed by using Tensorflow backend Keras [37].

Table 2. The parameter settings of SAEs in MLAE

	Number of input layer nodes	Number of hidden layer nodes	Number of output layer nodes (reconstruction)	Number of output layer nodes (supervised)
Obj-based SAE 1	100	80	100	1
Obj-based SAE 2	80	60	80	
AP-based SAE 1	60	50	60	8
AP-based SAE 2	50	30	50	

ADRM completes its job once it recommends an algorithm for the new problem. In spite of this, it is essential to conduct a comprehensive evaluation of the performance of ADRM when it comes to algorithm selection. It is important to choose the measurement metrics carefully. We choose two metrics, Spearman's rank correlation coefficient (SRCC) [38] and success rate (SR) to evaluate the performance of ADRM. SRCC is one of the non-parametric tests to measure the correlation between two variables. When the object is ADRM, SRCC calculates the correlation between the real ranking and the predicted ranking of the algorithms. SRCC is calculated by Eq. (1).

$$\rho_i = 1 - \frac{6 * \sum_{i=1}^{N} d_i^2}{N(N^2 - 1)} \tag{1}$$

In this equation, d_i represents the distance between the true and predicted ranking of a particular algorithm on problem i. N is the number of candidate algorithms in the algorithm repository. The value of ρ_i ranges from -1 to 1. The smaller d_i is, the more ρ_i converges to 1. That indicates there is a high consistency of the two rankings, in other words, ADRM has a good recommendation performance. Conversely, if ρ_i converges to -1, indicating that the consistency between the two rankings is low, ADRM doesn't perform well.

Another metric is SR, which should be presented as a percentage. Equation (2) shows how SR is calculated, it represents the accuracy rate of the match between the algorithm recommended by ADRM and the best algorithm in reality.

$$SR = \frac{\#well - predicted \ ideal \ algorithm \ number}{\#sampled \ toatal \ number \ of \ problems} \qquad (2)$$

4.2 Experiment Result

The results of each benchmark function were tested using 8 algorithms, as listed in Table 3. Following that, the flexibility of ADRM is demonstrated by the performance of its performance on previously unknown problems. In this experiment, we use 'Leave-one-out' cross-validation. Only six benchmark functions are used to train ADRM each time, and the left one is used for testing. Considering that the test function is completely new to ADRM, the performance reflects its flexibility. By completing 7 rounds of training, it can be guaranteed that all benchmark functions are used to train and test ADRM.

Table 3. Results of algorithms on benchmark functions

	CMAES	CSO	FEP	GA
f_1	6.42e+3 ± 2.64e+3	**3.82e−18 ± 2.91e−18**	3.44e+4 ± 8.25e+3	6.10e−1 ± 6.91e−2
f_2	5.12e+1 ± 1.03e+1	5.55e+1 ± 4.47e+0	8.71e+1 ± 1.01e+1	4.12e+1 ± 6.94e+0
f_3	6.34e+8 ± 3.24e+8	2.31e+3 ± 3.31e+3	4.47e +9 ± 3.45e+9	1.64e+3 ± 1.99e+3
f_4	2.58e+2 ± 8.54e+1	9.79e+1 ± 1.04e+1	7.48e+2 ± 1.53e+2	**1.02e+0 ± 5.28e−1**
f_5	6.35e+1 ± 2.01e+1	**0.00e+0 ± 0.00e+0**	3.23e+2 ± 9.07e+1	2.95e−1 ± 4.52e−2
f_6	8.81e+0 ± 4.70e 0	**4.01e−10 ± 1.29e−10**	1.68e+1 ± 2.04e+0	1.07e−1 ± 1.60e−2
f_7	−1.32e+6 ± 1.41e+5	**−1.87e+6 ± 5.07e+3**	−8.51e+ 5 ± 1.29e+5	−1.84e+6 ± 1.81e+4
	GPSO	IMODE	OFA	SHADE
f_1	8.99e−8 ± 2.83e−10	6.14e−3 ± 3.72e−3	3.16e+3 ± 4.71e+2	1.96e−9 ± 8.22e−10
f_2	4.78e+1 ± 3.92e+0	6.17e+1 ± 3.61e+0	8.59e+1 ± 1.12e+0	**2.22e+1 ± 2.66e+0**
f_3	1.24e+9 ± 2.25e+8	1.22e+3 ± 6.32e+2	4.79e+8 ± 6.78e+7	**3.19e+2 ± 1.56e+2**
f_4	2.21e+3 ± 1.38e+2	2.84e+1 ± 7.04e+0	1.18e+3 ± 3.96e+1	4.70e+2 ± 2.54e+1
f_5	3.07e−2 ± 4.39e−2	3.43e−3 ± 4.33e−3	2.73e+1 ± 4.29e+0	7.38e−3 ± 1.28e−2
f_6	1.98e+1 ± 7.72e−2	1.43e+0 ± 2.75e−1	1.54e+1 ± 5.10e−1	1.81e+0 ± 3.24e−1
f_7	−1.73e+5 ± 3.21e+4	−1.87e+6 ± 4.18e+3	−5.84e+5 ± 1.28e+4	−1.53e+6 ± 1.75e+5

There are two types of common meta-learner, so KNN is chosen to represent instance-based meta-learner and ANN to represent model-based meta-learner in the experiment. Table 4 shows the results of SRCC for ANN and KNN. It is clear to see that KNN performs better compared to ANN, all SRCC values for KNN were above 0.7 and the average SRCC exceeded 0.8, this is a pleasant result. Although the SRCC for ANN is relatively worse, it still achieves an average SRCC of 0.74, which is in the acceptable range.

Table 4. SRCC of different meta-learners

Problem	ANN-SRCC	KNN-SRCC
f_1	0.833	**0.905**
f_2	**0.714**	**0.714**
f_3	0.619	**0.762**
f_4	0.738	**0.762**
f_5	0.810	**0.881**
f_6	0.762	**0.869**
f_7	**0.714**	**0.714**
(MEAN)	0.741	**0.801**

Table 5 shows SR values of eight algorithms. Clearly, among the algorithm repository, CSO has the highest SR and performs the best on more than half of the problems. Followed by SHADE and GA, their SR values are 28.6% and 14.3% respectively. Furthermore, the top algorithm CSO ranks first on the four problems f_1, f_5, f_6, and f_7. Both KNN and ANN did not recommend CSO on problem f_7 only, instead, KNN recommended GA and ANN recommended IMODE. We also find that KNN performs better than ANN with SRCC as the measurement metric. However, ANN beats KNN in terms of SR. Generally speaking, both meta-learners showed appreciable results.

Table 5. SR of algorithms

	CMAES	CSO	FEP	GA	GPSO
SR	0.00%	**57.10%**	0.00%	14.30%	0.00%
	IMODE	OFA	SHADE	ADRM-KNN	ADRM-ANN
SR	0.00%	0.00%	28.60%	42.90%	**57.10%**

5   Conclusion

The problem of selecting the optimal algorithm for new problems without prior knowledge has been a challenge for researchers. Traditional methods such as trial-and-error have limitations in terms of time and accuracy. As the popularity of meta-learning has grown, a meta-learning-based approach has become a priority for solving the algorithm selection problem. However, most existing approaches have not fully utilized the original data and are sensitive to meta-features. To address these limitations, this paper proposes a supervised stacked autoencoder, MLAE, composed of multiple supervised autoencoders. MLAE is used to extract deep meta-features from the original data of the problem and algorithm, which avoids over-reliance on meta-features and enables data reconstruction, thereby improving the utilization of original data. To assess the generalizability and extensibility of this model, experiments were conducted on both benchmark functions and unseen problems. The results indicate that MLAE is effective for both instance-based and model-based meta-learner, and ADRM can recommend algorithms for unseen problems. However, there is room for improvement, such as considering multiple criteria metrics like computational cost and robustness to enhance the performance of ADRM.

Acknowledgement. This work was partially supported by the National Natural Science Foundation of China (No. 71971142 and 71501132), the Natural Science Foundation of Guangdong Province (No. 2022A1515010278 and 2021A1515110595).

References

1. Chu, X., et al.: Learning–interaction–diversification framework for swarm intelligence optimizers: a unified perspective. Neural Comput. Appl. **32**(6), 1789–1809 (2018)
2. Wolpert, D.H.a.M., W.G.: No free lunch theorems for optimization.pdf. IEEE Trans. Evol. Comput. **1**, 67–82 (1997)
3. Kerschke, P., et al.: Automated algorithm selection: survey and perspectives. Evol. Comput. **27**(1), 3–45 (2019)
4. Lemke, C., Budka, M., Gabrys, B.: Metalearning: a survey of trends and technologies. Artif. Intell. Rev. **44**(1), 117–130 (2015)
5. Smith-Miles, K.A.: IEEE. Towards Insightful Algorithm Selection For Optimisation Using Meta-Learning Concepts. International Joint Conference on Neural Networks, pp. 4118–4124. Hong Kong, PEOPLES R CHINA (2008)
6. Chu, X., et al.: Adaptive recommendation model using meta-learning for population-based algorithms. Inf. Sci. **476**, 192–210 (2019)
7. Dantas, A.L., Pozo, A.T.R.: IEEE. A meta- learning algorithm selection approach for the quadratic assignment problem. IEEE Congress on Evolutionary Computation (IEEE CEC) as part of the IEEE World Congress on Computational Intelligence (IEEE WCCI), pp. 1284–1291. Rio de Janeiro, BRAZIL (2018)
8. Cui, C., et al.: A recommendation system for meta-modeling: A meta-learning based approach. Expert Syst. Appl. **46**, 33–44 (2016)
9. Peng, Y.H., et al.: Improved dataset characterisation for meta-learning (2002)
10. Chu, X., et al.: Empirical study on meta-feature characterization for multi-objective optimization problems. Neural Comput. Appl. **34**(19), 16255–16273 (2022)
11. Rice, J.R.: The Algorithm Selection Problem (1976)
12. Khan, I., et al.: A literature survey and empirical study of meta-learning for classifier selection. IEEE Access **8**, 10262–10281 (2020)
13. Wang, G.T., et al.: A generic multilabel learning-based classification algorithm recommendation method. ACM Trans. Knowl. Discovery Data **9**(1) (2014)
14. Rossi, A.L.D., et al.: MetaStream: a meta-learning based method for periodic algorithm selection in time-changing data. Neurocomputing **127**, 52–64 (2014)
15. Muñoz, M.A., Kirley, M., Halgamuge, S.K.: The algorithm selection problem on the continuous optimization domain. In: Moewes, C., Nürnberger, A. (eds.) Computational Intelligence in Intelligent Data Analysis, pp. 75–89. Springer Berlin Heidelberg, Berlin, Heidelberg (2013). https://doi.org/10.1007/978-3-642-32378-2_6

16. Takahashi, R., et al.: Recommendation of web service selection algorithm based on web application review. IEEE-Region-10 Conference (IEEE TENCON), pp. 1882–1887. IEEE Reg 10, SOUTH KOREA (2018)

17. Degroote, H.: Online Algorithm Selection. 26th International Joint Conference on Artificial Intelligence (IJCAI), pp. 5173–5174. Melbourne, AUSTRALIA (2017)

18. Misir, M., Gunawan, A., Vansteenwegen, P.: Algorithm selection for the team orienteering problem. In: 22nd European Conference on Evolutionary Computation in Combinatorial Optimisation (EvoCOP) Held as Part of EvoStar Conference. 13222, pp. 33–45. Madrid, SPAIN (2022)

19. Baldi, P., Hornik, K.: Neural networks and principal component analysis: learning from examples without local minima. Neural Netw. **2**(1), 53–58 (1989)

20. Liu, G., Bao, H., Han, B.: A stacked autoencoder-based deep neural network for achieving gearbox fault diagnosis. Math. Probl. Eng. **2018**, 1–10 (2018)

21. Wang, L., et al.: A computational-based method for predicting drug-target interactions by using stacked autoencoder deep neural network. J. Comput. Biol. **25**(3), 361–373 (2018)

22. Chen, Y., et al.: deep learning-based classification of hyperspectral data. IEEE J. Select. Top. Appl. Earth Observ. Remote Sens. **7**(6), 2094–2107 (2014)

23. Tao, S., et al.: Bearing fault diagnosis method based on stacked autoencoder and softmax regression. In: 2015 34th Chinese Control Conference (CCC), pp. 6331–6335. IEEE (2015)

24. Smith-Miles, K.A.: Cross-disciplinary perspectives on meta-learning for algorithm selection. ACM Comput. Surv. **41**(1), 1–25 (2009)

25. Ferrari, D.G., de Castro, L.N.: Clustering algorithm selection by meta-learning systems: a new distance-based problem characterization and ranking combination methods. Inf. Sci. **301**, 181–194 (2015)

26. Tang, K., et al.: Benchmark functions for the CEC'2008 special session and competition on large scale global optimization. Nat. Inspired Comput. Appl. Lab. USTC, China **24**, 1–18 (2007)

27. Hansen, N., Ostermeier, A.: Completely derandomized self-adaptation in evolution strategies. Evol. Comput. **9**(2), 159–195 (2001)

28. Cheng, R., Jin, Y.: A competitive swarm optimizer for large scale optimization. IEEE Trans. Cybern. **45**(2), 191–204 (2014)

29. Yao, X., Liu, Y., Lin, G.: Evolutionary programming made faster. IEEE Trans. Evol. Comput. **3**(2), 82–102 (1999)

30. Holland, J.H.: Adaptation in Natural and Artificial Systems: An Introductory Analysis with Applications to Biology, Control, and Artificial Intelligence. MIT Press (1992)

31. Noel, M.M.: A new gradient based particle swarm optimization algorithm for accurate computation of global minimum. Appl. Soft Comput. **12**(1), 353–359 (2012)

32. Sallam, K.M., et al.: Improved multi-operator differential evolution algorithm for solving unconstrained problems. In: 2020 IEEE Congress on Evolutionary Computation (CEC), pp. 1–8. IEEE (2020)

33. Zhu, G.-Y., Zhang, W.-B.: Optimal foraging algorithm for global optimization. Appl. Soft Comput. **51**, 294–313 (2017)

34. Tanabe, R., Fukunaga, A.: Success-history based parameter adaptation for differential evolution. In: 2013 IEEE Congress on Evolutionary Computation, pp. 71–78. IEEE (2013)

35. Tian, Y., et al.: PlatEMO: a MATLAB platform for evolutionary multi-objective optimization [educational forum]. IEEE Comput. Intell. Mag. **12**(4), 73–87 (2017)

36. Pedregosa, F., et al.: Scikit-learn: machine learning in Python. J. Mach. Learn. Res. **12**, 2825–2830 (2011)

37. Gulli, A.: and S. Packt Publishing Ltd, Pal. Deep learning with Keras (2017)

38. Neave, H.R., Worthington, P.L.: Distribution-Free Tests. Unwin Hyman, (1988)

Optimization Design of Photovoltaic Power Generation System Under Complex Lighting Conditions

Wenke Lu[1], Huibin Han[1], Sining Hu[2($\boxtimes$)], and Jiachuan Shi[2]

[1] Design Institute of Jinan, China Railway Engineering Design Consulting Group Co., Ltd., Jinan, China
[2] School of Information and Electrical Engineering, Shandong Jianzhu University, Jianzhu, China
2606061116@qq.com

Abstract. Modeling and analyzing the electrical output characteristics of photovoltaic arrays under complex lighting conditions, and conducting research on the optimization design scheme of photovoltaic arrays and photovoltaic electrical systems. Modeling and analyzing the electrical output characteristics of photovoltaic arrays under complex lighting conditions, and conducting research on the optimization design scheme of photovoltaic arrays and photovoltaic electrical systems. Local shadow issues in photovoltaic modules can lead to uneven power generation, severe power reduction, and even hot spot effects. TCT configuration and micro inversion can effectively solve such problems, reducing the impact of shadow occlusion on the power generation efficiency of photovoltaic systems. Simulated the effectiveness of TCT configuration and confirmed the effectiveness of the TCT method. Through the discussion of TCT configuration and micro inversion, it is concluded that selecting a suitable solution based on the actual situation can significantly optimize the power generation efficiency of distributed photovoltaic power generation systems in complex lighting environments.

Keywords: PV array · complex lighting · mathematical modeling · output characteristic · TCT configuration · micro inversion

1 Introduction

Against the backdrop of the national "carbon peaking and carbon neutrality goals", the power system is transitioning towards low-carbon development, and new energy sources will inevitably gradually replace fossil fuels as the primary energy source in the future [1]. Solar energy, as the most promising new energy source in the future, has been widely used in photovoltaic power generation technology [2]. Photovoltaic Building Integration (BIPV), as a new technology that perfectly combines modern buildings with photovoltaic systems, is gradually becoming a research hotspot in the field of photovoltaic applications, and is also the main way to widely promote green buildings to achieve the "carbon peaking and carbon neutrality goals" [3].

© The Author(s), under exclusive license to Springer Nature Singapore Pte Ltd. 2023
H. Zhang et al. (Eds.): NCAA 2023, CCIS 1869, pp. 249–260, 2023.
https://doi.org/10.1007/978-981-99-5844-3_18

Through modeling and analyzing the output characteristic curve of photovoltaic cells, it is found that traditional photovoltaic arrays have only one peak in the power output curve under uniform light conditions, which is the maximum power point [2]. But now, due to the aesthetic requirements of BIPV buildings, the design conditions of photovoltaic modules are becoming more complex, which can also have a certain degree of impact on the modeling, analysis, design, and other work of photovoltaic arrays. If all photovoltaic modules are no longer designed with a single orientation, resulting in uneven illumination of the photovoltaic array, it will to some extent affect power generation efficiency; The local shadow problem of photovoltaic modules can lead to uneven power generation, severe power reduction, and even hot spot effects; If photovoltaic modules are connected in series under complex light conditions, the output curve of the photovoltaic array will change and the output power will also decrease [2]. Therefore, traditional mathematical models of photovoltaic arrays are no longer able to meet the current development needs.

This article models and simulates the S configuration and TCT configuration of photovoltaic arrays under uniform illumination and complex lighting conditions, respectively, and analyzes and studies the electrical output characteristic curves. Propose to distinguish and identify different areas of photovoltaic arrays under uniform light and shadow conditions, and adopt different methods according to actual situations. For economic benefits, traditional S-configuration method is adopted for photovoltaic arrays under uniform light, while efficiency is considered. For areas affected by shadows, micro inverters and TCT are combined to coordinate and cooperate, Ultimately, achieving the best efficiency of the entire project.

2 Photovoltaic Cell Modeling and Analysis

2.1 Mathematical Model of Photovoltaic Cell

The three common models for photovoltaic cells are: Rs model based on a single diode, Rs - Rp model, and D - diode model based on two diodes [4].

The single diode photovoltaic cell model is one of the simplest and most common mathematical models for photovoltaic cells. It models a photovoltaic cell as a diode and considers parameters such as photogenerated current, series resistance, and parallel resistance. The equivalent circuit model of a single diode photovoltaic cell is established as shown in Fig. 1.

In Fig. 1, the I_{ph}, denotes the current of the DC current source for simulating the photogenerated current, A; the diode is the equivalent P-N junction inside the cell, and I_d is the diode bypass current, A; the R_s and R_{sh} denote the equivalent series and shunt resistances inside the PV cell, Ω; I_{sh} is the shunt resistor bypass current, A; I is the PV cell output current, A.

According to Kirchhoff's current law, the expression of PV cell output current is:

$$I = I_{ph} - I_d - I_{sh} \tag{1}$$

The expression of photogenerated current I_{ph} is:

$$Iph = [Isc + (Ki * (T - 298))] * \frac{G}{1000} \tag{2}$$

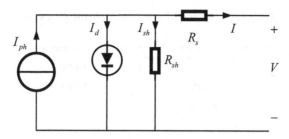

Fig. 1. Equivalent Circuit Model of Single Diode Photovoltaic Cells.

The mathematical expression of diode bypass current I_d is:

$$I_d = I_0 * \left[e^{\left((v+IRs) * \frac{q}{nKTNS} \right)} - 1 \right] \tag{3}$$

where I_0 is the saturation current of the diode, and the expression is:

$$I_0 = I_{rs} * \left(\frac{T}{T_n} \right)^3 * e^{\frac{q*Eg0*\left(\frac{1}{Tn} - \frac{1}{T} \right)}{nk}} \tag{4}$$

I_{rs} is the reverse saturation current of the diode, and the expression is:

$$I_{RS} = \frac{I_{sc}}{e^{\left(q * \frac{Voc}{nKTNs} \right)} - 1} \tag{5}$$

The expression of shunt resistor bypass current I_{sh} is:

$$I_{sh} = \frac{(V + IR_s)}{R_{sh}} \tag{6}$$

In Eqs. (1) to (6), the K_i denotes the current temperature coefficient (25 °C); G denotes solar radiation energy, W/m²; and T denotes the operating temperature, K; T_n denotes the nominal temperature, K; q denotes the electronic charge, C; V_{oc} indicates open-circuit voltage, V; I_{sc} denotes short-circuit current, A; n denotes the ideal factor of the diode; K denotes the Boltzmann factor, J/K; $Eg0$ denotes the band gap energy of the semiconductor, J; N_s denotes the number of series-connected cells.

2.2 Selection of Photovoltaic Cells

This paper uses Matlab/Simulink software to build a simulation model based on the formulas (1)–(6), and simulates and analyzes the output characteristics of photovoltaic cells under different light intensities.

For the comprehensive consideration of economy and characteristics, Shanghai Aerospace Automotive Electrical HT60-156P-235, a domestically produced photovoltaic cell, is selected in Simulink to study the output characteristics under different lighting conditions. The parameters of this type of photovoltaic cell are shown in Table 1.

Table 1. Parameters of single diode photovoltaic cells.

Name	Parameter	Name	Parameter
P_{max}	235.081W	V_{mp}	30.1V
V_{oc}	37.4V	I_{mp}	7.81A
I_{sc}	8.42A		

2.3 Output Characteristics of Photovoltaic Cells

The output characteristics of this type of photovoltaic cell are simulated using Matlab/Simulink. The I-V and P-V curves, namely the volt ampere characteristic curve and power voltage characteristic curve, are shown in Fig. 2. All the three curves are obtained at a standard temperature of 25 °C, with light intensity G of 1000 W/m², 700 W/m², and 400 W /m², respectively.

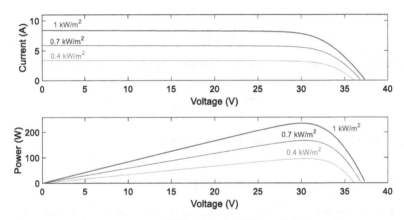

Fig. 2. Output curve of PV cell under different light intensity.

From the I-V and P-V characteristic curves in the figure, it can be seen that the output of the photovoltaic array at this time has only one peak, with a "single peak" output characteristic.

From the I-V curve, it can be seen that with the increase of light intensity, the short-circuit current of photovoltaic cells will show a significant increasing trend. This is due to the increase in light intensity, which leads to more thorough absorption of light energy and further conversion into electrical energy, resulting in higher current output. Therefore, with the improvement of lighting conditions, photovoltaic cells can achieve higher current output and provide greater power production capacity.In addition to the aforementioned effects, the open circuit voltage of photovoltaic cells will also slightly increase. This phenomenon is due to the increase in light intensity leading to an increase in photo generated charge carriers, which in turn increases the voltage output.

According to the P-V curve, as the light intensity increases, the output power of the photovoltaic cell will also increase. This is because the output power is the product of the output current and output voltage. When the light intensity increases, the output current and output voltage of the photovoltaic cell will both increase, leading to an increase in output power; The corresponding maximum power point will also increase. However, under different light intensities, there is almost no difference in the voltage corresponding to the maximum power point. Under each light intensity, the photovoltaic array has only one peak point, which is the maximum power point. The traditional MPPT algorithm can track the maximum power point [5].

In summary, light intensity is one of the main factors affecting the electrical output characteristics of photovoltaic arrays, and has a wide and important impact in photovoltaic systems. The change in light intensity directly affects the current and voltage output of photovoltaic cells, and thus has a significant impact on the power output and energy conversion efficiency of photovoltaic arrays. Therefore, in the design and optimization of photovoltaic systems, it is necessary to fully consider the non-uniformity of light intensity in order to develop reasonable strategies to optimize the power output and energy conversion efficiency of photovoltaic arrays.

3   Simulation of Photovoltaic Array Modeling Under Complex Light Intensity

The common topological structures of photovoltaic arrays mainly include: series (S) photovoltaic array configuration, series parallel (SP) photovoltaic array configuration, cross connected (TCT) photovoltaic array configuration, etc. [6].

In reference [6], the author concluded through research on the series configuration of photovoltaic arrays that the S configuration photovoltaic array will exhibit a multi peak state under complex light conditions, and the maximum output power is significantly lower than the maximum output power under uniform light conditions. In reference [7], the author concluded through research on photovoltaic arrays configured with TCT that the TCT configuration method of photovoltaic cells can effectively overcome the multi peak phenomenon caused by partial shadows and the impact of power reduction issues.

This paper will compare the photovoltaic arrays of S configuration and TCT configuration, and use Matlab/Simulink software for modeling and simulation analysis. Four photovoltaic modules will be used for simulation analysis and comparison of output characteristic curves under uniform lighting conditions and shadow occlusion lighting conditions, respectively. Among them, the uniform lighting conditions are: all four photovoltaic panels are 1000 W/m^2; The lighting conditions with shadow occlusion are: 1000 W/m^2, 800 W/m^2, 600 W/m^2, and 400 W/m^2.

3.1   Simulation Analysis of Output Curve of S-configuration Photovoltaic Array

The S configuration photovoltaic array simulation model built using Matlab/Simulink software is shown in Fig. 3.

Under uniform lighting conditions (all four photovoltaic panels are 1000 W/m^2), the output characteristic curve of the photovoltaic array is shown in "Scenario 1"; Under complex light conditions (1000 W/m^2, 800 W/m^2, 600 W/m^2, 400 W/m^2), the output characteristic curve of the photovoltaic array is shown in "Scenario 2". The I-V and P-V output characteristic curves of the S-configured photovoltaic array in two scenarios are shown in Fig. 4.

From the comparison between "Scenario 1" and "Scenario 2", it can be seen that when the S configuration photovoltaic array is subjected to complex lighting conditions, the electrical output characteristic curve of the photovoltaic array under uniform lighting conditions will undergo significant changes. Not only does the output power decrease and the maximum power point decrease, but it also produces multi peak phenomena. This is very detrimental to the practical application of photovoltaic technology.

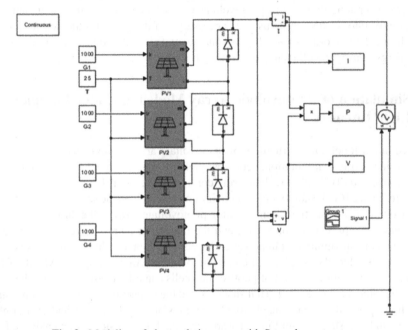

Fig. 3. Modeling of photovoltaic arrays with S-topology structure.

3.2 Mathematical Model of TCT Photovoltaic Configuration

In the TCT configuration of photovoltaic cells, each cell has two output paths, which can be roughly understood as a first in parallel and then in series connection method, and the required voltage can be obtained through this method. The TCT configuration of photovoltaic cells can effectively overcome the influence of partial shadows in the SP connected photovoltaic array configuration. 2 × 2 TCT photovoltaic configuration is shown in Fig. 5.

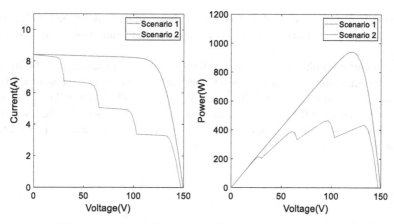

Fig. 4. I-V and P-V output characteristic curves of photovoltaic array under S configuration.

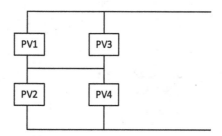

Fig. 5. TCT structure topology diagram.

The mathematical model of TCT photovoltaic array configuration is shown in formula (7).

$$\begin{cases} I_{pv} = I_i + I_{i+2}, i = 1, 2 \\ V_{Ri} = V_i = V_{i+2}, i = 1, 2 \\ V_{pv} = V_{R1} + V_{R2} = \sum_{i=1}^{2} V_{Ri} \\ P_{PV} = I_{pv} \times V_{PV} \end{cases} \tag{7}$$

In the equation: $I_1, I_2 \ldots, I_4$ are the output current of each photovoltaic module, A; $V_1, V_2 \ldots, V_4$ are the output voltage of each photovoltaic module, V; I_{PV}, V_{PV}, P_{PV} are the total output current, total output voltage, and total output power of the photovoltaic array.

3.3 Simulation Analysis of TCT Configuration Photovoltaic Array Output Curve

This section conducts simulation research on the mathematical model of TCT configuration photovoltaic array. Taking four sets of photovoltaic modules as an example, changing the illumination on the photovoltaic array at a standard temperature of 25 °C, using different light receiving forms, using Matlab/Simulink software to simulate and design the TCT configuration of the four sets of photovoltaic modules, and studying their electrical output characteristic curves. The Simulink simulation model built is shown in Fig. 6.

Under uniform lighting conditions (all four photovoltaic panels are 1000 W/m^2), the output characteristic curve of the photovoltaic array is shown in "Scenario 1"; Under complex light conditions (1000 W/m^2, 800 W/m^2, 600 W/m^2, 400 W/m^2), the output characteristic curve of the photovoltaic array is shown in "Scenario 2". The I-V and P-V output characteristic curves of the TCT configured photovoltaic array in two scenarios are shown in Fig. 7.

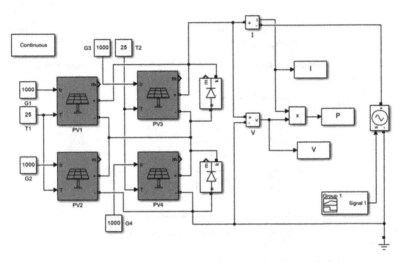

Fig. 6. Modeling of TCT Topological Structure for Photovoltaic Array.

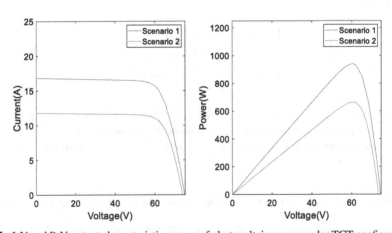

Fig. 7. I-V and P-V output characteristic curves of photovoltaic arrays under TCT configuration.

From the comparison between "Scenario 1" and "Scenario 2", it can be seen that when the TCT configuration photovoltaic array receives uneven light, its power decreases slightly, and its impact is negligible. Although the multi peak phenomenon still exists, the S configuration of the photovoltaic array has been significantly improved. The electrical output characteristics of the TCT configuration of the photovoltaic array greatly improve the performance of the photovoltaic power generation system, which is an optimization of the S configuration of the photovoltaic array.

4 Application of Micro Inverters in Solving Photovoltaic Arrays Under Complex Light Conditions

Although the TCT configuration of photovoltaic arrays can greatly alleviate the problems of power reduction and multi peak phenomenon under complex light conditions, there are still many shortcomings and shortcomings in practical applications. For example, efficiency loss, due to the presence of diodes and other components in the TCT configuration, additional connection lines and electrical contacts are used, which introduces losses and reduces the overall efficiency of the system; The cost increases, as TCT configuration requires more components and connecting lines, which not only increases system costs but also maintenance costs; Reliability issues include the additional components and connecting lines involved in TCT configuration, which increases the complexity of the system and increases the risk of faults and failures, thereby affecting the overall performance of the photovoltaic array and making maintenance more difficult.

Using micro inverters for photovoltaic grid connection is another effective method to solve the performance degradation of photovoltaic arrays under complex lighting conditions. The most economically valuable core equipment of the entire solar photovoltaic grid connected power generation system is the inverter. With the rapid development of photovoltaic power generation technology, diversified inverters have also emerged. The function of solar photovoltaic cells is to convert light energy into direct current, while the function of inverters is to convert direct current into 50 Hz, 220 V sine wave alternating current and integrate it into the power grid [8].

With the commercialization of building integrated photovoltaic systems (BIPV), a micro photovoltaic grid connected inverter (Micro inverter) has emerged. Micro photovoltaic grid connected inverters are small photovoltaic grid connected inverters installed on the back of each solar panel. Compared to traditional centralized or string photovoltaic grid connected inverters, micro photovoltaic inverters have many advantages: they will not affect the efficiency of the entire system due to the efficiency of a single solar photovoltaic cell plate, can naturally achieve component level MPPT, are simple and flexible to install and use, and are easy to adjust in scale The system has a compact structure and high reliability.

Overall, the introduction of micro inversion provides an innovative solution to the problems faced by photovoltaic arrays such as shadow occlusion and uneven lighting. Compared with traditional centralized inverters, using micro inverters enables us to achieve precise control of each photovoltaic cell, maximizing the energy conversion efficiency and power output of the system. This novel approach brings broader prospects for the application and development of photovoltaic energy, and promotes the continuous innovation and application of photovoltaic technology in the field of renewable energy.

5   Configuration Scheme of Photovoltaic Array

The complex lighting conditions of photovoltaic arrays mainly consider the impact of shadow occlusion, which can seriously affect the efficiency of the entire photovoltaic array. Therefore, it is necessary to clearly divide the uniform lighting area and shadow occlusion area, and choose appropriate methods for different areas in different situations to solve the problem of reduced efficiency of photovoltaic arrays under complex lighting conditions.

5.1   Uniform Light Receiving Area

For areas with uniform light reception, due to the absence of obstruction, there will be no hot spot effect due to uneven light reception, and the output characteristics will not exhibit multi peak phenomenon.

Therefore, there is no need to use the TCT configuration of photovoltaic arrays or use micro inverters to connect to the power grid, but instead use the traditional series topology of photovoltaic arrays.

5.2   Unevenly Illuminated Areas

When the photovoltaic array receives uneven light, there will be a problem of multiple peaks for traditional S-configuration photovoltaic arrays, and the power will be significantly reduced, seriously affecting the performance of photovoltaic cells. Therefore, under uneven lighting conditions, the traditional S configuration method is no longer applicable.

When there is shadow occlusion in the photovoltaic array, two methods can be chosen: using the TCT configuration of the photovoltaic array and using a micro inverter to connect to the power grid for the shaded part. The configuration of micro reverse and TCT needs to be analyzed and evaluated based on specific circumstances.

Based on the Scope and Degree of Shadow Occlusion. If the shadow only affects a small part of the photovoltaic array, choosing a micro inverter is more appropriate because the TCT configuration requires a centralized inverter. However, if using a micro inverter, only micro inverters need to be installed separately for the obstructed photovoltaic modules, which is more flexible and convenient, and can maximize the power generation efficiency in the shadow area. If the shadow range is large, choosing the TCT configuration of the photovoltaic array is more suitable because TCT configuration is a topology connection method for photovoltaic modules, where multiple modules are connected together and a centralized inverter is used to connect to the power grid, which is more suitable for dealing with situations of large-scale shadow occlusion. On the contrary, in this case, using micro inversion to connect to the power grid may cause cumbersome and high-cost problems.

Based on Space and Cost. If there is more reserved space and sufficient funds, it is more appropriate to choose micro reverse access to achieve better results, because if micro reverse access is used, each photovoltaic module needs to be connected to micro reverse, and the cost is higher. Due to limited reserved space and insufficient funding, choosing a TCT configuration for photovoltaic arrays is more appropriate to ensure effectiveness, as the TCT topology only requires one inverter to handle energy conversion. Therefore, for the number of photovoltaic modules, the TCT configuration takes up less space and has a relatively low cost.

However, it should be noted that in practical engineering, the specific combination and selection methods of TCT configuration and micro inversion need to be analyzed on a case-by-case basis, based on practical applications, taking into account various factors such as technical effectiveness and economic benefits, and ultimately making a relatively optimal choice. Proposal of a new scheme for complex light exposure problems of PV arrays.

6 Conclusion

This paper aims to analyze the optimization design of photovoltaic power generation system under complex lighting conditions. The study involves the utilization of Matlab/Simulink for modeling and simulation, followed by an analysis of the electrical output characteristics of photovoltaic arrays configured in both S and TCT configurations. The analysis considers the impact of uniform illumination as well as complex lighting conditions, allowing for a comprehensive understanding of the array's behavior under different scenarios.

In order to address the issue of uneven light reception, this article proposes two potential solutions: the utilization of TCT configuration and the application of micro inverters. By implementing TCT configuration, the shadow location can be accurately determined, and based on the specific circumstances, different plans can be chosen or even combined to achieve optimal results. This approach takes into account various factors and aims to maximize the overall benefits of the entire project.

Nevertheless, it is worth noting that practical applications present challenges when it comes to managing the relationship between micro inverters and centralized inverters. Finding an effective and efficient way to integrate these two components remains an ongoing challenge that requires further in-depth research and exploration by the scientific community. The exploration of this topic is crucial for the successful implementation of photovoltaic systems and the optimization of their performance in real-world settings.

In summary, this study contributes to the understanding of the electrical output characteristics of photovoltaic cells and arrays, while also proposing potential solutions to address the issue of uneven light reception. By considering factors such as shadow location and utilizing appropriate configurations, researchers aim to achieve the best comprehensive benefits for photovoltaic projects. However, the integration of micro inverters and centralized inverters poses a significant challenge, highlighting the need for continued research in this area to overcome this obstacle and further improve the practical applications of photovoltaic systems.

References

1. Ding, H., Zhang, B., Tang, W., et al.: Evaluation of distributed photovoltaic consumption capacity of distribution station area considering source-network-load-storage collaboration. Distrib. Utilization **40**(3), 2–8 (2023)

2. Li, G., Han, H., Fu, R., et al.: Modeling and analysis of photovoltaic array output characteristic sunder complex light conditions. Comput. Era **7**, 21–25 (2022)
3. Li, C., Li, G., Wang, Q., et al.: Simulation and optimization design of BIPV system of a regional railway station based on PVsyst. J. Nation Vocat. Univ. **36**(2), 95–100 (2022)
4. Wei, C., Shi, H., Xu, W., et al.: Modeling and research of photovoltaic array based on single diode model. Electron. Des. Eng. **25**(15), 141–144 (2017)
5. Hua, Y., Zhu, W., Guo, Q., et al.: Summary of MPPT algorithm research in photovoltaic power generation systems. Chin. J. Power Sour. **44**(12), 1855–1858 (2020)
6. Li, G.: Research on Optimal Design Method of Building Photovoltaic Power Generation System. Shandong Jianzhu University (2022)
7. Yang, Z.: Static reconfiguration strategy of photovoltaic arrays under partial shadow conditions. Shenyang Agricultural University (2022)
8. Chen, J.: Research on photovoltaic micro inverter. Anhui University of Science and Technology (2019)

Analysis of the Impact of Regional Customer Charging on the Grid Under the Aggregator Model

Chongyi Tian[1,2], Yubing Liu[1,2], Xiangshun Kong[1,2], and Bo Peng[1,2(✉)]

[1] School of Information and Electrical Engineering, Shandong University of Construction, Jinan 250101, China
631864503@qq.com
[2] ShandongKey: Laboratory of Intelligent Building Technology, Jinan 250101, China

Abstract. This paper studies the impact of regional user charging on the grid under the aggregator model. Firstly, the EV charging station is taken as the aggregator, the participants and the form of cooperation in this scenario are clarified, a three-party cooperation method between the aggregator, the user and the grid is designed, the travel characteristics of users in the region are analyzed, the daily residual SOC of EVs and the user trustworthiness are taken as the evaluation indexes of the dispatchable potential of users, and the aggregator-operator optimal dispatch model is established. The aggregator's revenue maximization is used as the objective function, and the aggregation of EV users in the region is carried out by the Monte Carlo method. It is verified that the aggregator's model can calm the grid load by guiding users to charge in an orderly manner, and at the same time can satisfy users' charging demand and mobilize their enthusiasm.

Keywords: Aggregator · Electric Vehicles · Optimal Dispatch · Monte Carlo

1 First Section

With the continuous growth of electric vehicle (EV) ownership, the EV load is highly stochastic, and the disorderly charging of a large number of EVs will have an impact on grid security. EVs have their unique characteristics of mobile energy storage, which can interact with the grid in both directions, and at the same time, EVs can smooth out the load curve for the grid through their mobile energy storage nature, and participate in grid auxiliary services through the aggregator's aggregation.

The impact on the grid can be reduced by guiding EV users through aggregators for orderly charging. The literature [1] proposes a kind of mixed integer linear programming (MILP) model to establish a model predictive energy management system for EVs. The literature [2] developed a divisional planning model by analysing two real distribution areas, showing that the losses to the grid from the scale of EVs differ under different charging strategies. In [3], a decentralised scheduling algorithm for EV charging is proposed, and a multi-intelligent system (MAS) architecture is constructed in which the aggregator is responsible for designing an appropriate virtual pricing strategy

H. Zhang et al. (Eds.): NCAA 2023, CCIS 1869, pp. 261–272, 2023.
https://doi.org/10.1007/978-981-99-5844-3_19

based on accurate electricity demand and generation forecasts to maximise the aggregator's profit. A distributed EV charging control method is proposed in the literature [4]. The convergence of the method when the EV control factors are decoupled and weakly coupled is discussed and the efficiency of the proposed management system is evaluated. The above literature only proposes corresponding control strategies for one role of EV alone, without considering it from the aggregator perspective. The literature [5] proposes a new optimisation framework based on a multiplier alternating direction approach based on the aggregator perspective, which is applied to the relevant aggregation objective to demonstrate its performance and generality, forming a linear relationship with the number of electric vehicles to achieve an optimisation criterion. The literature [6] proposes a new handshake approach, along with a novel distributed consensus-based power recovery method for EV user uncertainty, which uses the average power contribution of the rest of the CS to withstand the handshake in case of any CS contingency failure, verifying the effectiveness of the proposed control scheme in different scenarios. The literature [7] proposes an aggregator profit maximisation algorithm to simulate 10,000 electric vehicles in a region and verifies that the algorithm improves aggregator profits while reducing system load impact and user costs. In [8], a decentralised scheduling algorithm for EV charging is proposed, and a charging control model following the architecture of a multi-intelligent system (MAS) is developed. The aggregator is responsible for designing a pricing strategy based on accurate electricity demand and generation forecasts to maximise the aggregator's profit. The literature [9] proposes a modular approach to control algorithms in the aggregator model, exploring the advantages of separating the aggregator control algorithm in a V2G network into several modules that access a mutual database. The use of parked electric vehicles as distributed energy storage to support the grid is described. The literature [10] proposes a new multi-stage optimisation-based approach for routing and charging each individual electric vehicle, taking into account battery degradation, acceleration and speed-dependent power consumption, delivery delay penalties, tolls, fixed charging prices and incentives for available time. The literature [11] proposes a new business model framework for the design and development of viable business models for aggregators, additional flexibility for system and network operators, and lower charging costs for electric vehicle users. The literature [12] proposes a data-driven approach to optimise the existing charging station network by eliminating redundant charging stations and identifying congested areas of charging stations within the original charging network to provide suggestions for further solutions to electric vehicle charging.

Previously, EV participation in grid regulation was often disorderly participation or cluster orderly participation, without incorporating the role of aggregator. This paper uses EV charging stations as EV aggregators, specifies the participants and the form of cooperation among them, clarifies their respective requirements and obligations, combines daily residual SOC and user trustworthiness to establish user dispatchable potential indicators, establishes an aggregator collaborative user dispatch model, and takes The aggregator's revenue is maximised as the objective, and the user's load state and the peak-to-valley difference of the grid are taken as constraints. The results show that the introduction of the aggregator role can smooth out the load on the grid, and also enable customers to obtain some of the benefits, ultimately achieving a tripartite profit.

2   Operation Model

2.1   Participant

The EV aggregator acts as an intermediary between EV users and the grid, and can communicate with the grid on behalf of users to participate in demand response. There are three parties involved in the whole operation model, namely the user, the grid and the EV aggregator.

For EV users, the price of charging is an important factor in whether to participate in regulation. For the EV aggregator, the aggregation of EV user demand response resources and participation in grid peaking and other services maximises its own revenue. For the grid, through the EV aggregator's dispatch, it can provide feedback on tariff and compensation information to the EV aggregator, and through the aggregator's regulation, it can achieve the effect of load smoothing. The participant tasks and benefits are shown in Table 1.

Table 1. Tasks and benefits for each participant

Participants	Aggregators	User	Electricity network
Mission	1. Aggregation of users 2. Control of vehicle charging and discharging	Acceptance of aggregator dispatch	Provide aggregators with compensation information such as electricity prices
Revenue	1. Customer charging costs 2. Grid ancillary services revenue	Participation in regulated charging at a price lower than the normal charging price	Smoothing of peak-to-valley load differentials

2.2   Form of Cooperation

In this scenario, the EV charging station is used as an EV aggregator. When an EV user accesses the charging pile, the system will first judge whether the user's vehicle meets the requirements for participation in regulation. If you choose "participate in regulation" the charging price will be lower than the normal charging price, if you choose "not participate in regulation" the charging will be done at the normal price, at the same time the system will categorise the users who meet the conditions to participate in regulation into different controllable vehicles according to their vehicle parameter information. The user can also choose the charging mode, which includes fast charging and slow charging, i.e. there are four charging options for the user: fast charging for those who meet the conditions for participation in regulation, slow charging for those who meet the conditions for participation in regulation, fast charging for those who do not participate in regulation and slow charging for those who do not participate in regulation.

If the user finishes charging early within the specified participation period, the charging price will be settled at the normal price. If the EV aggregator does not meet the user's charging needs during the user's participation period, the EV aggregator will compensate the user at the normal price.

The electric vehicle aggregator can participate in auxiliary services such as peaking on the grid while regulating, and can provide feedback to the grid on the regulation curve of electricity consumption characteristics, while the grid can purchase electricity from the electric vehicle aggregator, and can also provide compensation information such as electricity prices to the electric vehicle aggregator. The EV aggregator will receive peak shaving and demand response compensation through the grid. The form of cooperation is shown in Fig. 1.

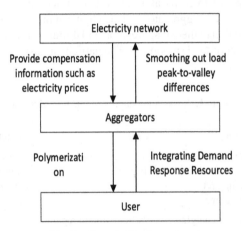

Fig. 1. Cooperation form chart

3   User Travel Characteristics and User Dispatchable Potential Indicators in the Region

3.1   User Travel Characteristics in the Region

Distribution of daily mileage:

The daily mileage starts from a log-normal distribution, and the probability density function is Eq. (1).

$$f_n(x) = \frac{1}{x\sigma_n\sqrt{2\pi}} \exp\left[-\frac{(\ln-\mu_h)^2}{2\sigma_n^2}\right] \tag{1}$$

where x is the mileage travelled, μ_h is the mean value of daily mileage travelled by tram users, and σ_n is the standard deviation, where μ_h is 3.2 and σ_n is 0.88.

The probability distribution of daily trip start mileage can be derived as shown in Fig. 2.

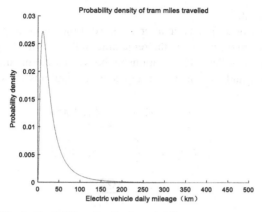

Fig. 2. Probability distribution of daily trip start mileage

The daily trip end time distribution follows a normal distribution, and the probability density function is Eq. (2).

$$f_t(x) = \begin{cases} \frac{1}{\sigma_s\sqrt{2\pi}}\exp\left(-\frac{(x-\mu_s)^2}{2\sigma_s^2}\right) & \mu_s - 12 < x < 24 \\ \frac{1}{\sigma_s\sqrt{2\pi}}\exp\left(-\frac{(x+24-\mu_s)^2}{2\sigma_s^2}\right) & 0 < x < \mu_s - 12 \end{cases} \tag{2}$$

where the mean value $\mu_s = 17.5$ and $\sigma_s = 3.2$.

The daily trip start mileage probability distribution can be derived as shown in Fig. 3.

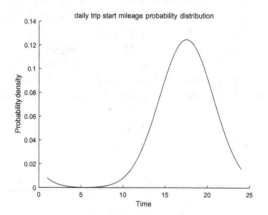

Fig. 3. Daily trip start mileage probability distribution

3.2 User Dispatchable Potential Indicators

In this paper, daily residual SOC and user trustworthiness are used as indicators of user dispatchable potential.

(1) Daily residual SOC

The majority of users charge after work, so if they want to participate in regulation, they need to discriminate the remaining SOC, and the daily remaining SOC indicator is proposed here. The relationship between the remaining daily SOC and the daily mileage and the time to participate in regulation is shown by Eqs. (3)-(5).

$$Q2 = [Q1 - \frac{S}{F_{max}}] \times 100\% \tag{3}$$

$$F_{max} = \frac{C}{\beta} - \frac{C}{\beta} \times 10\% \tag{4}$$

$$Tc = \frac{Q2}{\eta \cdot P_{ev}} \tag{5}$$

where: Q_2 is the remaining SOC of the EV for one day, Q_1 is the starting SOC for one day, S is the total starting mileage for one day, F_{max} is the maximum starting kilometres for one day, C is the battery capacity, β is the power consumption of the EV per kilometre, here it is taken as 0.1 kwh/km, here 10% of the mileage is taken as the mileage to the charging station and the maximum driving mileage is not taken into account for the time being, T_c is the remaining SOC that can participate in regulation time, η is the EV charging efficiency, and P_{ev} is the EV charging power. From Eq. (5), it can be seen that the larger the number of remaining SOCs on a day, the longer the time the user can participate in regulation.

(2) User trustworthiness

Due to the uncertainty of user charging, some users may terminate charging in the process of charging, which may also lead to the termination of participation in regulation, for the users themselves, their charging price will be charged according to the normal price, but for the EV aggregator, it may affect the whole regulation process, the EV aggregator can query the past user charging information through the station scheduling. Assume that a user has charged N times, T_{start} is the charging start time, T_{end} is the charging end time and T_{out} is the actual charging end time. The user trustworthiness index is judged by the process of Eqs. (6)–(8).

$$K_1 = \frac{\sum_{i=1}^{N} (T_{end} - T_{start})}{N} \tag{6}$$

$$K_2 = \frac{\sum_{i=1}^{N} (T_{out} - T_{start})}{N} \tag{7}$$

$$M = \frac{K_2}{K_1} \tag{8}$$

where K_1 is the due trustworthiness, K_2 is the actual trustworthiness and M is the trustworthiness ratio, the higher the M value the higher the trustworthiness of the user. The higher the M value, the higher the trustworthiness of the user. In extreme cases, the trustworthiness of the visible user is zero.

4 Optimal Scheduling Model for Aggregator Collaborative Users

4.1 Objective Function

This paper is mainly from the aggregator's perspective. For the aggregator, the objective is to maximize the overall revenue and the objective function is as follows:

$$f_{\max} = f_d + f_h \tag{9}$$

where f_d is the charging fee collected from users and f_h is the revenue from participation in grid auxiliary services.

The first part of f_d includes the collection of user charging fees, user service fees, user parking fees and the aggregator's compensation costs to users:

$$f_d = \sum_{i=1}^{m} Q_i^{out} \alpha_i^c + \sum_{i=1}^{m} \gamma_i^w + \sum_{i=1}^{m} \lambda_i^h - \sum_{i=1}^{m} \beta_i^{bu} \tag{10}$$

where: i is the electric vehicle number, assuming an m electric vehicles participate in the regulation, Q_i^{out} is the actual charged electricity for a day, a_i^c is the charging price for the day, γ_i^w is the user service fee charged for m vehicles in a day, λ_i^h is the parking fee charged for m vehicles in a day, and β_i^{bu} is the compensation amount paid to the user by the aggregator.

$$\beta_i^{bu} = \varphi^{bu} \cdot Q_i^{out} \tag{11}$$

where: φ^{bu} is the compensation tariff paid by the aggregator to the user for charging.

The second component f_h represents the aggregator's participation in grid auxiliary service revenue, including demand response compensation and peak shaving revenue.

$$f_h = \sum_{i=1}^{m} P_i^{out} \cdot \delta + \sum_{i=1}^{m} X_i^f \cdot \alpha_f + \sum_{i=1}^{m} X_i^g \cdot \alpha_g \tag{12}$$

where: P_i^{out} is the actual response of m electric vehicles after participation in regulation, is the service charge coefficient, here taken as 0.8, X_i^f is the peak load cut, αf is the "peak" tariff, X_i^g is the increased valley load, α_g is the "valley" tariff.

4.2 Constraints

For aggregators:

Constraint on the number of EVs that can be aggregated at one time by an EV aggregator.

$$N_{act} \leq N_{max} \tag{13}$$

as the EV charging stations are set as EV aggregators in this paper, N_{act} is the number of EVs actually charged and participating in regulation, and N_{max} is the maximum number of EVs allowed to participate in regulation by EV aggregators.

For EV users:

EV charging charge state constraint:

$$SOC_{min} \leq SOC_i \leq SOC_{max} \tag{14}$$

EV daily residual SOC constraint:

$$Q_{min} \leq Q_2 \leq Q_{max} \tag{15}$$

where, SOC_{min} is the minimum power value to meet the user's demand when the user is charging, SOC_i is the power value to meet the user's demand and SOC_{max} is the maximum power value to meet the user's demand when the user is charging. Q_{min} is the lowest SOC state allowed to participate in regulation and Q_{max} is the highest SOC state allowed to participate in regulation.

For the grid:

Peak-to-valley differential constraint:

$$X_g^{min} \leq X_g \leq X_g^{max} \tag{16}$$

$$X_f^{min} \leq X_f \leq X_f^{max} \tag{17}$$

where, X_g^{min} is the minimum valley load allowed for the grid after participating in regulation, X_g^{max} is the maximum valley load allowed for the grid after participating in regulation, X_g^{min} is the minimum peak load allowed for the grid after participating in regulation, X_g^{max} is the maximum valley load allowed for the grid after participating in regulation.

5 Example Analysis

In order to verify the impact of regional user charging on the grid under the aggregator model, a certain park is selected as the research scenario. Assuming that the aggregator and the user reach a cooperation agreement in the region, the actual number of EVs that can participate in the dispatch will be greatly reduced by screening the EVs in the selected region according to the EV user dispatchable potential index, and 1000 EVs are selected for simulation using the Monte Carlo method in this paper. The impact on the grid is analysed. During the simulation, the charging and discharging power is constant and the car is charged and discharged according to its rated charging and discharging power. The electric vehicle charging and discharging time-of-use tariffs are shown in the table. As can be seen from Table 2, the price of charging an electric vehicle at night is much less than during the day, while discharging can only be profitable during peak load hours on the grid. In order to better guide users to discharge, the discharge price is generally set higher than the charging price. In the course of life, users are motivated to charge during low hours and discharge during peak hours so as to gain revenue. So in order to motivate users, this paper sets the discharge price lower than the charging price.

It is assumed that the electricity purchased by the aggregator is at a time-sharing tariff: 7:00–22:00 is the peak hour tariff, RMB 1.20/(kW·h); 22:00–7:00 is the trough

Table 2. Electric vehicle charging and discharging prices

Projects	Charging		Discharge	
	Valley hours	Peak hours	Valley hours	Peak hours
Time period	21:00–08:00	08:00–21:00	\	12: 00–20: 00
Electricity prices/yuan	0.45	1.20	\	1.10

hour tariff, RMB 0.45/(kW·h); the compensation tariff for customers participating in the regulation is RMB 0.5/(kW·h); the day is divided into two main controllable peak hours: after work in the morning and before work in the afternoon (8:00–17:00); after work in the evening and before work in the next morning (18:00–7:00). (8:00–17:00); after returning home from work in the evening to the next morning before work (18:00–7:00). The process is described below.

Firstly, the total load of 1000 EVs in the region is aggregated using the Monte Carlo method, and the aggregation process is shown in Fig. 4.

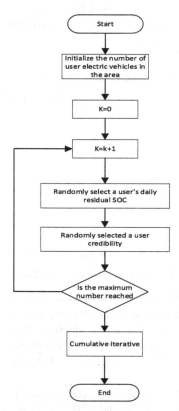

Fig. 4. Aggregation process using Monte Carlo method

The total load of 1000 electric vehicles is shown in Fig. 5.

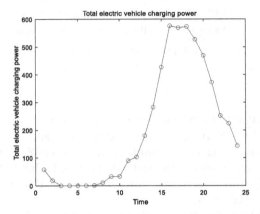

Fig. 5. The total load of 1000 electric vehicles

Secondly, using 24 h as the study period, the autonomous charging and discharging curves of electric vehicles and the charging and discharging curves after aggregator control are derived, as shown in Fig. 6.

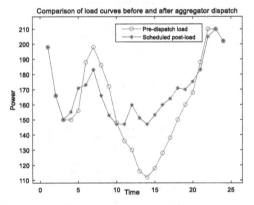

Fig. 6. Comparison of load curves before and after aggregator dispatch

When users charge autonomously, the charging peak is reached around 7:00 pm, which intensifies the load peak and is not conducive to the stable operation of the grid. Since the low valley tariff is applied after 21:00, the night peak occurs around 22:00 pm, when the charging demand of users decreases and the required minimum demand is reached.

With the participation of the aggregator, electric vehicles have the ability to reverse peak regulation between 11:00–14:00 and 17:00–20:00 at night, helping to smooth out the load on the grid, with 14:00–17:00 being the peak charging hours of the day. The results show that the charging and discharging control of the aggregator is able to cut the peaks and fill the valleys of the grid, relieving the pressure on the grid. Figure 6 shows that at 7:00 am the aggregator's control can cut the "peak" load by 150,000 kW·h, and at 14:00 noon the "valley" load can be filled by 350,000 kW·h, thus according to the aggregator's collaborative user optimisation dispatch model, the aggregator can calculate The aggregator can make a profit of 113,700 RMB, while the user can meet the charging demand and receive a compensation fee of 4.72 RMB if the demand is not met, so that all three parties can make a profit at the same time.

6 Conclusion

In order to maximise the dispatchable potential of EVs as a mobile energy source, increase aggregator revenue and user motivation, and reduce the negative impact of large-scale EV charging on the grid, this paper analyses the impact of regional EV user charging on the grid under the aggregator model. The paper firstly analyses the parties involved in this model, and secondly identifies the three forms of cooperation between users, the grid and the aggregator, analyses the travel characteristics of users and their dispatchable potential indicators, and constructs a model for the aggregator to collaborate with users and the grid to maximise the benefits of the aggregator; at the same time, it assists the grid to achieve peak and valley reduction. The analysis shows that increasing the role of aggregators can enable EV users and the grid to gain certain benefits, and is important for the economic operation of aggregators and the effective use of EVs. In the future, EVs can be clustered and processed for multiple regions, with all EVs in the region first stratified and then aggregated through Monte Carlo after stratification, so that the aggregator can meet load aggregation for multiple scenarios at the same time, with a broader scope.

Acknowledgements. This work is supported in part by the Natural Science Youth Foundation of Shandong Province, China under Grant ZR2021QE240, and in part by PhD Research Fund of Shandong Jianzhu University, China under Grant X21040Z.

References

1. Mouli, G.R.C., Kefayati, M., Baldick, R., et al.: Integrated PV charging of EV fleet based on energy prices, V2G, and offer of reserves. IEEE Trans. Smart Grid 10(2), 1313–1325 (2019)
2. Fernandez, L.P., San Román, T.G., Cossent, R., Domingo, C.M., Frias, P.: Assessment of the impact of plug-in electric vehicles on distribution networks. IEEE Trans. Power Syst. 26(1), 206–213 (2011)
3. Xydas, E., Marmaras, C., Cipcigan, L.M.: A multi-agent based scheduling algorithm for adaptive electric vehicles charging. Appl. Energy 177, 354–365 (2016)
4. Perez-Diaz, A., Gerding, E., McGroarty, F.: Coordination and payment mechanisms for electric vehicle aggregators. Appl. Energy 212, 185–195 (2018)
5. Rivera, J., Goebel, C., Jacobsen, H.A.: Distributed convex optimization for electric vehicle aggregators. IEEE Trans. Smart Grid, 1–12 (2016)
6. Sahoo, S., Prakash, S., Mishra, S.: A novel handshaking V2G strategy for grid connected PV assisted charging station. IET Renew. Power Gen. 11(11), 1410–1417 (2017)

7. Sortomme, E., El-Sharkawi, M.A.: Optimal charging strategies for unidirectional vehicle-to-grid. IEEE Trans. Smart Grid **2**(1), 131–138 (2011)
8. Karfopoulos, E.L., Hatziargyriou, N.D.: A multi-agent system for controlled charging of a large population of electric vehicles. IEEE Trans. Power Syst. **28**(2), 1196–1204 (2013)
9. Hk, A., Ac, A.: Modular strategy for aggregator control and data exchange in large scale vehicle-to-Grid (V2G) applications. Energy Procedia **151**, 7–11 (2018)
10. Diaz-Cachinero, P., Muñoz-Hernandez, J.I., Contreras, J.: Integrated operational planning model, considering optimal delivery routing, incentives and electric vehicle aggregated demand management. Appl. Energy **304**, 117698 (2021)
11. Afentoulis, K.D., Bampos, Z.N., Vagropoulos, S.I., Keranidis, S.D., Biskas, P.N.: Smart charging business model framework for electric vehicle aggregators. Appl. Energy **15**(328), 120179 (2022)
12. Wan, Z., Li, H., He, H., et al.: A data-driven approach for real-time residential EV charging management. In: 2018 IEEE Power & Energy Society General Meeting (PESGM). IEEE, pp. 1–5 (2018)

Indicators Directed Multi-strategy Artificial Bee Colony Algorithm

Jiali Wu[1], Jingwu Wang[2], Wenjun Wang[3], Jiawen Liu[1], Hui Wang[1(✉)],
Hai Zhang[1], and Min Hu[1]

[1] School of Information Engineering, Nanchang Institute of Technology,
Nanchang 330099, China
huiwang@whu.edu.cn, hujy0510@sina.com
[2] School of Electrical Engineering, Nanchang Institute of Technology,
Nanchang 330099, China
wu041031@sina.com
[3] School of Business Administration, Nanchang Institute of Technology,
Nanchang 330099, China
wangwenjun@nit.edu.cn

Abstract. Artificial bee colony (ABC) algorithm is commonly used to solve various optimization problems. Though ABC shows strong exploration capability, its weak exploitation may easily result in slow convergence. To tackle this issue, several multi-strategy ABC variants were proposed. Employing multiple search strategies with distinct features facilitates an appropriate balance between exploration and exploitation. However, choosing an appropriate strategy for the current search is a difficult task. This article suggests a new ABC variant named indicators directed multi-strategy ABC (IDMABC) to address this issue. Three evaluation indicators are designed to help ABC adaptively select suitable search strategies during the evolution. To validate the optimization capability of IDMABC, 22 classical problems are tested. IDMABC is evaluated against five other ABC variants. Results show the competitiveness of IDMABC.

Keywords: artificial bee colony · multi-strategy · multiple search strategies · indicators · adaptive

1 Introduction

Optimization issues occur frequently in real world. With the rapid development of society, many optimization issues are multi-modal, non-linear or even non-convex. Traditional optimization methods have some difficulties in solving those problems. Recently, several swarm intelligence optimization algorithms (SIOAs) were proposed, such as firefly algorithm [7,23], cuckoo Search [14], grey wolf optimizer [11,25], artificial bee colony algorithm [1,2,13,24], whale optimization algorithm [4,5], and shuffled frog leaping algorithm [3]. Among those SIOAs,

© The Author(s), under exclusive license to Springer Nature Singapore Pte Ltd. 2023
H. Zhang et al. (Eds.): NCAA 2023, CCIS 1869, pp. 273–285, 2023.
https://doi.org/10.1007/978-981-99-5844-3_20

ABC receives much attention because of its excellent search ability and few control parameters. Therefore, ABC has been effectively used for a wide range of hard optimization issues.

Although ABC presents good optimization efficiency, it still has some deficiencies, i.e. slow convergence rate and weak exploitation. Then, a concept of multi-strategy was introduced into ABC to boost exploitation and speed up convergence. In [19], a multi-strategy ensemble ABC (MEABC) was proposed, in which three different search strategies constitute a strategy pool. When the current search fails, another alternative search strategy is picked to generate offspring at random. In [26], the multi-strategy ABC employed five distinct approaches to find offspring. The amount of successful updates is kept using a counter. Then, the successful rate is defined for each search strategy. According to the successful rate, a roulette selection method will determine a appropriate search strategy. Sun et al. [21] randomly divided the employed bees into three sub-populations. Each sub-population is assigned a different search strategy. If a solution fails to generate a better solution, it is randomly added into another sub-population. It is expected that the different search equation can discover potentially better individuals. From the above analysis, it can be seen that selecting an appropriate search strategy in the optimization procedure remains a challenge. In this article, a novel multi-strategy ABC variant named indicators directed multi-strategy ABC (IDMABC) is proposed. In IDMABC, three evaluation indicators are designed to help ABC adaptively select suitable search strategies in the optimization procedure. To validate the optimization effectiveness of IDMABC, 22 classical problems are utilized. IDMABC is evaluated against five well-known ABC variants.

The remainder of this work will be structured as below. In Part 2, a brief discussion of the standard ABC is discussed. We describe the IDMABC that has been suggested in the third part. In the fourth part, study outcomes and comparisons are included. At last, our task is summarized in part 5.

2   Artificial Bee Colony Algorithm

The standard ABC switches between employed bees, onlooker bees, and scout bees in order to explore and exploit a great nectar source [12]. Various roles have various duties respectively, the whole process of the standard ABC consists of the following four stages.

(1) The initialization stage

This period generates SN solutions within the search area in a stochastic manner, where SN indicates the size of the population. Every solution (X_i) contains D dimensions: $X_i = \{x_{i,1}, x_{i,2}, \ldots, x_{i,D}\}$. The positions of X_i are obtained within x_{min} (lower bound) and x_{max} (upper bound) in the following method:

$$x_{i,j} = x_{min,j} + rand(0,1) \cdot (x_{max,j} - x_{min,j}) \tag{1}$$

where $i = 1, 2, 3, \ldots, SN$, $j = 1, 2, \ldots, D$, and the factor $rand(0,1)$ can be considered a stochastic value in the range of 0 and 1.

(2) Employed bees stage

This phase enters the formal search process, it will look for a individual close to each X_i. If the new individual is of higher quality, it will replace the previous one. The below is the formula for generating the candidate solution (V_i):

$$v_{i,j} = x_{i,j} + \phi_{i,j} \cdot (x_{i,j} - x_{k,j}) \tag{2}$$

where $\phi_{i,j}$ indicates a stochastic variable within the interval of $[-1, 1]$, X_k is randomly picked from the entire group, and $i \neq k$.

(3) Onlooker bees stage

The onlooker bees also uses the same search formula (Eq. (2)). Compares to the employed bee, it needs to select the better location by selection probability. The formula used to calculate the selection probability P_i of each X_i is as follows:

$$p_i = \frac{fit(X_i)}{\sum_{i=1}^{SN} fit(X_i)} \tag{3}$$

Where the determination of the quality of X_i is dependent on the fitness value of the solution ($fit(X_i)$), with a greater solution having a higher fitness value. The specific fitness value is generated by the following formula:

$$fit(X_i) = \begin{cases} \frac{1}{1+f(X_i)}, & \text{if } f(X_i) \geq 0 \\ 1 + |f(X_i)|, & \text{if } f(X_i) < 0 \end{cases} \tag{4}$$

where $f(X_i)$ denotes the function value of i-th solution.

(4) Scout bees stage

Abandoning overused individuals and finding new ones is the mission of the scout bees. Thus, when the solution remains unchanged for a long period, a new individual is generated randomly in the target area to replace the low-quality individual. The same formula is used as in the initialization phase.

3 Proposed Approach

As mentioned before, ABC shows strong exploration and weak exploitation. To effectively achieve a balance between the two, many improved methods were proposed. Recent studies demonstrated that ABC with multiple search methods are superior to that with a single method [6,8,15,17]. Multi-strategy ABCs need to dynamically adjust the search strategy depending on the present search condition. However, selecting an appropriate strategy is a difficult task. In this article, a novel ABC variant named indicators directed multi-strategy ABC (IDMABC) is proposed. There are two major modifications in IDMABC: 1) a strategy pool is made up of three unique kinds of search machines; and 2) three evaluation indicators are designed to adaptively choose suitable search strategies during the evolution.

3.1   Multiple Search Strategies

The search strategy is a key factor in determining the direction of the search. Different search strategies show different search features. Then, three different search strategies including GABC [28], modified MABC [21], and CABC [10] are selected to constitute a strategy pool.

The first search strategy (GABC) is a compromise, which does not prefer exploration or exploitation. GABC introduced the search information of the global optimal individual into the basic ABC. It is expected that Gbest can lessen the impacts of random selection and boost exploitation. The definition of GABC is given below.

$$v_{i,j} = x_{i,j} + \phi_{i,j} \cdot (x_{i,j} - x_{k,j}) + \psi_{i,j} \cdot (x_{best,j} - x_{i,j}) \tag{5}$$

where X_{best} is the global optimal individual, and $\psi_{i,j}$ denotes a stochastic value from the range $[0, 1.5]$.

The second search strategy (modified MABC) prefers exploitation. Depended on the mutation operation of DE, the modified ABC (MABC) was proposed [9], in which Gbest is utilised to lead search. In [21], a modified MABC search strategy is designed to further enhance the exploitation. Here is how to define the strategy of MABC.

$$v_{i,j} = x_{best,j} + \phi_{i,j} \cdot (x_{best,j} - x_{k,j}) \tag{6}$$

where $\phi_{i,j}$ presents a stochastic value within $[-1, 1]$.

The third search strategy (CABC) with strong exploration. The oscillation on the search causes inefficiency in ABC. Then, Gao et al. [10] created CABC method by using two randomly selected solutions. Below is a definition of the CABC search formula.

$$v_{i,j} = x_{r1,j} + \phi_{i,j} \cdot (x_{r1,j} - x_{r2,j}) \tag{7}$$

where $r1$ and $r2$ are two distinct values picked within the range $[1, SN]$ ($r1 \neq r2 \neq i$).

3.2   Evaluation Indicators

In Sect. 3.1, a strategy pool is made up of three unique kinds of search machines is established. During the search, suitable search strategies are required to cope with different search statuses. In some cases, the search should prefer exploitation. Perhaps, the search should prefer exploration in the next generation. To help ABC choose a suitable strategy in the search process, three evaluation indicators are designed in this section.

The first evaluation indicator (EI_1) compares the candidate solution generated in the previous iteration (V_i^{t-1}) with the current one (X_i^t). It aims to indicate whether the solution is updated successfully in the previous iteration. The indicator EI_1 is defined by

$$EI_1\left(X_i^t\right) = f\left(V_i^{t-1}\right) - f\left(X_i\right) \tag{8}$$

where t denotes iteration index. As seen, the indicator EI_1 can be used to determine whether the current iteration needs to change the search strategy. When $EI_1(X_i^t) = 0$, it means the last used strategy is effective and the current situation is defined as the sustainable state. Then, the current search strategy keeps unchangeable. When $EI_1(X_i^t) > 0$, it means the current search strategy is unsuccessful because the offspring cannot update the parent solution. Then, we try to check the quality of solutions generated in previous iterations.

The second evaluation indicator (EI_2) compares the quality of two previously generated solutions V_i^{t-1} and (V_i^{t-1-L}). The indicator EI_2 is defined by

$$EI_2\left(X_i^t\right) = f\left(V_i^{t-1}\right) - f\left(V_i^{t-1-L}\right) \tag{9}$$

where L is a parameter and its sensitive analysis is given in Sect. 4.2.

The second indicator EI_2 is used to determine whether the current strategy needs a adjustment. When $EI_1(X_i^t) > 0$ and $EI_2(X_i^t) < 0$, it shows that the last strategy is unsuccessful and the solution generated in the last iteration is superior to the one generated in the last L-th iteration. Therefore, this present strategy does not have a large deviation in the direction and it is still effective. In this case, the current situation is defined as a semi-sustained state. Then, the weighting factors ($\phi_{i,j}$ and $\psi_{i,j}$) of the current search strategy are doubled to expand the search range. When $EI_2(X_i^t) \geq 0$, the strategy is in an unsustainable state. By observing the third evaluation indicator (EI_3), appropriate search strategies can be selected.

The third evaluation indicator (EI_3) is defined as follows.

$$EI_3\left(X_i^t\right) = f\left(V_i^{t-1}\right) - \bar{f}(X^t) \tag{10}$$

where $\bar{f}(X^t) = \frac{\sum_{i=1}^{SN} f(X_i^t)}{SN}$. When $EI_3 \leq 0$, it means that the solution may be trapped into local minima and needs to use an exploration strategy. For this case, CABC is a suitable strategy. $EI_3 > 0$ indicates that the quality of the current solution is poor and the modified MABC is used to enhance the exploitation.

3.3 Framework of Proposed Approach

In IDMABC, there are two modifications. Firstly, three search strategies including GABC, modified MABC, and CABC are used to construct a strategy pool. Different strategies have different search features. Exploration and exploitation might be more evenly distributed throughout this phase of evolution. Then, three evaluation indicators (EI_1, EI_2, and EI_3) are defined based on the current search situation. This is helpful to select suitable search strategies.

Algorithm 1 explains how our IDMABC operates, where FEs denotes the number of function evaluations, $MAXFEs$ means the upper limit of function evaluations, $iter$ is iteration index, $SC(i)$ represents the strategy used in the current iteration, and $SL(i)$ indicates the strategy used in the previous iteration. Lines 7 to 38 are the selection process for the current strategy. As seen, GABC is used as the search strategy for each individual in the first iteration, followed by

---

Algorithm 1: Framework of IDMABC

```
 1 Randomly generate SN initial solutions X_i (i = 1, 2, ··· , SN);
 2 Compute each f(X_i) and f̄(X);
 3 Set FEs = SN, iter = 0, SC(i) = ∅ and SL(i) = ∅, where i = 1, 2, ··· , SN;
 4 while FEs ≤ MAXFEs do
 5 │   iter = iter + 1;
 6 │   for i = 1 to SN do
   │   │   /* The process of choosing a strategy                    */
 7 │   │   if iter == 1 then
 8 │   │   │   Set SC(i)=GABC;
 9 │   │   end
10 │   │   else
11 │   │   │   Compute EI_1 by Eq. (8);
12 │   │   │   if EI_1 == 0 then
13 │   │   │   │   Set SC(i) = SL(i) ;
14 │   │   │   end
15 │   │   │   else
16 │   │   │   │   Compute EI_2 by Eq. (9);
17 │   │   │   │   if iter > L and EI_2 ≥ 0 then
18 │   │   │   │   │   Compute EI_3 by Eq. (10);
19 │   │   │   │   │   if EI_3 > 0 then
20 │   │   │   │   │   │   Set SC(i)=modified MABC;
21 │   │   │   │   │   end
22 │   │   │   │   │   else
23 │   │   │   │   │   │   Set SC(i)=CABC;
24 │   │   │   │   │   end
25 │   │   │   │   end
26 │   │   │   │   else
27 │   │   │   │   │   if SL(i)==GABC then
28 │   │   │   │   │   │   Set SC(i)=GABC and expand the weighting factors φ_{i,j} and
   │   │   │   │   │   │   ψ_{i,j};
29 │   │   │   │   │   end
30 │   │   │   │   │   if SL(i)==modified MABC then
31 │   │   │   │   │   │   Set SC(i)=modified MABC and expand the weighting factor φ_{i,j};
32 │   │   │   │   │   end
33 │   │   │   │   │   if SL(i)==CABC then
34 │   │   │   │   │   │   Set SC(i)=CABC and expand the weighting factor φ_{i,j};
35 │   │   │   │   │   end
36 │   │   │   │   end
37 │   │   │   end
38 │   │   end
39 │   │   Generate V_i by the strategy SC(i);
40 │   │   Calculate f(V_i) and set FEs = FEs + 1;
41 │   │   if f(V_i) < f(X_i) then
42 │   │   │   Replace X_i with V_i;
43 │   │   end
44 │   │   Replace SL(i) with SC(i);
45 │   end
46 │   Update f̄(X);
47 end
```

---

the adjustment of the strategy according to different situations. The algorithm keeps looping until it reaches an upper limit. Compared with the standard ABC, it has better balance and performance by adjusting strategy during the search process.

4 Experimental Study

4.1 Test Problems and Variables Settings

In subsequent trials, we utilize 22 classic issues [22]. IDMABC is tested against five different ABCs to ensure its efficacy. Here is a summary of all the algorithms we compared.

- ABC [12].
- GABC [28].
- MABC [9].
- Opposition learning based ABC (O-ABC) [16].
- ABC based on based on multiple neighborhood topologies (ABC-MNT) [27].
- Our approach IDMABC.

The experiments consist of two parts: 1) parameter analysis of L; and 2) performance comparison of IDMABC against five ABC variants. We use values of 50 for SN, 100 for $limit$, and $5000 \cdot D$ for $MAXFEs$, respectively. After 30 iterations, the mean best function values are captured for each algorithm.

Table 1. Comparison results obtained by IDMABC with different L values.

Problems	$L=5$	$L=6$	$L=7$	$L=8$	$L=9$	$L=10$
f_1	6.76E-55	1.16E-55	**5.27E-56**	1.62E-50	5.95E-47	1.83E-42
f_2	1.44E-52	1.31E-51	**8.96E-53**	1.59E-50	1.32E-42	6.83E-40
f_3	**2.83E-56**	1.28E-55	6.85E-56	5.35E-51	3.81E-46	2.71E-43
f_4	**2.91E-141**	1.90E-139	8.20E-141	7.31E-138	9.27E-141	2.74E-128
f_5	1.55E-29	2.16E-29	**1.22E-29**	8.37E-24	1.78E-22	2.10E-21
f_6	1.66E+00	1.62E+00	**1.08E+00**	1.66E+00	1.56E+00	1.55E+00
f_7	**0**	**0**	**0**	**0**	**0**	**0**
f_8	**7.18E-66**	**7.18E-66**	**7.18E-66**	**7.18E-66**	**7.18E-66**	**7.18E-66**
f_9	**1.33E-02**	1.70E-02	1.44E-02	2.03E-02	1.96E-02	2.07E-02
f_{10}	1.66E+00	1.65E+01	**1.23E-01**	4.04E-01	1.40E+00	2.36E+00
f_{11}	**0**	1.99E-01	**0**	**0**	**0**	**0**
f_{12}	**0**	**0**	**0**	**0**	**0**	**0**
f_{13}	**0**	**0**	**0**	**0**	**0**	**0**
f_{14}	**−12569.5**	**−12569.5**	**−12569.5**	**−12569.5**	**−12569.5**	**−12569.5**
f_{15}	6.57E+00	2.80E+00	**2.93E-14**	3.29E-14	3.36E+00	4.74E+00
f_{16}	**1.57E-32**	**1.57E-32**	**1.57E-32**	**1.57E-32**	**1.57E-32**	**1.57E-32**
f_{17}	**1.35E-32**	**1.35E-32**	**1.35E-32**	**1.35E-32**	**1.35E-32**	**1.35E-32**
f_{18}	1.28E-15	4.22E-16	1.63E-31	5.88E-16	**4.45E-32**	1.22E-16
f_{19}	**1.35E-31**	**1.35E-31**	**1.35E-31**	**1.35E-31**	**1.35E-31**	**1.35E-31**
f_{20}	**0**	**0**	**0**	**0**	**0**	**0**
f_{21}	**−7.83E+01**	**−7.83E+01**	**−7.83E+01**	**−7.83E+01**	**−7.83E+01**	**−7.83E+01**
f_{22}	**−2.96E+01**	**−2.96E+01**	**−2.96E+01**	**−2.96E+01**	**−2.96E+01**	**−2.96E+01**

Table 2. Average ranking values of IDMABC in various L values.

L values	Average ranking
5	3.32
6	3.61
7	**2.52**
8	3.73
9	3.61
10	4.20

4.2 Parameter Analysis of L

When the present method cannot find an alternative individual, the second indicator EI_2 (Eq. (9)) is used to compare the quality of two previously generated solutions V_i^{t-1} with (V_i^{t-1-L}). As seen, the parameter L may greatly influence the value of EI_2. A large L is not meaningful. A small L may not check the effectiveness of the current search strategy. Thus, the parameter L should not

Table 3. Comparison outcomes obtained by IDMABC and 5 ABC variants at $D = 30$.

Problems	ABC	GABC	MABC	O-ABC	ABC-MNT	IDMABC
f_1	3.27E-17	5.21E-16	5.56E-22	9.42E-49	4.08E-41	**5.27E-56**
f_2	5.75E-08	5.44E-16	2.90E-19	1.92E-43	1.69E-37	**8.96E-53**
f_3	9.99E-19	5.12E-16	9.66E-24	1.30E-49	2.61E-41	**6.85E-56**
f_4	4.77E-27	1.04E-16	1.19E-47	1.48E-60	2.79E-85	**8.20E-141**
f_5	5.71E-11	1.38E-15	2.27E-12	1.31E-26	1.54E-21	**1.22E-29**
f_6	3.94E+01	1.64E+01	2.09E+01	2.59E+01	5.43E+00	**1.08E+00**
f_7	**0.00E+00**	**0.00E+00**	**0.00E+00**	**0.00E+00**	**0.00E+00**	**0.00E+00**
f_8	1.25E-22	1.01E-17	**7.18E-66**	**7.18E-66**	**7.18E-66**	**7.18E-66**
f_9	1.79E-01	**1.01E-01**	6.33E-02	5.46E-02	3.47E-02	1.44E-02
f_{10}	1.49E-02	**6.51E-04**	6.49E+00	4.26E-02	1.33E+00	1.23E-01
f_{11}	7.11E-15	**0.00E+00**	**0.00E+00**	**0.00E+00**	**0.00E+00**	**0.00E+00**
f_{12}	1.95E-14	**0.00E+00**	**0.00E+00**	**0.00E+00**	**0.00E+00**	**0.00E+00**
f_{13}	2.22E-16	4.44E-16	**0.00E+00**	6.77E-14	**0.00E+00**	**0.00E+00**
f_{14}	-12451.0	**−12569.5**	**−12569.5**	**−12569.5**	**−12569.5**	**−12569.5**
f_{15}	9.13E-06	5.06E-14	9.69E-11	3.29E-14	**2.93E-14**	**2.93E-14**
f_{16}	7.43E-18	7.65E-18	5.04E-23	**1.57E-32**	**1.57E-32**	**1.57E-32**
f_{17}	2.22E-16	5.44E-16	4.80E-22	**1.35E-32**	**1.35E-32**	**1.35E-32**
f_{18}	1.17E-06	4.40E-08	3.00E-14	2.23E-27	2.03E-22	**1.63E-31**
f_{19}	6.02E-15	1.18E-15	4.53E-22	**1.35E-31**	**1.35E-31**	**1.35E-31**
f_{20}	3.00E-01	**0.00E+00**	**0.00E+00**	**0.00E+00**	**0.00E+00**	**0.00E+00**
f_{21}	**−7.83E+01**	**−7.83E+01**	**−7.83E+01**	**−7.83E+01**	**−7.83E+01**	**−7.83E+01**
f_{22}	−2.95E+01	**−2.96E+01**	**−2.96E+01**	**−2.96E+01**	**−2.96E+01**	**−2.96E+01**
$w/t/l$	20/2/0	13/7/2	13/9/0	11/11/0	9/13/0	-/-/-

be too small or too large. To choose a good L, different L values are tried out. During this tests, L is assigned to 5, 6, 7, 8, 9, and 10, respectively.

Table 1 shows computational outcomes obtained by IDMABC in distinct L numbers on the test bench. Results demonstrate that $L = 7$ helps IDMABC obtain better solutions than other settings on all test problems except f_3, f_4, f_9, and f_{18}. IDMABC with $L = 5$ outperforms other cases on problems f_3, f_4, and f_9. For f_{18}, $L = 9$ is the best choice. For f_7, f_8, f_{12}-f_{14}, f_{16}, f_{17}, and f_{19}-f_{22}, the L value does not impact the capability of IDMABC.

For the purpose of determining which value of L is the most often preferred among all benchmarks, a Friedman test is run [18, 20]. The mean ranking results of IDMABC for various values of L is shown in Table 2. As shown, $L = 7$ ranks highest among a set of six distinct L numbers. It demonstrates that $L = 7$ is preferable to other options for IDMABC.

4.3   Tests on the Benchmark Issues

Five distinct ABC algorithms are contrasted against IDMABC in this subsection. Table 3 details the outcomes of computational runs of IDMABC, ABC,

Table 4. Comparison outcomes obtained by IDMABC and 5 ABC variants at $D = 100$.

Problems	ABC	GABC	MABC	O-ABC	ABC-MNT	IDMABC
f_1	3.69E-15	3.88E-15	2.28E-19	2.61E-45	1.18E-36	**7.59E-50**
f_2	2.31E-06	2.99E-15	7.18E-16	1.02E-40	4.50E-33	**8.33E-46**
f_3	2.53E-16	2.94E-15	8.92E-20	6.45E-47	9.77E-37	**2.05E-50**
f_4	2.34E-11	1.09E-11	1.62E-42	3.66E-40	8.56E-86	**1.92E-142**
f_5	1.23E-09	7.42E-15	1.64E-10	1.03E-25	2.11E-19	**1.98E-26**
f_6	7.98E+01	8.51E+01	7.58E+01	8.11E+01	4.31E+01	**3.19E+01**
f_7	1.00E+00	**0.00E+00**	**0.00E+00**	**0.00E+00**	**0.00E+00**	**0.00E+00**
f_8	6.74E-40	1.87E-46	**7.12E-218**	**7.12E-218**	**7.12E-218**	**7.12E-218**
f_9	1.35E+00	7.83E-01	3.47E-01	6.43E-01	1.45E-01	**1.29E-01**
f_{10}	7.51E-02	2.28E-01	1.02E+02	4.15E-01	2.66E-01	8.34E+00
f_{11}	6.63E-12	1.60E-14	**0.00E+00**	**0.00E+00**	**0.00E+00**	**0.00E+00**
f_{12}	1.00E+00	2.66E-14	**0.00E+00**	**0.00E+00**	**0.00E+00**	**0.00E+00**
f_{13}	1.55E-15	3.00E-15	**0.00E+00**	1.05E-14	**0.00E+00**	**0.00E+00**
f_{14}	−41053.7	−41861.2	**−41898.3**	**−41898.3**	**−41898.3**	**−41898.3**
f_{15}	8.06E-05	6.69E-10	8.10E-07	**1.71E-13**	1.82E-13	3.72E-12
f_{16}	9.62E-16	3.35E-15	1.18E-20	**4.71E-33**	**4.71E-33**	**4.71E-33**
f_{17}	3.49E-14	3.22E-15	1.71E-21	**1.35E-32**	**1.35E-32**	**1.35E-32**
f_{18}	1.48E-05	1.45E-05	2.76E-08	4.44E-27	2.22E-19	**1.82E-27**
f_{19}	7.91E-15	1.76E-14	6.36E-20	**1.35E-31**	**1.35E-31**	**1.35E-31**
f_{20}	3.14E+00	**0.00E+00**	**0.00E+00**	**0.00E+00**	**0.00E+00**	**0.00E+00**
f_{21}	**−7.83E+01**	**−7.83E+01**	**−7.83E+01**	**−7.83E+01**	**−7.83E+01**	**−7.83E+01**
f_{22}	−9.69E+01	−9.73E+01	−9.90E+01	−9.95E+01	**−9.96E+01**	−9.95E+01
$w/t/l$	20/1/1	18/3/1	14/8/0	9/11/2	8/11/3	-/-/-

Table 5. Friedman test for different six ABC variants.

Algorithms	$D = 30$	$D = 100$
	Mean ranking	Mean ranking
ABC	5.23	5.30
GABC	4.32	4.75
MABC	3.75	3.57
O-ABC	2.89	2.82
ABC-MNT	2.73	2.48
IDMABC	**2.09**	**2.09**

Table 6. Wilcoxon test for different five ABC variants compared with IDMABC.

IDMABC vs.	$D = 30$	$D = 100$
	p-value	p-value
ABC	**8.91E-04**	**7.95E-04**
GABC	**8.97E-03**	**1.70E-03**
MABC	**1.47E-03**	**9.82E-04**
O-ABC	**4.09E-02**	1.82E-01
ABC-MNT	**7.69E-03**	5.34E-01

GABC, MABC, O-ABC, and ABC-MNT for $D = 30$. The complete evaluation outcomes between IDMABC along with distinct ABCs are summarised in the final line of Table 3 with $w/t/l$. Listed below are explanations for the characters w, t, and l. IDMABC outperforms the compared ABC on w issues. IDMABC and both versions ABC provide t similar answers. On l issues, ABC compares well to IDMABC. As we shown, IDMABC and the original ABC obtain the same results in f_7 and f_{21}. On the remaining of 20 functions, IDMABC attains greater accuracy than the traditional ABC. IDMABC performs better than GABC and MABC on 13 issues. Only on two issues does GABC perform superior to IDMABC. MABC and IDMABC achieve similar performance on nine problems. Compared with O-ABC and ABC-MNT, IDMABC does not obtain worse results.

Fig. 1 displays four curves graphs of ABC, GABC, MABC, O-ABC, ABC-MNT, and IDMABC on four problems f_1, f_2, f_4, and f_{16} for $D = 30$. As seen, IDMABC obtains the fastest convergence speed than other ABCs. For problems f_1 and f_2, O-ABC converges more quickly than other algorithms in the beginning stage. As the iteration increases, IDMABC gradually converges faster than OABC and other ABC variants. For f_{16}, GABC is faster than other ABCs.

Table 4 displays the computational outcomes of all algorithms at $D = 100$. From these results, ABC outperforms IDMABC on only one problem f_{10}. On the other 21 issues, IDMABC is not inferior to ABC. IDMABC achieves higher

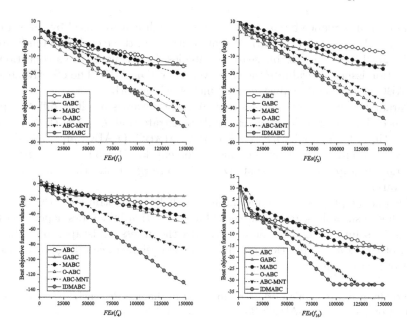

Fig. 1. Convergence graphs of six ABC variants with f_1, f_2, f_4, and f_{16} for $D = 30$.

quality answers than GABC on 18 issues. However, GABC performs better than IDMABC on f_{10}. Compared to the MABC, IDMABC obtains better results on 14 problems. For the remaining problems, both of them achieve the same results. IDMABC, O-ABC, and ABC-MNT obtain the same performance on 11 problems. O-ABC and ABC-MNT surpasses IDMABC on two and three problems, respectively. IDMABC remains superior to five various ABC algorithms despite having 100 dimensions across most issues.

To further compare the effectiveness of six ABC variants on the test set, two nonparametric statistical methods are conducted. Table 5 lists six mean ranks obtained by the Friedman test. A greater overall performance in the Friedman test is indicated by a lower average rank. As can be observed, IDMABC ranks highest for $D = 30$ and $D = 100$. It demonstrates that IDMABC outperforms the remaining five compared ABC variants. Comparing IDMABC to various ABC algorithms, the p-values are listed in table 6. For $D = 30$, IDMABC is superior to contrasting ABC variants. IDMABC significantly outperforms ABC, GABC, and MABC at $D = 100$.

5 Conclusion

To help multi-strategy ABC choose appropriate search strategies, this paper proposes indicators directed multi-strategy ABC (IDMABC). A strategy pool is built using GABC, modified MABC, and CABC, three different search machines. Exploration and exploitation are not mutually exclusive, rather, we need to find an appropriate balance by using the various search features used by the various strategies. According to the current search situation, three evaluation indicators are designed to select suitable search strategies. Experiments use a collection of 22 function issues to evaluate the effectiveness of IDMABC, The effectiveness of IDMABC is measured against five distinct ABC algorithms.

Parameter analysis of L shows that different L values result in different performance. The average ranking results demonstrate that $L = 7$ is the relatively best settings. Compared with ABC, GABC, MABC, O-ABC, and ABC-MNT, IDMABC obtains better results on most problems for $D = 30$ and $D = 100$. The convergence graphs illustrate that IDMABC converges faster than the other five ABC variants.

This paper proposes a simple method to help multi-strategy ABC dynamically choose suitable search strategies. The evaluation indicators depend on the changes of solution, and they do not consider the fitness landscape. In the future work, other strategy selection machine will be designed according to the viewpoint of fitness landscape.

Acknowledgments. This work was supported by Jiangxi Provincial Natural Science Foundation (Nos. 20212BAB202023 and 20212BAB202022).

References

1. Akay, B., Karaboga, D.: A modified artificial bee colony algorithm for real-parameter optimization. Inf. Sci. **192**, 120–142 (2012)
2. Bansal, J.C., Sharma, H., Jadon, S.S.: Artificial bee colony algorithm: a survey. Int. J. Adv. Intell. Paradigms **5**(1–2), 123–159 (2013)
3. Cai, J., Zhou, R., Lei, D.: Dynamic shuffled frog-leaping algorithm for distributed hybrid flow shop scheduling with multiprocessor tasks. Eng. Appl. Artif. Intell. **90**, 103540 (2020)
4. Chakraborty, S., Saha, A.K., Chakraborty, R., Saha, M.: An enhanced whale optimization algorithm for large scale optimization problems. Knowl.-Based Syst. **233**, 107543 (2021)
5. Chen, H., Xu, Y., Wang, M., Zhao, X.: A balanced whale optimization algorithm for constrained engineering design problems. Appl. Math. Model. **71**, 45–59 (2019)
6. Du, Z., Chen, K.: Enhanced artificial bee colony with novel search strategy and dynamic parameter. Comput. Sci. Inf. Syst. **16**(3), 939–957 (2019)
7. Fister, I., Fister, I., Jr., Yang, X.S., Brest, J.: A comprehensive review of firefly algorithms. Swarm Evol. Comput. **13**, 34–46 (2013)
8. Gao, W.F., Huang, L.L., Liu, S.Y., Chan, F.T., Dai, C., Shan, X.: Artificial bee colony algorithm with multiple search strategies. Appl. Math. Comput. **271**, 269–287 (2015)
9. Gao, W.F., Liu, S.Y.: A modified artificial bee colony algorithm. Comput. Oper. Res. **39**(3), 687–697 (2012)
10. Gao, W.F., Liu, S.Y., Huang, L.L.: A novel artificial bee colony algorithm based on modified search equation and orthogonal learning. IEEE Trans. Cybern. **43**(3), 1011–1024 (2013)
11. Hu, P., Pan, J.S., Chu, S.C.: Improved binary grey wolf optimizer and its application for feature selection. Knowl.-Based Syst. **195**, 105746 (2020)

12. Karaboga, D., et al.: An idea based on honey bee swarm for numerical optimization. Technical report, Technical report-tr06, Erciyes university, engineering faculty, computer ... (2005)
13. Kaya, E., Gorkemli, B., Akay, B., Karaboga, D.: A review on the studies employing artificial bee colony algorithm to solve combinatorial optimization problems. Eng. Appl. Artif. Intell. **115**, 105311 (2022)
14. Mareli, M., Twala, B.: An adaptive cuckoo search algorithm for optimisation. Appl. Comput. Inform. **14**(2), 107–115 (2018)
15. Peng, H., Wang, C., Han, Y., Xiao, W., Zhou, X., Wu, Z.: Micro multi-strategy multi-objective artificial bee colony algorithm for microgrid energy optimization. Futur. Gener. Comput. Syst. **131**, 59–74 (2022)
16. Sharma, T.K., Gupta, P.: Opposition learning based phases in artificial bee colony. Int. J. Syst. Assur. Eng. Manag. **9**, 262–273 (2018)
17. Song, X., Zhao, M., Xing, S.: A multi-strategy fusion artificial bee colony algorithm with small population. Expert Syst. Appl. **142**, 112921 (2020)
18. Wang, H., Rahnamayan, S., Sun, H., Omran, M.G.: Gaussian bare-bones differential evolution. IEEE Trans. Cybernet. **43**(2), 634–647 (2013)
19. Wang, H., Wang, W., Xiao, S., Cui, Z., Xu, M., Zhou, X.: Improving artificial bee colony algorithm using a new neighborhood selection mechanism. Inf. Sci. **527**, 227–240 (2020)
20. Wang, H., Wu, Z., Rahnamayan, S., Liu, Y., Ventresca, M.: Enhancing particle swarm optimization using generalized opposition-based learning. Inf. Sci. **181**(20), 4699–4714 (2011)
21. Wang, H., Wu, Z., Rahnamayan, S., Sun, H., Liu, Y., Pan, J.S.: Multi-strategy ensemble artificial bee colony algorithm. Inf. Sci. **279**, 587–603 (2014)
22. Xiao, S., Wang, W., Wang, H., Zhou, X.: A new artificial bee colony based on multiple search strategies and dimension selection. IEEE Access **7**, 133982–133995 (2019)
23. Yang, X.S.: Firefly algorithm, stochastic test functions and design optimisation. Int. J. Bio-inspired Comput. **2**(2), 78–84 (2010)
24. Ye, T., et al.: Artificial bee colony algorithm with efficient search strategy based on random neighborhood structure. Knowl.-Based Syst. **241**, 108306 (2022)
25. Zamfirache, I.A., Precup, R.E., Roman, R.C., Petriu, E.M.: Policy iteration reinforcement learning-based control using a grey wolf optimizer algorithm. Inf. Sci. **585**, 162–175 (2022)
26. Zeng, T., et al.: Artificial bee colony based on adaptive search strategy and random grouping mechanism. Expert Syst. Appl. **192**, 116332 (2022)
27. Zhou, X., Wu, Y., Zhong, M., Wang, M.: Artificial bee colony algorithm based on multiple neighborhood topologies. Appl. Soft Comput. **111**, 107697 (2021)
28. Zhu, G., Kwong, S.: Gbest-guided artificial bee colony algorithm for numerical function optimization. Appl. Math. Comput. **217**(7), 3166–3173 (2010)

Energy-Efficient Cellular Offloading Optimization for UAV-Aided Networks

Ge Yang[✉], Hongyuan Zheng, Xiangping Bryce Zhai, and Jing Zhu

Nanjing University of Aeronautics and Astronautics, Nanjing, China
{yangge,blueicezhaixp}@nuaa.edu.cn

Abstract. Unmanned aerial vehicle (UAV) plays an important role in wireless systems. It can be deployed flexibly to help improve communication quality. To solve the problem of energy limitation of UAV, we propose an energy-efficient offloading scheme with UAV. By jointly optimizing the energy consumption, user scheduling, UAV location and offloading ratio, we can minimize the total power consumption of the system under the premise of delay. We use K-means algorithm and user priority queue to deal with the real-time random mobility of users. Considering the complexity of the problem, we propose an offloading algorithm based on Deep Deterministic Policy Gradient (DDPG). In our framework, we can obtain the best computational offload policy in an complex and changeable environment. In addition, our energy-saving solution can extend the service time of the UAV. The simulation results show that our framework has better performance than conventional DDPG algorithm and other base solutions.

Keywords: Unmanned aerial vehicle(UAV) · DDPG · energy-efficient · cellular offloading

1 Introduction

Unmanned aerial vehicle (UAV) has great concern and application because its flexibility and convenience. With the reduction of cost, UAV is expected to provide more possibilities in the field of wireless communication. Some mobile applications usually require a large number of computing resources [1], which puts forward new requirements for UAV research [2]. Research shows that UAV can enhance the connectivity, throughput and coverage of the ground network in some scenarios.

Researchers have done a lot of research in this field. In [3], the authors highlighted that UAV can provide better service for cellular networks in multiple scenarios. In [4], the authors investigated the impact of UAV location on quality of service and proposed an artificial swarm algorithm-based method for UAV based station deployment. In order to improve the efficiency of offloading, researchers proposed an optimal mission server algorithm by UAV. In [5], the authors maximized the system throughput by optimizing UAV trajectory and user scheduling.

© The Author(s), under exclusive license to Springer Nature Singapore Pte Ltd. 2023
H. Zhang et al. (Eds.): NCAA 2023, CCIS 1869, pp. 286–301, 2023.
https://doi.org/10.1007/978-981-99-5844-3_21

Due to the complexity of the environment and the multidimensional nature of the state, it is an inevitable trend to apply reinforcement learning to UAV communication. In [6], the authors introduced the application of various RL algorithms in UAV networks to solve key problems such as trajectory control and resource management. In [7], the authors determined the optimal layout of UAV through Q-learning. In [8], the authors designed a positioning scheme of UAV based on deep Q-network (DQN) to improve throughput efficiency. Because the action space of DQN is discrete, the depth deterministic gradient algorithm has been proved to be an effective solution to the continuity problem [9].

However, most researches only focus on the advantages of UAV and ignore its short endurance. In [10], the authors improved the offloading efficiency and reduced the offloading energy consumption through the optimal matching modeling of UAV and edge nodes. Some reseachers jointly optimized UAV locus and user actions to minimize system energy consumption.

In general, researchers pursue the best system efficiency and ignore the energy saving of the whole system. In addition, researchers mainly consider the energy consumption of the system calculation and ignore the energy consumption of the UAV itself. For solving user actions, we first propose a K-means based clustering algorithm. Considering the complexity of the state and the unknown environment, we propose a DDPG algorithm.

The main contributions of this paper are summarized as follows:

- Considering time-varying channels, random tasks and high dimensionality of variables, we propose a cellular offloading framework based on the DDPG algorithm to obtain the optimal policy by combining user scheduling, UAV mobility and resource allocation.
- Considering the random distribution and demand of users, we propose an improved K-means clustering algorithm to optimize the user distribution of the system and define a priority queue of users to optimize user services.
- Considering the limitation of UAV energy consumption, we propose the idea of deploying periodic rather than real-time campaigns and balance the impact of system power consumption on rewards, thus minimizing system power consumption with basic delay requirements.

The rest of the paper is organized as follows. Section 2 describes the system model of energy-saving cellular offloading framework based on DDPG algorithm and the problem formula. Section 3 introduces the framework design in detail, including user clustering based on K-means algorithm, user priority queue and energy-saving offloading framework based on DDPG algorithm. Section 4 analyzes the numerical results. Finally, Sect. 5 presents our conclusions.

2   System Model

As shown in Fig. 1, we consider an outdoor downlink user intensive scenario, in which there is a central BS and a large number of mobile users. Due to the limited

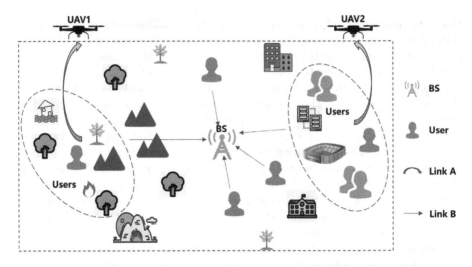

Fig. 1. Framework of UAV-aided cellular network.

capacity of BS, when multiple users request access, it may not meet the computing needs. In order to reduce the burden of BS and improve the communication quality of users, we propose a UAV-aided cellular offload framework. Users who cannot meet the requirements can offload the task to UAV for service. In order to simplify and subdivide the user distribution, the users are clustered according to their spatial positions and associated with the UAV, so that UAV can maintain a large distance to reduce interference. In addition, in order to reduce co-channel interference, it is assumed that UAV and BS use different frequency bands. We denote the set of users as $m \in \mathcal{M} = \{1, 2, 3, \dots, M\}$ and the set of UAVs can be denoted as $u \in \mathbf{U} = \{1, 2, 3, \dots, U\}$. The users are partitioned into m cells. Each user m is only served by UAV u.

In this paper, we set up a user model that moves randomly. Users will move aimlessly, their movement direction and speed are completely random in any discrete time slot t. During each t, an angle $\theta \in (0, \pi)$ will be randomly generated and the user will move a certain distance according to the angle. Due to continuous user roaming, the BS offloads may be variable. Although UAV can approach users, the initial user cluster may become worse with the change of user location, which leads to serious interference and low offloading efficiency. Therefore, regular re-clustering of users is a necessary condition to maintain the optimal data rate. UAVs re-cluster at a certain interval of T_r.

2.1 Signal Model

The channel model includes USERS to UAV and USERS to BS.

USERS-UAV: The air-to-ground channel model between each UAV and related users [11] is provided in the 3GPP Releases. The path loss depends on the line-of-sight (LoS) and non-line-of-sight (NLoS) link states between user m and UAV u can be expressed as (1), where h_u represents the flight altitude of UAV, f_c represents the carrier frequency, and the distance from UAV u to user m at time t can be represented as (2):

$$L(t) = \begin{cases} L_{LoS} = 30.9 + (22.5 - 0.5\log_{10} h_u)\log_{10} d_m^u(t) + 20\log_{10} f_c \\ L_{NLoS} = max\{L_{LoS}, 32.4 + (43.2 - 7.6\log_{10} h_u)\log_{10} d_m^u(t)\} \end{cases} \quad (1)$$

$$d_m^u(t) = \sqrt{h_u^2 + [x_u(t) - x_m(t)]^2 + [y_u(t) - y_m(t)]^2} \quad (2)$$

where $x_u(t)$ and $y_u(t)$ represents the coordinates of UAV u, $x_m(t)$ and $y_m(t)$ represents the coordinates of user m. Therefore, the average path loss between UAV u and user m can be calculated by the following formula:

$$L_m^u(t) = P_{LoS}{\cdot}L_{Los} + P_{NLoS}{\cdot}L_{NLos} \quad (3)$$

Considering small-scale fading, the channel gain from UAV u to user k at time t can be calculated as [12]:

$$g_m^k(t) = H_m^k(t){\cdot}10^{-L_m^u(t)/10} \quad (4)$$

where $H_m^k(t)$ represents the fading coefficient between UAV u and user m [13]. The wireless transmission rate can be given as:

$$r_m^k(t) = B\log_2\left(1 + \frac{P_{send}{\cdot}g_m^k(t)}{\sigma^2}\right) \quad (5)$$

where B represents the communication bandwidth, P_{send} represents the transmission power of user in the uplink link, and σ^2 represents the noise power.

USERS-BS: The coordinates of the BS are fixed. In the whole scene, the channel between users and BS is only related to the user's position and the obstacles between them. The channel gain between BS and the user m can be expressed as

$$g_m^b(t) = \frac{\mu}{[x_m(t) - x_b]^2 + [y_m(t) - y_b]^2} \quad (6)$$

where μ represents the channel gain at a reference distance of one meter. (x_b, y_b) represents the coordinates of BS. Due to the blockage from obstacles, the wireless transmission rate can be given as:

$$r_m^b(t) = B\log_2\left(1 + \frac{P_{send}{\cdot}g_m^b(t)}{\sigma^2 + q_m(t){\cdot}P_{NLoS}}\right) \quad (7)$$

where P_{NLoS} represents transmission loss [14], $q_m(t)$ indicating whether there is any blocking between BS and User m (0 indicates no blocking, 1 indicates blocking).

2.2   Computation Model

In the offloading framework, each slot t will offload the task to BS or UAV. $R_m^u(t) \in [0, 1]$ indicates the ratio of offloading tasks from user m to uav u.

Delay. In the system, the waiting delay of users can be divided into two parts. One part is the delay generated by calculation, and the other part is the transmission delay. According to the user's wireless transmission rate and task offloading ratio, we can calculate the calculation delay between user m and UAV u [15].

$$T_m^u(t) = \frac{R_m^u(t) \cdot Task_m(t) \cdot s}{f_u} \tag{8}$$

where $Task_m(t)$ represents the computing task size of user m, s represents the CPU cycle required to process each unit byte, and f_u represents the computing capacity of UAV. The transmission delay between user m and UAV u can be given as:

$$T_{tr(m,u)}(t) = \frac{R_m^u(t) \cdot Task_m(t)}{r_m^u(t)} \tag{9}$$

Similarly, the calculation and transmission delay between BS and user m can be expressed as:

$$T_m^b(t) = \frac{(1 - R_m^u(t)) \cdot Task_m(t) \cdot s}{f_{bs}} \tag{10}$$

$$T_{tr(m,bs)}(t) = \frac{(1 - R_m^u(t)) \cdot Task_m(t)}{r_m^b(t)} \tag{11}$$

where f_{bs} represents the computing capacity of BS. Therefore, the maximum waiting delay of user m in slot t can be expressed as:

$$T_{total}(t) = \max\{T_m^b(t) + T_{tr(m,bs)}(t), \\ T_m^u(t) + T_{tr(m,u)}(t)\} \tag{12}$$

Power. In the whole system, the energy consumption of the BS is negligible, so the energy consumption mainly includes the calculation and motion of the UAV. The calculated energy consumption of UAV u is determined by the number of tasks at each t. It can be expressed as [16]:

$$E_u^m(t) = \xi \cdot f_u^3 \cdot T_m^u(t) = \xi \cdot f_u^2 \cdot R_m^u(t) \cdot Task_m(t) \cdot s \tag{13}$$

UAV motion includes hovering and flying, and their energy consumption is different. Ideally, the energy consumed by UAV hovering E_h at each t is fixed. The energy consumed by UAV flight is related to the distance and speed of UAV flight. According to the principle of energy conservation, the electric energy consumed by UAV flight is converted into kinetic energy. The flight speed of

UAV is constant at v_u. The energy consumed by UAV flight at flight time t can be expressed as:

$$E_{fly}(t) = \int_t^{t+1} \frac{1}{2} m_u \cdot v_u^2 \cdot dt = \frac{1}{2} m_u \cdot v_u^2 \tag{14}$$

Therefore, the calculation formula of UAV energy in time t can be expressed as:

$$E_{total}(t) = \begin{cases} E_u^m(t) + E_h, & q_u(t) = 0 \\ E_u^m(t) + E_h + E_{fly}(t), & q_u(t) = 1 \end{cases} \tag{15}$$

where $q_u(t)$ represents the flight status of the UAV in time t (0 means hovering and 1 means flying).

2.3 Problem Formulation

Based on the proposed model, we formulated the optimization problem of UAV assisted energy-saving unloading. To ensure the effective utilization of limited resources, we aim to minimize the total power consumption while meeting the delay requirements by jointly optimizing user scheduling, UAV mobility and resource allocation. The optimization problem can be expressed as:

$$\min_{R_m^u(t), L_u(t), T_{total}(t)} \sum_{t=0}^{T} E_{total}(t) \tag{16a}$$

$$0 \leq R_m^u(t) \leq 1, \forall t, m, u \tag{16b}$$

$$L_u \in \{x_u(t), y_u(t) | x_u(t) \in [-L, L], y_u(t) \in [L, L]\} \tag{16c}$$

$$L_m \in \{x_m(t), y_m(t) | x_m(t) \in [-L, L], y_m(t) \in [-L, L]\} \tag{16d}$$

$$q_m(t) =\in \{0, 1\}, \forall t, m \tag{16e}$$

$$\sum_{t=0}^{T} E_{total}(t) \leq E_{all} \tag{16f}$$

$$T_{total}(t) \leq T_0. \forall t \tag{16g}$$

where (17b) represents the value range of offloading ratio of the offloading task. Constraints (17c) and (17d) ensure that users and UAVs move within the right area. Constraint (17e) indicates the blockage of the wireless channel between users and the BS. The constraint (17f) ensures that the UAV flight and calculation energy consumption satisfies the maximum battery capacity. The constraint (17g) specifies the maximum wait delay over the entire time period.

3 Framework Design

This section describes the framework design, including two steps. In the first section, users can be clustered by K-means based clustering algorithm. Offloaded users can be divided into multiple clusters to suppress interference between clusters. The second part introduces the design of offloading strategy based on DDPG algorithm.

3.1   Optimization of K-Means Clustering Algorithm

The location of UAV and user is uncertain, which affects the service quality of communication, so it is necessary to determine the best service location of UAV. On the other hand, there may be interference between UAVs, so they should be kept as far away as possible.

K-Means algorithm is well performed in this area for user clustering in wireless communication [17]. It has low complexity and is flexible to implement. In the system, the position $L = \{l_1, l_2...l_m\}$ of user is as the observation set. The goal of the algorithm is to divide the user set into U clusters according to the distance between user samples, so that the users in the cluster can be connected closer to each other, and the cluster centers of different clusters can be as far as possible. The conventional algorithm first randomly initializes the positions of U centroids, then assigns each user to the nearest centroid, recalculates the centroids of each cluster, and repeats the process until the centroids of all clusters don't change. Finally, users are divided into U clusters($C = \{C_1, C_2...C_u\}$) according to the number of UAVs and their own positions. However, the

---

Algorithm 1. Optimized K-means algorithm for user clustering

1: Initialize desired cluster number $\{C_1...C_u\}$, the maximum number of iterations N, maximum UAV load λ.
2: Input users location $L = \{l_1, l_2...l_m\}$ as observation set.
3: Randomly select a node U_1 from the user set L as the first cluster center.
4: **for** $l \in L$ **do**
5:     Calculate $d_m^u = \|l_m - U_1\|$
6:     Random select a new center using a weighted probability distribution according to d_m^u
7:     **if** number of center nodes equals u **then**
8:         End loop
9:     **end if**
10: **end for**
11: **for** $n = 1, 2...N$ **do**
12:     **for** $l \in L$ **do**
13:         Calculate $d_m^u = \|l_m - U_u\|$
14:         Allocate l_m to C_u with minimum d_m^u
15:     **end for**
16:     **if** $U_u(n) = U_u(n-1)$ **then**
17:     **end if**
18: **end for**
19: **while** $|Cu| > \lambda$ **do**
20:     Remove l_m with $max\{d_k^u\}$ from C_u
21:     Add l_m to $C_i \neq C_u$ with minimum d_k^i
22: **end while**
23: Output $\{C_1...C_u\}$

---

algorithm is sensitive to special points and points with edges, and the number of users that UAVs can serve has an upper limit. So we propose an improved K-means algorithm. When selecting the cluster center, it is assumed that the $n$initial cluster centers $(0 < n < U)$ have been selected. when selecting the $n+1$ cluster center, the point farther away from the current n cluster centers will have a higher probability of being selected. When selecting the first cluster center $(n=1)$, the random method is also used. In addition, if there are too many users in the cluster, users farthest from the hub will be split into another cluster. Repeat this operation until all clusters have legitimate users. The whole algorithm are clearly given in Algorithm 1:

3.2   Power Allocation Based on DDPG Algorithm

In the cellular offload scenario, the design of multiple UAVs is highly complex. The multi-UAVs to multi-users problem can be transformed into a single UAV to multi-users problem through the clustering algorithm mentioned above.

DDPG is proposed to solve such complex problems [18]. The DDPG algorithm mainly includes the following three key technologies:

- Replay Buffer: The agent puts the trained data into a fixed-size container class and Update network parameters with batch sampling.
- Target network: Target Actor network and Target Critical network for target estimation are used outside the Actor network and Critical network. When updating the target network, in order to avoid the parameter updating too fast, the soft update method is adopted.
- Noise exploration: The actions output by deterministic strategies are deterministic actions that lack exploration of the environment. During the training phase, noise is added to the actions output by the Actor network to provide the agent with exploration capabilities.

The DDPG algorithm is on the basis of the previous AC structure. The algorithm contains Actor and Critic networks. Each network has its corresponding target network, so the DDPG algorithm includes four networks: Actor network $\mu(\cdot \mid \theta^{\mu})$, Critic network $Q(\cdot \mid \theta^{Q})$, Target Actor network $\mu'(\cdot \mid \theta^{\mu'})$ and Target Critic network $Q'(\cdot \mid \theta^{Q'})$. The update of this algorithm is mainly to update the parameters of Actor and Critical network. Actor network updates θ^{μ} by maximizing the cumulative expected return and Critical network updates θ^{Q} by minimizing the error between the evaluation value and the target value.

In the training phase, we sample a batch of data from the Replay Buffer. Suppose the sampled data is $(s, a, r, s', done)$. For the Critical network, the Target Actor network is used to calculate the actions under the status s'. Then the Target Critic network is used to calculate the target value of the state action (s, a). The Critic network is used to calculate the evaluation value of the state action (s, a). Finally, the gradient descent algorithm is used to minimize the difference $L_c = (y - q)^2$ between the evaluation value and the expected value to update the parameters in the Critical network.

For the Actor network, first the Actor network is used to calculate the action in state s. Then the Critic network is used to calculate the evaluation value of the state action (s, a_{new}) (i.e. the cumulative expected return). Finally, the gradient descent algorithm is used to maximize the cumulative expected return q_{new} to update the parameters in the Actor network.

For the Target network, the algorithm adopts the soft update method, which can also be called Exponential Moving Average (EMA). A learning rate $\tau \in (0, 1)$ is introduced, the old target network parameters and the new corresponding network parameters are weighted and averaged, and then assigned to the target network:

$$\theta^{\mu'} = \tau\theta^\mu + (1 - \tau)\theta^{\mu'}$$
$$\theta^{Q'} = \tau\theta^Q + (1 - \tau)\theta^{Q'} \tag{17}$$

State Space: In the UAV-aided cellular offload, the state space in a cluster is determined by M users, one UAV and their environment. The input state S at time slot t can be expressed as:

$$S(t) = \{L_u(t), L_{bs}, L_{1...m}(t), P(t)$$
$$Task_{1...m}(t), q_{1...m}(t), E_{remain}(t), q_u(t)\} \tag{18}$$

where $L_u(t)$ represents the position of UAV u at time t, L_{bs} represents the position of BS, $L_{1...m}(t)$ represent the position of users at time t, $P(t)$ represents User's service order, $Task_{1...m}(t)$ represent the task size of User M at time t, $q_{1...m}(t)$ indicate the communication conditions between User M and BS, E_{remain} indicates the remaining power of the UAV u, $q_u(t)$ indicates whether UAV u flies or hovers.

Action Space: Based on the current state of the system and the observed environment, conventional DRL based cellular offloading algorithm of UAV usually considers the trajectory optimization of UAV. However, this small distance movement has little impact on user communication and consumes a lot of power for UAV. Therefore, the action space of the algorithm is only composed of task offloading ratio. The action a_t for each slot t can be expressed as:

$$a_t = \{R_m^u(t)\} \tag{19}$$

DDPG can efficiently reduce the high dimensionality of discrete action space by optimizing the power distribution in the continuous action space.

Reward Function: In order to learn the dynamic computing offload strategy of the proposed model, we consider minimizing energy consumption when completing tasks within acceptable computing delays. Therefore, the cost of agent will be calculated by the task delay and the total energy cost. The reward function $r_m^u(t)$ of each user m at time slot t is defined as:

$$r_m^u(t) = -\eta \cdot T_{total}(t) - (1 - \eta) \cdot E_{total}(t) \tag{20}$$

where $\eta \in [0,1]$ is a weighting factor. The reward $r_m^u(t)$ is a negative weighted sum of instantaneous power consumption and calculation delay. By setting different values of η, we can balance energy consumption and delay for dynamic of offloading strategy.

It is worth noting that although the trajectory movement of the UAV is not considered in the action space, we set the UAV to update its position every 50 iterations and fly to the best position through the algorithm. In addition, UAV serves users in the order of user priority queue, and the number of users' tasks arrives randomly. The detailed algorithm is listed in Algorithm 2.

Algorithm 2. Offload ratio distribution based on DDPG

https://www.overleaf.com/project/637c8395e0c59844d7fa667a

1: Initialize episode length E,Replay Buffer R;
2: Randomly initialize the actor network $\mu(s \mid \theta^\mu)$ and critic network $Q(s, a \mid \theta^Q)$
3: Initialize the corresponding target network Q' and μ'
4: **for** each episode $e = 1...E$ **do**
5:     Initialize positions of UAV and users as observation
6:     **for** $t = 1..T$ **do**
7:         **if** $t \mid 50 == 0$ **then**
8:             Update the position of UAV according to Algorithm 1
9:         **end if**
10:         Select action a_t according to current policy and exploration noise
11:         Execute action a_t , observe reward $r_m^u(t)$ and observe new state s_{t+1}
12:         Store transition (s_t, a_t, r_t, s_{t+1}) in R
13:         Sample N tuples from R
14:         calculate $y_i = r_i + \gamma Q'(s_{i+1}, \mu'(s_{i+1} \mid \theta^{\mu'}) \mid \theta^{Q'})$
15:         Update Critic network by minimizing the loss:$L = \frac{1}{N}\sum_{i=1}^N (y_i - Q(s_i, a_i \mid \theta^Q))^2$
16:         Update Actor network by sampled policy gradient
17:         Update the target networks according to (20)
18:     **end for**
19: **end for**

---

4   Numerical Results and Analysis

In this section, we prove the effectiveness of DDPG based computational offloading framework in UAV-aided cellular offloading system through numerical simulation. First, we introduce the setting of simulation parameters. Then, the performance of the algorithm based on DDPG is verified in different scenarios and compared with other baseline schemes.

4.1 Simulation Setup

In the cellular offload scenario, the entire scenario is assumed to be a rectangular coordinate system with BS as the origin. Users to be served are randomly distributed in each quadrant. The boundary of the scene is a square with a side length of 200 m centered. At each moment, one UAV serves only one user. The flight height of the UAV is fixed at h_u =100 m. The total mass of the UAV is set to M_u =10 kg, and the flight speed of the UAV is V_u =20 m/s. Time is slotted by $t_0 = 1s$ and the whole time period $T = 100s$. The computing capability of BS and UAV is set to $f_{BS} = 0.6$ GHz and $f_{UAV} = 1.2$ GHz. The AWGN is assumed to be $\theta = 100$ dBm [16]. The penetration loss between the user and BS under non-line-of-sight is $P_{NLoS} = 20$ dbm, The rest of the default simulation parameters are listed in Table 1.

Table 1. Simulation Parameters

Parameter	Description	Value
U	Number of uavs	4
M	Number of users	4
T	Total period	100 s
$L_b s$	Location of the BS	(0,0)
D_0	Service area boundary	200 m
M_u	Total mass of UAV	10 kg
h_u	Flight height of UAV	100 m
V_u	Flight speed of the UAV	20 m/s
V_m	Movement speed of User	0.5 m/s
B	Bandwidth of transmission	1 MHz
$P_s end$	Power of transmission	0.1 W
θ	AWGN power	−100 dBm
$P_N LoS$	Signal penetration loss	20 dBm
$E_a ll$	Battery of uav	500 kj
s	CPU cycles per bit	1000 cycles/bit
F_{bs}	Computing power of the BS	0.6 GHz
F_u	Computing power of the UAV	1.2 GHz

4.2 Results and Discussions

For comparison, four baseline approaches are described as follows:

Local-Only Algorithm: Without the assistance of UAV, all computing tasks of user are offloaded to the BS.

Random-Choose Algorithm: The user randomly chooses to offload the tasks.

Offloading Algorithm of DQN: In order to evaluate the performance of the proposed DDPG algorithm, DQN algorithm based on discrete action space is also implemented for the dynamic offloading problem. The offloading ratio levels can be set as $R_m^u = \{0, 0.2, 0.4...1\}$.

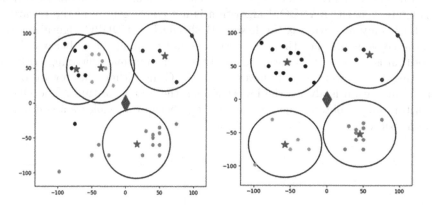

Fig. 2. Comparison between convention and optimization K-means algorithm.

DDPG with UAV moving: In order to reflect the energy saving of our framework, the real-time movement of UAV is added to the action space of DDPG. The action a_t for each slot t can be expressed as:

$$a_t = \{R_m^u(t), \Theta(t)\} \tag{21}$$

where $\Theta(t)$ represents the angle of UAV movement. The distance of UAV movement can be calculated by trigonometric function and flight speed V_u after transformation.

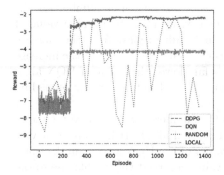
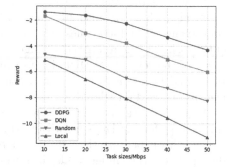

Fig. 3. Comparison between different algorithms.

Fig. 4. Performance of different algorithms under different tasks.

Figure 2 shows the initial distribution of BS and users. It shows the difference between the optimized clustering algorithm and the convention clustering algorithm. It can be seen that the convention clustering algorithm is vulnerable to the influence of edge points, and the distance between UAVs is relatively short. The optimized algorithm can achieve better clustering effect.

Figure 3 shows the performance comparison between different algorithms. We trained the algorithm for 1400 sets and set $\eta = 0.6$. It can be observed from this figure that the RANDOM algorithm is not convergent, and the reward of the Local-only algorithm is the lowest. Both DQN and DDPG algorithms can achieve convergence. For the same task size and training times, because DQN algorithm is discrete, it cannot accurately find the optimal offloading strategy, resulting in its reward slightly lower than DDPG algorithm. The reward of the DDPG algorithm is always the lowest among the four algorithms. The DDPG algorithm finally obtains the optimal strategy because it explores a continuous action space.

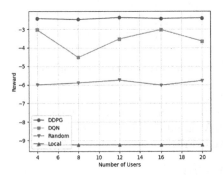

Fig. 5. Performance of different algorithms under different number of users.

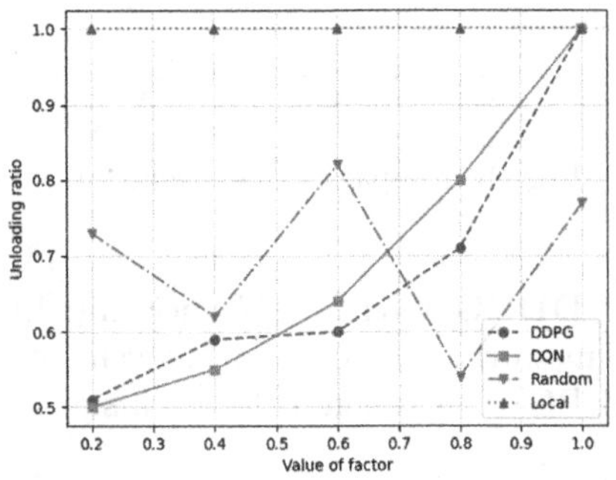

Fig. 6. Performance of Different Algorithms under Different Tasks.

Figure 4 shows the influence of different task size and user number on different algorithms. To visualize the result display, we take the average value of convergence after 1400 times of training as the result display. With the increase of the task size, the rewards of all algorithms gradually decrease, but the increase speed of the rewards of DDPG algorithm is far slower than that of DQN algorithm and Local-only algorithm, which shows the advantages of the offloading algorithm based on DDPG.

Figure 5 shows that with the increase of the number of users, the result shows that except for the DQN algorithm, the changes of all algorithms are relatively stable. This is because the output action values of the DQN algorithm fluctuate during different training, while the output values are relatively stable due to the continuity of the action space of the DDPG algorithm. In general, the results of the cellular offload framework based on the DDPG algorithm are optimal.

Figure 6 shows the influence of different reward factors on task offloading ratio. Higher offload rates indicate greater power computational consumption. The results show that as the reward factor increases, the effect of power consumption on the reward becomes smaller and the offloading ratio becomes larger. Under the same calculation frequency, if the power consumption has a greater impact on the reward, the algorithm will be more able to offload tasks to the BS.

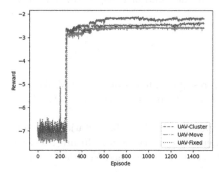

Fig. 7. Performance of different UAV strategies.

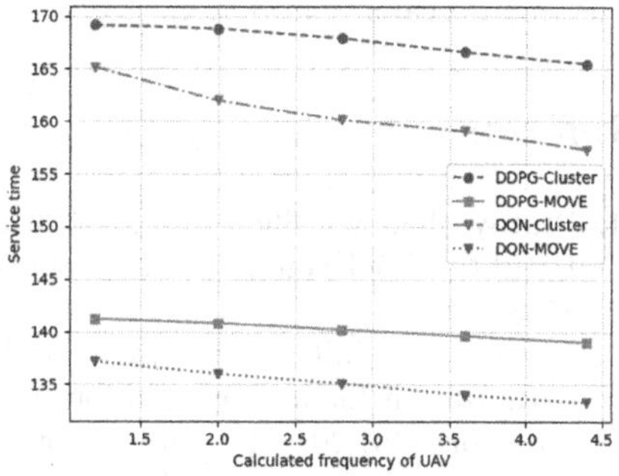

Fig. 8. The influence of different algorithms on UAV service time with the increase of UAV calculation frequency.

Figure 7 shows the change of reward under different UAV strategies of DDPG. When the reward factor $\eta = 0.6$, the results show that real-time update of UAV will consume a lot of power consumption, leading to the reward result even lower than the fixed UAV strategy. Because of the consideration of task delay, the result of every 50 clusters of UAVs is better than the fixed strategy of UAVs, just because each cluster of UAVs will find the best service location, thus reducing the delay.

Figure 8 compares the impact of the increase of UAV computing frequency on the UAV service time under the same conditions, it is obvious that the UAV clustering strategy based on DDPG algorithm is superior to the clustering strategy based on DQN algorithm. In addition, the result of clustering strategy is better than that of UAV real-time update strategy, because it reduces the cost of UAV flight. Finally, with the increase of UAV calculation frequency, the service time of UAV changes slowly.

5   Conclusion

In this paper, we describe the scenario of UAV-aided cellular offloading. By jointly optimizing UAV mobility, user scheduling and task offloading ratio, we can minimize system power consumption while delay is ensured. We are able to maximize the service time of UAV with guaranteed basic delay. Simulation results show that our framework can achieve better performance and longer UAV service time than the baseline algorithms.

Acknowledgement. This work was supported in part by the Natural Science Foundation of Jiangsu Province of China under Grant No. BK20222012, in part by the Fundamental Research Funds for the Central Universities, under Grant NS2023052, in part by the China National Key R&D Program during the 14th Five-year Plan Period under Grant No. 2022YFB2901600, and in part by the Fund of Prospective Layout of Scientific Research for NUAA.

References

1. Prathyusha, Y., Sheu, T.L.: Coordinated resource allocations for EMBB and URLLC in 5G communication networks. IEEE Trans. Veh. Technol. **71**, 8717–8728 (2022)
2. Cicek, C.T., Gultekin, H., Tavli, B., Yanikomeroglu, H.: UAV base station location optimization for next generation wireless networks: overview and future research directions. In: 2019 1st International Conference on Unmanned Vehicle Systems-Oman (UVS), pp. 1–6. IEEE (2019)
3. Xiao, Z., Zhu, L., Xia, X.G.: UAV communications with millimeter-wave beamforming: potentials, scenarios, and challenges. China Commun. **17**(9), 147–166 (2020)
4. Li, J., Lu, D., Zhang, G., Tian, J., Pang, Y.: Post-disaster unmanned aerial vehicle base station deployment method based on artificial bee colony algorithm. IEEE Access **7**, 168327–168336 (2019)
5. Ahmed, S., Chowdhury, M.Z., Jang, Y.M.: Energy-efficient UAV-to-user scheduling to maximize throughput in wireless networks. IEEE Access **8**, 21215–21225 (2020)
6. Hu, J., Zhang, H., Song, L., Han, Z., Poor, H.V.: Reinforcement learning for a cellular internet of UAVs: protocol design, trajectory control, and resource management. IEEE Wirel. Commun. **27**(1), 116–123 (2020)
7. Al-Ahmed, S.A., Shakir, M.Z., Zaidi, S.A.R.: Optimal 3D UAV base station placement by considering autonomous coverage hole detection, wireless backhaul and user demand. J. Commun. Netw. **22**(6), 467–475 (2020)
8. Koushik, A., Hu, F., Kumar, S.: Deep Q-learning-based node positioning for throughput-optimal communications in dynamic UAV swarm network. IEEE Trans. Cogn. Commun. Netw. **5**(3), 554–566 (2019)
9. Yu, Y., Tang, J., Huang, J., Zhang, X., So, D.K.C., Wong, K.K.: Multi-objective optimization for UAV-assisted wireless powered IoT networks based on extended DDPG algorithm. IEEE Trans. Commun. **69**(9), 6361–6374 (2021)
10. Dai, M., Su, Z., Li, J., Zhou, J.: An energy-efficient edge offloading scheme for UAV-assisted internet of things. In: 2020 IEEE 40th International Conference on Distributed Computing Systems (ICDCS), pp. 1293–1297. IEEE (2020)
11. Muruganathan, S.D., et al.: An overview of 3GPP release-15 study on enhanced LTE support for connected drones. IEEE Commun. Stan. Mag. **5**(4), 140–146 (2021)
12. Zhong, R., Liu, X., Liu, Y., Chen, Y.: Multi-agent reinforcement learning in NOMA-aided UAV networks for cellular offloading. IEEE Trans. Wireless Commun. **21**(3), 1498–1512 (2021)

13. Lyu, J., Zeng, Y., Zhang, R.: UAV-aided offloading for cellular hotspot. IEEE Trans. Wireless Commun. **17**(6), 3988–4001 (2018)
14. Coldrey, M., Berg, J.E., Manholm, L., Larsson, C., Hansryd, J.: Non-line-of-sight small cell backhauling using microwave technology. IEEE Commun. Mag. **51**(9), 78–84 (2013)
15. Wang, Y., Fang, W., Ding, Y., Xiong, N.: Computation offloading optimization for UAV-assisted mobile edge computing: a deep deterministic policy gradient approach. Wireless Netw. **27**(4), 2991–3006 (2021)
16. Hu, Q., Cai, Y., Yu, G., Qin, Z., Zhao, M., Li, G.Y.: Joint offloading and trajectory design for UAV-enabled mobile edge computing systems. IEEE Internet Things J. **6**(2), 1879–1892 (2018)
17. Arthur, D., Vassilvitskii, S.: k-means++: the advantages of careful seeding. Technical report, Stanford (2006)
18. Lillicrap, T.P., et al.: Continuous control with deep reinforcement learning. arXiv preprint: arXiv:1509.02971 (2015)

Artificial Bee Colony Based on Adaptive Search Strategies and Elite Selection Mechanism

Jiawen Liu[1], Wenjun Wang[2], Jiali Wu[1], Hui Wang[1(✉)], Hai Zhang[1], and Min Hu[1]

[1] School of Information Engineering, Nanchang Institute of Technology, Nanchang 330099, China
huiwang@whu.edu.cn, hujy0510@sina.com
[2] School of Business Administration, Nanchang Institute of Technology, Nanchang 330099, China
wangwenjun@nit.edu.cn

Abstract. In the field of optimization algorithms, artificial bee colony algorithm (ABC) shows strong search ability on many optimization problems. However, ABC still has a few shortcomings. It exhibits weak exploitation and slow convergence. In the late search stage, the original probability selection for onlooker bees may not work. Due to the above deficiencies, a modified ABC using adaptive search strategies and elite selection mechanism (namely ASESABC) is presented. Firstly, a strategy pool is created using three different search strategies. A tolerance-based strategy selection method is used to select a sound search strategy at each iteration. Then, to choose better solutions for further search, an elite selection means is utilized in the stage of onlooker bees. To examine the capability of ASESABC, 22 classical benchmark functions are tested. Results show ASESABC surpasses five other ABCs according to the quality of solutions.

Keywords: artificial bee colony · swarm intelligence · strategy selection · elite selection · optimization

1 Introduction

In order to efficiently address different optimization problems, strong optimization methods are required. Some swarm intelligence optimization algorithm(SIOAs) has gradually entered our field of vision, for instance particle swarm optimization (PSO) [10,19,21,28], firefly algorithm (FA) [1,17,23], and artificial bee colony (ABC) [2,4,9,11,16]. These SIOAs feature easier implementations, simpler principles, and better search capabilities than pure mathematical techniques. Among the above SIOAs, ABC is popular.

Although ABC has achieved outstanding results on several optimisation problems, it also has certain shortcomings, including slow convergence and poor exploitation. Especially, the probability selection for onlooker bees may not choose a certain sound solution in the late search stage. To tackle the above issues, different versions of ABC were proposed [3,6,15,26]. In [29], encouraged by PSO, the search method was changed by add the third item containing the optimal solution. Gao et al. [7] created a novel search technique, it is really very similar to the mutation operation of DE. In [25], an evolutionary ratio that adequately reflected the performance of each search strategy was defined. Then, based on the evolutionary ratio, an adaptive mechanism was proposed. In [13], Karaboga et al. proposed a novel definition of the neighborhood to enhance the search method, it can significantly improve the exploitation capability of onlooker bees. In. [4], Cui et al. defined a DFS framework and used elite solutions to change the search formula. Kiran and Hakli [14] introduced five search strategies in ABC, choose one of the five search strategies that is more suitable for the current search stage with the success rate. Cui et al. [5] selected good solutions based on a ranking method. Solutions with higher rankings have more chances to be selected. In [22], Wang et al. proposed a novel selection mechanism.

There are various means for ABC to improve its convergence performance, this work suggests a unique ABC (named ASESABC) with adaptive search strategies and an elite selection mechanism to ABC. Firstly, This study puts three search methods with kind of characteristics into the strategy pool. Then, to select an appropriate search strategy at each iteration, a tolerance-based strategy selection mechanism is designed. Moreover, an elite selection approach can effectively improve the working efficiency of onlooker bees, so we take advantage of it instead of probability selection. a set of 22 traditional problems with $D = 30$ and 100 are examined to evaluate ASESABC's performance, and Five modified ABC algorithms are compared to ASESABC's performance.

The essay next sections are hereinafter. The original ABC is plainly explained in Sect. 2. Section 3 offers the suggested strategy ASESABC. Section 4 presents experimental findings. In Sect. 5, conclusions are drawn.

2   Artificial Bee Colony Algorithm

Bees's foraging behavior is a very interesting natural phenomenon, According to this phenomenon, Karaboga et al. [12] puts forward a mathematical model called ABC. ABC has three different types of bees, which can complete their own simple works and cooperate with each other. The employed bees explore existing food sources in search of potentially superior ones. The onlooker bees identify some promising nectar sources and scour food source's neighbourhoods based on the search experiences of upper stage. When the employed bees could not search for a preferable solution, after the food source is abandoned, it transforms into a scout bee and look for a novel nectar source.

When ABC is put in place to an a specific case optimization problem [12], each nectar source is a promising solution. The initialized population contains SN solutions, denoted as X_1, X_2, ..., X_{SN}, where $X_i = (x_{i,1}, x_{i,2}, \ldots, x_{i,D})$ is the i-th individual, and D is problem's dimension. The standard ABC [12] are hereinafter.

Initialization--At this stage, initial SN solutions are produced at random according to the model below.

$$x_{i,j} = x_{min,j} + rnd \cdot (x_{max,j} - x_{min,j}), \tag{1}$$

where $i = 1, 2, \ldots, SN$, $[x_{min,j}, x_{max,j}]$ is the j-th dimension's border restriction, $j = 1, 2, \ldots, D$, and $rnd \in [0, 1]$ is an arbitrary value.

Employed bee phase--By using Eq. (2), a prospective individual V_i around each X_i is engender.

$$v_{i,j} = x_{i,j} + \phi_{i,j} \cdot (x_{i,j} - x_{k,j}), \tag{2}$$

where X_k is chosen stochastic from SN solutions $(k \neq i)$, $j \in \{1, D\}$ is an arbitrary integers, $\phi_{i,j}$ is an arbitrary value in [-1,1]. Based on the avaricious selection, if the candidate's individual V_i is more fit than X_i, V_i will replace X_i.

Onlooker bee phase--The last kind of bee ask the onlooker bees of their experiences. The onlooker bees then decide on some viable options and research their neighbourhoods. The selection likelihood of the i solution is expressed by p_i, its calculation process as the next shown

$$p_i = \frac{fit(X_i)}{\sum_{i=1}^{SN} fit(X_i)}, \tag{3}$$

where $fit(X_i)$ represents the fitness value of the i-th solution, and its calculation process as the next shown

$$fit(X_i) = \begin{cases} \frac{1}{1+f(X_i)}, & \text{if } f(X_i) \geq 0 \\ 1 + |f(X_i)|, & \text{if } f(X_i) < 0 \end{cases}, \tag{4}$$

where $f(X_i)$ is the goal value for the i-th individual. For each onlooker bee, a roulette wheel method is utilized to choose a good solution. Then, the onlooker bee still uses Eq. (2). To update the chosen solution, An avaricious selection is made.

Scout bee phase-- The previous two search phases, a solution may be updated by the greedy selection. When the search falls into a local minimum, by the neighborhood search, it is difficult to update solutions. In the scout bee phase, every solution X_i has a counter $trial_i$. If X_i is updated, the related $trial_i$ is set to 0; otherwise $trial_i$ is increased by 1. When the maximum $trial_i$ value reaches the $limit$ threshold value, the corresponding X_i maybe trapped into local minimum. Then, X_i is no longer used. An employed bee transforms into a scout bee, which employs Eq. (1) to create a novel solution.

3 Proposed Approach

3.1 Tolerance-Based Strategy Selection Method

In numerous research articles [8,27], researchers sought to use multiple strategies to maintain equilibrium ABC's exploration and exploitation, which is turning out

---

Algorithm 1: Tolerance-based strategy selection

---

1 **if** $T_s > M$ **then**
2 | Randomly select a new strategy SP_h from the strategy pool and $h \neq s$;
3 | Set SP_h as the next search strategy, and $s = h$;
4 | Set $T_s = 0$;
5 **end**

---

to be an effective tactic. In [24], it was suggested to use a multi-strategy ensemble of ABC (MEABC). The 3 methods of searching that make up MEABC's pool of strategies. Then, in accordance with the quality of offspring, a straightforward random technique is utilized to alter the search approach. if unable to discover a superior solution, the current search strategy is swapped out with a different stochastic search method culled from the strategy pool. It is transparent that the search tactics might be modified often. That will result in the oscillation of the search.

This paper makes reference to MEABC and utilises the same strategy pool(SP), which includes the search methods ABC,GABC,and improved ABC/best/1. The SP is defined as below.

$$SP_s = \begin{cases} v_{i,j} = x_{i,j} + \phi_{i,j} \cdot (x_{i,j} - x_{k,j}) & (5) \\ v_{i,j} = x_{i,j} + \phi_{i,j} \cdot (x_{i,j} - x_{k,j}) + \psi_{i,j} \cdot (gbest_j - x_{i,j}) & (6) \\ v_{i,j} = gbest_j + \phi_{i,j} \cdot (gbest_j - x_{k,j}) & (7) \end{cases}$$

where $s = 1, 2, 3$, SP_s is the s-th technique in the SP, $k \in \{1, 2, ..., SN\}$ is a stochastic index $(k \neq i)$, $gbest_j$ is the j-th dimension of the best individual, $j \in \{1, 2, ..., D\}$ is a stochastic index, $\phi_{i,j}$ is a stochastic element in [-1,1], and $\psi_{i,j}$ is a stochastic outcome in $[0, C]$.

In trying to look for an appropriate solution, a suitable technique for the current search can effectively improve performance. Therefore, it is crucial to choose a suitable search technique. During the search, to choose appropriate search strategies, a tolerance-based adaptive method is proposed. At the beginning of every generation, a search technique is selected from the SP. Then, the entire population uses the selected strategy to generate offspring. When the search situation changes, a new search strategy may be selected to replace the current one.

For each search strategy SP_s, it is assigned a counter T_s to record the number of unsuccessful searches ($s = 1, 2, 3$). Supposing that the current search strategy is SP_s. If a parent solution X_i is unable to be updated through its offspring V_i, the current search is unsuccessful and the corresponding counter T_s is added by 1. when every bee has completed their work, the counter T_s is compared with a preset threshold M. If T_s is larger than M, the current strategy, SP_s, is replaced by an new one, SP_h, which was chosen at random from the strategy pool. According to the above analysis, the tolerance-based strategy selection method described in Algorithm 1.

Algorithm 2: Elite selection method

---

1 Sort the population;
2 Select the top m solutions to form a temporary set Q;
3 Add $gbest$ in the collection $eSet$;
4 Calculate $mdist$ according to Eq. (8);
5 **for** each $X_i \in Q$ **do**
6 $\quad$ $flag = true$;
7 $\quad$ **for** each $X_j \in eSet$ **do**
8 $\quad\quad$ **if** $dist(i,j) > mdist$ **then**
   $\quad\quad\quad$ /* Assign the flag to false and exit the loop   */
9 $\quad\quad\quad$ $flag = false$;
10 $\quad\quad\quad$ break;
11 $\quad\quad$ **end**
12 $\quad$ **end**
   $\quad$ /* Update the elite set   */
13 $\quad$ **if** $flag == true$ **then**
14 $\quad\quad$ $eSet \leftarrow eSet \bigcup X_i$;
15 $\quad$ **end**
16 **end**

3.2   Elite Selection for the Onlooker Bee Phase

Individuals are primarily looking for good solutions near good nectar sources on the onlooker bee stage. Then, a selection probability p_i by Eq. (3) is assigned to every individual. The $fit(X_i)$ determines the probability p_i. Solutions with better fitness values increase the probability of selecting a solution for further search. However, we may not choose better solutions based on fitness values. By the suggestions of [24], the fitness definition Eq. (4) is not available in the later search stage. When the func cost of the solution is sufficiently minimal, a solution's fitness value $fit(X_i)$ runs to be 1. Then, all solutions may have the same fitness values in this situation. As a result, we cannot use the probability p_i to choose better solutions for the onlooker bees.

In this section, to choose some excellent solutions for the onlooker bees, an elite selection strategy is designed. Firstly, according to the fitness values of solutions, the population is sorted. A temporary set Q is created using the top m solutions, where $m = \rho \cdot SN$ is a parameter and lots of scientists usually define $\rho \in (0,1)$. For each solution $X_i \in Q$, the Euclidean distance $dist_i$ amid X_i and $gbest$ is calculated. Then, the mean Euclidean distance of Q is computed as below.

$$mdist = \frac{\sum_{i=1}^{m} dist_i}{m-1} \tag{8}$$

The $gbest$ is added to the elite set $eSet$. For a solution $X_i \in Q$, the Euclidean distance $dist(i,j)$ between X_i and each solution $X_j \in eSet$ are calculated. If $dist(i,j)$ is less than $mdist$ for all $X_j \in eSet$, the solution X_i is added to the collection $eSet$. The detailed steps are listed in Algorithm 2.

3.3   Framework of Proposed Approach

This study proposes a modified ABC (called ASESABC) that includes adaptive search strategies and an elite selection mechanism. Firstly, a strategy pool con-

---

Algorithm 3: Proposed Approach (ASESABC)

---

1 Based on Eq. (1), randomly construct SN solutions;
2 Randomly choose a strategy to initialize SP_s, and set $T_s = 0$;
3 Initialize $eSet$ according to Algorithm 2;
4 **while** $FEs \leq MaxFEs$ **do**
      /* Employ bee phase   */
5 | **for** $i = 1$ to SN **do**
6 | | Based on the search strategy SP_s, construct a new solution V_i;
7 | | Calculate $f(V_i)$ and $fit(V_i)$, and set FEs++;
8 | | **if** $f(V_i) < f(X_i)$ **then**
9 | | | Update X_i by V_i, and set $trial_i = 0$;
10 | | **end**
11 | | **else**
12 | | | $trial_i$++ and T_s++;
13 | | **end**
14 | **end**
      /* Onlooker bee phase   */
15 | **for** $i = 1$ to SN **do**
16 | | Randomly select an elite solution X_e from the set $eSet$;
17 | | Based on the search strategy SP_s construct a new solution V_e;
18 | | Calculate $f(V_e)$ and $fit(V_e)$, and set FEs++;
19 | | **if** $f(V_e) < f(X_e)$ **then**
20 | | | Update X_e by V_e, and set $trial_e = 0$;
21 | | **end**
22 | | **else**
23 | | | $trial_e$++ and T_s++;
24 | | **end**
25 | **end**
      /* Scout bee phase   */
26 | **if** max $\{trial_i\} > limit$ **then**
27 | | Based on Eq. (1)initialize X_i;
28 | | Calculate $f(X_i)$ and $fit(X_i)$;
29 | | Set FEs++ and $trial_i = 0$;
30 | **end**
31 | Execute the strategy update by Algorithm 1;
32 | Update $eSet$ by Algorithm 2;
33 **end**

---

tains 3 search rules. A tolerance-based strategy selection method is designed to choose suitable search strategies. Then, an elite selection tactic is utilised to aid the onlooker bee greater opt a good solution.

In Algorithm 3, the ASESABC framework is described, where FEs denotes the amount of problem assessments, $limit$ is a predetermined value, $MaxFEs$ is the greatest number of FEs, and M is a threshold value. The main operations of ASESABC consist of six parts: initialization (Lines 1–3), employed bee phase (in rows 5–14), onlooker bee phase (in rows 15–25), Scout bee phase (Lines 26–30), strategy adjustment (Line 31), and elite set updating (Line 32).

4 Experimental Study

4.1 Benchmark Problems and Parameter Settings

This chapter shows what is the outcome of the algorithm, it take 22 benchmark tasks to validate the ASESABC's capacity in the following experiments [20]. The parameter of Dimensional D is 30 and 100. We take ABC and four ABC

Table 1. Results of ASESABC under different M values when $D = 30$.

Problems	$M=40$	$M=50$	$M=60$	$M=70$	$M=80$
f_1	5.39E-57	5.27E-57	3.66E-68	**2.39E-90**	8.06E-74
f_2	3.48E-53	4.39E-55	8.52E-66	**5.49E-85**	1.83E-71
f_3	7.78E-58	1.04E-58	2.9E-70	**4.23E-91**	3.57E-50
f_4	9.55E-50	1.07E-49	1.23E-46	**2.44E-50**	1.58E-47
f_5	1.11E-29	2.9E-30	1.2E-36	**1.22E-46**	2.33E-43
f_6	7.39E+00	1.06+01	8.93E+00	**7.22E+00**	1.32E+01
f_7	**0.00E+00**	**0.00E+00**	**0.00E+00**	**0.00E+00**	**0.00E+00**
f_8	**7.18E-66**	**7.18E-66**	**7.18E-66**	**7.18E-66**	**7.18E-66**
f_9	3.98E-02	3.97E-02	3.47E-02	**3.42E-02**	5.88E-02
f_{10}	9.75E-04	8.75E-04	7.49E-04	**1.14E-04**	5.22E-02
f_{11}	**0.00E+00**	**0.00E+00**	**0.00E+00**	**0.00E+00**	**0.00E+00**
f_{12}	**0.00E+00**	**0.00E+00**	**0.00E+00**	**0.00E+00**	**0.00E+00**
f_{13}	**0.00E+00**	**0.00E+00**	**0.00E+00**	**0.00E+00**	**0.00E+00**
f_{14}	**−12569.5**	**−12569.5**	**−12569.5**	**−12569.5**	**−12569.5**
f_{15}	**2.62E-14**	2.93E-14	**2.22E-14**	**2.22E-14**	**2.22E-14**
f_{16}	**1.57E-32**	**1.57E-32**	**1.57E-32**	**1.57E-32**	**1.57E-32**
f_{17}	**1.35E-32**	**1.35E-32**	**1.35E-32**	**1.35E-32**	**1.35E-32**
f_{18}	1.58E-30	1.51E-30	**7.66E-32**	2.22E-16	2.22E-16
f_{19}	**1.35E-31**	**1.35E-31**	**1.35E-31**	**1.35E-31**	**1.35E-31**
f_{20}	**0.00E+00**	**0.00E+00**	**0.00E+00**	**0.00E+00**	**0.00E+00**
f_{21}	**−7.83E+01**	**−7.83E+01**	**−7.83E+01**	**−7.83E+01**	**−7.83E+01**
f_{22}	**−2.96E+01**	**−2.96E+01**	**−2.96E+01**	**−2.96E+01**	**−2.96E+01**

Table 2. The result of ASESABC on mean ranking values among different M.

M values	Mean ranking
40	3.20
50	3.23
60	2.82
70	**2.23**
80	3.52

variants as benchmark algorithms and based on it compare with ASESABC's result. Following is a list of the relevant algorithms.

- ABC [12].
- Quick ABC (qABC) [13].
- ABC with variable search strategy (ABCVSS) [14].

Table 3. The result of ASESABC compared with benchmark algorithms when $D = 30$.

Problem	ABC	qABC	ABCVSS	MEABC	TSaABC	ASESABC
F_1	7.4E-16	7.31E-16	6.34E-36	1.86E-40	4.67E-52	**3.11E-84**
F_2	1.53E-09	9.07E-11	1.58E-26	7.05E-37	5.23E-48	**1.13E-80**
F_3	4.97E-16	6.82E-16	9.35E-39	1.34E-41	1.58E-52	**1.80E-88**
F_4	1.11E-17	2.47E-21	5.92E-44	**4.46E-90**	4.38E-86	3.66E-64
F_5	2.02E-10	1.65E-10	6.04E-19	1.16E-21	1.59E-27	**4.13E-46**
F_6	1.29E+01	**9.93E-02**	9.15E-01	5.2E+00	1.42E+00	1.33E+00
F_7	**0.00E+00**	**0.00E+00**	**0.00E+00**	**0.00E+00**	**0.00E+00**	**0.00E+00**
F_8	8.72E-20	7.03E-66	**7.18E-66**	**7.18E-66**	**7.18E-66**	**7.18E-66**
F_9	7.17E-02	7.47E-02	4.29E-02	3.55E-02	2.71E-02	**1.0E-02**
F_{10}	5.74E-03	9.8E-03	7.86E-02	2.08E-01	4.3E-01	**2.13E-03**
F_{11}	1.78E-15	2.13E-14	**0.00E+00**	**0.00E+00**	**0.00E+00**	**0.00E+00**
F_{12}	1.44E-12	3.55E-15	**0.00E+00**	**0.00E+00**	**0.00E+00**	**0.00E+00**
F_{13}	5.55E-16	1.44E-16	2.33E-15	**0.00E+00**	**0.00E+00**	**0.00E+00**
F_{14}	−12554.6	−12451.0	−12451.0	**−12569.5**	**−12569.5**	**−12569.5**
F_{15}	4.27E-06	1.51E-06	**2.13E-14**	3.29E-14	2.81E-14	**2.13E-14**
F_{16}	7.31E-16	7.54E-16	4.87E-32	**1.57E-32**	**1.57E-32**	**1.57E-32**
F_{17}	3.1E-15	2.4E-15	1.67E-32	**1.35E-32**	**1.35E-32**	**1.35E-32**
F_{18}	3.43E-06	7.46E-07	1.72E-18	2.78E-22	1.75E-27	**2.22E-34**
F_{19}	1.63E-13	1.03E-15	2.30E-31	**1.35E-31**	**1.35E-31**	**1.35E-31**
F_{20}	6.87E-03	1.41E-03	**0.00E+00**	**0.00E+00**	**0.00E+00**	**0.00E+00**
F_{21}	−7.83E+01	−7.83E+01	−7.83E+01	−7.83E+01	−7.83E+01	−7.83E+01
F_{22}	−2.95E+01	−2.93E+01	2.96E+01	2.96E+01	2.96E+01	2.96E+01

- Multi-strategy ensemble ABC (MEABC) [24].
- Adaptive ABC with two search strategies (TSaABC) [18].
- Proposed approach ASESABC.

For the above six ABCs, the values for the common parameters swarm number (SN), maximum number of function evaluations ($MaxFEs$), and $limit$ are set to 50, $5000 \cdot D$, and $SN \cdot D$, respectively. Throughout the experiments, each problem is tested 30 times for every ABC and the average outcomes are reported.

4.2   Study on the Parameter M

In Sect. 3.1, a new parameter M is defined as a threshold value to control the tolerance of changing strategies. For a small M, the strategy may be changed frequently, and this tends to cause oscillation in the search. For a large M, an inappropriate strategy may be used for a long time. This is not beneficial for balancing exploitation and exploration. Therefore, the value of M may influent on the performance of ASESABC. To compare the effectiveness of various ASESABC, the best M can be obtained. In this section, M is set to 40, 50, 60, 70, and 80, respectively.

The computation of ASESABC for various M values is shown in Table 1. As seen, ASESABC with $M = 70$ is not worse than other cases except for problem f_{18}. $M = 60$ achieves the best result in f_{18}. To further determine the M value, Friedman experiment is tested. ASESABC's ranking values under various M values are shown in Table 2. According to the results, $M = 70$ has the highest rating value. It implies that the optimal setting is $M = 70$.

4.3 Comparison of ASESABC with Other ABC Algorithms

The proposal ASESABC is contrasted to five distinct ABCs with $D = 30$ and 100. Table 3 lists the comparative outcomes of ASESABC for $D = 30$. Six algorithms produced the best outcomes, which are displayed in bold. As can be seen from the outcomes, on all experiment issues, ASESABC is not worse than other ABCs, outside of f_4 and f_6. On f_4, MEABC obtains the best solution. qABC outperforms other ABCs on f_6. Compared with ABC, ABCVSS, and TSaABC and ASESABC achieve similar performance on 2, 8, and 12 problems, respectively. On 22 test functions, ASESABC's results are better than or equal to

Table 4. The result of ASESABC compared with benchmark algorithms when $D = 100$.

Problem	ABC	qABC	ABCVSS	MEABC	TSaABC	ASESABC
F_1	6.1E-15	1.69E-15	4.54E-31	1.58E-36	5.38E-49	**1.01E-78**
F_2	4.5E-09	1.16E-10	7.47E-23	3.53E-33	1.35E-40	**2.87E-71**
F_3	2.8E-15	3.43E-15	3.18E-29	1.14E-36	3.44E-45	**2.09E-80**
F_4	5.4E-17	5.31E-22	2.86E-47	1.41E-85	**9.84E-94**	6.46E-44
F_5	2.3E-09	2.21E-09	1.57E-16	2.23E-19	1.00E-22	**1.43E-40**
F_6	5.3E+01	5.1E+01	**1.7E+01**	3.9E+01	3.0E+01	7.9E+01
F_7	**0.00E+00**	**0.00E+00**	**0.00E+00**	**0.00E+00**	**0.00E+00**	**0.00E+00**
F_8	5.9E-28	**7.1E-218**	**7.1E-218**	**7.1E-218**	**7.1E-218**	**7.1E-218**
F_9	2.25E-01	2.98E-01	2.17E-01	1.68E-01	1.26E-01	**1.19E-01**
F_{10}	1.93E-02	4.48E-02	8.69E-02	2.8E-01	2.02E+00	**8.9E-03**
F_{11}	1.8E-15	1.78E-15	**0.00E+00**	**0.00E+00**	**0.00E+00**	**0.00E+00**
F_{12}	1.01E+00	8.7E-14	**0.00E+00**	**0.00E+00**	**0.00E+00**	**0.00E+00**
F_{13}	1.9E-15	1.72E-15	1.67E-15	**0.00E+00**	**0.00E+00**	**0.00E+00**
F_{14}	−41412	−41048.9	−41898.3	−41898.3	−41898.3	−41898.3
F_{15}	1.6E-05	3.39E-06	**3.06E-14**	3.95E-13	1.29E-13	1.86E-13
F_{16}	3.4E-15	2.76E-16	6.73E-32	**4.71E-33**	**4.71E-33**	**4.71E-33**
F_{17}	7.2E-15	1.71E-15	2.69E-32	**1.35E-32**	**1.35E-33**	**1.35E-32**
F_{18}	5.22E-03	3.72E-05	1.96E-13	1.91E-19	1.02E-22	**7.96E-28**
F_{19}	3.6E-13	4.10E-15	2.97E-31	**1.35E-31**	**1.35E-31**	**1.35E-31**
F_{20}	4.09E-02	3.28E-02	**0.00E+00**	**0.00E+00**	**0.00E+00**	**0.00E+00**
F_{21}	**−7.83E+01**	**−7.83E+01**	**−7.83E+01**	**−7.83E+01**	**−7.83E+01**	**−7.83E+01**
F_{22}	−9.79E+01	−9.76E+01	−9.76E+01	**−9.96E+01**	**−9.96E+01**	**−9.96E+01**

Table 5. The result of ASESABC and benchmark algorithms' mean ranking values.

Algorithm	$D = 30$	$D = 100$
	Mean ranking	Mean ranking
ABC	5.14	5.32
qABC	4.73	4.50
ABCVSS	3.70	3.34
MEABC	2.75	2.98
TSaABC	2.50	2.48
ASESABC	**2.18**	**2.39**

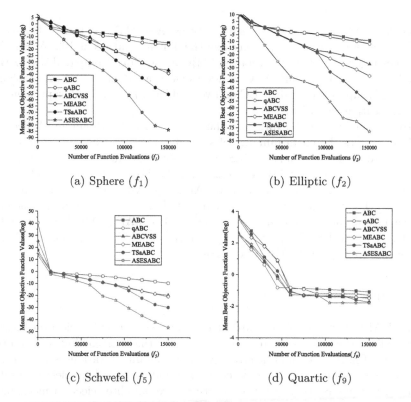

Fig. 1. The graph of convergence curves of ASESABC and benchmark algorithms on four test function (f_1, f_2, f_5, and f_9) when $D = 30$.

ABC, ABCVSS, TSaABC. Figure 1 shows the convergence curve of ASESABC and benchmark algorithms on four test functions f_1, f_2, f_5, and f_9. With other ABCs compared, ASESABC showed the fastest convergence. For f_9, TSaABC's convergence is better than ASESABC in the intermediate search process, but ASESABC is faster than TSaABC soon in later.

Table 6. The result of passing the Wilcoxon experiment among ASESABC and benchmark algorithms.

ASESABC vs.	$D = 30$	$D = 100$
	p-value	p-value
ABC	**8.9E-05**	**1.3E-03**
qABC	**1.5E-03**	**1.5E-03**
ABCVSS	**4.18E-03**	1.24E-01
MEABC	**3.29E-02**	2.48E-01
TSaABC	**3.29E-02**	7.9E-01

Table 4 lists the comparison results between ASESABC and five other ABCs for $D = 100$. For most test problems and the dimensional size is 100, ASESABC continues to perform better than other ABC algorithms. Compared with ABC, qABC, and MEABC, ASESABC is not worse than them on all 22 problems. ABCVSS is superior to ASESABC on two problems f_6 and f_{15}. For problem f_4, TSaABC surpasses ASESABC.

To compare six ABC algorithms, there are tests in Friedman and Wilcoxon. Table 5 provides the mean rankings of six compared algorithms on $D = 30$ and 100. According to the results, ASESABC receives the highest ranking value. It proves that, out of the six ABC algorithms, ASESABC is the most effective one. The Wilcoxon tests p-values amid ASESABC and ABC variants have emerged in Table 6. The bold of p-values is demonstrated in the situation under the 0.05 evident levels. For $D = 30$, in comparison to ABC, qABC, ABCVSS, MEABC, and TSaABC, ASESABC performs noticeably better. For $D = 100$, ASESABC significantly outperforms ABC and qABC.

5   Conclusion

A novel ABC with adaptive search techniques and an elite selection mechanism (called ASESABC) is presented to address the shortcomings of the original ABC. Firstly, we built a strategy pool containing various search rules. Then, a tolerance-based strategy selection method is created to select a suitable search technique at every iteration. Furthermore, an elite selection mechanism is employed to substitute the original probability selection. In the experiments, with $D = 30$ and 100, 22 well-known issues are utilized.

ABC, qABC, ABCVSS, MEABC, and TSaABC performance are compared with that of ASESABC. Outcomes demonstrate that ASESABC is surpass other ABC algorithms on most issues. Although the problem dimension is increased to 100, our approach still gets the best results among the six ABC variants. Study on the parameter M shows that different M values affect the performance of ASESABC. The statistical test demonstrates that $M = 70$ obtains the relatively best choice. In the experiments, ASESABC is only put to the test with some classical problems. In a further study, the performance of ASESABC on some complex benchmark sets will be investigated.

Acknowledgements. This work was supported by Jiangxi Provincial Natural Science Foundation (Nos. 20212BAB202023 and 20212BAB202022).

References

1. Arora, S., Kaur, R.: An escalated convergent firefly algorithm. J. King Saud Univ. Comput. Inf. Sci. **34**, 308–315 (2018)
2. Bajer, D., Zoric, B.: An effective refined artificial bee colony algorithm for numerical optimisation. Inf. Sci.: Int. J. **504**, 221–275 (2019)
3. Cui, L., Kai, Z., Li, G., Fu, X., Jian, L.: Modified Gbest-guided artificial bee colony algorithm with new probability model. Soft. Comput. **22**(2), 1–27 (2018)
4. Cui, L., et al.: A novel artificial bee colony algorithm with depth-first search framework and elite-guided search equation. Inf. Sci. **367–368**, 1012–1044 (2016)
5. Cui, L., et al.: A ranking-based adaptive artificial bee colony algorithm for global numerical optimization. Inf. Sci. **417**, 169–185 (2017)
6. Gao, W., Chan, F., Huang, L., Liu, S.: Bare bones artificial bee colony algorithm with parameter adaptation and fitness-based neighborhood. Inf. Sci. **316**, 180–200 (2015)
7. Gao, W.F., Liu, S.Y.: A modified artificial bee colony algorithm. Comput. Oper. Res. **39**(3), 687–697 (2012)
8. Gao, W., Huang, L., Liu, S., Chan, F., Dai, C., Shan, X.: Artificial bee colony algorithm with multiple search strategies. Appl. Math. Comput. **271**, 269–287 (2015)
9. Gao, W., Liu, S., Huang, L.: Enhancing artificial bee colony algorithm using more information-based search equations. Inf. Sci. **270**, 112–133 (2014)
10. Wang, H., Wu, Z., Rahnamayan, S., Liu, Y., Ventresca, M.: Enhancing particle swarm optimization using generalized opposition-based learning. Inf. Sci. **181**(20), 4699–4714 (2011)
11. Ji, J., Song, S., Tang, C., Gao, S., Tang, Z., Todo, Y.: An artificial bee colony algorithm search guided by scale-free networks. Inf. Sci. **473**, 142–165 (2019)
12. Karaboga, D.: An idea based on honey bee swarm for numerical optimization, Technical report - tr06 (2005)
13. Karaboga, D., Gorkemli, B.: A quick artificial bee colony (qABC) algorithm and its performance on optimization problems. Appl. Soft Comput. **23**, 227–238 (2014)
14. Kiran, M.S., Hakli, H., Gunduz, M., Uguz, H.: Artificial bee colony algorithm with variable search strategy for continuous optimization. Inf. Sci. **300**, 140–157 (2015)
15. Liang, Z., Hu, K., Zhu, Q., Zhu, Z.: An enhanced artificial bee colony algorithm with adaptive differential operators. Appl. Soft Comput. **58**, 480–494 (2017)
16. Peng, C.: Best neighbor-guided artificial bee colony algorithm for continuous optimization problems. Soft Comput.: Fusion Found., Methodol. Appl. **23**(18) (2019)
17. Peng, H., Zhu, W., Deng, C., Wu, Z.: Enhancing firefly algorithm with courtship learning. Inf. Sci. **543**, 18–42 (2020)
18. Song, Q.: A high-efficiency adaptive artificial bee colony algorithm using two strategies for continuous optimization. Swarm Evol. Comput. **50**, 100549 (2019)
19. Tao, X., Li, X., Chen, W., Liang, T., Qi, L.: Self-adaptive two roles hybrid learning strategies-based particle swarm optimization. Inf. Sci. **578**(8), 457–481 (2021)

20. Ty, A., et al.: Artificial bee colony algorithm with efficient search strategy based on random neighborhood structure. Knowl.-Based Syst. **241**, 108306 (2022)

21. Wang, F., Zhang, H., Li, K., Lin, Z., Yang, J., Shen, X.L.: A hybrid particle swarm optimization algorithm using adaptive learning strategy. Inf. Sci. **436–437**, 162–177 (2018)

22. Wang, H., Wang, W., Xiao, S., Cui, Z., Zhou, X.: Improving artificial bee colony algorithm using a new neighborhood selection mechanism. Inf. Sci. **527**, 227–240 (2020)

23. Wang, H., et al.: Firefly algorithm with neighborhood attraction. Inf. Sci. **382–383**, 374–387 (2016)

24. Wang, H., Wu, Z., Rahnamayan, S., Sun, H., Liu, Y., Pan, J.S.: Multi-strategy ensemble artificial bee colony algorithm. Inf. Sci. **279**, 587–603 (2014)

25. Xs, A., Ming, Z.A., Sx, B.: A multi-strategy fusion artificial bee colony algorithm with small population. Expert Syst. Appl. **142**, 112921 (2020)

26. Yu, W., Zhan, Z., Zhang, J.: Artificial bee colony algorithm with an adaptive greedy position update strategy. Soft. Comput. **22**, 437–451 (2018)

27. Yurtkuran, A., Emel, E.: An adaptive artificial bee colony algorithm for global optimization. Appl. Math. Comput. **271**, 1004–1023 (2015)

28. Zhang, X., Lin, Q.: Three-learning strategy particle swarm algorithm for global optimization problems. Inf. Sci.: Int. J. **593**, 289–313 (2022)

29. Zhu, G., Kwong, S.: Gbest-guided artificial bee colony algorithm for numerical function optimization. Appl. Math. Comput. **217**(7), 3166–3173 (2010)

Minimization of Active Power Loss Using Enhanced Particle Swarm Optimization

Samson Ademola Adegoke[1], Yanxia Sun[1]([✉]), and Zenghui Wang[2]

[1] Department of Electrical and Electronic Engineering Science, University of Johannesburg, Johannesburg, South Africa
ysun@uj.ac.za
[2] Department of Electrical Engineering, University of South Africa, Pretoria, South Africa

Abstract. Weak bus identification in power system networks is crucial for planning and operation since most generators operate close to their operating limits, resulting in generator failures. This work aims to identify the critical/weak node and reduce the system's actual power losses. The line stability index (L_{mn}) and fast voltage stability index (FVSI) were used to identify the critical node and lines close to instability in the power system networks. The enhanced particle swarm optimization (EPSO) was chosen because of its ability to communicate with better individuals, making it more efficient to obtain a prominent solution. EPSO and other PSO variants were used to minimize the system's actual/real losses. Nodes 8 and 14 were identified as the critical nodes of the IEEE 9 and 14 bus systems, respectively. The power losses of the IEEE 9 bus system were reduced from 9.842 MW to 7.543 MW, and for the 14 bus system, the losses were reduced from 13.775 MW of the base case to 12.253 MW for EPSO. The EPSO gives a better active power loss reduction and improves the node voltage profile than other PSO variants and algorithms in the literature. This suggests the feasibility and suitability of the EPSO to improve the grid voltage quality.

Keywords: Voltage stability · Identification of weak bus · FVSI and L_{mn} · Diminish power loss · PSO variants · EPSO

1 Introduction

Voltage stability (VS) is a major focus of power system (PS) utility companies. Therefore, VS is the ability of systems to keep the voltage profile when undergoing large or small disturbances. The increase in the infrastructure and load demand rate leads to the high utilization of PS energy equipment. This has made the system experience voltage instability that usually leads to blackouts in some parts of the world, destruction of some businesses and daily activities, and increased power loss. Based on this problem, this study is motivated by identifying the weak bus that could cause the blackout in the system and reducing the system transmission loss, which is caused by a shortage of reactive power (RP), which is significantly affecting the quality of energy delivery to the consumer end. Environmental and economic factors are two leading causes for establishing

© The Author(s), under exclusive license to Springer Nature Singapore Pte Ltd. 2023
H. Zhang et al. (Eds.): NCAA 2023, CCIS 1869, pp. 315–333, 2023.
https://doi.org/10.1007/978-981-99-5844-3_23

a new transmission line (TL). As the load increases, the system is heavily loaded, and maintaining stability becomes difficult; thus, the system operates close to the instability point [1–4]. Evaluation of system stability is based on the node/bus voltage profile. Presently, there are numerous techniques for finding system stability. The P-V and Q-V curves [5], Line collapse proximity index (LCPI) [6], line stability factor (LQP) [7], line voltage stability index (LVSI) [8], a simplified voltage stability index (SVSI) [9], NVSI [10], L-index [11], global voltage stability index (GVSI) [12], etc. Some classical voltage stability assessments and various techniques have been proposed for weak bus identification in power system networks. Some of them are Genetic algorithms based on support vector machine (GA-VSM) [13], "Artificial neural network (ANN) [14], Fuzzy logic" [15], Ant colony (AC) [16], electric cactus structure (ECS) [17], network response structural characteristic (NRSC) [18].

Shortage of RP in the PS network has caused a tremendous waste of electricity in the distribution system, resulting in extra emission of carbon production and power generation cost. Therefore, reducing losses in transmission line networks (TLN) is essential for system safety. However, the best way to reduce losses in TL of PS networks is RP optimization (RPO). Two methods used in solving the RPO problem are traditional and evolutionary algorithms. Traditional methods such as Newton-Rapson (NR) [19], interior point methods [20], quadratic programming [21], and linear programming [22]. Recently, evolution algorithms have been used to solve RPO problems such as particle swarm optimization (PSO) [23], hybrid pathfinder algorithm (HPFA) [24], Hybrid (HPSO-PFA) [25], modified PFA (mPFA) [26], chaotic krill herd [27], moth–flame optimizer (MFO) [28], ant lion optimizer (ALO) [29], tree seed algorithm (TSA) [30], and improved pathfinder algorithm (IPFA) [31]. However, PSO is good in search capacity and has less programming than others [32].

Harish et al. used fast voltage stability index (FVSI) and line stability index (Lmn) to identify the location of flexible alternating current (FACTS) devices along with PSO, artificial bee colony (ABC), and hybrid genetic algorithm (H-GA) to find the sizing of the FACTS devices [33]. Also, a novel method for strengthening PS stability was proposed by Jaramillo et al. [34]. FVSI was used to identify the node on which the SVC should be installed under N - 1 scenario. It was reported that the result obtained could reestablish the FVSI in each contingency before the outage [34]. Voltage collapse critical bus index (VCCBI) [35], "L-Index, Voltage Collapse Proximity Index (VCPI), and modal analysis" [36], which are part of voltage stability indices (VSI), were used to identify weak/vulnerable buses in electrical power system. Power loss reduction has been made using a hybrid loop-genetic-based algorithm (HLGBA) [37], Jaya algorithm (JAYA), diversity-enhanced (DEPSO), etc. [38].

This research considers the Identification of critical nodes and loss reduction, which is lacking in the previous work mentioned above. FVSI and L_{mn} were used to find the critical node in the system based on load flow (LF) results from MATLAB software. FVSI and L_{mn} are chosen because of their efficiency in identifying the weak/vulnerable bus (i.e., the fastness of FVSI and the accuracy of L_{mn}) [39]. The critical node was determined by the value of the indices (i.e., FVSI and L_{mn}). When the indices value reaches unity or is close to unity, that node is the critical node of the system. The reactive powers of all the load buses were increased one after the other to determine the maximum RP on each of the load buses/nodes. Also, each line's value of FVSI and L_{mn} was computed to determine the load-ability limit on each load bus. The ranking was done based on the indices value of each node; hence, the node with the highest indices values is the system's critical bus. This bus contained the smallest RP when the load bus was varied. The identified node needs reactive power support to avoid voltage collapse. Also, enhanced PSO (EPSO) was used to minimize the PS network loss along with PSO variants that have been developed by previous research, such as PSO-based time-varying acceleration coefficients (PSO-TVAC) [40], random inertia weight PSO (RPSO), PSO based on success rate (PSO-SR) [41]. To overcome the premature convergence of PSO, the chosen EPSO has been applied, and it uses neighborhood exchange to share more information with other best individual (neighborhood) to improve itself, which make it more efficient in getting a prominent solution (i.e., exploitation stage) to optimizing the objective function. However, the contribution of this work are 1) comparison of different PSO variants along with EPSO for power loss reduction. 2) Identified the PS critical node for the perfect operation of generators to avoid breakdown.

The rest of the paper is structured as follows: Sect. 2 presents the problem formulation of voltage stability indices and RPO, and Sect. 3 discusses the PSO, it is variants and EPSO. The result and discussion is presented in Sect. 4, and the conclusion and future work is the Sect. 5.

2 Problem Formulation

2.1 Formulation of Voltage Stability Indices

FVSI and L_{mn} are part of the VSI methods for identifying weak buses. They were formed based on two bus systems. For FVSI and L_{mn}, for a system to be stable, the value must be smaller than unity and unstable when it equals unity and above.

2.1.1 FVSI

FVSI was developed [42] based on the concept of a single line of power flow (PF). The FVSI is evaluated by:

$$FVSI = \frac{4Z^2 Q_1}{V_1^2 X} \leq 1 \tag{1}$$

where; Z is the impedance. Q_1 is the RP at sending ends, X is the reactance of the line, and V_1 is the voltage at the sending end.

2.1.2 L_{mn}

L_{mn} Was proposed by [43]. Using the concept of PF from a single-line diagram, the discriminant of the quadratic voltage equation is set to be higher than or equal to zero. The equation is given below.

$$L_{mn} = \frac{4X Q_2}{[V_1 \sin(\theta - \delta)]^2} \leq 1 \tag{2}$$

where; θ is the angle of the TL, δ is the power angle, and Q_2 is the RP at receiving ends.

2.2 Steps Involve in Identifying Critical Node in EPS

Identifying critical nodes in EPS is essential for delivering electricity supply to consumers. The following steps are involved.

1. Input the line and bus data of the IEEE test case.
2. Run the PF solution in MATLAB using the NR method at the base case
3. Calculate the stability values of FVSI and L_{mn} of the IEEE test system
4. Gradually increased the RP of the load bus until the values of FVSI and L_{mn} were closer to one (1)
5. The load bus with the highest FVSI and L_{mn} value is selected at the critical node.
6. Steps 1 to 5 are repeated for all the load buses.
7. The highest RP loading is selected and called maximum load-ability.
8. The voltage magnitude at the critical loading of a particular load bus is obtained and is called the critical voltage of a specific load bus.

2.3 Formulation of RPO

The main objective of RPO is to reduce the network's actual power loss.

2.3.1 Objective Function

The main goal of RPO is to reduce the network's actual power losses, which are described as follows.

$$\text{Minimize} f = P_{loss}(x, u) \tag{3}$$

which satisfying

$$\begin{cases} g(x, u) = 0 \\ h(x, u) \leq 0 \end{cases} \tag{4}$$

where; $f(x, u)$ is the objective function, $g(x, u)$ is the equality constraints, $h(x, u)$ is the inequality constraints, x is the vector of the state variables, u is the vector of the control variables.

The real power loss minimization in the TL is given below, and its purpose is to reduce the overall loss in the transmission networks

$$Minf = P_{loss} \sum_{K=1}^{N_L} G_k \left(v_i^2 + v_j^2 - 2V_i V_j \cos \theta_{ij} \right) \tag{5}$$

where; P_{loss} is the real total losses, K is the branch, G_k is the conductance of the branch k, Vj and V_i is the voltage at the ith and jth bus, N_L is the total number of TL, θ_{ij} is the voltage angle between bus i and j.

2.3.2 Equality Constraints

The PS network's active and reactive PF equation is called the equality constraints.

$$P_{gi} - P_{di} - V_i \sum_{j=1}^{N} V_j(g_{ij} \cos \theta_{ij} + b_{ij} \sin \theta_{ij}) = 0 \tag{6}$$

$$Q_{gi} - Q_{di} - V_i \sum_{j=1}^{N} V_j(g_{ij} \sin \theta_{ij} - b_{ij} \cos \theta_{ij}) = 0 \tag{7}$$

Here, g_{ij} *and* b_{ij} are conductance and susceptance, and θ_{ij} is the phase difference of voltages.

2.3.3 Inequality Constraints

Inequality constraints are operational variables that must be kept within acceptable limits.

1) Generator constraints

$$V_{gi}^{min} \leq V_{gi} \leq V_{gi}^{max} i = 1 \ldots, N_g \tag{8}$$

$$Q_{gi}^{min} \leq Q_{gi} \leq Q_{gi}^{max} i = 1 \ldots, N_g \tag{9}$$

2) Reactive compensation constraints

$$Q_{ci}^{min} \leq Q_{ci} \leq Q_{ci}^{max} i = 1 \ldots, N_C \tag{10}$$

3) Transformer tap ratio constraints

$$T_k^{min} \leq T_k \leq T_k^{max} i = 1 \ldots, N_T \tag{11}$$

where, V_{gi}^{min} *and* V_{gi}^{max} are voltage amplitude limits, Q_{gi}^{min} *and* Q_{gi}^{max} are the generation limits of reactive power, P_{gi}^{min} *and* P_{gi}^{max} are the limits of active power, Q_{ci}^{min} *and* Q_{ci}^{max} are the reactive compensation limits, and T_k^{min} *and* T_k^{max} are the transformer tap limits.

The penalty function is used to make optimization problems more straightforward and rigorous. In other to solve the optimization problem penalty function has to be selected. The primary function of the penalty function is to keep the system security within the acceptable limit.

$$f_T = f + \lambda_V \sum_{K=1}^{N_B} (V_i - V_i^{lim})^2 + \lambda_g \sum_{K=1}^{N_B} (Q_{gi} - Q_{gi}^{lim})^2 + \lambda_s \sum_{K=1}^{N_B} (S_i - S_i^{lim})^2 \tag{12}$$

where,

$$\lambda_V, \lambda_g, \text{ and } \lambda_s \text{ are the penalty factors.} \tag{13}$$

$$V_i^{lim} = \begin{cases} V_i^{lim}, & if \ V_i < V_i^{min} \\ V_i^{lim}, & if \ V_i > V_i^{max} \end{cases} \tag{14}$$

$$Q_{gi}^{lim} = \begin{cases} Q_{gi}^{lim}, & if \ Q_{gi} < Q_{gi}^{min} \\ Q_{gi}^{lim}, & if \ Q_{gi} > Q_{gi}^{max} \end{cases} \tag{15}$$

$$S_i^{lim} = \begin{cases} S_i^{lim}, & if \ S_i < S_i^{min} \\ S_i^{lim}, & if \ S_i > S_i^{max} \end{cases} \tag{16}$$

3 Particle Swarm Optimization (PSO)

3.1 PSO and It is Variants

3.1.1 Overview of PSO

PSO was created in 1995 by Kennedy and Eberhart [44]. The social behavior of birds and schooling fish was the basis for the population-based stochastic optimization method known as PSO. In the search space, PSO makes use of the promising area. Each particle moves and adjusts its position following its past behavior and the best particle within a decision time. Each particle is identified by a d-dimensional vector that depicts its location in the search space. The position of the vector is represented as a possible solution to the optimization issue. Whenever an iterative process is performed, the velocity is added to update each particle's position [45]. The best particle in the swarm (population) and the distance from the best cognitive both impact the particle's velocity. The formulae for velocity and position are shown below.

$$V_i^{k+1} = w_1 v_i^k + c_1 \times r_1(p_{best} - s_i^k) + c_2 \times r_2(g_{best} - s_i^k) \tag{17}$$

$$s_i^{k+1} = s_i^k + V_i^{k+1} \tag{18}$$

3.1.2 RPSO

In RPSO, the weighting factor was usually between 0.5 and 1 [46]. Roy Ghatak et al. claims that the random inertia weight component enhanced the initial objective function. Stocking to the local optimal at the end of the iteration may affect the accuracy of the solution [41]. The value of w is found in Eq. 19.

$$w = 0.5 + rand()/2 \tag{19}$$

3.1.3 PSO-SR

To find the best method for effective management of inertia weight (w), a novel adaptive w was developed based on success rate (SR) [46]. At each iteration, the swarm position is determined using the SR. Indicate that a big value of w is necessary to advance toward the optimal point when the SR is large. Additionally, for a low value, the particle oscillates around the ideal location and requires a small increase in the w value to reach the perfect result [41].

$$w(t) = \left(w^{max} - w^{min}\right) * SR + w^{min} \tag{20}$$

where; SR is the success rate and is chosen to be 1; otherwise, it is zero, w^{min} is the minimum limits of the w, and w^{max} is the maximum limits of the w [41].

3.1.4 PSO-TVAC

Due to the lack of diversity towards the end of the search area, PSO with PSOTV-w was utilized, which locates the optimal solution more quickly but is less effective at tuning the optimal solution. The accuracy and effectiveness of the PSO to obtain optimal solutions are significantly influenced by the tuning parameter [40, 47]. Based on this concept, a TVAC was proposed to enhance the global search at the start of optimization, allowing the particle to move to the global optimum at the end of the search space. As the search progresses, c_1 and c_2 alter over time, decreasing the cognitive components and increasing the social components. This indicates that the particle converges to the global optimum towards the end of the search process due to minor cognitive and greater social components. Additionally, the particle can roam throughout the search area rather than initially gravitating toward the best population due to enhanced cognitive and minor social components [40].

$$C_1 = (C_{1t} - C_{1k})\frac{z}{iter_{max}} + C_{1k} \tag{21}$$

$$C_2 = (C_{2t} - C_{2k})\frac{z}{iter_{max}} + C_{2k} \tag{22}$$

where; z is the current iteration, $iter_{max}$ is the maximum iteration, and $C_{1k}, C_{1t}, C_{2t},$ and C_{2k} are the initial and final values of the cognitive and social acceleration factors. The value for C_{1t} and C_{2k} is 0.5, and 2.5 for C_{1k} and C_{2t} is the most accurate value [48].

3.2 EPSO

In other to overcome the issue of falling into local optimal from the standard PSO, the chosen EPSO added some expansion to basic PSO, such as constriction factor and neighborhood model.

3.2.1 Exchange of Neighborhood

Since each particle learns from its own personal and global positions in the PSO algorithm in the social cognitive system, apart from personal experience and better information received from the search areas, it is advisable to share with better individuals to enhance or improve itself [49]. Therefore, using that concept, a new acceleration constant (c_3) is added to the original PSO equation, making it more efficient to obtain the best solution. The additional c_3 gives the swarm the capability to reach the exploitation stage.

$$V_i^{k+1} = \varphi\left(w_1 v_i^k + c_1 r_1\left(p_{best} - s_i^k\right) + c_2 r_2\left(g_{best} - s_i^k\right) + c_3 \times r_3\left(p_{best,t} - s_i^k\right)\right)$$

(23)

where, $p_{best,t}$ is the vector position for an excellent individual domain (i;e the overall best position), and r_3 is random numbers in the interval of (0,1)

3.2.2 Implementation of EPSO to RPO

The steps involved in solving the RPO problem are given below, and the flow chat is shown in Fig. 1

1. Initialize: Set the number of particles, initial velocity, the total number of iterations, generator voltages, the transformer tap settings, and accelerated constants.
2. Run load flows to determine the objective function (real power loss) and evaluate the penalty function concerning inequality constraint violation.
3. Counter updating: Update the iter = iter + 1
4. Evaluate each particle and save the global and personal best positions.
5. Update the velocity as given in Eq. (23)
6. Update the position as given in Eq. (18).

Check whether solutions in steps 3 and 4 are within the limit; if it is above the limit, apply Eq. (12) to keep the violation.

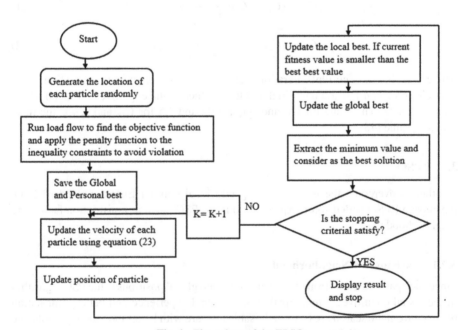

Fig. 1. Flow chart of the EPSO

1. The position of the local best should be updated if the current fitness value is smaller than the best one.
2. Update global best
3. Search for minimum value: The minimum value in all the individual iterations is considered the best solution.
4. Stopping criteria: If the stopping criteria have been satisfied, stop; if not, go back to step 5.

4 Result and Discussion

The algorithm settings used in this study are shown in Table 1. The simulation was executed using MATLAB 2018a. IEEE 9 and 14 bus systems were employed to evaluate the effectiveness of the techniques for actual power reduction. The best outcomes are documented after each test system has been run 30 times.

4.1 Voltage Stability Indices

4.1.1 IEEE 9 Bus System

The line and bus data are taken [50, 51]. The critical node of the system was selected by varying/increasing the RP of the load bus until one of the indices values approaches unity, and the bus contains the lowest permissible RP. The load bus's RP increases one after the other to find the maximum reactive power of each load bus. Table 2 illustrates the value of the indices, the voltage magnitude, and each load bus ranking. The bus with the highest values of the index and a smaller load-ability value of RP is ranked first and is the system's critical node. Node 8 is the critical node and ranks first in the system because it contains a small RP load variation of 240 MVar and a value of 1.028 for FVSI. The lines connected to it are 7–8 and 8–9. This bus needs urgent attention to avoid the breakdown of the generator.

Table 1. Optimum setting of the algorithm

Parameters	Value
Number of Iterations	200
Particle Number	50
Acceleration constant	C1 = C2 = C3 = 2.05
Maximum and minimum w	0.9 and 0.4
Constriction factor	0.729

Table 2. The indices value and ranking of the bus in the IEEE 9 bus system

Bus Numbers	Q (MVar)	L_{mn}	FVSI	Voltage magnitude (p.u)	Ranking
5	260	0.902	0.933	0.800	2
6	290	0.865	0.889	0.824	3
8	**240**	0.998	**1.028**	0.802	**1**

4.1.2 IEEE 14 Bus System

The IEEE 14 bus system contains 4 generators, a slack bus, 20 transmission lines, and 9 load buses located at buses 4, 5, 7, 9, 10, 11, 12, 13, and 14. The line and bus data are taken [38, 39].

Table 3 shows each load bus's indices value, ranking, and voltage magnitude. The RP of the load bus varies one bus at a time until the value of indices reaches unity or the LF solution fails to converge. When varying the RP of each load bus, one of the indices values makes the bus vulnerable to voltage collapse; at this stage, the indices values were noted, and the bus was selected as the critical bus. It also presents the ranking of each bus, which was done based on the highest values of the indices and the least RP injected into the load bus. The load node/bus with the highest indices value and the small permissible reactive load was selected as the critical node of the system and ranked first. It clearly shows that node 14 is the critical bus/node of the system because it has the highest values of the index of 1.023 for FVSI and a small permissible reactive load of 76.5 MVar. This node/bus has two connected lines, 9–14 and 13–14. The load connected to the line experience voltage instability. This was identified as the weakest bus that needs proper attention to avoid voltage collapse. The most stable bus are nodes/buses 4 and 5. They have the height value of RP injected into the bus, 360 MVar, and 352.5 MVar, respectively.

Table 3. The indices value and ranking of the bus in the IEEE 14 bus system.

Bus Number	Q (Var)	Voltage magnitude (p.u)	Ranking	L_{mn}	FVSI
4	360	0.833	9	0.944	0.892
5	352.5	0.997	8	0.998	0.999
7	160	0.771	7	0.929	0.928
9	150	0.712	5	0.981	0.970
10	120	0.663	4	0.942	0.904
11	103	0.748	3	0.912	0.974
12	78	0.790	2	0.865	0.868
13	151.8	0.747	6	0.923	0.993
14	**76.5**	0.693	**1**	0.966	**1.023**

4.2   Reactive Power Optimization

To test the chosen EPSO for RPO, the EPSO and other PSO variants are coded in MATLAB software. Table 4 shows the control variable limits of the two test systems.

Table 4. Control variable limits IEEE 9 and 14 bus system.

Voltage	14 bus system	9 bus system
	Limits	Limits
V_{max}	1.10	1.10
V_{min}	0.95	0.95
T_{max}	1.025	1.025
T_{min}	0.975	0.975
Q_{max}	0.20	0.90
Q_{min}	0.0	0.0

4.2.1   IEEE 9 Bus System

Table 4 illustrates the variable control limits of the algorithms used in this research. The system contains three generators (1, 2, and 3), three transformers, and six load buses (4, 5, 6, 7, 8, and 9).

Figure 2 illustrate the convergence curve of the system. The loss of PSO, EPSO, PSO-TVAC, PSO-SR, and RPSO are 7.608 MW, 7.543 MW, 7.589 MW, 7.600 MW, and 7.602 MW, respectively, from the base case of 9.842 MW. The optimized power loss is expected to be smaller than the base case result. Therefore, EPSO gives the smallest power loss of **7.543 MW**; this gives the significance of the chosen method in power loss reduction. Figure 3 illustrates the voltage profile at each bus before and after the optimization. It can be seen that EPSO offered the highest voltage profile. This show that EPSO is more suitable for improving the node voltage than the rest of the PSO variants. Table 5 compares the power loss of EPSO with other PSO variants and algorithms. The EPSO gives accurate results and outperforms all of them.

Table 5. Comparison of the IEEE 9 bus system with other algorithms

Algorithms	PSO	EPSO	PSO-TVAC	PSO-SR	RPSO	DA [52]	CA [52]
Best	7.6077	**7.543**	7.5894	7.600	7.6023	14.74	14.82
Worst	8.957	8.257	8.685	8.878	8.989	–	–
Mean	8.282	7.900	8.137	8.239	8.296	–	–
STD	0.954	**0.505**	0.775	0.902	0.957	–	–

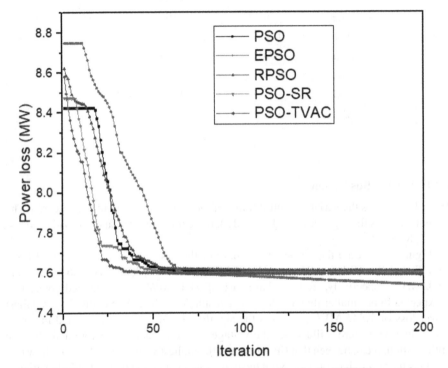

Fig. 2. The convergence curve minimizes real power loss for the IEEE 9 bus system

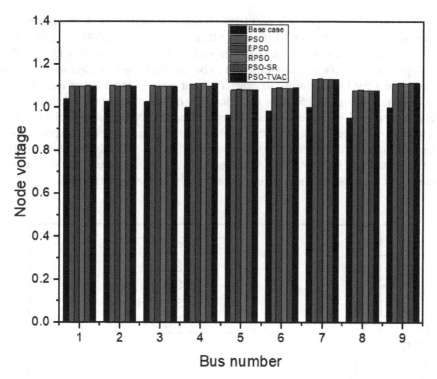

Fig. 3. Voltage magnitude before optimization (base case) and after optimization of IEEE 9 bus system

4.2.2 IEEE 14 Bus System

The control variable is given in Table 4. The N-R was used as the base case, while EPSO was chosen and compared with other PSO variants. The initial/base case system loads, total generation, and power losses of the test system from the LF solution by NR method are given below.

$$P_{load} = 259\,\text{MW}, P_{loss} = 13.775\,\text{MW}, \text{ and } P_G = 272.757\,\text{MW}$$

The curve of the real power loss of EPSO is demonstrated in Fig. 4. The power loss reduction of PSO, EPSO, PSO-TVAC, PSO-SR, and RPSO are 12.263 MW, 12.253 MW, 12.260 MW, 12.261 MW, and 12.264 MW, respectively from the base case of 13.775 MW. After optimization, power loss is expected to be less than the base case result. It can be seen that EPSO outperforms all other PSO variants, giving a lower reduction of **12.253 MW**. However, Fig. 5 shows the voltage magnitude of each bus/node before and after optimization. EPSO effectively increases each node's voltage and gives the smallest loss reduction. The superiority of EPSO was validated by comparing the real power loss, mean, and standard deviation (STD) of the result obtained with other algorithms, like DEPSO, PSO-CFA, HLGBA, MTLA, BA, and LCA, as presented in Table 6. Notably, it shows that the EPSO methods give excellent results and outperform them.

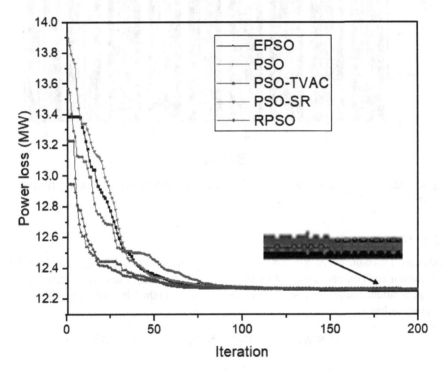

Fig. 4. Convergence minimization of real power loss

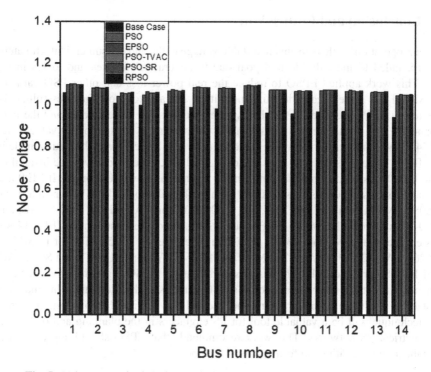

Fig. 5. Voltage magnitude before optimization (base case) and after optimization.

Table 6. Comparison of IEEE14 bus system with other algorithms

Algorithms	Best	Worst	Mean	STD
PSO	12.263	12.879	12.571	0.436
EPSO	**12.253**	12.311	12.282	0.041
PSO-TVAC	12.260	12.587	12.424	0.232
PSO-SR	12.261	12.762	12.512	0.354
RPSO	12.259	12.324	12.292	0.046
PSO-CFA [53]	12.416	–	–	–
DEPSO [38]	13.4086	–	–	–
JAYA [38]	13.466			
HLGBA [37]	13.1229	–	–	–
LCA [54]	12.9891	13.1638	–	$5.5283 * 10^{-3}$
PBIL [54]	13.0008	13.1947	–	$9.7075 * 10^{-4}$
JAYA [55]	12.281	–	–	–

5 Conclusion and Future Work

The PS operation's role is to ensure a stable voltage to the end consumer. Unfortunately, the PS failed to meet the desired goal due to generators' failures and losses in the TL. This work applied EPSO to reduce the real power loss and other PSO variants. FVSI and L_{mn} were used to identify the critical bus and to know the stressfulness of the lines in a PS. For the IEEE 9 bus system, bus 8 is the critical node, and the lines connected to it are the most stressful line of the system. It has the lowest value of RP of 240 MVar, and one of the indices reaches unity (1). Node 14 was the critical node in the IEEE 14 bus system, and the lines connected to it experienced voltage instability. The EPSO was used to reduce/diminish the actual/real power loss on the IEEE 9 and 14 bus systems. The loss was reduced from 9.842 MW to 7.543 MW for EPSO and 7.608 MW,7.602 MW, 7.589 MW, and 7.600 MW for PSO, RPSO, PSO-TVAC, and PSO-SR, respectively, for the IEEE 9 bus system. Also, the losses on the IEEE 14 bus system were reduced from 13.775 MW (the base case) to 12.253 MW for EPSO and 12.263 MW, 12.259 MW, 12.260 MW, and 12.261 MW for PSO, RPSO, PSO-TVAC, and PSO-SR, respectively. The result shows that the EPSO algorithm gives a better loss reduction than other techniques and PSO variants in the literature. This indicates that EPSO is suitable for improving the grid voltage quality, thereby suggesting that the technique will be a valuable tool for PS engineers in the planning and operations of electrical PS networks. This work recommends that EPSO should apply to other metaheuristic algorithms to form a hybrid method.

Acknowledgment. This work was supported in part by the South African National Research Foundation under Grants 141951 and 132797.

References

1. Subramani, C., Dash, S.S., Jagdeeshkumar, M., Bhaskar, M.A.: Stability index based voltage collapse prediction and contingency analysis (2009). https://doi.org/10.5370/JEET.2009.4.4.438
2. Adegoke, S.A., Sun, Y.: Power system optimization approach to mitigate voltage instability issues : a review. 1–40 (2023). https://doi.org/10.1080/23311916.2022.2153416
3. Zha, Z., Wang, Bo., Fan, H., Liu, L.: An improved reinforcement learning for security-constrained economic dispatch of battery energy storage in microgrids. In: Zhang, H., Yang, Z., Zhang, Z., Wu, Z., Hao, T. (eds.) NCAA 2021. CCIS, vol. 1449, pp. 303–318. Springer, Singapore (2021). https://doi.org/10.1007/978-981-16-5188-5_22
4. Mu, B., Zhang, Xi., Mao, X., Li, Z.: An optimization method to boost the resilience of power networks with high penetration of renewable energies. In: Zhang, H., Yang, Z., Zhang, Z., Wu, Z., Hao, T. (eds.) NCAA 2021. CCIS, vol. 1449, pp. 3–16. Springer, Singapore (2021). https://doi.org/10.1007/978-981-16-5188-5_1
5. Cañizares, C.: Voltage Stability Assessment: Concepts, Practices and Tools (2002)
6. Tiwari, R., Niazi, K.R., Gupta, V.: Line collapse proximity index for prediction of voltage collapse in power systems. Int. J. Electr. Power Energy Syst. **41**, 105–111 (2012). https://doi.org/10.1016/j.ijepes.2012.03.022
7. Mohamed, A., Jasmon, G.B., Yusof, S.: A static voltage collapse indicator. J. Ind. Technol. **7**, 73–85 (1998)

8. Ratra, S., Tiwari, R., Niazi, K.R.: Voltage stability assessment in power systems using line voltage stability index. Comput. Electr. Eng. **70**, 199–211 (2018). https://doi.org/10.1016/j.compeleceng.2017.12.046

9. Pérez-Londoño, S., Rodríguez, L.F., Olivar, G.: A simplified voltage stability index (SVSI). Int. J. Electr. Power Energy Syst. **63**, 806–813 (2014). https://doi.org/10.1016/j.ijepes.2014.06.044

10. Sarat, K.S., Suresh, S.R., Jayaram, V., Kumar, S.: New Voltage Stability Index (NVSI) for voltage stability analysis in power system. Int. J. Electr. Electron. Eng. Res. (IJEEER) **2**(4), 13–20 (2012)

11. And, P.K., H.G.: Estimating the voltage stability of a power system. IEEE Trans. Power Deliv. PWRD-1(3) (1986)

12. Sakthivel, S., Mary, D., Ezhilan, C.: Global voltage stability limit improvement by real and reactive power optimization through evolutionary programming algorithm. Int. J. Adv. Sci. Tech. Res. **1**, 88–102 (2012)

13. Sajan, K.S., Kumar, V., Tyagi, B.: Genetic algorithm based support vector machine for online voltage stability monitoring. Int. J. Electr. Power Energy Syst. **73**, 200–208 (2015). https://doi.org/10.1016/j.ijepes.2015.05.002

14. Devaraj, D., Preetha Roselyn, J.: On-line voltage stability assessment using radial basis function network model with reduced input features. Int. J. Electr. Power Energy Syst. **33**, 1550–1555 (2011). https://doi.org/10.1016/j.ijepes.2011.06.008

15. Satpathy, P.K., Das, D., Gupta, P.B.D.: A novel fuzzy index for steady state voltage stability analysis and Identification of critical busbars. Electr. Power Syst. Res. **63**, 127–140 (2002). https://doi.org/10.1016/S0378-7796(02)00093-7

16. Hamid, Z.A., Musirin, I., Rahim, M.N.A., Kamari, N.A.M.: Application of electricity tracing theory and hybrid ant colony algorithm for ranking bus priority in power system. Int. J. Electr. Power Energy Syst. **43**, 1427–1434 (2012). https://doi.org/10.1016/j.ijepes.2012.07.010

17. Yang, D.S., Sun, Y.H., Zhou, B.W., Gao, X.T., Zhang, H.G.: Critical nodes identification of complex power systems based on electric cactus structure. IEEE Syst. J. **14**, 4477–4488 (2020). https://doi.org/10.1109/JSYST.2020.2967403

18. Adebayo, I., Jimoh, A.A., Yusuff, A.: Voltage stability assessment and Identification of important nodes in power transmission network through network response structural characteristics. IET Gener. Transm. Distrib. **11**, 1398–1408 (2017). https://doi.org/10.1049/iet-gtd.2016.0745

19. Ambriz-Perez, H., Acha, E., Fuertc-Esquivel, C.R.: Advanced SVC models for newton-raphson load how and newton optimal power flow studies. IEEE Power Eng. Rev. **19**, 46 (1999)

20. Ramos, J.M., Exposito, A.G., Quintana, V.H.: Transmission power loss reduction by interior-point methods: implementation issues and practical experience. IEE Proc. Gen. Trans. Distrib. **152**(1), 90-98 (2005). https://doi.org/10.1049/ip-gtd:20041150

21. Grudinin, N.: Reactive power optimization using successive quadratic programming method. IEEE Trans. Power Syst. **13**, 1219–1225 (1998). https://doi.org/10.1109/59.736232

22. Alsac, O., Bright, J., Prais, M., Stott, B.: Further developments in LP-based optimal power flow. IEEE Trans. Power Syst. **5**(3), 697-711 (1990)

23. Singh, R.P., Mukherjee, V., Ghoshal, S.P.: Optimal reactive power dispatch by particle swarm optimization with an aging leader and challengers. Appl. Soft Comput. J. **29**, 298–309 (2015). https://doi.org/10.1016/j.asoc.2015.01.006

24. Suresh, V., Senthil Kumar, S.: Research on hybrid modified pathfinder algorithm for optimal reactive power dispatch. Bull. Polish Acad. Sci. Tech. Sci. **69**, 1–8 (2021). https://doi.org/10.24425/bpasts.2021.137733

25. Adegoke, S.A., Sun, Y.: Optimum reactive power dispatch solution using hybrid particle swarm optimization and pathfinder algorithm, 403–410 (2022). https://doi.org/10.47839/ijc.21.4.2775

26. Yapici, H.: Solution of optimal reactive power dispatch problem using pathfinder algorithm. Eng. Optim. **53**, 1946–1963 (2021). https://doi.org/10.1080/0305215X.2020.1839443
27. Mukherjee, A., Mukherjee, V.: Chaotic krill herd algorithm for optimal reactive power dispatch considering FACTS devices. Appl. Soft Comput. J. **44**, 163–190 (2016). https://doi.org/10.1016/j.asoc.2016.03.008
28. Ng Shin Mei, R., Sulaiman, M.H., Mustaffa, Z., Daniyal, H.: Optimal reactive power dispatch solution by loss minimization using moth-flame optimization technique. Appl. Soft Comput. J. **59**, 210–222 (2017). https://doi.org/10.1016/j.asoc.2017.05.057
29. Mouassa, S., Bouktir, T., Salhi, A.: Ant lion optimizer for solving optimal reactive power dispatch problem in power systems. Eng. Sci. Technol. an Int. J. **20**, 885–895 (2017). https://doi.org/10.1016/j.jestch.2017.03.006
30. Üney, M.Ş., Çetinkaya, N.: New metaheuristic algorithms for reactive power optimization. Teh. Vjesn. **26**, 1427–1433 (2019). https://doi.org/10.17559/TV-20181205153116
31. Adegoke, S.A., Sun, Y.: Diminishing active power loss and improving voltage profile using an improved pathfinder algorithm based on inertia Weight. Energies. **16**(3), 1270, (2023). https://doi.org/10.3390/en16031270
32. AlRashidi, M.R., El-Hawary, M.E.: A survey of particle swarm optimization applications in electric power systems. IEEE Trans. Evol. Comput. **13**, 913–918 (2009). https://doi.org/10.1109/TEVC.2006.880326
33. Harish Kiran, S., Dash, S.S., Subramani, C.: Performance of two modified optimization techniques for power system voltage stability problems. Alexandria Eng. J. **55**, 2525–2530 (2016). https://doi.org/10.1016/j.aej.2016.07.023
34. Jaramillo, M.D., Carrión, D.F., Muñoz, J.P.: A novel methodology for strengthening stability in electrical power systems by considering fast voltage stability index under N − 1 scenarios. 1–23 (2023)
35. Adebayo, I.G., Sun, Y.: Voltage stability based on a novel critical bus identification index. In: 2019 14th IEEE Conference on Industrial Electronics and Applications (ICIEA), pp. 1777–1782 (2019). https://doi.org/10.1109/ICIEA.2019.8834091
36. Adebayo, I.G., Sun, Y.: A comparison of voltage stability assessment techniques in a power system. (2018)
37. Alam, M.S., De, M.: Optimal reactive power dispatch using hybrid loop-genetic based algorithm. In: 2016 National Power Systems Conference (NPSC) 2016, pp. 1–6 (2017). https://doi.org/10.1109/NPSC.2016.7858901
38. Vishnu, M., Sunil Kumar, T.K.: An improved solution for reactive power dispatch problem using diversity-enhanced particle swarm optimization. Energies. **13** (2020). https://doi.org/10.3390/en13112862
39. Samuel, I.A., Katende, J., Awosope, C.O.A., Awelewa, A.A.: Prediction of voltage collapse in electrical power system networks using a new voltage stability index. Int. J. Appl. Eng. Res. **12**, 190–199 (2017)
40. Chaturvedi, K.T., Pandit, M., Srivastava, L.: Particle swarm optimization with time varying acceleration coefficients for non-convex economic power dispatch. Int. J. Electr. Power Energy Syst. **31**, 249–257 (2009). https://doi.org/10.1016/j.ijepes.2009.01.010
41. Roy Ghatak, S., Sannigrahi, S., Acharjee, P.: Comparative performance analysis of DG and DSTATCOM using improved pso based on success rate for deregulated environment. IEEE Syst. J. **12**, 2791–2802 (2018). https://doi.org/10.1109/JSYST.2017.2691759
42. Musiri, I., Abdul Rahman, T.K.: On-line voltage stability based contingency ranking using fast voltage stability index (FVSI). In: IEEE/PES Transmission and Distribution Conference and Exhibition, vol. 2, pp. 1118–1123 (2002). https://doi.org/10.1109/TDC.2002.1177634
43. Moghavvemi, M., Faruque, M.O.: Power system security and voltage collapse: a line outage based indicator for prediction. Int. J. Electr. Power Energy Syst. **21**, 455–461 (1999). https://doi.org/10.1016/S0142-0615(99)00007-1

44. Imran, M., Hashim, R., Khalid, N.E.A.: An overview of particle swarm optimization variants. Procedia Eng. **53**, 491–496 (2013). https://doi.org/10.1016/j.proeng.2013.02.063
45. Poli, R., Kennedy, J., Blackwell, T., Freitas, A.: Particle swarms: the second decade. J. Artif. Evol. Appl. **2008**, 1–3 (2008). https://doi.org/10.1155/2008/108972
46. Vinodh Kumar, E., Raaja, G.S., Jerome, J.: Adaptive PSO for optimal LQR tracking control of 2 DoF laboratory helicopter. Appl. Soft Comput. J. **41**, 77–90 (2016). https://doi.org/10.1016/j.asoc.2015.12.023
47. Neyestani, M., Farsangi, M.M., Nezamabadipour, H., Lee, K.Y.: A modified particle swarm optimization for economic dispatch with nonsmooth cost functions. IFAC Proc. **42**, 267–272 (2009). https://doi.org/10.3182/20090705-4-SF-2005.00048
48. Ratnaweera, A., Halgamuge, S.K., Watson, H.C.: Self-organizing hierarchical particle swarm optimizer with time-varying acceleration coefficients. IEEE Trans. Evol. Comput. **8**, 240–255 (2004). https://doi.org/10.1109/TEVC.2004.826071
49. Cao, S., Ding, X., Wang, Q., Chen, B.: Opposition-based improved pso for optimal reactive power dispatch and voltage control. Math. Probl. Eng. (2015). https://doi.org/10.1155/2015/754582
50. Asija, D., Choudekar, P., Soni, K.M., Sinha, S.K.: Power flow study and contingency status of WSCC 9 bus test system using MATLAB. In: 2015 International Conference on Recent Developments in Control, Automation and Power Engineering (RDCAPE), pp. 338–342 (2015)
51. Parmar, R., Tandon, A., Nawaz, S.: Comparison of different techniques to identify the best location of SVC to enhance the voltage profile. ICIC Express Lett. **14**, 81–87 (2020). https://doi.org/10.24507/icicel.14.01.81
52. Khan, I., Li, Z., Xu, Y., Gu, W.: Distributed control algorithm for optimal reactive power control in power grids. Int. J. Electr. Power Energy Syst. **83**, 505–513 (2016). https://doi.org/10.1016/j.ijepes.2016.04.004
53. Roy, R., Das, T., Mandal, K.K.: Optimal reactive power dispatch using a novel optimization algorithm. J. Electr. Syst. Inf. Technol. **8** (2021). https://doi.org/10.1186/s43067-021-00041-y
54. Suresh, V., Kumar, S.S.: Optimal reactive power dispatch for minimization of real power loss using SBDE and DE-strategy algorithm. J. Ambient Intell. Humaniz. Comput. (2020). https://doi.org/10.1007/s12652-020-02673-w
55. Barakat, A.F., El-Sehiemy, R.A., Elsayd, M.I., Osman, E.: Solving reactive power dispatch problem by using JAYA optimization algorithm. Int. J. Eng. Res. Africa. **36**, 12–24 (2018). https://doi.org/10.4028/www.scientific.net/JERA.36.12

Preference Weight Vector Adjustment Strategy Based Dynamic Multiobjective Optimization

Jiamin Yu[✉] and Cuili Yang

Beijing University of Technology, Beijing 100124, China
jmyu0220@163.com, clyang5@bjut.edu.cn

Abstract. The multiple objective functions or constraints of dynamic multi-objective optimization problems (DMOPs) generally vary over time. The multi-objective evolutionary algorithm based on decomposition (MOEA/D) not only tracks the entire Pareto optimal front (POF) but also has good performance. In effect, a decision maker (DM) is only cared about part of POF in different environments, which is called region of interest (ROI). To solve this problem, preference weight vector adjustment strategy based dynamic MOEA/D (MOEA/D-DPWA) is proposed. Firstly, the reference vector adjustment strategy is put forward to bring in the preference information of DM into DMO. Then, the dynamic response strategy is implemented, which is designed to regenerate the population near to ROI when environmental dynamics change. Finally, experimental results indicate that MOEA/D-DPWA is valid in tackling DMOPs.

Keywords: Dynamic multiobjective optimization · Preference weight vector · Decision maker

1 Introduction

A great deal of visible MOPs are dynamic multiobjective optimization problems (DMOPs) in nature [1], such as the process of shop scheduling [2] and transportation routing optimization [3]. Consequently, a number of dynamic multi-objective evolutionary algorithms (DMOEAs) have been applied with effect to solve DMOPs in recent years. Due to dynamic characteristics [4], how to design DMOEAs to effectively tackle DMOPs remains a great challenge.

Generally speaking, the final goal of solving DMOPs in practical applications is to supply a small amount of Pareto optimal solutions (POS) that satisfies the preference of DM [5]. An effective DMOEA is generally designed to approximate a whole POF with each limited time interval [6]. However, in practical use, DM

This work was supported by the National Key Research and Development Project under Grant 2021ZD0112002, the National Natural Science Foundation of China under Grants 61973010, 62021003 and 61890930-5.

is only concerned with some areas of POF, which is referred to as ROI [7]. Providing the approximate solution of the whole POF for DM not only increases the workload, but also brings difficulties to the decision-making process of DM [8]. Accordingly, how to design a DMOEA to find POS that meet DM preference is a issue that needs to be addressed in DMO. It is a natural opinion that preference information of DM and DMOEA are combined with to accurately search for ROI [9].

There are some problems in solving DMOPs. Firstly, due to the dynamic nature of DMOP, the preference of DM may be variant with time. Secondly, searching POF approximation with DMOEA is conducted in a limited time [10]. Therefore, although numerous MOEAs have been put forward to handle static MOPs based on preferences [11], they may not be effective in dynamic MOPs. Finally, solutions other than ROI may not be useful to the decision making process, or even the high computing costs are caused. In recent years, some researches [12] have combined DM preferences with DMOEAs. One of the challenges of DMO is how to effectively incorporate DM preferencesindi into the dynamic search process has been clearly indicated in [13]. The novel DMOEA has been proposed in [12]. The algorithm shows that DM uses one or more preference points as preference information, and these points can be modified interactively in the optimization process [12]. The knee-guided prediction approach has been designed [14] in which the knee point is deemed the preferred solution of DM. In short, there are still many challenges in integrating DM preferences into the evolution process.

To solve the above problems, perference weight vector adjustment strategy based DMOEA (MOEA/D-DPWA) is designed. The major contributions of this article are as below. To start with, the reference vector adjustment strategy aims to integrate DM's preference into DMO, and guide the entire population towards ROI. Next, the corresponding dynamic response strategy is executed to handle the change of environment. It is allowed that some solutions to evolve independently in a short time to obtain a promising direction to more accurately guide the evolution. Finally, the experiment results evidence that the algorithm is competitive in disposing of preference based DMOPs.

2   Background

Only the unconstrained DMOPs are considered, which are described as,

$$\min F(x,t) = (f_1(x,t), f_2(x,t), ..., f_m(x,t))^T$$
$$s.t.\quad x \in \Omega$$

(1)

where Ω represents the decision space, $x = (x_1, x_2, ..., x_m)^T$ is the decision variable vector, in which m is the number of objectives, $t = 0, 1, 2...$ represents discrete time instants, $f(x,t)$ is the objective function, and $f_i(x,t)$ is the ith objective function.

Definition 1 (*Dynamic Pareto Dominance*) [15]: *Suppose x_1 and x_2 are two different decision vectors, x_1 is dynamic Pareto-dominates x_2 at time t, indicated as $x_1 \prec_t x_2$, if and only if*

$$\begin{cases} \forall k = 1, \cdots, m, f_k(x_1, t) \leq f_k(x_2, t) \\ \exists k = 1, \cdots, m, f_k(x_1, t) \leq f_k(x_2, t) \end{cases} \tag{2}$$

Definition 2 (*Dynamic Pareto Optimal Set, PS_t*): *If a decision vector x^* at time t satisfies*

$$\left\{ PS_t = \{ x^* \mid \neg \exists x \in \Omega : x \prec_t x^* \} \right. \tag{3}$$

then x^* is called Pareto optimal solution, and the set composed of all the Pareto optimal solutions at time t is called PS_t).

Definition 3 (*Dynamic Pareto Optimal Front, PF_t*): *The dynamic Pareto optimal front (PF_t) is the corresponding objective vectors of the PS_t at time t, and the PF_t can be defined as*

$$\left\{ PF_t = \{ F(x^*, t) \mid x^* \in PS_t \} \right. \tag{4}$$

3   Proposed Algorithm

To solve the DMOPs, a DMOEA based preference information is developed, which is simply named as MOEA/D-DPWA. The MOEA/D-DPWA includes three techniques to acquire a new population that respond to the environmental changes. Firstly, if the environment and preference point changes, the evenly distributed weight vector is moved to nearby position of preference vector, which is provided by the decision maker (DM). Then, a dynamic response strategy is implemented. The strategy aims to acquire a new initial population, which is closed to ROI. Finally, MOEA/D is used to optimize and obtain the preference solution that satisfies the preference of DM. Details are described as follows.

3.1   MOEA/D

In the circumstances, MOEA/D [16] is selected and combined with the proposed strategies, whose working mechanism is introduced as follows. The MOEA/D decomposes the original MOP into several subproblems and optimises them simultaneously by means of an aggregation function. It is worth noting that each subproblem is optimised using only the information from the solutions of neighbouring subproblems, so the computational complexity is low. The Tchebycheff aggregation function [17] is used in this article, defined as,

$$\begin{aligned} min \quad & g(\mathbf{x} \mid \lambda, \mathbf{z}^*) = \max_{1 \leq p \leq m} \left| f_p(x) - z_p^* \right| / \lambda_p \\ subject \quad & to \quad x \in \Omega \end{aligned} \tag{5}$$

where $z^* = (z_1^*, \cdots, z_p^*)$ is the ideal point, $p = 1, \cdots, m$, and $z_p^* = min\{f_p(x), x \in \Omega\}$ is the minimum value of each objective vector, $\lambda = (\lambda_1, \cdots, \lambda_p)$ is reference vector.

Generally speaking, MOEA/D is to find the approximation of entire POF. However, this article aims to search for a partial solution set of POF in the light of DM'preference. Therefore, how to update reference vectors be described explicitly in the following section. Distribution vectors can be built around a preference vector, so as to quickly track ROI within a limited time interval.

3.2   Reference Vector Adjustment Strategy

It is difficult for DM to provide a reasonable preference in DMOPs because it is necessary to find POF/POS that varies with time for a limited time. To solve this problem, the reference vector movement policy is designed to enable the solutions to quickly converge to ROI. In particular, the reference vector is the weight vector.

Firstly, the position of the preference point has a certain influence on the effectiveness of algorithm. So as to prevent the impact of the location from the preference point, the preference point is converted into a preference vector. The preference vector is defined as starting from the origin and ending at the preference point. After the preference point is achieved, the corresponding mapping point is determined by DM in line with the preference point. Therefore, the mapping point R' is the crossover point of preference vector and the hyperplane $f_1 + f_2 + \cdots + f_m = 1$. Given reference point $R = (f_1, f_2, \cdots, f_m)$, the corresponding $R' = (f_1', f_2', \cdots, f_m')$ can be calculated as follows,

$$\begin{cases} \frac{f_2}{f_2'} = \frac{f_2}{f_2'} = \cdots = \frac{f_m}{f_m'} \\ f_1' + f_2' + \cdots + f_m' = 1 \end{cases} \tag{6}$$

$$f_j' = \frac{f_j}{\sum_{j=1}^m f_j}, j = 1, 2, \cdots, m \tag{7}$$

The weight vector is adjusted after the mapping point is obtained and the preference area is determined. Aroused by the MOEA/D, the original weight vectors are evenly distributed in objective space, then moved them to preference region in the light of the preference vector. The exact adjustment process is visually illustrated in Fig. 1, here it is assumed that the preference point supplied by DM is (0.5, 0.5), meanwhile, the neighborhood size is constant. Here are the specific procedures.

Step 1: A group of evenly weight vectors is manufactured by the means proposed by Das and Dennies [18], in which $N = \binom{H+M-1}{M-1}$ points are evenly sampled on objective space.

Step 2: Given the preference point, the preference vector is acquired by connecting with the origin and the preference point. Then, the mapping point is determined and biased distribution of the weight vectors are obtained. The weight vector is moved in line with following formula,

$$\lambda' = \lambda + \mu * (\lambda_{ref} - \mu) \tag{8}$$

where λ' is the weight vector after moving, λ is a uniformly distributed weight vector, λ_{ref} is the preference weight vector, $\mu \in (0,1)$ is the moving step size.

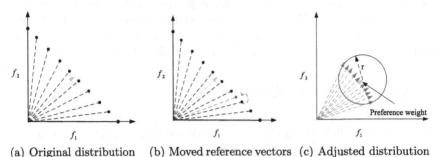

(a) Original distribution   (b) Moved reference vectors   (c) Adjusted distribution

Fig. 1. Illustration of the Moving Process of Evenly Distributed Weight Vector.

3.3   Dynamic Response Strategy

Changes in environment and preferences will cause the convergence losses of the population. The dynamic response strategy is applied to address this issue. Specifically, the preference of DM should be unknown during the optimization process. Therefore, in this paper, preference points are randomly set. All preference points are stored in a set A to search for approximate ROIs in different environments. When the environment and preference point changes, some strategies are adopted to respond the change. The detailed procedures of dynamic response are as below:

Firstly, the center point X_t of the current population PS_t is calculated by,

$$X_t = \frac{1}{|PS_t|} \sum_{x_t \in PS_t} x_t \tag{9}$$

where PS_t is the nondominated solution set of the population under the time t environment, and $|PS_t|$ is the size of PS_t.

Secondly, the partial solution is selected for satisfying DM preference under the current environment to form a new set G with the size of K. The basis for selection is that in the current population, the partial solution closest to the Euclidean distance of the preference point A_{t-1} given by DM is selected, and the small part of the selected decomposition is independently evolved in a short time through crossover and mutation. The total number of evolutions is defined as N. The new set G is obtained by independent inheritance, and the central point X_G of the new set G is calculated by,

$$X_G = \frac{1}{|G|} \sum_{x_t \in G} x_t \tag{10}$$

where $|G|$ is the size of G.

Then, the direction of movement that leads the population to evolve to the preferred region is calculated by,

$$D = X_G - X_t \tag{11}$$

Finally, the solutions in the overall are updated as below,

$$x_{t+1} = x_t + D \tag{12}$$

Particularly, the process of MOEA/D-DPWA is expressed in Algorithm 1.

Algorithm 1. MOEA/D-DPWA

Input: $F(x, t)$: A DMOP; PS_t: current population; T: size of neighborhood; m: the number of objectives; t: current time instant; N_e: the number of environmental changes; N: the population size.

Output: PS_{t+1}: solutions of new environment $t + 1$; A: preference point set.

1: Initialization a empty preference point set A, t=1;
2: **for** $p = 1 : N_e$ **do**
3:    **while** $q < m$ **do**
4:       $A_p^q = rand$,
5:       $q = q + 1$;
6:    **end while**
7: **end for**
8: **while** *the stop criterion is not met* **do**
9:    **if** *there is a change in environment* **then**
10:       $t = t + 1$;
11:       Connect the preference point with the origin to obtain the preference vector.
12:       Adjust weight vectors by weight vector adjustment strategy.
13:       Select some solutions and Revolves alone over several generations;
14:       Compute X_t and X_G by Eq. (9) and Eq. (10), respectively;
15:       Compute the evolutionary direction D by Eq. (11);
16:       **for** $x_t \in PS_t \backslash G$ **do**
17:          Acquire the solutions x_{t+1} for the new environment through Eq. (12)
18:       **end for**
19:    **else**
20:       $PS_t = \mathbf{MOEA}(F(x, t), PS_t)$;
21:    **end if**
22: **end while**

---

4   Simulation and Evaluation

In this section, the FDA [19], dMOP [23] benchmark test suites are used to evaluate the performance of algorithm, and a total of six test problems are selected. The time instance t is defined as $t = (1/n_t) \lfloor \tau/\tau_t \rfloor$, where n_t, τ_t and τ represent the severity of change, the frequency of change, and the maximum generation, respectively.

4.1 Performance Indicators

As the algorithms incorporate the preference of DM, in an effort to better measure the effectiveness of algorithms, the generational distance(GD) and its variant MGD [20] are used. The GD is calculated by,

$$GD = \frac{\sqrt{\sum_{i=1}^{n} d_i^2}}{n} \tag{13}$$

where n is the number of Pareto optimal solutions, and d_i is the minimum Euclidean distance from the ith solution to POF in the algorithm. MGD is defined as the average value of GD in some time steps of run:

$$MGD = \frac{1}{|T|} \sum_{t \in T} GD(t) \tag{14}$$

4.2 Compared Algorithms

Three DMOEAs are considered for comparison in the experiment. Additionally, the mainly aim of DNSGA-II [21] is search for the whole POF rather than the approximate solution subset of POF. For the sake of fairness, the r-dominance relationship is applied in DNSGA-II [9]. Each compared algorithm is briefly introduced as follows.

r-DNSGA-II-A: Whenever the environment changes, 20% of the new population is taken the place of randomly created solutions.

r-DNSGA-II-B: Different from creating random solutions, 20% of the new solution is substituted for mutated solutions of the current solutions.

DMOEA/D: To adapt dynamic environments, MOEA/D is revised as the dynamic version, in which 20% of the population are randomly regenerated.

4.3 Parameter Settings

In this article, the parameters of the contrasted algorithms are set according to [21, 22]. Some important parameters are in summary as below.

1) The size of population is 100 and the dimension of decision variable is 10. The severity of changes is set to $n_t = 5, 10$ and the frequency of changes is set to $\tau_t = 10, 20, 30$.

2) The neighborhood size supplied by DM is set to 0.3. The step size of the reference point is 0.5. Each test problem is run independently for 10 times, and 50 environmental changes have been simulated.

4.4 Comparative Study

The MGD values of all comparison algorithms are displayed in Table 1. On FDA1, FDA3, dMOP1, dMOP2, it is indicate that MOEA/D-DPWA obviously outperforms other comparison algorithms. This implies that in most cases,

MOEA/DPWA has outstanding ability to track dynamically changing POF and POS. Compared with other algorithms, MOEA/D-DPWA is slightly lag behind in one or two dynamic settings on FDA2. The dynamic characteristics of FDA2 change rapidly, so tracking ROI at limited time intervals do not perform well.

Table 1. Mean values of MGD obtained by compared algorithms.

Problems	(n_t, τ_t)	r-DNSGA-II-A	r-DNSGA-II-B	DMOEA/D	MOEA/D-DPWA
FDA1	(5,10)	2.802e-1(+)	1.760e-1(+)	3.102e-1(+)	**2.192e-3**
	(10,10)	2.509e-1(+)	1.748e-1(+)	2.969e-1(+)	**1.675e-3**
	(10,20)	1.872e-1(+)	6.646e-2(+)	2.563e-1(+)	**7.107e-4**
	(10,30)	1.438e-1(+)	1.885e-2(+)	2.040e-1(+)	**6.489e-4**
FDA2	(5,10)	3.936e-1(+)	4.001e-2(+)	**1.837e-2(−)**	3.626e-2
	(10,10)	3.747e-1(+)	3.827e-2(+)	**1.817e-2(−)**	3.448e-2
	(10,20)	3.348e-2(−)	3.377e-2(+)	**1.887e-2(−)**	3.270e-2
	(10,30)	3.048e-2(−)	3.049e-2(−)	**1.895e-2(−)**	3.232e-2
FDA3	(5,10)	2.944e-2(+)	3.924e-2(+)	1.472e-2(+)	**1.407e-2**
	(10,10)	8.125e-3(+)	1.312e-2(+)	1.394e-2(+)	**1.568e-3**
	(10,20)	2.289e-3(+)	3.661e-3(+)	4.023e-3(+)	**1.439e-3**
	(10,30)	1.936e-3(+)	6.195e-3(+)	1.561e-3(+)	**1.405e-3**
dMOP1	(5,10)	5.045e-4(+)	4.210e-4(+)	1.903e-4(+)	**1.536e-4**
	(10,10)	3.614e-4(+)	3.447e-4(+)	1.757e-4(+)	**1.281e-4**
	(10,20)	1.549e-4(+)	1.489e-4(+)	1.126e-4(+)	**1.102e-4**
	(10,30)	1.115e-4(+)	1.093e-4(+)	1.039e-4(+)	**1.018e-4**
dMOP2	(5,10)	4.096e-1(+)	4.021e-1(+)	4.141e-1(+)	**6.863e-4**
	(10,10)	4.001e-1(+)	3.987e-1(+)	4.045e-1(+)	**3.310e-4**
	(10,20)	3.010e-1(+)	3.009e-1(+)	3.919e-1(+)	**1.614e-4**
	(10,30)	1.753e-1(+)	1.757e-1(+)	3.051e-1(+)	**1.232e-4**
dMOP3	(5,10)	5.552e-3(+)	2.093e-2(+)	**1.134e-3(−)**	2.279e-3
	(10,10)	3.091e-3(+)	4.929e-3(+)	**8.565e-4(+)**	1.647e-3
	(10,20)	1.161e-3(+)	1.056e-3(+)	6.809e-4(+)	**5.476e-4**
	(10,30)	1.562e-4(−)	**1.543e-4(−)**	2.425e-4(−)	5.091e-4

At the same time, it can be seen that the smaller MGD values acquired by the MOEA/D-DPWA. It is indicated that the POS and POF achieved by MOEA/D-DPWA are nearer to the preference region given by DM. When the severity of the change is set from small to large, it means that the environment changes from more intense to slower, and all algorithms have better dynamic tracking capabilities. Meanwhile, when the change severity is constant, the more slower change frequency, the performance of all algorithms are better.

In general, decomposition-based algorithms perform well. In an effort to intuitively display the dynamic tracking capabilities of different algorithms, POFs of four algorithms at different times are drawn on four test cases, FDA1 and dMOP1-3, as shown in Figs. 2, 3, 4 and 5. Compared with other algorithms, it can be apparently noticed that the proposed approach converges to the preference region defined by DM and the changing POF and POS is quickly tracked.

Meanwhile, it has the best convergence. It is proved that the performance of MOEA/D-DPWA is superior to the other comparison methods mentioned above when dealing with DMOPs.

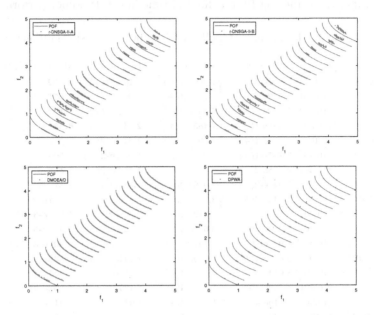

Fig. 2. Obtained POFs for FDA1 with $n_t = 10$ and $\tau_t = 10$.

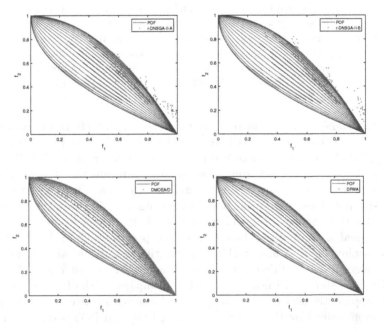

Fig. 3. Obtained POFs for dMOP1 with $n_t = 10$ and $\tau_t = 10$.

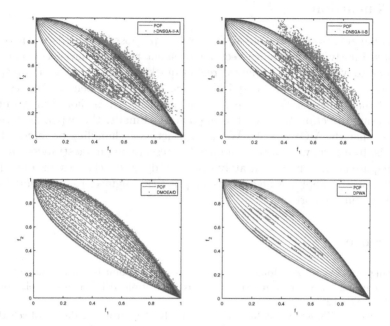

Fig. 4. Obtained POFs for dMOP2 with $n_t = 10$ and $\tau_t = 10$.

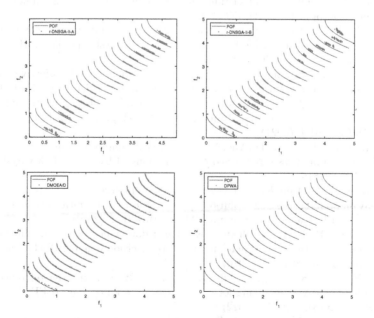

Fig. 5. Obtained POFs for dMOP3 with $n_t = 10$ and $\tau_t = 10$.

5   Conclusion

In this paper, a preference based DMOEA is proposed, which is used to solve DMOPs. Firstly, the reference vector movement policy is devised to make the solutions converge to the preferred region quickly. Then the dynamic response strategy is executed aims to quickly track the environmental changes. It is allowed that some solutions to evolve independently in a short time to obtain a promising direction to guide the population. Finally, the experiment results indicate that the MOEA/D-DPWA is satisfactory in handling preference based DMOPs. In future, we will be committed to researching more strategies to meet DM requirements while accurately tracking dynamic changes in POS/POF. Simultaneously, applying algorithms to practical engineering problems is also one of our future work.

References

1. Jiang, M., Wang, Z., Hong, H., Yen, G.G.: Knee point-based imbalanced transfer learning for dynamic multiobjective optimization. IEEE Trans. Evol. Comput. **25**(1), 117–129 (2020)
2. Nguyen, S., Zhang, M., Johnston, M., Tan, K.C.: Automatic design of scheduling policies for dynamic multi-objective job shop scheduling via cooperative coevolution genetic programming. IEEE Trans. Evol. Comput. **18**(2), 193–208 (2013)
3. Peng, X., Xu, D., Zhang, F.: UAV online path planning based on dynamic multiobjective evolutionary algorithm. In: Proceedings of the 30th Chinese Control Conference, pp. 5424–5429. IEEE (2011)
4. Cao, L., Xu, L., Goodman, E.D., Bao, C., Zhu, S.: Evolutionary dynamic multiobjective optimization assisted by a support vector regression predictor. IEEE Trans. Evol. Comput. **24**(1), 305–319 (2019)
5. Hu, Y., Zheng, J., Zou, J., Jiang, S., Yang, S.: Dynamic multi-objective optimization algorithm based decomposition and preference. Inf. Sci. **571**, 175–190 (2021)
6. Zhao, Q., Yan, B., Shi, Y., Middendorf, M.: Evolutionary dynamic multiobjective optimization via learning from historical search process. IEEE Trans. Cybern. **52**(7), 6119–6130 (2021)
7. Li, K., Chen, R., Savić, D., Yao, X.: Interactive decomposition multiobjective optimization via progressively learned value functions. IEEE Trans. Fuzzy Syst. **27**(5), 849–860 (2018)
8. Li, K., Liao, M., Deb, K., Min, G., Yao, X.: Does preference always help? A holistic study on preference-based evolutionary multiobjective optimization using reference points. IEEE Trans. Evol. Comput. **24**(6), 1078–1096 (2020)
9. Said, L.B., Bechikh, S., Ghédira, K.: The r-dominance: a new dominance relation for interactive evolutionary multicriteria decision making. IEEE Trans. Evol. Comput. **14**(5), 801–818 (2010)
10. Xie, Y., Yang, S., Wang, D., Qiao, J., Yin, B.: Dynamic transfer reference point-oriented MOEA/D involving local objective-space knowledge. IEEE Trans. Evol. Comput. **26**(3), 542–554 (2022)
11. Tomczyk, M.K., Kadziński, M.: Decomposition-based interactive evolutionary algorithm for multiple objective optimization. IEEE Trans. Evol. Comput. **24**(2), 320–334 (2019)

12. Nebro, A.J., Ruiz, A.B., Barba-González, C., García-Nieto, J., Luque, M., Aldana-Montes, J.F.: InDM2: interactive dynamic multi-objective decision making using evolutionary algorithms. Swarm Evol. Comput. **40**, 184–195 (2018)
13. Helbig, M., Deb, K., Engelbrecht, A.: Key challenges and future directions of dynamic multi-objective optimisation. In: 2016 IEEE Congress on Evolutionary Computation (CEC), pp. 1256–1261. IEEE (2016)
14. Zou, F., Yen, G.G., Tang, L.: A knee-guided prediction approach for dynamic multi-objective optimization. Inf. Sci. **509**, 193–209 (2020)
15. Zhang, K., Shen, C., Liu, X., Yen, G.G.: Multiobjective evolution strategy for dynamic multiobjective optimization. IEEE Trans. Evol. Comput. **24**(5), 974–988 (2020)
16. Li, H., Zhang, Q.: Multiobjective optimization problems with complicated Pareto sets, MOEA/D and NSGA-II. IEEE Trans. Evol. Comput. **13**(2), 284–302 (2008)
17. Li, K., Zhang, Q., Kwong, S., Li, M., Wang, R.: Stable matching-based selection in evolutionary multiobjective optimization. IEEE Trans. Evol. Comput. **18**(6), 909–923 (2013)
18. Zhang, Q., Li, H.: MOEA/D: a multiobjective evolutionary algorithm based on decomposition. IEEE Trans. Evol. Comput. **11**(6), 712–731 (2007)
19. Farina, M., Deb, K., Amato, P.: Dynamic multiobjective optimization problems: test cases, approximations, and applications. IEEE Trans. Evol. Comput. **8**(5), 425–442 (2004)
20. Jiang, S., Yang, S.: A steady-state and generational evolutionary algorithm for dynamic multiobjective optimization. IEEE Trans. Evol. Comput. **21**(1), 65–82 (2016)
21. Deb, K., Rao N, U.B., Karthik, S.: Dynamic multi-objective optimization and decision-making using modified NSGA-II: a case study on hydro-thermal power scheduling. In: Obayashi, S., Deb, K., Poloni, C., Hiroyasu, T., Murata, T. (eds.) EMO 2007. LNCS, vol. 4403, pp. 803–817. Springer, Heidelberg (2007). https://doi.org/10.1007/978-3-540-70928-2_60
22. Cao, L., Xu, L., Goodman, E.D., Li, H.: A first-order difference model-based evolutionary dynamic multiobjective optimization. In: Shi, Y., et al. (eds.) SEAL 2017. LNCS, vol. 10593, pp. 644–655. Springer, Cham (2017). https://doi.org/10.1007/978-3-319-68759-9_52
23. Zhou, A., Jin, Y., Zhang, Q.: A population prediction strategy for evolutionary dynamic multiobjective optimization. IEEE Trans. Cybern. **44**(1), 40–53 (2013)

Complementary Environmental Selection for Evolutionary Many-Objective Optimization

Zichen Wei, Hui Wang$^{(\boxtimes)}$, Shuai Wang, Shaowei Zhang, and Dong Xiao

School of Information Engineering, Nanchang Institute of Technology,
Nanchang 330099, China
huiwang@whu.edu.cn

Abstract. In many-objective evolutionary algorithms (MaOEAs), environmental selection is an important operation, which can greatly affect the performance of convergence and distribution of solutions. However, different environmental selection strategies have different preferences. It is difficult to design an appropriate environment selection strategy to balance the convergence and population diversity. To address this issue, this paper proposes a complementary environmental selection strategy for evolutionary many-objective optimization (called CES-MaOEA). Firstly, a dual-population mechanism is utilized. The first population uses the environmental selection of NSGA-III, and the second population employs the environmental selection of radial space division based evolutionary algorithm (RSEA). Through complementary cooperation of two populations, the proposed strategy can make full use of the advantages of the two environmental selection methods. In order to verify the effectiveness of our approach, two well-known benchmark sets including DTLZ and MaF are tested. Performance of CES-MaOEA is compared with five state-of-the-art MaOEAs. Experimental results show that CES-MaOEA achieves competitive performance in terms of convergence and population diversity.

Keywords: evolutionary many-objective optimization · Pareto dominance · radial space projection · complementary environmental selection

1 Introduction

There are many practical problems with at least two optimization objectives, which are usually conflicting with each other. These problems are called multi-objective optimization problems (MOPs) [7]. MOPs exist in many real-world engineering applications, e.g., reservoir resource scheduling [3,12], car model optimization [15,19] and feature selection [14,18], etc. Unlike single-objective optimization problems [25–27], MOPs have many non-inferior solutions at the same time, which form the real Pareto front (PF) [2]. Multi-objective evolutionary algorithms (MOEAs) [13] have been successfully used to solve various MOPs. When a MOP has more than three objectives, it is also called many-objective

H. Zhang et al. (Eds.): NCAA 2023, CCIS 1869, pp. 346–359, 2023.
https://doi.org/10.1007/978-981-99-5844-3_25

optimization problems (MaOPs) [5]. Compared with MOPs, MaOPs are more complex and challenging. Most existing MOEAs show difficulties when dealing with MaOPs, because most solutions become incomparable and the population can hardly converge to the true PF.

To effectively solve MaOPs, different versions of many-objective evolutionary algorithms (MaOEAs) were proposed [1,6,22,23]. In MaOEAs, environmental selection is used to select fitter individuals as parents entering the next generation. So, environmental selection is an important step, which can greatly influence the performance of convergence and population diversity. Different environmental selection strategies have different preferences. It is not easy to design an environment selection strategy to suit various MaOPs. Though the reference point method used in NSGA-III [6] can maintain population diversity, it weakens the ability to deal with irregular PFs. In addition, the Pareto dominance cannot provide sufficient selection pressure with growth of objectives. In [8], a radial space division based evolutionary algorithm (RSEA) was proposed. In RSEA, though the radial space projection used for environmental selection easily caused the lost curvature information of the PFs, but it can effectively strengthen the selection pressure and ignore the features of irregular PFs.

Based on the above analysis, combining the superiorities of different environmental selection strategies may be a good choice for MaOEAs. Therefore, this paper proposes a complementary environmental selection strategy (called CES-MaOEA) for evolutionary many-objective optimization. Firstly, A dual-population mechanism is employed. The first population uses the environmental selection of NSGA-III, and the second population employs the environmental selection of RSEA. The environmental selection of NSGA-III depends on the Pareto-dominance and reference points. The Pareto-dominance cannot provide sufficient selection pressure, and the uniform reference points may reduce the performance on irregular PFs. The above deficiencies of NSGA-III can be enhanced by the radial space projection of RSEA. In RSEA, the lost curvature information of the PFs is caused by the dimensionality reduction during the radial space projection. This can be made up by the environmental selection of NSGA-III. Through complementary cooperation of two populations, the proposed strategy can make full use of the advantages of the two environmental selection methods. To verify the effectiveness of the proposed CES-MaOEA, two well-known benchmark sets including DTLZ and MaF are used to test the performance of CES-MaOEA. Results of CES-MaOEA are compared with five state-of-the-art MaOEAs.

The rest of the paper is organized as follows. In Sect. 2, the definitions of MaOPs and some existing environmental selection strategies are briefly introduced. In Sect. 3, the proposed approach CES-MaOEA is described. Simulation experiments and results are presented in Sect. 4. Finally, the work is summarized in Sect. 5.

2   Background

2.1   Problem Descriptions

A general MOP can be defined as follows:

$$\min F(\mathbf{x}) = [f_1(\mathbf{x}), f_2(\mathbf{x}), \cdots, f_M(\mathbf{x})],$$
$$\text{s.t.} \quad \mathbf{x} \in \Omega, \tag{1}$$

where $\mathbf{x} = (x_1, x_2, \ldots, x_D)$ represent the decision vector, Ω is the decision space, D is the number of decision variables, $f_i(\mathbf{x})$ denotes the i-th objective, and M is the number of objectives. When $M > 3$, the MOPs are called as many-objective optimization problems (MaOPs). Optimizing one objective leads to the deterioration of at least other objectives. So, it is expected to find a set of non-dominated solutions called Pareto-optimal solution set [10].

2.2   Environmental Selection

Withe the growth of objectives, the proportion of non-dominated solutions in the population sharply increases. The selection pressure is weakened and the population can hardly converge to the true PF. To tackle these issues, numerous MaOEAs were designed. In MaOEAs, environmental selection is an important part, which can greatly affect the performance of convergence and distribution of solutions.

NSGA-II is one of the most popular MOEAs and its environment selection depends on Pareto dominance and crowding degree [7]. For MaOPs, the Pareto dominance cannot provide enough selection pressure. Then, Deb and Jain [6] proposed an extended version called NSGA-III, which uses reference points to replace the crowding degree. In [28], decomposition-based MOEA (MOEA/D) transforms an MaOP into several single-objective subproblems. In the environmental selection, it uses reference vectors to maintain the population diversity. Bader and Zitzler [1] proposed a fast hypervolume-based many-objective optimization (HyPE), which employs a hypervolume indicator in environmental selection to assess the diversity of each solution. Tian et al. [20] proposed an indicator-based MOEA with reference point adaptation (AR-MOEA). In the environmental selection, AR-MOEA combines an enhanced inverted generational distance indicator and reference point adaptation to balance population diversity and convergence. In [11], an MOP is decomposed into several simple MOPs by dividing the objective space into several subspaces. Then, a diversity-first-and-convergence-second strategy is adopted to perform the environmental selection. In [8], radial space projection was used for the environmental selection. Solutions are projected from a high-dimensional objective space into a two-dimensional radial space.

Although the above environmental selection strategies have achieved promising performance on many benchmark problems, they still have some deficiencies in solving MaOPs. For example, the Pareto-dominance based methods easily suffer from dominant resistance (DR) [9], diversity promotion (ADP) [16] and falling

---

Algorithm 1: Framework of CES-MaOEA

---

1 Randomly initialize the populations $P1$ and $P2$;
2 **while** *The stopping condition is not reached* **do**
3     $Temp \leftarrow P1 \cup P2$;
4     $MP1 \leftarrow$ Use the mating selection to choose solutions from $Temp$;
5     $MP2 \leftarrow$ Use the mating selection to choose solutions from $Temp$;
6     $P1' \leftarrow$ Conduct simulated binary crossover and polynomial mutation on $MP1$;
7     $P2' \leftarrow$ Conduct simulated binary crossover and polynomial mutation on $MP2$;
8     $R1 \leftarrow P1 \cup P1'$;
9     $R2 \leftarrow P2 \cup P2'$;
10     $P1 \leftarrow$ Select N solutions from $R1$ based on the environmental selection of RSEA;
11     $P2 \leftarrow$ Select N solutions from $R2$ based on the environmental selection of NSGA-III;
12 **end**
13 Output the population $P1$;

---

into local optima. The hypervolume indicator needs much computing time. The convergence and diversity of MOEA/D are easily degraded. The radial space projection loses the curvature information of the PFs. As seen, all environmental selection methods are not perfect. Combining the superiorities of different environmental selection strategies may be a good choice for MaOEAs.

3 Proposed Approach

As mentioned before, environmental selection is an important operation for MaOPs, which can greatly influence the performance of convergence and population diversity. However, it is difficult to design an environment selection strategy to suit various MaOPs. Therefore, combining the advantages of different environmental selection strategies may be suitable for MaOPs. In this paper, complementary environmental selection based many-objective evolutionary algorithm (called CES-MaOEA) is proposed. Firstly, CES-MaOEA employs a dual-population mechanism. Then, the environmental selection strategies of NSGA-III and RSEA are used in the two populations, respectively.

3.1 Basic Framework

The framework of the proposed CES-MaOEA is described in Algorithm 1. Firstly, two populations $P1$ and $P2$ are randomly initialized, respectively (Lines 1). $P1$ uses the environmental selection of RSEA, and $P2$ employs the environmental selection NSGA-III. During the iterative operation, $P1$ and $P2$ are merged into a temporary population $Temp$ (Line 3). Based on mating selection, two mating pools $MP1$ and $MP2$ are chosen from $Temp$, respectively (Lines 4–5). Then, two offspring population $P1'$ and $P2'$ are obtained by $MP1$ and $MP2$, respectively (Line 6–7). The parent population $P1$ and and its offspring population $P1'$ are merged into $R1$. Similarly, $P2$ and $P2'$ are merged into $R2$ (Line 8–9). When the search operations on $P1$ and $P2$ are completed, the environmental selection on each population is conducted (Lines 10–11). Based on the population $R1$, an offspring population $P1$ is obtained by using the environmental selection of RSEA. Based on the population $R2$, an offspring population $P1$ is achieved by using the environmental selection of NSGA-III. When the stopping condition is satisfied, the final population $P1$ is output.

3.2 Complementary Environmental Selection

In CES-MaOEA, the environmental selection strategies of NSGA-III and RSEA are used in the two populations, respectively. The environmental selection of NSGA-III depends on the Pareto-dominance and reference points. With increasing number of objectives, most solutions in the population become incomparable. The Pareto-dominance cannot provide sufficient selection pressure to obtain better solutions. Though the reference point method can maintain population diversity, it weakens the ability to deal with irregular PFs.

Like NSGA-III, RSEA also uses the Pareto dominance in environmental selection, but it employs another new strategy called radial space projection. Solutions are projected from a high-dimensional objective space into a two-dimensional radial space. Though the curvature information of the PFs is missing due to the dimensionality reduction, it can effectively strengthen the selection pressure and ignore the features of irregular PFs.

It is obvious that the deficiencies of NSGA-III can be enhanced by the radial space projection of RSEA. Similarly, the lost curvature information of the PFs can be made up by the reference points of NSGA-III. Through complementary cooperation of two populations, the proposed strategy can make full use of the advantages of the two environmental selection methods.

4 Experimental Study

4.1 Benchmark Problems and Parameter Settings

In order to verify the performance of our approach CES-MaOEA, two well-known benchmark sets including DTLZ and MaF are utilized in the following experiments [23,24]. The DTLZ contains seven problems (DTLZ1-DTLZ7), and the MaF has 12 problems (MaF1-MaF12). These problems contain various properties such as multi-modal, irregular PF, and deceptiveness. For each problem, the number of objective functions M is set to 3, 5, 10, 15, and 20. Performance of CES-MaOEA is compared with five state-of-the-art MaOEAs. The involved algorithms are listed as below.

- NSGA-III [6].
- RSEA [8].
- FDEA-II [17].
- KnEA [29].
- MOEA/D-DE [13].
- Proposed approach CES-MaOEA.

Table 1. HV values obtained by six MaOEAs on the DTLZ benchmark.

Problem	M	NSGA-III	RSEA	FDEA-II	KnEA	MOEA/D	DECES-MaOEA
DTLZ1	3	**8.3861E-1** =	8.3719E-1 -	8.1945E-1 -	7.2499E-1 -	6.4422E-1 -	8.3837E-01
	5	**9.7055E-1** +	9.5449E-1 -	9.4772E-1 -	6.7167E-1 -	7.9832E-1 -	9.6245E-01
	10	9.7377E-1 -	9.9625E-1 -	9.9857E-1 -	0.0000E+0 -	5.9558E-1 -	**9.9907E-01**
	15	9.9836E-1 =	9.8820E-1 -	9.3486E-1 -	0.0000E+0 -	9.4061E-1 -	**9.9937E-01**
	20	7.9388E-1 -	9.6281E-1 -	6.4288E-1 -	0.0000E+0 -	8.3054E-1 -	**9.8270E-01**
DTLZ2	3	**5.5948E-1** +	5.5501E-1 -	5.3011E-1 -	5.3914E-1 -	5.2511E-1 -	5.5578E-01
	5	**7.7967E-1** +	7.4862E-1 =	6.9836E-1 -	7.6101E-1 +	6.1731E-1 -	7.5221E-01
	10	**9.6322E-1** +	9.2476E-1 -	9.2663E-1 -	9.5683E-1 -	5.4612E-1 -	9.5942E-01
	15	9.8117E-1 =	9.6369E-1 -	9.3605E-1 -	8.1386E-1 -	3.4125E-1 -	**9.8492E-01**
	20	8.3252E-1 -	9.6845E-1 -	5.6008E-1 -	7.4597E-1 -	3.2388E-1 -	**9.9425E-01**
DTLZ3	3	3.9675E-1 -	4.6255E-1 =	3.3896E-1 -	3.5481E-1 -	1.1089E-1 -	**4.8175E-01**
	5	3.3769E-1 -	5.4182E-1 -	3.0736E-1 -	4.2408E-1 -	4.0564E-1 -	**7.0275E-01**
	10	2.4148E-1 -	4.9189E-1 -	0.0000E+0 -	0.0000E+0 -	7.3122E-2 -	**7.6254E-01**
	15	4.4475E-1 -	2.2236E-1 -	1.5142E-2 -	0.0000E+0 -	3.4205E-1 -	**7.8852E-01**
	20	0.0000E+0 -	1.5722E-1 =	2.3034E-1 =	0.0000E+0 -	2.2371E-1 =	**3.3334E-01**
DTLZ4	3	5.3217E-1 -	**5.5554E-1** -	5.3443E-1 -	5.4023E-1 -	5.1178E-1 -	5.5509E-01
	5	7.6491E-1 +	7.5596E-1 =	7.1300E-1 -	**7.6535E-1** +	6.6092E-1 -	7.5440E-01
	10	9.6173E-1 -	9.1680E-1 -	9.1846E-1 -	9.5565E-1 -	6.1310E-1 -	**9.6224E-01**
	15	9.8648E-1 =	9.4363E-1 -	9.2446E-1 -	**9.9224E-1** +	4.8889E-1 -	9.8842E-01
	20	9.8362E-1 =	9.5239E-1 -	6.8591E-1 -	9.9601E-1 -	4.2510E-1 -	**9.9639E-01**
DTLZ5	3	1.9419E-1 -	1.7448E-1 -	1.8081E-1 -	1.9220E-1 -	1.9436E-1 -	**1.9862E-01**
	5	1.0003E-1 -	**1.2272E-1** +	1.0623E-1 -	8.0456E-2 -	1.2106E-1 +	1.1501E-01
	10	7.5854E-2 =	8.7799E-2 +	0.0000E+0 -	4.9594E-2 -	**9.7377E-2** +	7.1016E-02
	15	7.0575E-2 =	8.5772E-2 +	0.0000E+0 -	1.9739E-2 -	**9.4039E-2** +	6.9515E-02
	20	0.0000E+0 -	**9.1700E-2** +	0.0000E+0 -	2.1757E-2 -	9.0915E-2 +	5.2431E-02
DTLZ6	3	1.9085E-1 -	1.8322E-1 -	1.7683E-1 -	1.9115E-1 =	1.9477E-1 -	**1.9912E-01**
	5	9.0840E-2 -	1.1491E-1 +	1.0895E-1 -	7.8597E-2 -	**1.2135E-1** +	9.6975E-02
	10	1.9231E-2 +	6.6280E-2 -	0.0000E+0 -	0.0000E+0 -	**9.7572E-2** +	1.4780E-02
	15	5.6818E-3 -	7.0570E-2 -	0.0000E+0 -	5.6818E-3 -	**9.4059E-2** +	1.9014E-02
	20	0.0000E+0 -	7.3068E-2 -	0.0000E+0 -	0.0000E+0 -	**9.0904E-2** +	2.2780E-02
DTLZ7	3	2.6701E-1 -	2.6988E-1 =	2.5359E-1 -	**2.6995E-1** -	2.0909E-1 -	2.6777E-01
	5	2.3712E-1 -	**2.4736E-1** -	1.5821E-1 -	2.4246E-1 =	7.1654E-3 -	2.4602E-01
	10	1.5332E-1 -	1.9130E-1 -	1.2322E-1 -	6.5244E-2 -	1.6347E-4 -	**2.0150E-01**
	15	7.0261E-2 -	1.4964E-1 -	1.2412E-1 -	3.2578E-5 -	0.0000E+0 -	**1.7899E-01**
	20	1.0390E-1 -	1.2792E-1 -	9.0723E-2 -	1.2936E-7 -	1.1689E-15 -	**1.5554E-01**
+/-/=		6/22/7	8/20/7	0/32/3	3/29/3	8/26/1	

To have a fair comparison, all the above six MaOEAs are run the PlatEMO platform [21]. For M=3, 5, 10, 15, and 20, the corresponding population size N is set to 91, 105, 230, 240, and 210, respectively. And the maximum number of evaluations ($MaxFEs$) is set to 30000, 50000, 100000, 150000, and 200000, respectively. For each problem, each algorithm is run 30 times and the mean results are reported.

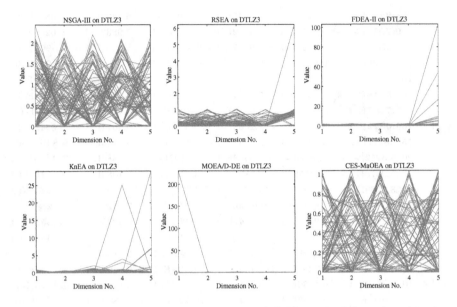

Fig. 1. Distribution of solutions obtained by six MaOEAs on DTLZ3 with 5 objectives

In the experiments, two popular performance indicators including inverse generation distance (IGD) [4] and hypervolume (HV) [30] are utilized. IGD is a widely used metric for evaluating the convergence and diversity of a population. It is calculated by selecting a set of reference points on the actual PF surface and then calculating the average of the distance from each reference point to the nearest solution. A smaller IGD value indicates a better performance of the algorithm. HV is a classical evaluation index to assess the diversity of the population. It calculates the volume of the region in the objective space enclosed by the set of non-dominated solutions and reference points obtained by the algorithm. A larger HV means a better performance of the algorithm.

4.2   Results and Discussions

Table 1 shows the HV values obtained by NSGA-III, RSEA, FDEA-II, KnEA, MOEA/D-DE, and CES-MaOEA on the DTLZ test suite. In the last row of Table 1, the comparison results between CES-MaOEA and other MaOEAs are summarized by the symbols "+/ − / =", where "+", "−", and "=" means that CES-MaOEA is worse, better, and similar than to the compared algorithm, respectively. It can be seen from Table 1 that CES-MaOEA outperforms NSGA-III, RSEA, FDEA-II, KnEA, and MOEA/D-DE on 22, 20, 32, 29, and 26 test instances, respectively. On DTLZ1-DTLZ3 and DTLZ7, CES-MaOEA obtains

Table 2. IGD values obtained by six MaOEAs on the DTLZ benchmark.

Problem	M	NSGA-III	RSEA	FDEA-II	KnEA	MOEA/D-DE	CES-MaOEA
	3	**2.0972E-2** +	2.2283E-2 -	3.1809E-2 -	6.1032E-2 -	9.9869E-2 -	2.1839E-02
	5	**6.3082E-2** +	7.9936E-2 -	8.6923E-2 -	1.7961E-1 -	1.3492E-1 -	7.4863E-02
DTLZ1	10	1.5388E-1 -	1.3531E-1 =	1.5355E-1 -	4.0872E+0 -	3.0399E-1 -	**1.3010E-01**
	15	**1.4377E-1** =	2.0699E-1 -	2.3310E-1 -	5.1004E+0 -	1.7726E-1 -	1.4663E-01
	20	3.0127E-1 =	2.7109E-1 -	4.5040E-1 -	5.3510E+0 -	2.7336E-1 -	**2.3757E-01**
	3	**5.4476E-2** +	5.7659E-2 =	8.8495E-2 -	7.1330E-2 -	7.5600E-2 -	5.7514E-02
	5	**1.9562E-1** +	2.5984E-1 -	2.7720E-1 -	2.2837E-1 +	3.0966E-1 -	2.3472E-01
DTLZ2	10	4.6073E-1 =	5.1268E-1 -	4.9259E-1 -	**4.4760E-1** +	6.8992E-1 -	4.5529E-01
	15	5.6686E-1 =	6.2444E-1 -	6.5122E-1 -	6.3373E-1 =	8.2708E-1 -	**5.5234E-01**
	20	7.8685E-1 =	7.1076E-1 -	9.7803E-1 -	7.3098E-1 =	1.0528E+0 -	**6.2922E-01**
	3	2.5190E-1 -	1.6732E-1 =	3.8454E-1 -	3.4986E-1 -	1.1232E+1 -	**1.3023E-01**
	5	9.7528E-1 =	4.5607E-1 -	7.9424E-1 -	6.2960E-1 -	4.9443E+0 -	**2.3403E-01**
DTLZ3	10	2.4790E+0 -	9.1839E-1 -	1.0837E+1 -	3.8876E+2 -	6.1123E+0 -	**5.7338E-01**
	15	1.1587E+0 -	1.4301E+0 -	1.2640E+1 -	8.2385E+2 -	9.3253E-1 -	**7.1301E-01**
	20	7.2220E+1 -	1.5353E+0 =	1.4256E+1 -	6.0109E+2 -	1.5934E+0 =	**1.4324E+00**
	3	1.1535E-1 -	**5.7595E-2** +	8.5412E-2 -	7.0586E-2 -	1.4364E-1 -	5.7794E-02
	5	2.2511E-1 +	2.5775E-1 -	2.8321E-1 -	**2.2455E-1** +	3.7898E-1 -	2.3637E-01
DTLZ4	10	4.6987E-1 -	5.4039E-1 -	5.0139E-1 -	**4.4785E-1** +	7.4217E-1 -	4.6072E-01
	15	5.5520E-1 =	6.5743E-1 -	6.4939E-1 -	**5.4791E-1** +	8.8846E-1 -	5.6159E-01
	20	6.7293E-1 -	7.1637E-1 -	7.7776E-1 -	**6.1842E-1** +	1.0282E+0 -	6.2587E-01
	3	1.2610E-2 -	1.9946E-1 -	4.2138E-2 -	1.0975E-2 -	1.4335E-2 -	**6.1270E-03**
	5	1.2002E-1 -	5.9830E-2 +	9.0212E-2 -	1.7313E-1 -	**3.4258E-2** +	6.4114E-02
DTLZ5	10	2.5180E-1 -	9.9561E-2 +	1.5733E+0 -	3.2454E-1 -	**5.0174E-2** +	1.4423E-01
	15	3.9079E-1 -	1.1332E-1 =	2.0524E+0 -	5.0041E-1 -	**8.9057E-2** =	9.3222E-02
	20	5.8967E-1 -	2.2426E-1 -	1.6541E+0 -	6.7784E-1 -	2.4006E-1 -	**1.3169E-01**
	3	1.8285E-2 -	1.1621E-1 -	4.3554E-2 -	1.4465E-2 -	1.4491E-2 -	**5.7090E-03**
	5	2.7043E-1 -	8.1219E-2 +	8.0276E-2 +	3.0988E-1 -	**3.0365E-2** +	1.6881E-01
DTLZ6	10	9.5590E-1 -	1.7501E-1 =	1.4756E+0 -	1.6402E+0 -	**4.3615E-2** +	2.2284E-01
	15	1.1814E+0 -	2.2158E-1 +	8.7561E+0 -	1.8702E+0 -	**5.8996E-2** +	4.5274E-01
	20	7.2252E+0 -	**3.3427E-1** =	1.0248E+1 -	1.4176E+0 -	3.4787E-1 +	4.2651E-01
	3	**7.7437E-2** +	1.1933E-1 -	1.1217E-1 =	1.2278E-1 =	1.8866E-1 -	1.3755E-01
	5	3.4298E-1 +	5.1946E-1 -	6.5826E-1 -	**3.2761E-1** +	9.6583E-1 -	4.8615E-01
DTLZ7	10	1.1816E+0 -	2.3431E+0 -	2.6370E+0 -	**9.0450E-1** +	1.3130E+0 =	1.3196E+00
	15	4.2964E+0 +	7.7988E+0 -	3.6458E+0 +	2.3030E+0 +	**1.9241E+0** +	5.4077E+00
	20	8.2841E+0 +	1.3862E+1 -	8.5165E+0 -	3.5788E+0 +	**2.2330E+0** +	1.0094E+01
+/-/=		9/19/7	5/21/9	3/31/1	10/22/3	8/24/3	

very competitive results. Especially for $M = 15$ and 20, CES-MaOEA surpasses other MaOEAs. However, CES-MaOEA shows poor performance on DTLZ5 and DTLZ6. On these two problems, MOEA/D-DE achieves better results than other algorithms. It is noted that CES-MaOEA is superior to five other MaOEAs on

Table 3. HV values obtained by six MaOEAs on the MaF benchmark.

Problem	M	NSGA-III	RSEA	FDEA-II	KnEA	MOEA/D-DE	CES-MaOEA
	3	2.0496E-1 -	2.1837E-1 -	1.8483E-1 -	2.1259E-1 -	1.9857E-1 -	2.1874E-1
	5	4.8897E-3 -	8.7815E-3 =	4.5778E-3 -	9.1800E-3 +	4.8106E-3 -	8.6222E-03
MaF1	10	4.5529E-7 -	5.3506E-7 =	1.4408E-7 -	3.4024E-7 -	6.1031E-8 -	5.6751E-07
	15	5.5146E-12 -	5.5307E-12 -	1.2139E-12 -	0.0000E+0 -	0.0000E+0 -	1.0534E-11
	20	2.8879E-17 +	4.7080E-17 =	6.3281E-18 -	0.0000E+0 =	0.0000E+0 =	1.6533E-17
	3	2.3765E-1 -	2.4335E-1 -	2.3702E-1 -	2.4352E-1 =	2.2664E-1 -	2.4317E-01
	5	1.5750E-1 -	1.7760E-1 -	1.5399E-1 -	1.8361E-1 +	1.4062E-1 -	1.7944E-01
MaF2	10	2.0760E-1 -	2.3479E-1 -	1.0063E-1 -	1.6169E-1 -	1.6570E-1 -	2.3706E-01
	15	1.3442E-1 -	2.2308E-1 =	8.3709E-2 -	1.2477E-1 -	1.5522E-1 -	2.2179E-01
	20	1.3665E-1 -	1.6509E-1 -	5.8689E-2 -	9.9585E-2 -	1.2672E-1 -	2.0616E-01
	3	7.3881E-1 -	6.4024E-1 -	1.7698E-1 -	7.3872E-1 -	2.3191E-1 -	8.8033E-01
	5	7.5770E-1 -	7.0477E-1 -	3.2147E-1 -	5.4328E-1 -	8.6281E-1 -	9.3396E-01
MaF3	10	0.0000E+0 =	0.0000E+0 =	0.0000E+0 =	0.0000E+0 =	8.4219E-1 +	6.0053E-02
	15	0.0000E+0 =	0.0000E+0 =	0.0000E+0 =	0.0000E+0 =	9.9977E-1 +	1.1751E-01
	20	0.0000E+0 =	0.0000E+0 =	3.7091E-3 -	0.0000E+0 =	9.9918E-1 +	0.0000E+00
	3	3.3032E-1 -	4.7089E-1 =	3.3849E-1 -	4.4666E-1 +	1.6420E-1 -	4.2063E-01
	5	4.0880E-2 -	9.6031E-2 =	4.8810E-2 -	9.4153E-2 -	1.0825E-2 -	9.3793E-02
MaF4	10	1.8653E-4 -	3.8411E-4 +	8.7038E-5 -	2.6982E-5 -	2.5463E-7 -	3.4445E-04
	15	2.1614E-7 -	4.0430E-7 +	5.0711E-8 -	2.8925E-11 -	4.9143E-11 -	4.0105E-07
	20	1.3267E-10 -	1.9973E-10 =	1.8901E-12 -	4.5714E-17 -	7.2224E-15 -	2.1099E-10
	3	5.1826E-1 -	5.5507E-1 =	4.9916E-1 -	5.4056E-1 -	4.9584E-1 -	5.5520E-01
	5	7.5533E-1 +	7.5367E-1 +	6.0050E-1 -	7.6528E-1 +	5.5404E-1 -	7.5004E-01
MaF5	10	9.6764E-1 +	9.2470E-1 -	7.8029E-1 -	9.5410E-1 -	4.2204E-1 -	9.6230E-01
	15	9.9103E-1 +	9.4973E-1 -	7.8224E-1 -	9.9072E-1 +	3.8134E-1 -	9.8810E-01
	20	9.9532E-1 =	9.6340E-1 -	6.1524E-1 -	9.9594E-1 -	3.4913E-1 -	9.9657E-01
	3	1.9295E-1 -	1.9807E-1 -	1.8127E-1 -	1.7441E-1 -	1.9408E-1 -	1.9880E-01
	5	1.2569E-1 -	1.1613E-1 -	1.1297E-1 -	1.2703E-1 -	1.2151E-1 -	1.2891E-01
MaF6	10	0.0000E+0 -	6.7683E-2 -	5.9765E-2 =	6.2953E-3 -	9.7421E-2 +	8.4590E-02
	15	1.4932E-3 -	3.1928E-2 -	0.0000E+0 -	0.0000E+0 -	9.4123E-2 +	5.5771E-02
	20	1.3033E-26 -	6.2090E-2 -	0.0000E+0 -	0.0000E+0 -	9.0879E-2 +	6.6199E-02
	3	2.6580E-1 -	2.7207E-1 -	2.4968E-1 -	2.7635E-1 =	2.1127E-1 -	2.6999E-01
	5	2.3544E-1 -	2.4500E-1 -	1.7541E-1 -	2.4697E-1 =	1.1951E-2 -	2.3995E-01
MaF7	10	1.4784E-1 -	1.8640E-1 -	1.2130E-1 -	5.6301E-2 -	1.0096E-3 -	2.0084E-01
	15	6.7582E-2 -	1.5064E-1 -	1.2599E-1 -	1.5938E-4 -	2.0104E-8 -	1.7805E-01
	20	1.0652E-1 -	1.2931E-1 -	8.4808E-2 -	0.0000E+0 =	6.6209E-10 -	1.5435E-01
	3	2.4683E-1 -	2.5973E-1 -	2.4197E-1 -	1.5034E-1 -	2.7027E-1 =	2.6983E-01
	5	9.3662E-2 -	1.1537E-1 -	6.9813E-2 -	9.5310E-2 -	1.1273E-1 -	1.1772E-01
MaF8	10	9.1179E-3 -	1.0985E-2 =	4.6448E-3 -	1.0187E-2 -	9.3015E-3 -	1.0961E-02
	15	5.0794E-4 -	6.5373E-4 =	1.6018E-4 -	5.7121E-4 -	5.4610E-4 -	6.5581E-04
	20	2.4077E-5 -	3.7398E-5 -	6.0997E-6 -	3.0824E-5 -	3.3156E-5 -	3.8961E-05
	3	8.2441E-1 -	7.9689E-1 -	8.1115E-1 -	3.9036E-1 -	7.9041E-1 -	8.2925E-01
	5	1.6476E-1 -	1.9840E-1 -	2.3019E-1 +	1.3335E-1 -	2.9662E-1 +	1.6312E-01
MaF9	10	8.3609E-3 -	1.5723E-2 +	1.2294E-2 =	7.3768E-5 -	1.2060E-2 -	1.2077E-02
	15	8.0805E-4 -	9.0821E-4 -	5.6851E-4 -	8.2391E-4 -	6.2270E-4 -	1.1899E-03
	20	1.1436E-5 -	6.3402E-5 =	3.7483E-5 -	0.0000E+0 -	1.4418E-5 -	6.2650E-05
	3	9.1441E-1 -	9.3153E-1 =	9.1836E-1 -	9.2585E-1 -	4.0614E-1 -	9.2684E-01
	5	9.7911E-1 =	9.9305E-1 +	9.8633E-1 +	9.8996E-1 +	6.2993E-1 -	9.8433E-01
MaF10	10	9.9015E-1 -	9.9821E-1 =	9.9997E-1 +	9.9784E-1 =	9.5012E-1 -	9.9809E-01
	15	9.9932E-1 +	9.9913E-1 +	1.0000E+0 +	9.9828E-1 =	7.1264E-1 -	9.9861E-01
	20	9.9974E-1 +	9.9987E-1 +	1.0000E+0 +	9.9701E-1 -	9.9958E-1 +	9.9928E-01
	3	9.2682E-1 =	9.2933E-1 +	9.1353E-1 -	9.2646E-1 =	8.8169E-1 -	9.2682E-01
	5	9.9064E-1 =	9.9052E-1 =	9.8892E-1 -	9.8811E-1 -	9.3568E-1 -	9.9083E-01
MaF11	10	9.9474E-1 -	9.9623E-1 =	9.9911E-1 +	9.9225E-1 -	9.8808E-1 -	9.9600E-01
	15	9.9429E-1 -	9.9775E-1 =	9.9813E-1 =	9.9485E-1 -	9.9531E-1 -	9.9712E-01
	20	9.9521E-1 +	9.9616E-1 +	9.9469E-1 +	9.7868E-1 =	9.9879E-1 +	9.8894E-01
	3	5.2520E-1 -	5.2674E-1 -	5.0436E-1 -	5.2555E-1 -	4.7672E-1 -	5.3297E-01
	5	6.9793E-1 -	7.1268E-1 -	6.3859E-1 -	7.3053E-1 +	5.3729E-1 -	7.1075E-01
MaF12	10	8.5816E-1 -	8.6524E-1 -	8.0928E-1 -	9.0088E-1 +	4.4433E-1 -	8.8799E-01
	15	8.9161E-1 -	9.0318E-1 -	9.0140E-1 =	9.0372E-1 -	5.6826E-1 -	9.1501E-01
	20	7.6657E-1 -	9.1968E-1 -	8.5019E-1 -	8.4013E-1 -	3.7244E-1 -	9.2980E-01
+/-/=		7/40/13	6/24/30	7/46/7	8/38/14	9/48/3	

Table 4. IGD values obtained by six MaOEAs on the MaF benchmark.

Problem	M	NSGA-III	RSEA	FDEA-II	KnEA	MOEA/D-DE	CES-MaOEA
	3	6.1009E-2 -	4.3306E-2 +	7.6342E-2 -	4.9529E-2 -	6.7669E-2 -	4.3578E-02
	5	1.9722E-1 -	1.5546E-1 +	2.6493E-1 -	1.3213E-1 +	1.7048E-1 -	1.5869E-01
MaF1	10	2.9312E-1 -	2.5556E-1 -	3.7227E-1 -	2.3673E-1 +	2.7285E-1 -	2.4589E-01
	15	3.2101E-1 -	3.1582E-1 -	4.2740E-1 -	3.1969E-1 =	4.0541E-1 -	3.1036E-01
	20	4.4127E-1 -	4.3105E-1 -	5.5878E-1 -	4.4034E-1 -	4.8405E-1 -	4.2613E-01
	3	3.5716E-2 -	3.0365E-2 =	4.4450E-2 -	3.4987E-2 -	4.5245E-2 -	3.0245E-02
	5	1.2915E-1 =	1.3117E-1 -	1.6851E-1 -	1.3870E-1 -	1.4332E-1 -	1.2727E-01
MaF2	10	2.0864E-1 +	3.1886E-1 -	6.4712E-1 -	1.6578E-1 +	3.2466E-1 -	2.6280E-01
	15	2.2274E-1 +	2.9723E-1 -	8.1973E-1 -	1.8130E-1 +	3.6863E-1 -	2.4457E-01
	20	2.5637E-1 =	4.2039E-1 -	8.7670E-1 -	1.8335E-1 +	6.0370E-1 -	2.6434E-01
	3	1.2330E+0 -	7.1820E-1 -	3.9648E+0 -	1.8734E+0 -	2.8293E+2 -	2.1004E-01
	5	5.4263E-1 -	1.4609E+0 -	4.2694E+0 -	3.5056E+0 -	6.3624E+0 -	1.8123E-01
MaF3	10	2.7777E+3 =	3.9038E+2 =	1.3151E+2 =	3.6926E+7 -	2.5579E+0 +	5.9054E+02
	15	1.5830E+2 =	1.9922E+3 -	1.4107E+2 -	7.4231E+8 -	1.4043E-1 +	5.2009E+01
	20	2.1262E+3 -	2.4412E+3 -	2.4880E+2 =	9.6667E+9 -	1.9391E-1 +	4.9946E+02
	3	1.0046E+0 -	5.1079E-1 +	1.3269E+0 -	6.6118E-1 +	1.5726E+1 -	6.9940E-01
	5	4.1265E-1 -	3.3086E+0 -	4.3533E+0 -	2.8002E+0 -	3.5291E+1 -	2.7997E+00
MaF4	10	1.0320E+2 -	1.0225E+2 -	1.4184E+2 -	1.3796E+2 -	8.7732E+2 -	9.1347E+01
	15	3.9122E+3 -	3.3160E+3 -	3.3830E+3 -	1.6363E+3 +	3.6698E+3 -	3.3215E+03
	20	1.3204E+5 =	1.3205E+5 -	1.1960E+5 -	4.5030E+4 +	6.0485E+5 -	1.3657E+05
	3	5.2526E-1 -	2.6072E-1 =	5.2839E-1 -	3.2102E-1 -	5.8863E-1 -	2.6123E-01
	5	2.4058E+0 -	2.5725E+0 -	3.8798E+0 -	2.6067E+0 -	5.9416E+0 -	2.3857E+00
MaF5	10	8.6991E+1 -	6.7366E+1 -	1.1304E+2 -	7.7207E+1 -	3.0518E+2 -	6.2945E+01
	15	2.4989E+3 -	2.1416E+3 -	2.6473E+3 -	1.9526E+3 -	7.3258E+3 -	1.5641E+03
	20	7.1370E+4 -	4.8620E+4 -	9.5937E+4 -	2.5015E+4 =	1.7095E+5 -	2.9835E+04
	3	1.4541E-2 -	1.3039E-2 -	3.4690E-2 -	3.7238E-2 -	1.4322E-2 -	6.0723E-03
	5	4.3612E-2 -	4.0530E-1 -	6.3046E-2 -	1.0199E-2 +	2.5703E-2 -	1.5522E-02
MaF6	10	8.6432E-1 -	2.8226E-1 =	2.3996E-1 +	9.7593E+0 -	2.7372E-2 +	2.5063E-01
	15	6.9641E-1 -	3.4799E-1 =	9.5280E+1 -	4.4048E+1 -	2.1935E-2 +	3.6751E-01
	20	7.6203E-1 -	5.9254E-1 =	7.1862E+1 -	6.7706E+1 -	1.4859E-1 +	5.0123E-01
	3	9.3855E-2 +	1.0265E-1 =	1.1765E-1 +	6.9710E-2 +	2.0960E-1 -	1.2084E-01
	5	3.4753E-1 +	5.4890E-1 -	5.7132E-1 -	3.3031E-1 +	9.3894E-1 -	5.7857E-01
MaF7	10	1.1497E+0 +	2.4060E+0 -	2.3908E+0 -	9.0379E-1 +	1.3292E+0 -	1.3752E+00
	15	4.3336E+0 +	7.7687E+0 -	4.6109E+0 +	2.4416E+0 +	1.9275E+0 +	5.5526E+00
	20	8.2811E+0 +	1.3372E+1 -	7.7479E+0 +	3.5923E+0 +	2.2173E+0 +	9.9290E+00
	3	1.3282E-1 -	1.0089E-1 -	1.3722E-1 -	4.6447E-1 -	9.3536E-2 -	7.9291E-02
	5	2.2636E-1 -	1.7420E-1 -	4.6065E-1 -	2.9614E-1 -	1.3157E-1 =	1.4211E-01
MaF8	10	4.1981E-1 -	1.7593E-1 -	7.6371E-1 -	1.5157E-1 +	1.4880E-1 +	1.6648E-01
	15	3.8211E-1 -	2.1941E-1 =	1.0993E+0 -	1.7027E-1 +	1.7345E-1 +	2.1440E-01
	20	4.3117E-1 -	2.7446E-1 =	1.3549E+0 -	2.1646E-1 +	1.7585E-1 +	2.6933E-01
	3	7.9441E-2 =	1.1730E-1 -	9.9409E-2 -	2.1471E+0 -	8.9396E-2 -	6.8390E-02
	5	5.4114E-1 =	4.3216E-1 =	2.8512E-1 +	7.4729E-1 -	1.4628E-1 +	5.5557E-01
MaF9	10	6.1930E-1 -	1.6337E-1 +	3.9338E-1 -	3.8991E+1 -	2.7038E-1 =	2.7720E-01
	15	1.0704E+0 +	3.4819E-1 -	5.7687E-1 -	4.4640E-1 -	5.4932E-1 -	8.1152E-01
	20	4.4432E+0 +	1.7772E-1 -	4.3886E-1 -	8.2273E+1 -	5.7300E-1 -	1.6882E-01
	3	2.1666E-1 -	1.7987E-1 =	2.7024E-1 -	2.1308E-1 -	1.2160E+0 -	1.8459E-01
	5	4.6891E-1 -	5.0629E-1 -	7.1917E-1 -	5.0710E-1 -	1.5966E+0 -	4.7266E-01
MaF10	10	1.1302E+0 -	1.0540E+0 -	1.4151E+0 -	1.0123E+0 -	2.2698E+0 -	9.7985E-01
	15	1.4746E+0 -	1.5836E+0 -	2.0340E+0 -	1.5696E+0 -	2.6429E+0 -	1.4592E+00
	20	4.5132E+0 -	3.4706E+0 -	4.1855E+0 -	3.4848E+0 -	4.8937E+0 -	3.3071E+00
	3	1.6296E-1 +	1.6599E-1 =	2.6350E-1 -	1.9715E-1 -	3.3836E-1 -	1.6592E-01
	5	4.5840E-1 +	5.2173E-1 -	7.0316E-1 -	5.7527E-1 -	1.1132E+0 -	4.9396E-01
MaF11	10	1.3272E+0 -	1.0737E+0 -	1.4680E+0 -	1.2351E+0 -	1.8351E+0 -	1.0231E+00
	15	1.5522E+0 +	1.9270E+0 -	2.3026E+0 -	1.7112E+0 -	2.2316E+0 -	1.6215E+00
	20	3.9752E+0 =	4.2383E+0 -	4.4325E+0 -	3.7776E+0 +	4.8402E+0 -	4.0877E+00
	3	2.2435E-1 -	2.3363E-1 -	3.5877E-1 -	2.4339E-1 -	3.3463E-1 -	2.2511E-01
	5	1.1172E+0 -	1.3455E+0 -	1.7559E+0 -	1.2364E+0 +	2.0077E+0 -	1.2871E+00
MaF12	10	4.5471E+0 =	4.7463E+0 -	5.3504E+0 -	4.3724E+0 +	7.1883E+0 -	4.5123E+00
	15	7.9594E+0 +	8.8697E+0 -	9.5923E+0 -	6.9080E+0 +	1.1709E+1 -	8.3117E+00
	20	1.4348E+1 -	1.3923E+1 -	1.6288E+1 -	1.0786E+1 +	2.9025E+1 -	1.3179E+01
+/-/=		12/34/14	4/36/20	5/49/6	21/35/4	12/44/4	

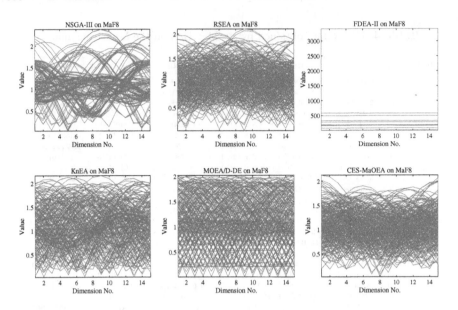

Fig. 2. Distribution of solutions obtained by six MaOEAs on MaF8 with 15 objectives

DTLZ3 with different number of objectives. CES-MaOEA is not worse than FDEA-II on all 35 test instances. Compared with NSGA-III and RSEA, CES-MaOEA achieves better results on most test instances. It demonstrates that the proposed complementary environment selection is effective.

Figure 1 lists the distribution of the solutions obtained by six MaOEAs on DTLZ3 with 5 objectives. As seen, CES-MaOEA achieves the best population diversity and convergence than five other MaOEAs. NSGA-III and CES-MaOEA have better distributions than RSEA, FDEA-II, KnEA, and MOEA/D-DE, but NSGA-III is worse than CES-MaOEA on the convergence.

Table 2 shows the IGD values of six MaOEAs on the DTLZ test suite. From the results, CES-MaOEA outperforms NSGA-III, RSEA, FDEA-II, KnEA, and MOEA/D-DE on 19, 21, 31, 22, and 24 test instances, respectively. Like the HV indicator, CES-MaOEA performs better than other algorithms on DTLZ1-DTLZ3 and DTLZ7 with 15 and 20 objectives. For DTLZ3, CES-MaOEA obtains the best results on five different objectives. On DTLZ4-DTLZ6, the performance of CES-MaOEA is a little poor. From the results of IGD and HV, CES-MaOEA is superior to NSGA-III and RSEA regarding the population diversity and convergence. This validates the effectiveness of complementary environmental selection. CES-MaOEA is better than FDEA-II on almost all problems in terms of HV and IGD values.

The HV values of six MaOEAs on the MaF benchmark are given in Table 3. As shown, CES-MaOEA outperforms NSGA-III, RSEA, FDEA-II, KnEA, and MOEA/D-DE on 40, 24, 46, 38, and 48 test instances, respectively. On MaF2-MaF4, MaF6-MaF9, and MaF12, CES-MaOEA is not worse than NSGA-III on five different objectives. RSEA is not better than CES-MaOEA on MaF1-MaF3, MaF5-MaF8, and MaF12 with different test cases. FDEA-II is better than CES-MaOEA on seven test instances, which are mainly distributed in two problems MaF10 and MaF11. For MaF3 and MaF6, MOEA/D-DE outperforms CES-MaOEA on six test instances. Compared with NSGA-III and RSEA, CES-MaOEA obtains worse results on seven and six test instances, respectively. Through the complementary of NSGA-III and RSEA, the proposed CES-MaOEA can obtain good performance.

Table 4 shows the IGD values of six algorithms on the MaF benchmark. It can be see that CES-MaOEA surpasses NSGA-III, RSEA, FDEA-II, KnEA, and MOEA/D-DE on 34, 36, 49, 35, and 44 test instances, respectively. On MaF5 and MaF10, CES-MaOEA is not worse than five other MaOEAs on all five different objectives. NSGA-III is better than CES-MaOEA on 12 test instances, which are distributed in MaF2, MaF7, MaF11, and MaF12. For the rest of eight problems, CES-MaOEA is not worse than NSGA-III. RSEA outperforms CES-MaOEA on only four test instances. From the comparison among CES-MaOEA, NSGA-III, and RSEA, CES-MaOEA can better balance the convergence and population diversity. It confirms the effectiveness of the proposed the complementary environmental selection strategy.

Figure 2 shows the distribution of the solutions obtained by six MaOEAs on MaF8 with 15 objectives. From the graphs, CES-MaOEA, RSEA, KnEA, and MOEA/D-DE obtains better popular diversity than NSGA-III and FDEA-II. For the convergence, CES-MaOEA is the best one among six MaEAs.

On the DTLZ and MaF benchmark sets, CES-MaOEA obtains competitive performance in terms of HV and IGD values. The proposed approach can make a good balance between convergence and population diversity. Compared with NSGA-III and RSEA, CES-MaOEA achieves better results on most test instances. It demonstrates the effectiveness of the proposed complementary environmental selection strategy.

5   Conclusion

This paper proposes a complementary environmental selection based many-objective evolutionary algorithm (CES-MaOEA). In CES-MaOEA, a dual-population mechanism is utilized. The environmental selection methods of NSGA-III and RSEA are employed in two populations, respectively. Through complementary cooperation of two populations, the proposed strategy can make full use of the advantages of the two environmental selection methods. To verify the effectiveness of CES-MaOEA, two benchmark sets DTLZ and MaF with 3, 5, 10, 15, and 20 objectives are tested.

Simulation results show CES-MaOEA obtains better performance than NSGA-III, RSEA, FDEA-II, KnEA, and MOEA/D-DE in terms of IGD and HV indicators. Compared with NSGA-III and RSEA, CES-MaOEA achieves better results on most test instances. It demonstrates that the complementary environmental selection is effective. In this work, we make an attempt to combine

NSGA-III and RESA to improve the environmental selection. More investigations on different environmental selection strategies will be studied in the future work.

Acknowledgements. This work was supported by the National Natural Science Foundation of China (No. 62166027), and Jiangxi Provincial Natural Science Foundation (No. 20212ACB212004).

References

1. Bader, J., Zitzler, E.: HypE: an algorithm for fast hypervolume-based many-objective optimization. Evol. Comput. **19**(1), 45–76 (2011)
2. Balachandran, M., Gero, J.: A comparison of three methods for generating the pareto optimal set. Eng. Optim. **7**(4), 319–336 (1984)
3. Chang, F.J., Lai, J.S., Kao, L.S.: Optimization of operation rule curves and flushing schedule in a reservoir. Hydrol. Process. **17**(8), 1623–1640 (2003)
4. Coello, C.A.C., Cortés, N.C.: Solving multiobjective optimization problems using an artificial immune system. Genet. Program Evolvable Mach. **6**, 163–190 (2005)
5. Cui, Z., Zhang, M., Wang, H., Cai, X., Zhang, W.: A hybrid many-objective cuckoo search algorithm. Soft. Comput. **23**, 10681–10697 (2019)
6. Deb, K., Jain, H.: An evolutionary many-objective optimization algorithm using reference-point-based nondominated sorting approach, part i: solving problems with box constraints. IEEE Trans. Evol. Comput. **18**(4), 577–601 (2013)
7. Deb, K., Pratap, A., Agarwal, S., Meyarivan, T.: A fast and elitist multiobjective genetic algorithm: NSGA-II. IEEE Trans. Evol. Comput. **6**(2), 182–197 (2002)
8. He, C., Tian, Y., Jin, Y., Zhang, X., Pan, L.: A radial space division based evolutionary algorithm for many-objective optimization. Appl. Soft Comput. **61**, 603–621 (2017)
9. Ishibuchi, H., Matsumoto, T., Masuyama, N., Nojima, Y.: Effects of dominance resistant solutions on the performance of evolutionary multi-objective and many-objective algorithms. In: Proceedings of the 2020 Genetic and Evolutionary Computation Conference, pp. 507–515 (2020)
10. Jafaryeganeh, H., Ventura, M., Soares, C.G.: Application of multi-criteria decision making methods for selection of ship internal layout design from a pareto optimal set. Ocean Eng. **202**, 107151 (2020)
11. Jiang, S., Yang, S.: A strength pareto evolutionary algorithm based on reference direction for multiobjective and many-objective optimization. IEEE Trans. Evol. Comput. **21**(3), 329–346 (2017)
12. Laborie, P.: Algorithms for propagating resource constraints in AI planning and scheduling: Existing approaches and new results. Artif. Intell. **143**(2), 151–188 (2003)
13. Li, H., Zhang, Q.: Multiobjective optimization problems with complicated pareto sets, MOEA/D and NSGA-II. IEEE Trans. Evol. Comput. **13**(2), 284–302 (2008)
14. Li, J., et al.: Feature selection: a data perspective. ACM Comput. Surv. (CSUR) **50**(6), 1–45 (2017)
15. Mei, L., Thole, C.A.: Data analysis for parallel car-crash simulation results and model optimization. Simul. Model. Pract. Theory **16**(3), 329–337 (2008)
16. Purshouse, R.C., Fleming, P.J.: On the evolutionary optimization of many conflicting objectives. IEEE Trans. Evol. Comput. **11**(6), 770–784 (2007)

17. Qiu, W., Zhu, J., Wu, G., Fan, M., Suganthan, P.N.: Evolutionary many-objective algorithm based on fractional dominance relation and improved objective space decomposition strategy. Swarm Evol. Comput. **60**, 100776 (2021)
18. Sharawi, M., Zawbaa, H.M., Emary, E.: Feature selection approach based on whale optimization algorithm. In: 2017 Ninth International Conference on Advanced Computational Intelligence (ICACI), pp. 163–168. IEEE (2017)
19. Song, G., Yu, L., Geng, Z.: Optimization of Wiedemann and Fritzsche car-following models for emission estimation. Transp. Res. Part D: Transp. Environ. **34**, 318–329 (2015)
20. Tian, Y., Cheng, R., Zhang, X., Cheng, F., Jin, Y.: An indicator-based multiobjective evolutionary algorithm with reference point adaptation for better versatility. IEEE Trans. Evol. Comput. **22**(4), 609–622 (2017)
21. Tian, Y., Cheng, R., Zhang, X., Jin, Y.: PlatEMO: A MATLAB platform for evolutionary multi-objective optimization. IEEE Comput. Intell. Mag. **12**(4), 73–87 (2017)
22. Wang, H., Wei, Z., Yu, G., Wang, S., Wu, J., Liu, J.: A two-stage many-objective evolutionary algorithm with dynamic generalized pareto dominance. Int. J. Intell. Syst. **37**, 9833–9862 (2022)
23. Wang, S., Wang, H., Wu, J., Liu, J., Zhang, H.: Many-objective artificial bee colony algorithm based on decomposition and dimension learning. In: Zhang, H., et al. (eds.) NCAA 2022. Communications in Computer and Information Science, vol. 1638, pp. 150–161. Springer, Singapore (2022). https://doi.org/10.1007/978-981-19-6135-9_12
24. Wei, Z., et al.: Many-objective evolutionary algorithm based on dominance and objective space decomposition. In: Zhang, H., et al. (eds.) NCAA 2022. Communications in Computer and Information Science, vol. 1638, pp. 205–218. Springer, Singapore (2022)
25. Ye, T., Wang, H., Wang, W., Zeng, T., Zhang, L., Huang, Z.: Artificial bee colony algorithm with an adaptive search manner and dimension perturbation. Neural Comput. Appl. **34**, 1–15 (2022)
26. Zeng, T., Wang, H., Wang, W., Ye, T., Zhang, L., Zhao, J.: Data-driven artificial bee colony algorithm based on radial basis function neural network. Int. J. Bio-Inspired Comput. **20**(1), 1–10 (2022)
27. Zeng, T., et al.: Artificial bee colony based on adaptive search strategy and random grouping mechanism. Expert Syst. Appl. **192**, 116332 (2022)
28. Zhang, Q., Li, H.: MOEA/D: a multiobjective evolutionary algorithm based on decomposition. IEEE Trans. Evol. Comput. **11**(6), 712–731 (2007)
29. Zhang, X., Tian, Y., Jin, Y.: A knee point-driven evolutionary algorithm for many-objective optimization. IEEE Trans. Evol. Comput. **19**(6), 761–776 (2014)
30. Zitzler, E., Thiele, L.: Multiobjective evolutionary algorithms: a comparative case study and the strength pareto approach. IEEE Trans. Evol. Comput. **3**(4), 257–271 (1999)

An Investigation on the Effects of Exemplars Selection on Convergence and Diversity in Large-Scale Particle Swarm Optimizer

Minchong Chen, Xuejing Hou, Qi Yu, Dongyang Li$^{(\boxtimes)}$, and Weian Guo$^{(\boxtimes)}$

Tongji University, 4800 Cao'an Road, Shanghai, China
{2232930,xuejing_hou,2233440,1710333,guoweian}@tongji.edu.cn

Abstract. The selection of candidate exemplars significantly influences the performance of particle swarm optimizer (PSO) variants. Motivated by this phenomenon, this paper makes an investigation on how the selection of candidate exemplars affects convergence acceleration and diversity preservation. In this proposed investigation, two aspects are taken into consideration, including the selection model and the number of candidate exemplars. On the one hand, three different selection models are utilized so that the general quality of selected exemplars can be adjusted. On the other hand, the number of candidate exemplars changes from small to large so as to improve the exemplars' dominance among the swarm. In order to respectively and independently analyze the effects of the two aspects, comprehensive experiments are conducted to test the roles of exemplars numbers and exemplars selection model to optimization performance. The conclusions exhibit a guiding and instructive significance for exemplars selection in particle swarm optimizers.

Keywords: Particle swarm optimizer · Convergence · Diversity · Exemplar selection

1 Introduction

Exemplars play an important role in various kinds of swarm optimizers. For example, particle swarm optimizer (PSO) [1], ant colony algorithm (ACO) [2], artificial bee colony algorithm (ABC) [3], firefly algorithm (FA) [4], cuckoo search algorithm (CS) [5] and many others [6–9]. In these algorithms, the selection of exemplars has a direct impact on the search performance. This phenomenon is particularly prominent in the particle swarm optimizer. Furthermore, due to the representation and significant influence of PSO in swarm optimizers, the PSO algorithm is chosen as the carrier of the proposed investigation.

However, cPSO shows deteriorated performance on large-scale optimization problems (LSOPs). In cPSO, specifically, the two learning exemplars for each updated particle are the global best position *gbest* and the historically best position found by individuals *pbest*$_i$, the *gbest* may remain unchanged for several iterations, which brings damage to the swarm diversity and makes the incompetence of cPSO on LSOPs.

H. Zhang et al. (Eds.): NCAA 2023, CCIS 1869, pp. 360–375, 2023.
https://doi.org/10.1007/978-981-99-5844-3_26

To improve the performance of cPSO, numerous efforts have been made, many of which are focused on devising novel exemplar selections. These researches can be mainly divided into two categories:

First, coupled control on convergence and diversity. In this category, the selected exemplars have superiority on both convergence and diversity for updated particle to learn from. For example, Yang et al. propose a level-based learning swarm optimizer (DLLSO) [10] where particles are grouped into different levels according to their fitness, two learning exemplars of corresponding updated particle are randomly selected from two different higher levels. Following this exemplar selection mechanism, Yang et al. put forward another PSO variants named SDLSO [11]. In this optimizer, two better exemplars are randomly selected from the swarm. Thanks to the exemplar selection mechanism of the two optimizers, the selected exemplars possess favorable information about convergence, while the diversity mainly from the randomness during exemplar selection. Although a promising improvement has been shown by PSO variants in this category on LSOPs, however, the contributions that selected exemplars respectively make to convergence and diversity are ambiguous, which makes it difficult for exemplars to achieve direct and independent control over convergence and diversity.

Second, decoupled control on convergence and diversity. In this category, each updated particle learns from two exemplars guide convergence and diversity, respectively. For instance, Li et al. propose an adaptive particle swarm optimizer with decoupled exploration and exploitation (APSO-DEE) [12]. In this optimizer, convergence and diversity are measured by two different indicators, then updated individuals will learn from an exemplar with better convergence and an exemplar with better diversity, achieving a direct and independent control over convergence and diversity by selecting exemplars accordingly. However, it takes extra computing resources to measure the diversity of individuals. Consequently, PSO variants in this category are often computationally expensive.

From the experimental results presented by these aforementioned PSO variants, it can be recognized that, the selection of exemplar makes a great impact on the performance of PSO variants. Motivated by these findings, this paper makes an investigation on how the selection of candidate exemplars affects swarm convergence and diversity. The main contributions are listed as follows.

First, three different selection models of candidate exemplars are utilized, which aim to adjust the general quality of selected exemplars. Second, the number of exemplars varies from small to large, which is designed to improve the dominance of exemplars among the swarm. Third, a novel update strategy that ensures convergence and diversity is proposed for each updated particle to learn from selected exemplars.

The rest of this paper is organized as follows: Sect. 2 presents the related work on large-scale optimization, where several representative PSO variants in the aforementioned categories are listed. Following which is the design guideline in Sect. 3. Section 4 is the experiment study, where a great number of experimental phenomena and analyses are demonstrated. Conclusions and future work are given in the final Sect. 5.

2 Related Work

Without loss of generality, this paper considers the minimization problems defined as (1),

$$\min f(X), X = [x_1, x_2, ..., x_n] \tag{1}$$

where X is the decision variable vector, n is the number of variables.

LSOPs are prevalent in the real world, many existing PSO variants are found to be inefficient in handling them, especially for some previous PSO variants [13, 14]. Many studies have improved the performance of cPSO from the perspective of tuning parameters [15, 16], but they still cannot avoid the inherently negative impact of selected learning exemplars on convergence and diversity. It can be inferred that, in most cases, the selection of candidate exemplars plays a more important role in swarm convergence and diversity than parameters. Research that is focused on designing novel exemplar selection mechanisms can be mainly divided into the following two categories.

2.1 Coupled Control over Convergence and Diversity

Selecting exemplars to conduct a coupled control over convergence and diversity has become a widely used tool in LSOPs recently. PSO variants in this category pay attention to selecting exemplars with favorable information about both convergence and diversity to guide updated particles.

In earliest work of cPSO [1], the *gbest* and *pbest*$_i$, both of which are better or no worse than the corresponding updated particle, will be changed over generations, providing updated individuals with better convergence and diversity. Aiming to alleviate the negative effects that the *gbest* in cPSO brings, Chen et al. propose ALC-PSO [17], where the *gest* is replaced by the *Leader*, an exemplar that iterates more frequently. The swarm diversity in this exemplar selection mechanism is enhanced compared with that of cPSO. By adopting the competition mechanism, Cheng et al. propose CSO [18]. In this optimizer, the better half of individuals and a particle representing the mean position of the swarm are selected as the learning exemplars for corresponding updated particle. Thanks to the exemplar selection mechanism, the swarm convergence and diversity are both promoted. In the same year, Cheng et al. propose a social learning swarm optimizer (SL-PSO) [19]. In this algorithm, the learning exemplars are any better individuals of the corresponding updated particle, and the exemplars are various in different dimensions so that the swarm diversity can be maintained in a large search space.

However, in this category, the superiorities of selected exemplars in convergence and diversity are not clearly distinguished and thus remain difficult for targeted adjustment in convergence and diversity.

2.2 Decoupled Control over Convergence and Diversity

Controlling convergence and diversity in a decoupled method has recently emerged as a hot research topic. PSO variants in this category make efforts to measure the convergence and diversity of individuals independently. Then selecting different learning exemplars that targeted exploitation and exploration for according updated particle, respectively.

Li et al. propose an adaptive particle swarm optimizer with decoupled exploration and exploitation (APSO-DEE) [12], where the convergence and diversity of individuals are measured by two different indicators, then a particle with good convergence performance and a particle with great diversity capacity are selected as learning exemplars for convergence acceleration and diversity preservation, respectively. Following this idea, Li et al. propose a particle swarm optimizer with dynamic balance of convergence and diversity (PSO-DBCD) [20]. Both the two optimizers achieve decoupled control over convergence and diversity by selecting targeted exemplars to guide particle updating.

3 Design Guideline

3.1 Selection Model of Candidate Exemplars

The first aspect of the proposed investigation is the selection model of candidate exemplars. Consequently, three different exemplar selection models are utilized in the proposed investigation. First, random model, where candidate exemplars are randomly selected from the swarm without considering their fitness. In this case, the fitness of the selected exemplar is distributed with complete randomness.

Second, liner model. Before selecting candidate exemplars, individuals in the swarm are sorted according to their fitness, better particles have higher probabilities of being selected than worse individuals. The probability of each particle being selected as a candidate exemplar is calculated according to (2):

$$p_i = \frac{(\sum fitness) - fitness(i)}{\sum \left[(\sum fitness) - fitness(i) \right]} \tag{2}$$

where the $fitness(i)$ is the fitness value of the ith particle, $\sum fitness$ is the sum of the fitness of the swarm. p_i is the probability that the ith particle being selected as a candidate exemplar.

Third, quadric model. After sorting the swarm in accordance with fitness, candidate exemplars are selected from the swarm in the form of a quadratic. In this selection model, better individuals have higher probabilities of being selected compared with the linear model. The probability of each particle that being selected as a candidate exemplar is calculated according to (3):

$$p_i = \frac{[(\sum fitness) - fitness(i)]^2}{\sum \left[(\sum fitness) - fitness(i) \right]^2} \tag{3}$$

Denote that the general quality of the selected exemplars is different with different selection model. The schematic diagrams of the three selection models are shown in Fig. 1 as follows, where individuals with a blue background are particles that are being selected as exemplars.

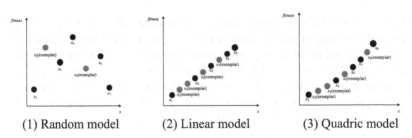

(1) Random model   (2) Linear model   (3) Quadric model

Fig. 1. Three selection models.

3.2 Number of Candidate Exemplars

The second aspect of the proposed investigation is the number of candidate exemplars, which is varied between 2 and 10, with an increasing step of 2. For one thing, when the number changes from small to large, the proportion of exemplars in the swarm increases as well. As a result, the influence of exemplars among swarm gradually becomes dominant. For anther thing, when the selection model is fixed unchanged while adjusting the number, the effect that the number on convergence and diversity can be studied independently of the selection model. Last but not least, if the increasing step is set to 1, the differences between experimental phenomena could be too subtle to analyze in some cases.

3.3 Update Strategy

After selecting M exemplars, the corresponding particle is updated only if at least one of the exemplars is better than it. Otherwise, it directly enters the next generation. When the update condition is met, the velocity and position of the current particle are updated according to (4) and (5), respectively.

$$v_i(t+1) = rv_i(t) + \sum_{k=1}^{m} r_k(x_{bk}(t) - x_i(t)) \tag{4}$$

$$x_i(t+1) = x_i(t) + v_i(t+1) \tag{5}$$

where $v_i(t)$ and $x_i(t)$ are the velocity and position vector of ith particle in tth generation. x_{bk} $(1 \leq k \leq m)$ are those in the selected exemplars that are better than the updated particle, which is also named used exemplars, the number of them is m, where $1 \leq m \leq M$ is satisfied. r and $r_k (1 \leq k \leq m)$ are values randomized within [0,1].

In the update strategy used, each updated particle learns from any selected exemplars that have better fitness, the reasons for such a design are listed as follows: First, when the number of exemplars is small, the exemplars provide information that is more conducive to convergence. As the number of exemplars gradually increases, the diversity of the exemplars themselves improves, which is also beneficial to enhancing the diversity of swarm. Second, to the best of our knowledge, the velocity update strategy in (4) is widely used in existing PSO variants. However, in this update strategy, the effects of exemplars on convergence and diversity are highly coupled. Consequently, the respective effects of changing the selection of exemplars on convergence and diversity are unclear and have yet to be investigated. In summary, the update strategy in (4) is adopted in this paper to investigate the effects of exemplars selection where exemplars have coupled affection on convergence and diversity.

4 Experiment Study

In order to verify how the selection model and the number of candidate exemplars affect swarm convergence and diversity, experiments are conducted with respect to large-scale benchmark suit posted by CEC 2013 [21]. The reasons why the large-scale benchmark suit is chosen for investigation are stated as follows: First, compared with small-scale benchmark optimization problems, LSOPs raise a higher demand for both convergence and diversity due to the exponentially growing search space. Second, the selection of exemplars can affect convergence and diversity in a more direct way.

In the experiments, first, the influences of the exemplar's number on convergence and diversity are researched by increasing the number from 2 to 10 with three given selection models. Second, the effects of the exemplar's selection model on convergence and diversity are studied by keeping the number unchanged while switching the selection model. For clarity, experimental results are presented mainly in graphic form.

In the proposed investigation, the maximum fitness evaluation FES max is set to 1E + 05 rather than 3E + 06. The reasons for such a design are stated as follows: First, the convergence and diversity are interacting, when convergence suffers, so dose the performance of diversity, and vice versa. Second, there are stage-level differences on the performance of convergence and diversity. Consequently, it is advisable for select a stage in the optimization process for analysis. All figure results are taken from the experimental data of the first 100 iterations.

To make fair and comprehensive comparisons, 30 independent runs on each function are conducted. The convergence and diversity in each generation are measured by the best fitness the entropy, respectively. All the experiments are conducted on a PC with an Inter Core i7-11400F 2.6 GHz CPU, 16G memory and a Microsoft Windows 11 Enterprise 64-bit operating system.

4.1 Experimental Results of Convergence

In Fig. 2, we are to investigate the effects of exemplar number on convergence, where the selection model is fixed unchanged while the number is increased from 2 to 10.

From Fig. 2, it can be pointed out that, in most functions, a small number of candidate exemplars will present an advantage of the swarm convergence within given iterations. However, the results of F_3, F_6 and F_{10} deserve extra attention. All of these functions are multimodal *Ackley Function*, a category of function featuring a large number of local optimal areas. From the optimization results, one can recognize that the better performance of swarm convergence takes place with a relatively large number of candidate exemplars, a condition that leads to higher swarm diversity (which will be discussed in the next section). It is worth noting that these phenomena mentioned above are shared

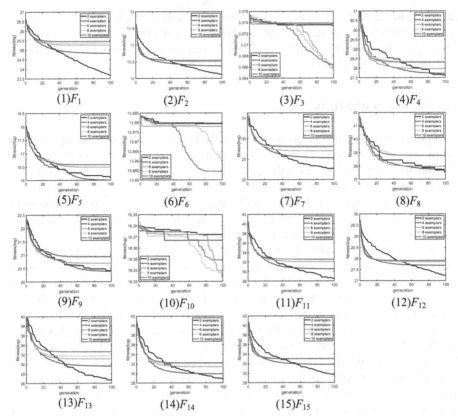

Fig. 2. Variation trend of convergence on the random selection model with different numbers of exemplars (where the increasing step of number is 2).

with the other two selection models. Interested readers can refer to Fig. 13 and Fig. 14 in the appendix.

In Fig. 3, we are to investigate the characteristics of convergence when the number of exemplars is relatively small. Consequently, the increasing step of exemplar number is reduced to 1.

As depicted in Fig. 3, in most cases, a discernible superiority of convergence occurs when the number is no more than 3. Similar phenomena are revealed in the other two selection models. Interested readers can refer to Fig. 15 and Fig. 16 in the appendix.

According to the experimental results above, the effects differ when the exemplar number is small or large. In order to make the difference more obvious, in Fig. 4, the increasing step is set to 4.

From Fig. 4, it can be found that, as the number of candidate exemplars gradually becomes large, the swarm is to be convergent at an increasing rate. This phenomenon is shared with the other two selection models. Interested readers can refer to Fig. 17 and Fig. 18 in the appendix.

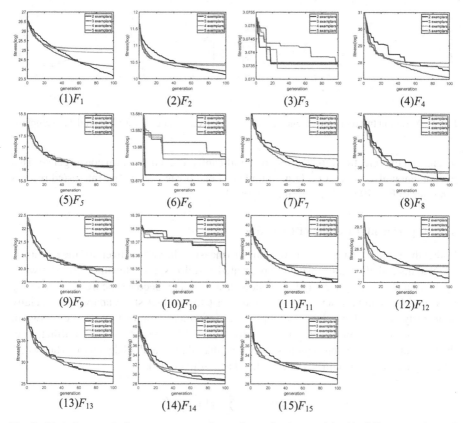

Fig. 3. Variation trend of convergence on the random selection model with different numbers of exemplars (where the increasing step of number is 1).

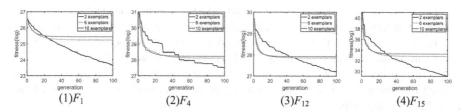

Fig. 4. Variation trend of convergence on the random selection model with different numbers of exemplars (where the increasing step of number is 4).

In Fig. 5, we are to investigate the effects of exemplar selection model on convergence, where the number is fixed unchanged.

In Fig. 5, four functions are selected (F_1, F_4, F_{12} and F_{15}), which represent four different kinds of functions: *Fully-separable Function, Partially Additively Separable Function, Overlapping Function* and *Non-separable Function*, respectively. First, the three curves that represent different selection models have many overlapping parts, which

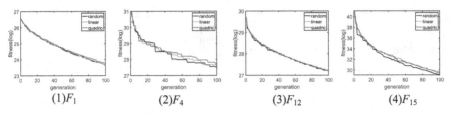

Fig. 5. Variation trend of convergence on the three selection models with 2 candidate exemplars.

means that the selection model does not significantly affect the swarm convergence. This phenomenon also holds for a larger number of candidate exemplars. Interested readers can refer to Fig. 19 and Fig. 20 in the appendix.

4.2 Experimental Results of Diversity

Rather than applying the standard deviation to measure the swarm diversity, the entropy is adopted instead, the main reasons are presented as follows: First, the interaction between diversity and convergence happens to be in a short stage. Second, the trend of diversity measured by standard deviation varies too fast to conduct stage analysis, especially at an early stage. Third, the entropy provides information on overall degree of swarm confusion, an indicator that can more intuitively reflect the swarm diversity. Furthermore, the trend of entropy changes more slowly than that of the standard deviation. The swarm diversity measured by entropy is calculated according to (6):

$$Ent(P) = -\sum_{i=1}^{NP} x_i^d \log_2 x_i^d \qquad (6)$$

where NP is the swarm size, x_i^d is the dth dimension component of the position of the ith particle, $Ent(P)$ is the measurement of the swarm diversity.

In Fig. 6, we are to investigate the effects of exemplar number on diversity, where the selection model is fixed unchanged while the number is increased from 2 to 10.

From Fig. 6, it turns out that, in most cases, increasing the number of candidate exemplars can significantly promote the swarm diversity. In other words, the swarm diversity exhibits a positive trend with the number of candidate exemplars. This phenomenon is shared with the other two selection models. Interested readers can refer to Fig. 21 and Fig. 22 in the appendix.

In Fig. 7, we are to investigate the differences between increasing the number of exemplars and increasing the population size in improving diversity.

From Fig. 7, one can recognize that, when the number of candidate exemplars is fixed unchanged while increasing the population size, the swarm diversity shows a limited improvement. However, when the population size remains unchanged while increasing the number of candidate exemplars, the swarm diversity is significantly enhanced. The phenomenon found is shared with the other two selection models. Interested readers can refer to Fig. 23 and Fig. 24 in the appendix.

In Fig. 8, we are to investigate the effects of exemplar selection model on divsersity, where the number is fixed unchanged.

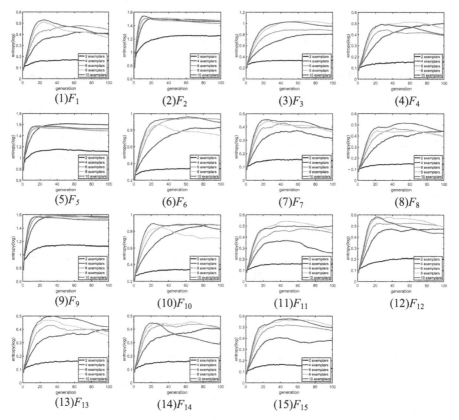

Fig. 6. Variation trend of diversity on the random selection model with different numbers of exemplars (where the increasing step of number is 2).

From Fig. 8, it can be revealed that, when the number of candidate exemplars is small, the three curves that represent different selection models have relatively more overlaps. In this case, the selection model has little influence on swarm diversity.

However, as the number of candidate exemplars becomes large, the overlapping areas between the three curves decrease, emerging branches since the middle or late stages. In this condition, the effect that the selection model on swarm diversity becomes more significant, which is shown in following Fig. 9 and Fig. 10.

4.3 Summary and Analysis

In this section, a summary of the aforementioned experimental results and some further analyses are presented, which are listed as follows:

- The swarm diversity is likely to be promoted with the exemplars' number, however, increasing the number of exemplars is to the disadvantage of swarm convergence. This is mainly because more exemplars will guide particles to explore diverse areas, however, due to the limited fitness evaluations, individuals in each local optimal space

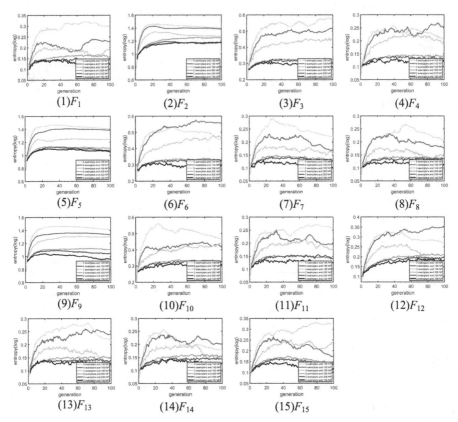

Fig. 7. Variation trend of diversity with the random selection model, the number of candidate exemplars varies from 2 to 5 and the population size varies from 100 to 300, respectively.

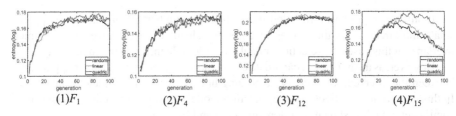

Fig. 8. Variation trend of diversity on the three selection models with 2 candidate exemplars.

cannot conduct sufficient exploitation. At the same time, this phenomenon verifies that accelerating convergence and maintaining diversity are two conflicting goals.

- As the number of exemplars increases, the swarm is more likely to encounter premature convergence. The reason is similar to that of the first point. In large-scale optimization, in the case of updated particles simultaneously learning from multiple exemplars in all dimensions, it is recommended that the number of candidate exemplars be preferably no more than 3. Moreover, the number of candidate exemplars will

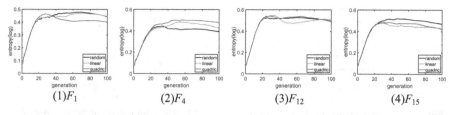

Fig. 9. Variation trend of diversity on the three selection models with 6 candidate exemplars.

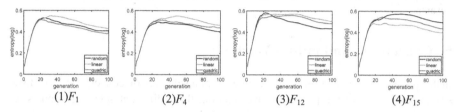

Fig. 10. Variation trend of diversity on the three selection models with 10 candidate exemplars.

significantly affect the performance of the swarm convergence, while the selection model will not.

- Compared with increasing the population size, raising the number of candidate exemplars is a more effective method to enhance swarm diversity. This can be interpreted by that, although increasing the population size brings inherently better diversity, however, within the given fitness evaluations, the number of iterations decreases because of a larger population size, leading to a relatively poor exploration ability of the swarm. While exemplars have a direct guiding function on updated individuals, they are more likely to guide particles to unexplored areas and thus can enhance the swarm diversity.

- According to the overall experimental results, there are no significant differences between the performance of convergence and diversity under the three selection models utilized. In fact, the effects of the three selection models on convergence and diversity are problem-dependent, making it difficult to draw a general conclusion.

- As for multimodal function optimization, good swarm diversity is strongly needed, which is conductive to swarm escapes from the local optimal areas. While convergence acceleration is more valued for unimodal functions, so that the swarm is able to exploit the optimal area with great efficiency and refine more promising solutions.

- Last but not least, as revealed in Fig. 11 and Fig. 12, the selection model of candidate exemplars does not significantly affect the number of used exemplars, while the number of candidate exemplars does.

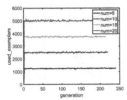

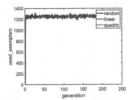

Fig. 11. The comparison of the number of used exemplars with the random selection model and different number of candidate exemplars.

Fig. 12. The comparison of the number of used exemplars with 5 candidate exemplars and three selection models.

5   Conclusions and Future Work

In this paper, an investigation of effects of exemplar selection on convergence and diversity is conducted. In the proposed investigation, three different selection models are adopted to adjust the general quality of selected exemplars; the number of candidate exemplars varies from small to large to enhance the exemplars' dominance among the swarm; a novel learning strategy that can ensure convergence and diversity is utilized for each updated particle. Experimental results demonstrate that the number of candidate exemplars has a considerable impact on swarm convergence and diversity, whereas the effects of the three selection models utilized are relatively weak. Some analyses and conclusions given in this paper may be beneficial to future research in exemplar selection for large-scale optimization.

In the future, our attention will be paid on putting the research findings of this paper into the design of large-scale optimizers. For example, how to design an effective selection model and cooperate the selection model with number to further enhance diversity or to achieve a balance between convergence and diversity are the direction for future investigation.

Acknowledgement. This work is supported by National Key R&D Program of China under Grant Number 2022YFB2602200; the National Natural Science Foundation of China under Grant Number 62273263, 72171172 and 71771176; Shanghai Municipal Science and Technology Major Project (2022–5-YB-09); Natural Science Foundation of Shanghai under Grant Number 23ZR1465400.

Appendix

Due to space limitations, only part of each group of figures is shown.

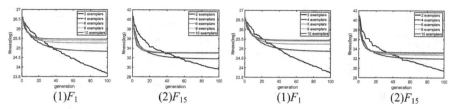

Fig. 13. Variation trend of convergence on the linear selection model with different numbers of exemplars.

Fig. 14. Variation trend of convergence on the quadric selection model with different numbers of exemplars.

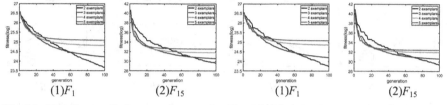

Fig. 15. Variation trend of convergence on the linear selection model with different numbers of exemplars.

Fig. 16. Variation trend of convergence on the quadric selection model with different numbers of exemplars.

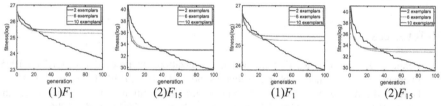

Fig. 17. Variation trend of convergence on the linear selection model with different numbers of exemplars.

Fig. 18. Variation trend of convergence on the quadric selection model with different numbers of exemplars.

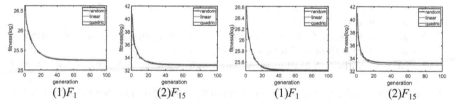

Fig. 19. Variation trend of convergence on the three selection models with 6 candidate exemplars.

Fig. 20. Variation trend of convergence on the three selection models with 10 candidate exemplars.

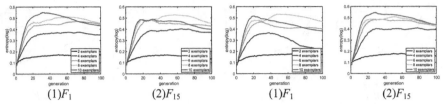

$(1)F_1$ $\qquad$ $(2)F_{15}$ $\qquad\qquad$ $(1)F_1$ $\qquad$ $(2)F_{15}$

Fig. 21. Variation trend of diversity on the linear selection model with different numbers of exemplars.

Fig. 22. Variation trend of diversity on the quadric selection model with different numbers of exemplars.

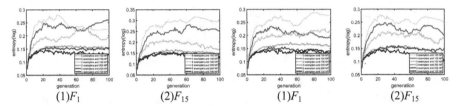

$(1)F_1$ $\qquad$ $(2)F_{15}$ $\qquad\qquad$ $(1)F_1$ $\qquad$ $(2)F_{15}$

Fig. 23. Trend of diversity with the linear selection model, the number of candidate exemplars varies from 2 to 5 and the population size varies from 100 to 300, respectively.

Fig. 24. Trend of diversity with the quadric selection model, the number of candidate exemplars varies from 2 to 5 and the population size varies from 100 to 300, respectively.

References

1. Kennedy, J., Eberhart, R.: Particle swarm optimization. In: Proceedings of ICNN 1995-International Conference on Neural Networks, pp. 1942–1948. IEEE, Location (1995)
2. Dorigo, M., Maniezzo, V., Colorni, A.: Ant system: optimization by a colony of cooperating agents. IEEE Trans. Syst. Man Cybern. Part B (Cybernetics) **26**(1), 29–41 (1996)
3. Karaboga, D., Akay, B.: A comparative study of artificial bee colony algorithm. Appl. Math. Comput. **214**(1), 108–132 (2009)
4. Yang, X.S., He, X.: Firefly algorithm: recent advances and applications. Int. J. Swarm Intell. **1**(1), 36–50 (2013)
5. Yang, X.S., Deb, S.: Engineering optimisation by cuckoo search. Int. J. Math. Model. Numer. Optimisation **1**(4), 330–343 (2010)
6. Yang, X.S., Hossein, G.A.: Bat algorithm: a novel approach for global engineering optimization. Eng. Comput. **29**(5), 464–483 (2012)
7. Mirjalili, S., Mirjalili, S.M., Lewis, A.: Grey wolf optimizer. Adv. Eng. Softw. **69**, 46–61 (2014)
8. Mirjalili, S., Lewis, A.: The whale optimization algorithm. Adv. Eng. Softw. **95**, 51–67 (2016)
9. Mirjalili, S., Gandomi, A, H., Mirjalili, S, Z., Saremi, S., Faris, H., Mirjalili, S, M.: Salp swarm algorithm: a bio-inspired optimizer for engineering design problems. Adv. Eng. Softw. **113**, 163–191 (2017)
10. Yang, Q., Chen, W., Deng, J., Li, T.: A level-based learning swarm optimizer for large-scale optimization. IEEE Trans. Evol. Comput. **22**(4), 578–594 (2017)
11. Yang, Q., Chen, W., Gu, T., Jin, H., Mao, W., Zhang, J.: An adaptive stochastic dominant learning swarm optimizer for high-dimensional optimiztion. IEEE Trans. Cybern. **52**(3), 1960–1976 (2020)

12. Li, D., Guo, W., Lerch, A., Li, L., Wang, L., Wu, Q.: An adaptive particle swarm optimizer with decoupled exploration and exploitation for large scale optimization. Swarm Evol. Comput. **60**, 100789 (2021)
13. Mahadevan, K., Kannan, B.S.: Comprehensive learning particle swarm optimization for reactive power dispatch. Appl. Soft Comput. **10**(2), 641–652 (2010)
14. Qu, B.Y., Suganthan, P.N., Das, S.: A distance-based locally informed particle swarm model for multimodal optimization. IEEE Trans. Evol. Comput. **17**(3), 387–402 (2012)
15. Ratnaweera, A., Halgamuge, S.K., Watson, H.C.: Self-organizing hierarchical particle swarm optimizer with time-varying acceleration coefficients. IEEE Trans. Evol. Comput. **8**(3), 240–255 (2004)
16. Jordehi, A.R., Jasni, J.: Parameter selection in particle swarm optimization: a survey. J. Exper. Theor. Artif. Intell. **25**(4), 527–542 (2013)
17. Chen, W., et al.: Particle swarm optimization with an aging leader and challengers. IEEE Trans. Evol. Comput. **17**(2), 241–258 (2012)
18. Cheng, R., Jin, Y.: A competitive swarm optimizer for large scale optimization. IEEE Trans. Cybern. **45**(2), 191–204 (2014)
19. Cheng, R., Jin, Y.: A social learning particle swarm optimization algorithm for scalable optimization. Inf. Sci. **291**, 43–60 (2015)
20. Li, D., Wang, L., Guo, W., Zhang, M., Hu, B., Wu, Q.: A particle swarm optimizer with dynamic balance of convergence and diversity for large-scale optimization. Appl. Soft Comput. **132**, 109852 (2023)
21. Li, X., Tang, K., Omidvar, M.N., Yang, Z., Qin, K., China, H.: Benchmark functions for the CEC 2013 special session and competition on large-scale global optimization. Gene **7**(33), 8 (2013)

A LSTM Assisted Prediction Strategy for Evolutionary Dynamic Multiobjective Optimization

Guoyu Chen and Yinan Guo[✉]

China University of Mining and Technology, Xuzhou 100083, China
nanfly@126.com

Abstract. Dynamic multiobjective optimization problems (DMOPs) are widely spread in real-world applications. Once the environment changes, the time-varying Pareto-optimal solutions (PS) are required to be timely tracked. The existing studies have pointed out that the prediction based mechanism can initialize high-quality population, accelerating search toward the true PS under the new environment. However, they generally ignore the correlation between decision variables during the prediction process, insufficiently predict the future location under the complex problems. To solve this issue, this paper proposes a long short-term memory (LSTM) assisted prediction strategy for solving DMOPs. When an environmental change is detected, the population is divided into center point and manifold. As for center point, historical ones are utilized to train LSTM network and predict the future one. Subsequently, the manifold is estimated by Gaussian model in terms of two past ones. In this way, an initial population is generated at the new time by combining the predicted center point and manifold. The intensive experimental results have demonstrated that the proposed algorithm has good performance and computational efficiency in solving DMOPs, outperforming the several state-of-the-art dynamic multiobjective evolutionary algorithms.

Keywords: Dynamic multiobjective optimization · LSTM · prediction · evolutionary algorithm

1 Introduction

Many real-world optimization problems involve multiple and time-varying objective functions that are conflicted in nature [1, 2]. These problems are referred to as dynamic multiobjective optimization problems [3–5], which can be formulated as follows:

$$\min \ F(x, t) = (f_1(x, t), \dots, f_m(x, t))^T$$
$$\text{s.t.} x \in \Omega \tag{1}$$

where t is the time index. x represents an n-dimensional decision vector in decision space Ω. $f_i(x, t)$, $i = 1, \dots, m$, is the ith objective function at time t, and m is the number of objectives.

To solve DMOPs, various dynamic multiobjective evolutionary algorithms (DMOEAs) have been designed to track Pareto-optimal solutions (PS) varying over time under dynamic environments. In diversity introduction strategies, random or mutated solutions are generated to replace the ones of current population [6]. Following this idea, Deb et al. [7] developed two versions of D-NSGA-II, called D-NSGA-II-A and D-NSGA-II-B. Apparently, the simple introduction of diversity can destroy the structure of current population to track the changing Pareto optima, but it is weak to solve DMOPs with severe change. Additionally, Sahmoud and Topcuoglu [8] proposed a memory mechanism, which is embedded into NSGA-II. In the algorithm, an explicit memory was constructed to preserve non-dominated solutions during the evolution process, and provided the ones for a similar environment in the future. In this way, the method can well solve periodical problems.

Among the above two mainstream strategies, they generally lack full mining of historical knowledge, achieving insufficient performance in solving various DMOPs. Therefore, Hatzakis and Wallace [9] utilized autoregression (AR) model to construct a feedforward prediction strategy (FPS). The method extrapolated the location of new environment by learning historical ones. To achieve more robust prediction, Zhou et al. [10] proposed a novel strategy, called population prediction strategy (PPS). The population was divided into center point and manifold, and the future center point was estimated using AR model based on a sequence of past ones. Also, the manifold was estimated by past two ones, and an initial population under the new environment was generated by combing the predicted center point and manifold.

The existing prediction methods have shown good ability in handling with DMOPs. However, they predict center point by independently estimating one by one decision variable, neglecting the influence of correlation between different decision variables. Apparently, this gap can prevent the existing prediction methods from solving more scenarios. To solve the issue, this paper proposes a long short-term memory (LSTM) assisted prediction strategy (LP) based DMOEA, called LP-DMOEA. Once an environment changes, LP predicts the future center point by LSTM, and the manifold is represented by Gaussian model to estimate the one under the new environment. In this way, a high-quality population is initialized under the new environment, speeding up the convergence.

2   Related Work

2.1   Prediction Based DMOEAs

Prediction-based approach, as a popular one in DMOEAs, learns the changing rule of PS obtained from historical environments to generate a high-quality initial population, with the purpose of timely tracking the time-varying PS. Among the existing studies, various methods, e.g., transfer learning [11], autoregression [10], and grey model [12], are utilized as the learning model. Additionally, Rambabu et al. [13] proposed a mixture-of-experts-based ensemble framework. In the method, various predictors were managed by a gating network, and switched based on their prediction performance under dynamic environments. Apparently, the ensemble methods can fully utilize the advantage of different prediction methods, achieving better performance in solving DMOPs.

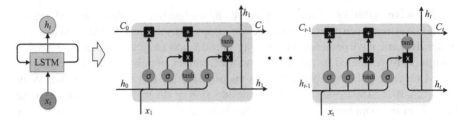

Fig. 1. The workflow of LSTM.

To achieve more stable change response, Liang et al. [14] developed a hybrid method, which combines memory and prediction strategies. Once a similar environment appeared, the memory strategy was executed. Otherwise, the prediction strategy was employed to initialize the population under the new environment. Recently, transfer learning method is also introduced to assist prediction. Jiang et al. [15] developed a trend prediction model to predict the knee points under the new environment. Following that, an imbalance transfer learning method was proposed to generate an initial population in terms of these estimated knee points under the new environment. Although the existing methods explore the prediction in different angels, they generally neglect the influence of correlation between different decision variables during prediction process, which may achieve worse estimation. Apparently, the dynamics of corresponding correlation makes DMOPs more difficult for problem-solver, weakening the performance of DMOEAs in adapting more scenarios [3].

2.2 Long Short-Term Memory

In recent years, the recurrent neural network (RNN) [16] has been pointed out that it is a good tool for processing multivariate time series. However, a gradient explosion may easily emerge in RNN, especially for long time series. To solve this issue, the long short-term memory (LSTM) [17], as a powerful variant of RNN, is developed, and gains wide applications in various practical scenarios. The structure of LSTM network is depicted in Fig. 1, such a memory block consists of a cell, an input gate, an output gate, and a forget gate. Following that, the calculation of them can be expressed as follows:

$$f_t = \sigma(w_f[h_{t-1}, x_t] + b_f) \tag{2}$$

$$i_t = \sigma(w_i[h_{t-1}, x_t] + b_i) \tag{3}$$

$$\tilde{c}_t = \tanh(w_c[h_{t-1}, x_t] + b_c) \tag{4}$$

$$o_t = \sigma(w_o[h_{t-1}, x_t] + b_o) \tag{5}$$

Algorithm 1: The framework of LP-DMOEA

Input: Dynamic optimization function $F(x,t)$, population size N

Output: A series of approximated populations P

1: Set time step $t = 0$;

2: Initialize a population P_0 with size N;

3: **while** *termination criterion not met*

4:   **if** an environmental change occurs

5:     $t = t + 1$;

6:     **if** $t - 1 < K$

7:       Perform Kalman filter prediction method;

8:     **else**

9:       $[P_t, LSTM] = LP(P, LSTM)$;

10:     **end if**

11:   **end if**

12:   Perform NSGA-II;

13: **end while**

$$c_t = f_t \odot c_{t-1} + i_t \odot \tilde{c}_t \tag{6}$$

$$h_t = o_t \odot \tanh(c_t) \tag{7}$$

where i_t, f_t, o_t, and c_t denote the outputs of input gate, forget gate, output gate, and cell at time step t. x_t indicates the input vector of memory block at time step t. h_{t-1} and h_t are output vector of memory block at time step $t-1$ and t, respectively. $\tilde{c}_t$ is the candidate information of input gate. σ and tanh denote activation function. Additionally, w_f, w_i, w_c and w_o are the learned weight vector, and bf, b_i, b_c and b_o are the corresponding bias vectors. During the training process, a LSTM unit is shared at each time step to store and transfer historical knowledge for future time step.

3 LSTM Assisted Prediction Strategy Based DMOEA

3.1 The Framework of LP-DMOEA

As the framework of DMOEA based on a LSTM assisted prediction strategy (LP-DMOEA) is shown in Algorithm 1, a population P_0 with size N is firstly initialized (line 2 of Algorithm 1). Then, $0.1N$ individuals are randomly selected from the current population, and re-calculate them. Apparently, the difference of objective values in two adjacent generations means that a new environment has appeared. Following that, the proposed LSTM assisted prediction strategy (LP) is performed to generate a new initial population (line 9 of Algorithm 1). Here, it is worth mentioning that the Kalman filter (KF) prediction method [18] is introduced once $t - 1 < K$ (line 7 of Algorithm 1), and K is the threshold. Otherwise, NSGA-II [19] is executed to find the PS under each environment (line 12 of Algorithm 1).

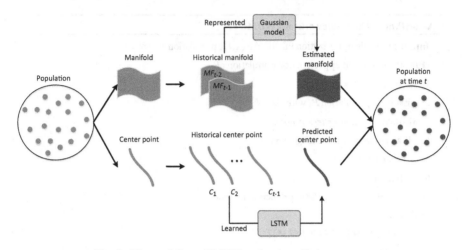

Fig. 2. The workflow of LSTM assisted prediction strategy.

3.2 LSTM Assisted Prediction Strategy

In the proposed LSTM assisted prediction strategy (LP) as presented in Fig. 2, population P (where it contains N individuals, and N is the population size) is predicted by separately estimating center point $C \in \mathbf{R}^{1 \times n}$ and manifold $MF \in \mathbf{R}^{N \times n}$, which can be calculated as follows:

$$C = \frac{1}{|P|} \sum_{x_i \in P} x_i \tag{8}$$

$$MF = P - C \tag{9}$$

Here, center point and manifold determine the location and distribution of population in decision space, respectively. For center point, traditional methods generally predict one by one decision variable, neglecting the influence of correlation between different decision variables. To overcome it, one layer LSTM (the number of hidden units is $2n$.) as hidden layer is introduced. Denoted $X = (C_1, \ldots, C_{K-1})$ and $Y = (C_2, \ldots, C_K)$ as the input and output for training data, respectively. Based on them, the LSTM can be achieved by iteratively optimizing Eq. (2)–(7). Following that, the center point at new time t can be predicted by inputting C_{t-1} as follows:

$$C_t = \text{LSTM}(C_{t-1}) \tag{10}$$

After obtaining C_t, the network state is subsequently updated and utilized for the future prediction.

As for the manifold, it is represented by Gaussian model, which can be formulated as follows:

$$MF \sim \text{N}(\vec{0}, \Sigma) \tag{11}$$

Apparently, the covariance matrix Σ can efficiently represent the distribution of population, and it can improve the diversity in certain extent during the subsequent generation process by Gaussian model. In order to obtain Σ_t at time t, two historical ones are introduced:

$$\Sigma_t = \frac{\Sigma_{t-1} + \Sigma_{t-2}}{2} \tag{12}$$

where Σ_{t-1} and Σ_{t-2} are covariance matrixes of PS_{t-1} and PS_{t-2}, respectively. In this way, the population at time t can be initialized as follows:

$$P_t = C_t + MF_t \tag{13}$$

In addition, all individuals $x \in P_t$ are repaired once they beyond the boundary of decision space:

$$x_{i,j} = \begin{cases} x_{i,j} & L_j < x_{i,j} < U_j \\ L_j & x_{i,j} < L_j \\ U_j & x_{i,j} > U_j \end{cases} \tag{14}$$

where $i = 1, \ldots, N, j = 1, \ldots, n$, L_j and U_j denote the lower and upper bounds of jth variable in decision space.

4 Experiment

4.1 The Framework of LP-DMOEA

In the experiment, nine benchmark functions, including DF1-DF9 [20], are introduced for comparisons, and the number of decision variables n is set to 10 for all test instances. The population size $N = 100$. Also, the change frequency $\tau_t = 10, 20$, the change severity $n_t = 2, 5$, and each run contains 100 environmental changes.

In order to demonstrate the performance of LP-DMOEA, four state-of-the-art DMOEAs are employed for the comparisons, i.e., D-NSGA-II-A [7], PPS [10], KF [18], and KT-MOEA/D [15]. For D-NSGA-II-A, 20% of random individuals are introduced. In PPS, the length of history mean point series $M = 23$, , and the order of AR(p) $p = 3$. In KF, the equal element values of $\mathbf{Q}$ and $\mathbf{R}$ diagonal matrices are set to 0.04 and 0.01, respectively. The number of predicting knee points is set to $p = 10$ in KT-MOEA/D. In LP-DMOEA, the threshold K is set to 30. Also, the total number of iterations is set to 300 in the Adam optimizer, $\beta_1 = 0.5$ and $\beta_2 = 0.999$ are used to train the LSTM network.

To evaluate the performance of algorithms, MIGD [21] and MS [22] are introduced as the performance indicators. MIGD can comprehensively reflect the diversity and convergence of obtained population, and larger MIGD means worse performance of algorithm. Denoted $R(t)$ as a set of uniformly distributed samples of true Pareto front at time t (PF(t)), MIGD can be calculated as follows:

$$\text{MIGD} = \frac{1}{T} \sum_{t=1}^{T} \frac{\sum_{i=1}^{|R(t)|} D(P(t), r_i)}{|R(t)|} \tag{15}$$

where T is the number of environments that occurred in a run. $D(P(t), r_i)$ denotes the minimum Euclidean distance between ith point in $R(t)$ and the $P(t)$ obtained by an algorithm in the objective space.

MS measures the extent degree of Pareto optima covering the true PF. Smaller MS value denotes the worse distribution of obtained PF, and the indicator can be formulated as follows:

$$
\text{MS} = \sqrt{\frac{1}{m} \sum_{k=1}^{m} \left[\frac{\min(\overline{f_k}, \overline{F_k}) - \max(\underline{f_k}, \underline{F_k})}{\overline{F_k} - \underline{F_k}} \right]}
\tag{16}
$$

where $\overline{f_k}$ and $\underline{f_k}$ be the maximum and minimum values of kth objective function in Pareto optima obtained by algorithms, $\overline{F_k}$ and $\underline{F_k}$ are the maximum and minimum values of kth objective function of true PF

Each experiment independently runs 20 times, and the best results among the compared algorithms on each test instance are highlighted in blue. Also, Wilcoxon rank-sum test [23] is introduced to point out the significance between different results at the 0.05 significance level, where $'+'$, $'-'$, and $'='$ denote that the results obtained by another DMOEA are significantly better, significantly worse, and statistically similar to that obtained by compared algorithm, respectively.

4.2 Analysis of Parameter K

In LP-DMOEA, K determines when using the LSTM assisted prediction strategy. As K varies from 20 to 40, every 5, and Fig. 3 depicts the MIGD values obtained by LP-DMOEA with different K. As observed from the statistical results, MIGD values are more stable as K is 30. Apparently, the smaller K provides LSTM with fewer training samples, but larger K may worsen the training data, causing the worse prediction accuracy. Therefore, $K = 30$ is more suitable to keep a good tradeoff between these issues, achieving the robust prediction.

4.3 Comparisons with Various DMOEAs

In order to demonstrate the effectiveness of LP-DMOEA, four popular change response strategies on DMOPs are introduced for comparisons, and the corresponding statistical results of MIGD and MS are summarized in Tables 1 and 2, respectively. As observed from the experimental results, the proposed LP-DMOEA outperforms the other competitors on the most test instances in terms of MIGD and MS indicators. Due to that D-NSGA-II-A introduces random solutions to improve the diversity of population, showing worse tracking ability than LP-DMOEA under more complex problems. Additionally, compared with PPS, KF, and KT-MOEA/D, the proposed LSTM assisted prediction strategy in the paper can achieve more accurate prediction, thus shows more robust performance in solving DMOPs.

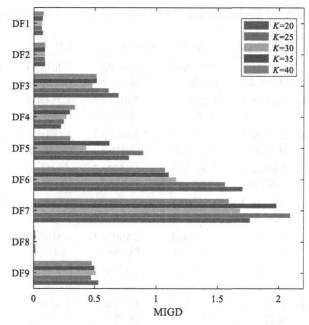

Fig. 3. MIGD values of different K

4.4 Running Time of Various DMOEAs

In order to analyze the computation efficiency, the running time of all comparative algorithms is depicted in Fig. 4. Apparently, LP-DMOEA consumes less running time than PPS, KF, and KT-MOEA/D. This is because that the LSTM assisted prediction strategy proposed in the paper only trains the prediction model once, and it can be updated by the future prediction. However, the former three prediction models, they all require the training of prediction model after each environmental change. In contrast, fewer computation costs have been paid by the D-NSGA-II-A, because it does not need to build the prediction model for initializing a population under the new environment. In general, the proposed LP-DMOEA achieves the promising tradeoff between algorithm performance and computational efficiency in solving DMOPs.

Table 1. Mean and standard deviation of MIGD obtained by comparative algorithms.

Problem	τ_t, n_t	LP-DMOEA	D-NSGA-II-A	PPS	KF	KT-MOEA/D
DF1	(10, 2)	**1.0521e-1** (**1.36e-2**)	2.0598e-1 (7.14e-3)−	4.6074e-1 (2.25e-2)−	5.3611e-1 (1.85e-2)−	4.5126e-1 (2.01e-2)−
	(10, 5)	**2.6940e-2** (**2.22e-3**)	1.2550e-1 (5.96e-3)−	2.0711e-1 (4.03e-2)−	1.2547e-1 (1.48e-2)−	1.3230e-1 (2.03e-2)−
	(20, 2)	**2.2969e-2** (**1.82e-3**)	7.4299e-2 (3.77e-3)−	2.6084e-1 (3.11e-2)−	2.6841e-1 (1.38e-2)−	2.6937e-1 (1.46e-2)−
	(20, 5)	**9.3011e-3** (**4.67e-4**)	4.3091e-2 (1.63e-3)−	5.4593e-2 (6.61e-3)−	4.3451e-2 (5.34e-3)−	4.4382e-2 (6.19e-3)−
DF2	(10, 2)	**9.5351e-2** (**8.01e-3**)	1.3620e-1 (3.01e-3)−	2.4143e-1 (9.24e-3)−	4.1330e-1 (1.98e-2)−	2.8894e-1 (1.53e-2)−
	(10, 5)	**8.6157e-2** (**8.09e-3**)	9.3350e-2 (3.14e-3)−	2.1902e-1 (5.88e-3)−	1.8180e-1 (6.91e-3)−	1.4897e-1 (6.71e-3)−
	(20, 2)	**3.0101e-2** (**4.37e-3**)	4.7943e-2 (2.13e-3)−	1.4274e-1 (6.71e-3)−	2.4071e-1 (6.23e-3)−	1.4607e-1 (5.72e-3)−
	(20, 5)	**3.0710e-2** (**4.17e-3**)	3.5871e-2 (1.83e-3)−	1.2314e-1 (5.30e-3)−	1.0729e-1 (5.31e-3)−	8.0856e-2 (2.54e-3)−
DF3	(10, 2)	6.0485e-1 (3.40e-2)	6.9986e-1 (2.92e-2)−	5.6035e-1 (9.88e-3)+	5.1968e-1 (8.41e-3)+	5.6984e-1 (1.62e-2)+
	(10, 5)	**3.5129e-1** (**5.27e-2**)	4.0003e-1 (1.09e-2)−	5.5274e-1 (1.38e-2)−	**5.9739e-1** (**6.73e-3**)−	5.7219e-1 (1.35e-2)−
	(20, 2)	**3.6617e-1** (**2.72e-2**)	4.0335e-1 (9.15e-3)−	5.4355e-1 (6.91e-3)−	4.7009e-1 (2.15e-2)−	5.5295e-1 (7.90e-3)−
	(20, 5)	**2.6259e-1** (**6.08e-2**)	3.1454e-1 (8.44e-3)−	4.6936e-1 (3.86e-2)−	5.5569e-1 (7.62e-3)−	5.2383e-1 (2.15e-2)−
DF4	(10, 2)	4.0444e-1 (3.08e-2)	**2.1138e-1** (**1.21e-2**)+	2.9606e-1 (1.33e-2)+	3.9087e-1 (4.10e-2) =	2.6174e-1 (8.89e-3)+
	(10, 5)	**1.2632e-1** (**6.33e-3**)	1.3364e-1 (2.41e-3)−	2.0820e-1 (6.72e-3)−	2.3623e-1 (8.63e-3)−	2.1678e-1 (7.15e-3)−
	(20, 2)	1.3428e-1 (7.92e-3)	**1.2458e-1** (**3.29e-3**)+	1.9823e-1 (6.69e-3)−	2.0959e-1 (1.03e-2)−	1.9166e-1 (5.59e-3)−
	(20, 5)	**8.9401e-2** (**2.60e-3**)	9.0773e-2 (1.55e-3) =	1.6861e-1 (3.15e-3)−	1.7766e-1 (4.09e-3)−	1.7171e-1 (4.55e-3)−
DF5	(10, 2)	1.9345e-1 (1.68e-1)	4.8898e-1 (2.87e-2)−	**2.9578e-2** (**3.63e-3**)+	1.1883e-1 (1.44e-2) =	1.2500e-1 (1.41e-2) =
	(10, 5)	6.6356e-1 (5.94e-1)	2.9448e-1 (1.02e-2) =	**1.5527e-2** (**1.16e-3**)+	4.7466e-2 (8.37e-3)+	4.7513e-2 (1.01e-2)+

(continued)

Table 1. (*continued*)

Problem	τ_t, n_t	LP-DMOEA	D-NSGA-II-A	PPS	KF	KT-MOEA/D
	(20, 2)	5.6515e-2 (2.54e-2)	1.6956e-1 (9.95e-3)−	**9.9116e-3 (5.42e-4)+**	3.0003e-2 (4.53e-3)+	3.4418e-2 (5.91e-3)+
	(20, 5)	1.3940e-1 (1.57e-1)	6.8454e-2 (2.89e-3) =	**6.9799e-3 (2.15e-4)+**	1.3147e-2 (6.77e-4)+	1.5470e-2 (1.96e-3)+
DF6	(10, 2)	1.4781e + 0 (1.93e-1)	1.1638e + 0 (4.75e-2) +	7.8352e-1 (6.53e-2)+	**6.3227e-1 (6.33e-2)+**	6.6110e-1 (3.59e-2)+
	(10, 5)	8.5017e-1 (5.64e-1)	1.0564e + 0 (3.10e-2) -	4.9753e-1 (9.01e-2) =	**3.9140e-1 (2.04e-2)+**	4.8643e-1 (7.96e-2) =
	(20, 2)	4.0388e-1 (8.71e-2)	**2.9595e-1 (1.17e-2) +**	5.7164e-1 (4.76e-2)−	3.6631e-1 (3.15e-2) =	5.0576e-1 (2.49e-2)−
	(20, 5)	**2.3143e-1 (1.82e-1)**	2.7523e-1 (9.67e-3)−	3.7730e-1 (2.84e-2)−	3.5226e-1 (1.59e-2)−	4.0758e-1 (4.99e-2)−
DF7	(10, 2)	**1.3394e + 0 (1.02e + 0)**	5.0631e + 0 (5.12e-2)−	9.3804e + 0 (4.95e-2)−	9.4480e + 0 (8.24e-2)−	9.0767e + 0 (1.42e-1)−
	(10, 5)	**2.0284e + 0 (6.48e-1)**	2.4152e + 0 (3.12e-2)−	3.9149e + 0 (2.34e-2)−	3.8522e + 0 (3.88e-2)−	3.6866e + 0 (5.56e-2)−
	(20, 2)	**1.0353e + 0 (2.03e-2)**	4.9097e + 0 (4.18e-2)−	9.6397e + 0 (8.16e-3)−	9.5356e + 0 (2.21e-2)−	9.3513e + 0 (8.83e-2)−
	(20, 5)	**1.0445e + 0 (7.94e-1)**	2.3564e + 0 (5.39e-2)−	3.9362e + 0 (1.05e-2)−	3.7804e + 0 (2.26e-2)−	3.7645e + 0 (2.62e-2)−
DF8	(10, 2)	7.6013e-3 (2.01e-4)	**6.6629e-3 (9.04e-5)+**	2.0907e-1 (5.42e-2)−	1.8571e-1 (1.66e-2)−	1.8558e-1 (1.15e-2)−
	(10, 5)	1.7596e-2 (2.06e-3)	**1.3536e-2 (1.02e-3)+**	1.8253e-1 (3.02e-2)−	1.9119e-1 (1.32e-2)−	1.8801e-1 (1.22e-2)−
	(20, 2)	6.4502e-3 (5.11e-5)	**6.3505e-3 (5.28e-5)+**	1.7595e-1 (3.37e-2)−	1.5379e-1 (9.54e-3)−	1.7005e-1 (3.43e-3)−
	(20, 5)	1.4708e-2 (1.53e-3)	**1.2098e-2 (6.68e-4)+**	1.6410e-1 (2.45e-3)−	1.5127e-1 (9.64e-3)−	1.6933e-1 (2.46e-3)−
DF9	(10, 2)	6.4265e-1 (2.27e-1)	7.3706e-1 (2.69e-2)−	**5.2260e-1 (3.88e-2) =**	1.0145e + 0 (5.80e-2)−	5.8995e-1 (5.72e-2) =
	(10, 5)	3.7411e-1 (6.51e-2)	5.7846e-1 (1.95e-2)−	6.3551e-1 (2.03e-2)−	5.5561e-1 (4.52e-2)−	**3.7002e-1 (5.06e-2) =**
	(20, 2)	**2.4432e-1 (3.65e-2)**	3.6161e-1 (1.14e-2)−	3.9917e-1 (1.26e-2)−	3.9748e-1 (3.19e-2)−	3.1658e-1 (3.65e-2)−
	(20, 5)	**1.9655e-1 (2.62e-2)**	3.1881e-1 (1.30e-2)−	5.6884e-1 (7.56e-2)−	3.0861e-1 (1.67e-2)−	2.5734e-1 (3.75e-2)−
+/ − / =			8 / 25 / 3	7 / 27 / 2	6 / 27 / 3	6 / 26 / 4

Table 2. Mean and standard deviation of MS obtained by comparative algorithms.

Problem	τ_t, n_t	LP-DMOEA	D-NSGA-II-A	PPS	KF	KT-MOEA/D
DF1	(10, 2)	**8.9882e-1** **(1.37e-2)**	7.8442e-1 (1.38e-2)−	2.8871e-1 (3.58e-2)−	4.6877e-1 (2.42e-2)−	3.2900e-1 (2.17e-2)−
	(10, 5)	**9.7574e-1** **(3.49e-3)**	8.8028e-1 (6.87e-3)−	5.8932e-1 (7.49e-2)−	6.9308e-1 (3.14e-2)−	6.8075e-1 (4.32e-2)−
	(20, 2)	**9.7574e-1** **(4.23e-3)**	9.0616e-1 (1.05e-2)−	4.7435e-1 (5.28e-2)−	6.4260e-1 (2.44e-2)−	4.7220e-1 (2.08e-2)−
	(20, 5)	**9.9460e-1** **(8.69e-4)**	9.5841e-1 (3.21e-3)−	8.7585e-1 (1.62e-2)−	8.7132e-1 (1.56e-2)−	8.7016e-1 (1.78e-2)−
DF2	(10, 2)	**8.5665e-1** **(7.77e-3)**	7.8745e-1 (1.28e-2)−	4.3843e-1 (1.08e-2)−	5.3145e-1 (1.77e-2)−	5.5958e-1 (1.41e-2)−
	(10, 5)	**8.8134e-1** **(6.61e-3)**	6.7915e-1 (9.42e-3)−	4.2470e-1 (1.49e-2)−	6.2410e-1 (1.23e-2)−	6.7700e-1 (9.73e-3) =
	(20, 2)	**9.3187e-1** **(5.52e-3)**	9.0604e-1 (1.01e-2)−	5.6813e-1 (1.58e-2)−	6.7427e-1 (8.62e-3)−	6.7235e-1 (8.61e-3)−
	(20, 5)	**9.4087e-1** **(5.01e-3)**	8.6389e-1 (9.36e-3)−	5.6783e-1 (1.10e-2)−	7.2654e-1 (1.10e-2)−	7.4818e-1 (1.05e-2)−
DF3	(10, 2)	2.2404e-1 (3.59e-2)	**3.1974e-1** **(1.52e-2)+**	8.6473e-2 (1.19e-2)−	2.2638e-1 (1.28e-2) =	7.4548e-2 (1.53e-2)−
	(10, 5)	**3.8322e-1** **(7.53e-2)**	3.8301e-1 (1.52e-2) =	8.7587e-2 (1.18e-2)−	5.5757e-2 (5.47e-3)−	7.1691e-2 (1.06e-2)−
	(20, 2)	**3.9745e-1** **(4.19e-2)**	3.5073e-1 (1.66e-2)−	9.9857e-2 (7.35e-3)−	2.1502e-1 (3.15e-2)−	8.5919e-2 (6.82e-3)−
	(20, 5)	**4.8618e-1** **(8.80e-2)**	3.7813e-1 (1.35e-2)−	1.6020e-1 (3.56e-2)−	8.9286e-2 (7.09e-3)−	1.1363e-1 (2.18e-2)−
DF4	(10, 2)	**9.4947e-1** **(8.89e-3)**	9.4154e-1 (9.35e-3)−	6.5185e-1 (1.62e-2)−	5.6357e-1 (3.00e-2)−	6.7974e-1 (8.56e-3)−
	(10, 5)	9.4192e-1 (4.91e-3)	**9.5196e-1** **(2.90e-3)+**	7.2127e-1 (8.71e-3)−	6.9074e-1 (9.91e-3)−	7.1887e-1 (6.04e-3)−
	(20, 2)	9.5272e-1 (2.72e-3)	**9.5880e-1** **(1.85e-3)+**	7.3325e-1 (6.46e-3)−	7.1050e-1 (1.33e-2)−	7.4027e-1 (4.42e-3)−
	(20, 5)	9.3814e-1 (2.22e-3)	**9.5378e-1** **(2.47e-3)+**	7.5295e-1 (2.77e-3)−	7.4481e-1 (4.59e-3)−	7.5574e-1 (3.72e-3)−
DF5	(10, 2)	**9.9987e-1** **(4.67e-4)**	9.9029e-1 (6.01e-3)−	9.7279e-1 (8.87e-3)−	7.8078e-1 (3.25e-2)−	7.5229e-1 (2.85e-2)−
	(10, 5)	**9.9999e-1** **(3.86e-5)**	9.9720e-1 (2.13e-3)−	9.8509e-1 (4.60e-3)−	8.1916e-1 (2.62e-2)−	8.3403e-1 (3.00e-2)−

(*continued*)

Table 2. (*continued*)

Problem	τ_t, n_t	LP-DMOEA	D-NSGA-II-A	PPS	KF	KT-MOEA/D
	(20, 2)	**1.0000e + 0** **(0.00e + 0)**	9.9612e-1 (4.78e-3)−	9.9111e-1 (2.59e-3)−	9.1011e-1 (1.67e-2)−	8.9092e-1 (2.17e-2)−
	(20, 5)	**1.0000e + 0** **(0.00e + 0)**	9.9978e-1 (7.07e-4)−	9.9550e-1 (1.52e-3)−	9.5336e-1 (6.14e-3)−	9.4269e-1 (1.01e-2)−
DF6	(10, 2)	9.8535e-1 (5.47e-3)	**9.9978e-1** **(1.61e-5)+**	1.7339e-1 (2.50e-2)−	2.2450e-1 (1.82e-2)−	2.6834e-1 (2.00e-2)−
	(10, 5)	9.9436e-1 (2.81e-3)	**9.9992e-1** **(9.24e-6)+**	1.8842e-1 (3.30e-2)−	2.2227e-1 (3.03e-2)−	2.2166e-1 (4.36e-2)−
	(20, 2)	9.9251e-1 (5.43e-3)	**9.9998e-1** **(2.84e-6)+**	2.7663e-1 (3.45e-2)−	2.6932e-1 (3.22e-2)−	3.3272e-1 (2.00e-2)−
	(20, 5)	9.9831e-1 (1.70e-3)	**9.9999e-1** **(1.18e-6)+**	2.0675e-1 (2.91e-2)−	2.3926e-1 (3.37e-2)−	2.4046e-1 (5.68e-2)−
DF7	(10, 2)	**5.8590e-1** **(4.45e-2)**	2.4411e-1 (1.60e-2)−	2.1143e-2 (3.14e-3)−	3.4355e-2 (4.21e-3)−	6.7148e-2 (1.39e-2)−
	(10, 5)	**3.5354e-1** **(7.09e-2)**	1.8722e-1 (4.19e-2)−	5.8420e-2 (1.28e-2)−	9.5658e-2 (8.39e-3)−	9.7170e-2 (2.05e-2)−
	(20, 2)	**6.2270e-1** **(5.88e-3)**	2.7199e-1 (1.56e-2)−	3.2731e-2 (2.73e-3)−	5.5684e-2 (5.04e-3)−	6.0580e-2 (8.07e-3)−
	(20, 5)	**5.0197e-1** **(1.06e-1)**	2.0069e-1 (4.16e-2)−	8.4050e-2 (1.26e-2)−	1.2634e-1 (5.85e-3)−	1.2533e-1 (1.56e-2)−
DF8	(10, 2)	9.9594e-1 (1.80e-3)	**9.9747e-1** **(2.65e-3) =**	5.0757e-1 (1.87e-1)−	5.5350e-1 (5.57e-2)−	6.0117e-1 (4.03e-2)−
	(10, 5)	9.6568e-1 (1.13e-2)	**9.9167e-1** **(5.03e-3)+**	5.9445e-1 (1.02e-1)−	5.6490e-1 (3.91e-2)−	5.9752e-1 (4.27e-2)−
	(20, 2)	9.9949e-1 (2.69e-4)	**9.9982e-1** **(6.17e-4) =**	6.0534e-1 (1.17e-1)−	6.5288e-1 (3.43e-2)−	6.3533e-1 (1.13e-2)−
	(20, 5)	9.8484e-1 (8.70e-3)	**9.9740e-1** **(2.81e-3)+**	6.4566e-1 (8.01e-3)−	6.6585e-1 (3.03e-2)−	6.4121e-1 (7.64e-3)−
DF9	(10, 2)	**1.0000e + 0** **(0.00e + 0)**	9.9106e-1 (7.04e-3)−	2.0128e-1 (1.96e-2)−	5.1322e-1 (3.58e-2)−	4.4798e-1 (2.19e-2)−
	(10, 5)	**1.0000e + 0** **(0.00e + 0)**	9.8286e-1 (8.84e-3)−	1.4671e-1 (1.74e-2)−	4.7009e-1 (3.25e-2)−	4.4883e-1 (4.55e-2)−
	(20, 2)	**1.0000e + 0** **(0.00e + 0)**	9.8262e-1 (1.55e-2)−	2.5992e-1 (1.78e-2)−	6.5675e-1 (1.98e-2)−	5.3587e-1 (4.48e-2)−
	(20, 5)	**1.0000e + 0** **(0.00e + 0)**	9.8267e-1 (7.46e-3)−	1.9136e-1 (7.01e-2)−	5.5318e-1 (2.98e-2)−	5.4720e-1 (4.60e-2)−
+/−/=			10 / 23 / 3	0 / 36 / 0	0 / 35 / 1	0 / 35 / 1

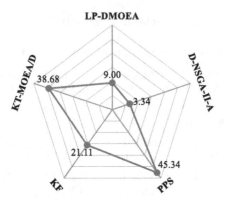

Fig. 4. Average running time of comparative algorithms (Unit: second).

5   Conclusion

In this paper, we propose a long short-term memory (LSTM) assisted prediction strategy based DMOEA, namely LP-DMOEA, to solve DMOPs. Population is firstly divided into center point and manifold. Following that, the historical center points are introduced to train LSTM network and predict the future one, and the manifold is represented by Gaussian model to estimate future one in terms of past two ones. In this way, a high-quality population is initialized, timely tracking the changing PS under dynamic environments. The intensive experimental results have demonstrated that the proposed LP-DMOEA has promising performance and computational efficiency in solving DMOPs.

However, LSTM can efficiently predict the location of population, but the estimation of manifold may be misled by simply utilizing historical information. Apparently, complex changes may arise in the correlation between different decision variables, wrong estimation can cause worse distribution of population. Therefore, how to efficiently estimate manifold needs further study.

References

1. Egea, J.A., Gracia, I.: Dynamic multiobjective global optimization of a waste water treatment plant for nitrogen removal. IFAC Proc. **45**, 374–379 (2012)
2. Wang, Z., Li, G., Ren, J.: Dynamic path planning for unmanned surface vehicle in complex offshore areas based on hybrid algorithm. Comput. Commun. **166**, 49–56 (2021)
3. Jiang, S., Yang, S.: Evolutionary dynamic multiobjective optimization: benchmarks and algorithm comparisons. IEEE Trans. Cybern. **47**, 198–211 (2017)
4. Azzouz, R., Bechikh, S., Ben Said, L.: Dynamic multi-objective optimization using evolutionary algorithms: a survey. In: Bechikh, S., Datta, R., Gupta, A. (eds.) Recent Advances in Evolutionary Multi-objective Optimization. ALO, vol. 20, pp. 31–70. Springer, Cham (2017). https://doi.org/10.1007/978-3-319-42978-6_2
5. Jiang, S., Zou, J., Yang, S., Yao, X.: Evolutionary dynamic multi-objective optimisation: a survey. ACM Comput. Surv. **55**, 1–47 (2023)
6. Raquel, C., Yao, X.: Dynamic multi-objective optimization: a survey of the state-of-the-art. In: Studies in Computational Intelligence, pp. 85–106 (2013)

7. Deb, K., Rao N., U.B., Karthik, S.: Dynamic multi-objective optimization and decision-making using modified NSGA-II: a case study on hydro-thermal power scheduling. In: Obayashi, S., Deb, K., Poloni, C., Hiroyasu, T., Murata, T. (eds.) EMO 2007. LNCS, vol. 4403, pp. 803–817. Springer, Heidelberg (2007). https://doi.org/10.1007/978-3-540-70928-2_60

8. Sahmoud, S., Topcuoglu, H.R.: A memory-based NSGA-II algorithm for dynamic multi-objective optimization problems. In: Squillero, G., Burelli, P. (eds.) EvoApplications 2016. LNCS, vol. 9598, pp. 296–310. Springer, Cham (2016). https://doi.org/10.1007/978-3-319-31153-1_20

9. Hatzakis, I., Wallace, D.: Dynamic multi-objective optimization with evolutionary algorithms. In: Proceedings of the 8th Annual Conference on Genetic and Evolutionary Computation – GECCO 2006, p. 1201. ACM Press, New York (2006)

10. Zhou, A., Jin, Y., Zhang, Q.: A population prediction strategy for evolutionary dynamic multiobjective optimization. IEEE Trans. Cybern. **44**, 40–53 (2014)

11. Guo, Y., Chen, G., Jiang, M., Gong, D., Liang, J.: A knowledge guided transfer strategy for evolutionary dynamic multiobjective optimization. IEEE Trans. Evol. Comput. **66**, 1 (2022)

12. Wang, C., Yen, G.G., Jiang, M.: A grey prediction-based evolutionary algorithm for dynamic multiobjective optimization. Swarm Evol. Comput. **56**, 100695 (2020)

13. Rambabu, R., Vadakkepat, P., Tan, K.C., Jiang, M.: A mixture-of-experts prediction framework for evolutionary dynamic multiobjective optimization. IEEE Trans. Cybern. **50**, 5099–5112 (2020)

14. Liang, Z., Zheng, S., Zhu, Z., Yang, S.: Hybrid of memory and prediction strategies for dynamic multiobjective optimization. Inf. Sci. (Ny) **485**, 200–218 (2019)

15. Jiang, M., Wang, Z., Hong, H., Yen, G.G.: Knee point-based imbalanced transfer learning for dynamic multiobjective optimization. IEEE Trans. Evol. Comput. **25**, 117–129 (2021)

16. Cho, K., et al.: Learning phrase representations using RNN encoder–decoder for statistical machine translation. In: Proceedings of the 2014 Conference on Empirical Methods in Natural Language Processing (EMNLP), pp. 1724–1734. Association for Computational Linguistics, Stroudsburg (2014)

17. Hochreiter, S., Schmidhuber, J.: Long short-term memory. Neural Comput. **9**, 1735–1780 (1997)

18. Muruganantham, A., Tan, K.C., Vadakkepat, P.: Evolutionary dynamic multiobjective optimization via kalman filter prediction. IEEE Trans. Cybern. **46**, 2862–2873 (2016)

19. Deb, K., Pratap, A., Agarwal, S., Meyarivan, T.: A fast and elitist multiobjective genetic algorithm: NSGA-II. IEEE Trans. Evol. Comput. **6**, 182–197 (2002)

20. Jiang, M., Wang, Z., Qiu, L., Guo, S., Gao, X., Tan, K.C.: A fast dynamic evolutionary multiobjective algorithm via manifold transfer learning. IEEE Trans. Cybern. **51**, 3417–3428 (2021)

21. Chen, G., Guo, Y., Huang, M., Gong, D., Yu, Z.: A domain adaptation learning strategy for dynamic multiobjective optimization. Inf. Sci. (Ny) **606**, 328–349 (2022)

22. Goh, C., Tan, K.C.: A competitive-cooperative coevolutionary paradigm for dynamic multiobjective optimization. IEEE Trans. Evol. Comput. **13**, 103–127 (2009)

23. Li, L., Lin, Q., Ming, Z., Wong, K.C., Gong, M., Coello, C.A.C.: An immune-inspired resource allocation strategy for many-objective optimization. IEEE Trans. Syst. Man Cybern. Syst. **PP**(99), 1–14 (2022)

An Adaptive Brain Storm Optimization Based on Hierarchical Learning for Community Detection

Wenya Shi[1], Yifei Sun[1(✉)], Shi Cheng[2], Jie Yang[1], Xin Sun[1], Zhuo Liu[1], Yifei Cao[1], Jiale Ju[1], and Ao Zhang[1]

[1] School of Physics and Information Technology, Shaanxi Normal University, Xi'an 710119, China
{wenyas,yifeis,jieyang2021,sunxin_,zhuoliu,yifeic,jujiale,
aozhang}@snnu.edu.cn
[2] School of Computer Science, Shaanxi Normal University, Xi'an 710119, China
cheng@snnu.edu.cn

Abstract. Community structure is an important feature of complex networks and it is essential for analyzing complex networks. In recent years, community detection based on heuristic algorithms has received much attention in various fields. To improve the effectiveness and accuracy of the algorithm in big data era, an adaptive brain storm optimization based on hierarchical learning (ABSO-HL) is proposed. Instead of the fixed probability, the adaptive probability is adopted in mutation and crossover operations of the proposed ABSO-HL. The proposed updated strategy selects one or two solutions for mutation and crossover to generate a new one. The proposed hierarchical learning strategy is used to accelerate the process by searching in the neighborhood of the new solution, and obtain the optimal partition in an efficient way. The usefulness and effectiveness of the proposed algorithm were demonstrated through a lot of experiments on both real-world and synthetic networks.

Keywords: Brain Storm Optimization · Community Detection · Hierarchical Learning · Complex Networks

1 Introduction

Numerous real-world systems can be represented as networks, for example, collaborative network, aviation network, biological network [1–3], etc. These networks are called complex networks, where the relationships are very complicated among the individuals in the network. Community structure is an important topological property of complex networks [4]. Community structure refers to nodes with similar characteristics or roles that are more densely connected to each other within the community, and more sparsely connected outside of it [5–7].

In recent years, with the efforts of many researchers in various fields, a lot of algorithms have been proposed for community detection, mainly divided into hierarchical clustering-based methods and modularity-based optimization methods. Hierarchical

H. Zhang et al. (Eds.): NCAA 2023, CCIS 1869, pp. 390–402, 2023.
https://doi.org/10.1007/978-981-99-5844-3_28

clustering-based methods include splitting and coalescing strategies. The famous GN (Girvan-Newman) algorithm [4] is a representative of the splitting method. Newman and Girvan [8, 9] proposed the modularity metric as a means of assessing the effectiveness of a network community partition. Finding a network division that maximizes the modularity is equivalent to solving the network community detection problem. Although optimizing the modularity is a non-deterministic polynomial hard (NP-hard) problems, many researchers have proposed heuristic algorithms (HAs) to solve it, including particle swarm optimization [10], simulated annealing algorithm [11], extreme value optimization algorithm [12], and genetic algorithm, etc. Community detection can also be regarded as multi-objective optimization problems (MOPs) [13], and thus, many scholars have proposed numerous classical multi-objective evolutionary algorithms (MOEAs) to solve MOPs in the past three decades [14].

In this study, we proposed an adaptive brain storm optimization algorithm based on hierarchical learning (ABSO-HL) for solving the community detection problems [16]. It consists of two main components. First, an update strategy is adopted in a mutation and crossover pool by using adaptive probability, and a better solution is generated by one or two existing solution from different clusters based on adaptive probability, resulting in a better population. Second, the hierarchical learning strategy operates on the optimal solution of the new population, searching the neighborhood of the solution to efficiently and rapidly identify potential optimal solutions and avoid becoming stuck in a local optimum. Comprehensive experiments on several real-world and synthetic networks show that ABSO-HL has effectiveness and practicality compared with other existing algorithms.

The structure of the rest of this article is as follows: an overview of complex networks and the classical approach for community detection are given in Sect. 2. Section 3 details the proposed algorithm ABSO-HL, while Sect. 4 provides experimental data and analyzes the results. Section 5 concludes our work and offers future directions for research.

2 Background Knowledge

A complex network is typically composed of edges connecting many nodes, represented in graph theory as $G = (V, E)$ [15]. Where V represents the set of nodes and E represents the set of edges. Generally, nodes in the network correspond to different entities in real-world systems, and edges denote the relationships between them. The adjacency matrix A is symmetric and its elements only have values 0 and 1, representing the adjacency relationship between points. When the element value A_{ij} equals 1, two nodes v_i and v_j are regarded as existing one link, otherwise, there is no connection between v_i and v_j.

2.1 Modularity

In 2004, Newman and Girvan proposed modularity as a measure for assessing the partition of a network community [9, 20]. The magnitude of modularity depends on the partition of the network, ranging from -0.5 to 1. A higher modularity value indicates a better division of the network. The formula for modularity is shown below:

$$Q = \sum_{i=1}^{c} \left[\frac{l_i}{m} - \left(\frac{k_i}{2m} \right)^2 \right] \tag{1}$$

where c is the number of communities, l_i is the number of edges in the ith community, m is the total number of edges in the network, and k_i is the sum of all node degrees in the ith community.

2.2 Normalized Mutual Information

Normalized mutual information (NMI) [17, 18] is used to measure the correlation between two variables. It is applied in the field of community detection to quantify the similarity between two distinct communities. The formula for NMI is shown below:

$$I(S_1, S_2) = \frac{-2 \sum_{i=1}^{N_1} \sum_{j=1}^{N_2} C_{ij} \log\left(C_{ij}N / C_{i.}C_{.j}\right)}{\sum_{i=1}^{N_1} C_{i.} \log(C_{i.}/N) + \sum_{j=1}^{N_2} C_{.j} \log\left(C_{.j}/N\right)} \tag{2}$$

where S1(S2) denotes the different partitions, and N1(N2) is the number of communities in S1(S2); C is a confusion matrix whose element C_{ij} denotes the number of identical nodes existing in both the community in partition S_1 and S_2; N is the number of nodes, and $C_{i.}$ ($C_{.j}$) represents the sum of the elements in row i (column j). The value of NMI ranges between 0 and 1. If $I(S_1, S_2) = 1$, partition S_1 is equal to S_2, otherwise, partition S_1 is not equal to S_2.

3 Proposed Algorithms

3.1 Framework of ABSO-HL

The Brain Storm Optimization (BSO) algorithm was proposed by Shi in 2011, and it simulates the process of human idea generation [16]. In BSO algorithm, the population is usually divided into multiple clusters, and the individual with the largest (smallest) fitness value in each cluster is regarded as the cluster center. In each iteration, either the common individual or the cluster center can be selected for variation, allowing the algorithm to search for the optimal solution as much as possible in the solution space. Although BSO can perform well in small networks compared to classical algorithms, it still has many shortcomings. For example, the BSO algorithm cannot get the maximum Q in large networks and it is unstable. To improve this drawback of the algorithm, we combine hierarchical learning with BSO so that the algorithm can be applied to large networks.

We proposed ABSO-HL for community detection by optimizing modularity in this study. Every solution can be considered as a division of the network. ABSO-HL consists of several parts: initialization, k-means clustering, mutation and crossover, and hierarchical learning. Figure 1 shows the overall process of ABSO-HL. This section will provide further details on the remaining aspects.

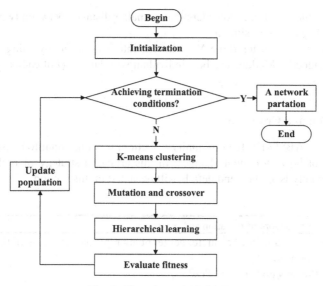

Fig. 1. Flowchart of ABSO-HL

3.2 Initialization and Representation

To maintain population diversity, a random coding initialization was used. Each solution can be expressed as $P_i = [x_i^1, x_i^2, ..., x_i^N]$, where $i \in \{1, 2, ..., N_p\}$, N is the total number of nodes, N_p represents the number of population, x_i^j indicates the community identifier of vertex v_i in solution P_i, the value of x_i^j is an integer ranges from 1 to N, $j \in \{1, 2, ..., N\}$. If $x_i^a = x_i^b$, it means the vertices v_a and v_b are in the same community, If $x_i^a \neq x_i^b$, it means they are in different communities. Figure 2 shows this expression of a network with 9 nodes.

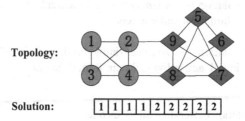

Fig. 2. Illustration of the representation scheme.

3.3 K-means Clustering

K-means [19] algorithm is a popular clustering technique that divides a dataset into k clusters, with each cluster represented by its centroid. By default, k-means uses the

Euclidean distance metric. It considers that a smaller distance between two individuals implies that they are more similar.

In this paper, we clustered the N_p solutions into N_c clusters by using the k-means function provided by Matlab. The best individual with largest Q of each cluster will be the center.

3.4 Mutation and Crossover

When applying ABSO-HL to community detection, it is important to consider solving the problem of how individual changes. In this article, a strategy of multiple points varying separately is applied and details will be stated in this section.

Algorithm1 : Framework of mutation and crossover

Input: The probability of a cluster center being replaced: P_1; the solutions: *POP*; the population size: N_p;

Output: the new population: *POP;*

1 **if** *random*(0,1) < P_1

2   randomly generate a solution take the place of a random cluster center;

3 **end**

4 **for** every solution

5   **if** *random*(0,1)< P_2 // select one cluster

6   **if** *random*(0,1)< 0.3

7   a new solution ← *mutation* (center, P_2); // use the center

8   **else**

9   a new solution ← *mutation* (individual, P_2); // use the individual randomly selected from the cluster

10   **end**

11   **else** // select two different clusters

12   **if** *random*(0,1)< 0.7

13   a new solution ← *crossover* (center1, center2, P_2); //center1 and center2 are from different clusters

14   **else**

15   a new solution; ← *crossover* (individual1, individual2, P_2); //individual1 and individual2 are from different clusters

16   **end**

17   **end**

18   **if** the new solution better than previous one

19   replace it;

20   **end**

21 **end**

Algorithm 1 presents the pseudo code for mutation and crossover. In line 1, to ensure the diversity of solutions, we allow solutions that are worse than existing solutions to be generated. A probability *r is* generated randomly, if $r < P_1$, a random solution will replace a center from any arbitrary cluster. In addition, clusters with more individuals are more likely to be selected, and cluster centers have more opportunities of being chosen than other solutions in each cluster to produce a new solution. The parameter P_2 is an adaptive probability and it will increase with the number of iterations. In the early stage of the algorithm, P_2 is relatively small, resulting in more preferable to choose two solutions to combine, which will expand the search range. Later in the algorithm, the probability of choosing one solution will be relatively larger, which makes the algorithm gradually stabilize. The probability P_2 is defined as follows:

$$p_2 = 0.8 * \frac{i}{i_{max}} \tag{3}$$

where i is the current iteration number, i_{max} is the maximum iteration number.

In line 5 of Algorithm 1 is used to select one or two clusters. The variable r is a number from 0 to 1, if $r < P_2$, one solution is chosen to mutate and generate a new one, otherwise two solutions from different clusters are selected to cross. In lines 6–10 of Algorithm 1, we set two strategies in the mutation pool, and the algorithm selects one of the update strategies from mutation poll based on the adaptive probability. We consider that nodes should be in the same community as the majority of its neighbors. At the early stage of the iteration, we are more likely to perform the operation of changing the community label of one-fifth of the nodes to their most neighbors'. However, its community label stays the same when it has no neighbors. The specific process of mutation is shown in Fig. 3, green and pink circles represent different community partitions. Node 3 is assigned to the community where its majority of neighboring nodes belong to, resulting in it's community label change to the same community as that of node 7. In the later iteration, we prefer to perform the operation of changing the community label of any one node to the neighbor's that makes Q the largest.

Fig. 3. Variation diagram of node 3

Lines 12–16 of Algorithm 1 are used to cross any two individuals or cluster centers from different clusters. The crossover process can be described as follows: at the early stage of the iteration, we want the algorithm to search as widely as possible in the solution space, making more community labels of nodes to change. The process of exchanging any segment community labels in two solutions is shown in Fig. 4.

However, in the later iteration, we make greater use of the two-way crossover operation, as described in Fig. 5. Firstly, two solutions x_a and x_b are randomly selected from the population, and a vertex v_i is arbitrarily chosen as the crossover point. The community label of vertex v_i in solution x_a and x_b is noted as x_a^i and x_b^i, respectively. Next, the community label of vertices in x_a with the same community label as x_a^i are assigned to the corresponding nodes in x_b (i.e. $x_b^j \leftarrow x_a^j, \forall j \in \{j \mid x_a^j = x_a^i\}$). Also, the community label of vertices in x_b with the same community label as x_b^i are distributed to the corresponding node in x_a (i.e. $x_a^j \leftarrow x_b^j, \forall j \in \{j \mid x_b^j = x_b^i\}$).

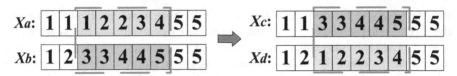

Fig. 4. Crossover of two solutions

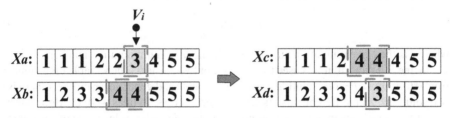

Fig. 5. Two-way crossover

3.5 Hierarchical Learning

To accelerate convergence and avoid getting trapped in local optima, a hierarchical learning strategy is proposed in this section to refine the optimal solution P_l of the new population. The hierarchical learning algorithm consists of two levels.

First-level learning is shown in Algorithm 2. Taking P_l as the initial solution, and taking $\{r_1, r_2, ..., r_n\}$ as the random sequence. A new solution P_n is obtained by replacing the label of node V_{r_i} to one of its neighbors'. If the fitness value Q is larger than before, we will update P_l with P_n.

Algorithm2 : First-Level Learning

    Input: Solution with largest Q : P_l;

    Output: A new solution: P_l;

1 creative a random series $\{r_1, r_2, ..., r_n\}$;

2 **for** every node V_{r_i} in P_l

3     $P_n \leftarrow$ update its community label x_i^j with one of its neighbors';

4   **if** $Q_{P_n} > Q_{P_l}$

5       $P_l \leftarrow P_n$;

6   **end**

7 **end**

The second-level learning applies to the results P_l and its process is described in Algorithm 3. A new solution $P_t = [v_1', v_2', ..., v_k']$ is generated, considering its nodes as each community in P_l. Assuming P_l has k communities $(c_1, c_2, ..., c_k)$, there are k notes in P_t. The total weight of link between communities in P_l is equal to the weight of link between the corresponding new nodes. Every new node will be assigned an unique label (e.g. $v'_i = i$, $1 < i < k$, where v'_i represents the community label). Then, employ first-level learning on P_t to search for an improved solution.

Algorithm3 : Second-Level Learning

Input: Solution with largest Q : P_l;

Output: A new solution $P_n = [x^1_n, x^2_n, ..., x^N_n]$;

1 generate a solution $P_t = [v'_1, v'_2, ..., v'_k]$, where k represents the total number of

   communities in P_l, v'_i represents the corresponding community s_j in P_l, $1 < i < k$.

   B is a adjacency matrix of P_t, denotes connections between new nodes. $B_{ij} \leftarrow$

   $\sum_{i \in s_i} \sum_{j \in s_j} A_{ij}$;

2 $v'_i = i$, $1 < i < k$;

3      $P_t \leftarrow$ first-level learning (P_l);

4 **While** $i \leq k$

5      $x^j_n \leftarrow v'_i$, $\forall j \in \{j | v_j \in s_j\}$

6 **end**

---

4 Experiments

4.1 Experimental Settings

This section evaluates the performance of ABSO-HL on five real-world networks and eleven synthetic networks. The parameters used in ABSO-HL are listed in Table 1. All experiments are simulated by MATLAB R2020b and run independently for 30 sessions. The computer's operating system is Microsoft (R) Windows 10 Home and the other configurations are 16 GB DDR4 RAM, Ryzen (TM), AMD (R), 7-5800U with Radeon Graphics CPU 1.90 GHZ.

Table 1. Set of Parameters for ABSO-HL Algorithm

N_p	N_c	i_{max}	p_1	p_2
200	5	100	0.7	$0.8 * \frac{i}{i_{max}}$

Experimental Networks. The classical GN benchmark network [4] has 128 nodes, which are divided into 4 communities with 32 nodes in each community. The mixing parameter γ is used to control the proportion of neighbors of each node in the other communities. As γ increases, community structure becomes increasingly difficult to discover.

Table 2. Topological characteristics of real-world networks

Network	Nodes	Edges	$\bar{k}$
Emails	1133	545	9.622
Jazz	198	2742	27.70
Football	115	613	10.66
Dolphins	62	159	5.129
Karate	34	79	4.588

The real-world networks are Emails within the University network (Emails) [21], the Jazz Musicians Network (Jazz) [22], the American College Football network (Football) [4], the Bottlenose Dolphins network (Dolphins) [23] and theZachary's Karate Club Network (Karate) [24]. Table 2 provides details of these real-world networks.

4.2 Experimental Results and Analysis

In this paper, we compared the performance of ABSO-HL with 10 existing algorithms. They are MODPSO [25], Meme-net [26], GA-net [27], GN [4], MOGA-net [28], MOCD [29], MOEA/D-net [30], MODPSO-CD-r [31], CNM [32] and Louvain [33].

Table 3 shows the experimental data of ABSO-HL, BSO, MODPSO, MOGA-net and Louvain in five real-world networks. Based on the experimental results, we can conclude that ABSO-HL shows relatively good performance on all evaluation metrics compared to other algorithms. BSO can find the Q-maximum network partition in relatively small networks like Karate, Dolphins and Football. However, the accuracy of BSO will decreases when the number of nodes in the network increases, such as in the jazz and Email networks, where BSO does not obtain the maximum Q. Therefore, we combine BSO with the strategy of hierarchical learning to improve the performance of the algorithm in large networks.

The proposed ABSO-HL can find the network partition with the largest Q among the four networks except the Email network. Although ABSO-HL does not get the best community division in Email networks, it has the largest Q of all algorithms and more stable compared with others. MODPSO can find the largest Q in Karate and Football networks, but it is not stable. Similarly, Louvain can find the largest Q in Football and Jazz networks, but it is not stable either. MOGA-net did not find the network division with the maximum Q in the 5 real-world networks, and the stability also needs to be improved.

Figure 6 shows the results of the experimental comparison between ABSO-HL and other algorithms on the GN benchmark network. The results demonstrate that when the mixing parameter $\gamma \leq 0.1$, all of the algorithms can obtain the maximum NMI, and as γ gradually becomes larger, the NMI of several algorithms subsequently become smaller. When $\gamma = 0.45$, only ABSO-HL and MODPSO can identify the right network partition. When $\gamma = 0.5$, the average NMI of ABSO-HL is 0.5026, only MODPSO and MOPSO-CD-r better than it.

Table 3. Results of ABSO-HL compare with real-world network

Network	Indexes	ABSO-HL	BSO	MODPSO	MOGA-net	Louvain
Karate	Q_{max}	**0.4198**	**0.4198**	**0.4198**	0.4159	0.4188
	Q_{ave}	**0.4198**	0.4186	0.4182	0.3945	0.4165
	Q_{std}	**0**	0.0002	0.0079	0.0089	0.0077
Dolphins	Q_{max}	**0.5222**	**0.5222**	0.5216	0.5034	0.5168
	Q_{ave}	**0.5212**	0.5187	0.5208	0.4583	0.5160
	Q_{std}	**0.0005**	0.0036	0.0062	0.0163	0.0029
Football	Q_{max}	**0.6046**	**0.6064**	**0.6046**	0.4325	**0.6046**
	Q_{ave}	**0.6045**	0.6031	0.6038	0.3906	0.6043
	Q_{std}	**0.0009**	0.001	0.0011	0.0179	0.0009
Jazz	Q_{max}	**0.4451**	0.4449	0.4421	0.2952	**0.4451**
	Q_{ave}	**0.4451**	0.4442	0.4419	0.2929	0.4443
	Q_{std}	**0**	0.0003	0.0001	0.0084	0.0023
Emails	Q_{max}	**0.5749**	0.5647	0.5193	0.3007	0.5412
	Q_{ave}	**0.5686**	0.5526	0.3493	0.2865	0.5392
	Q_{std}	**0.0051**	0.0109	0.0937	0.0075	0.0091

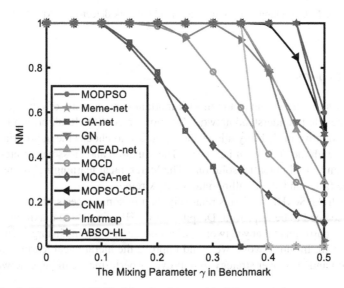

Fig. 6. The average NMI of the algorithms with different mixing parameters.

The community partition of the email network identified by ABSO-HL is depicted in Fig. 7. The network is divided into 9 communities, with 1133 nodes in total, and each community is denoted by a distinct color.

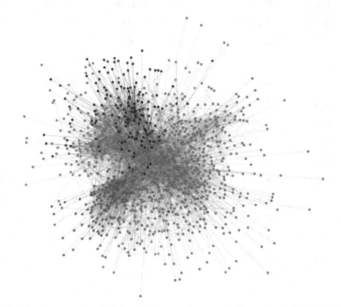

Fig. 7. Community division of the email network detected by ABSO-HL.

5 Conclusions

Community structure is one of the most important characteristic of complex networks that helps people to understand network function. In ABSO-HL algorithm proposed in this study, adaptive probability allows the algorithm to choose the appropriate strategy of variation with the number of iterations. The mutation and crossover strategy enables the algorithm to find potential solutions. The hierarchical learning helps the algorithm to converge faster and avoid falling into local extremes.

Although ABSO-HL performs relatively well in some aspects, there are still many performance need to be improved. Despite ABSO-HL can get the maximum Q and the best network partition in many networks, its accuracy and stability in large networks need to be improved. In future work, we could improve the hierarchical learning strategy or the mutation and crossover strategy to improve its performance in large networks.

Acknowledgements. This work was supported by the Natural Science Basic Research Plan in Shaanxi Province of China (Program No. 2022JM-381,2017JQ6070) National Natural Science Foundation of China (Grant No. 61703256), Foundation of State Key Laboratory of Public Big Data (No. PBD2022–08) and the Fundamental Research Funds for the Central Universities (Program No.GK202201014, GK202202003, GK201803020).

References

1. Strogatz, S.: Exploring complex networks. Nature **410**, 268–276 (2001)
2. Latora, V., Nicosia, V., Russo, G.: Complex Networks: Principles, Methods and Applications, 1st edn. Cambridge University Press, UK (2017)
3. Lyu, C., Shi, Y., Sun, L.: A Novel Local Community Detection Method Using Evolutionary Computation. IEEE Trans. Cybern. **51**(6), 3348–3360 (2021)
4. Girvan, M., Newman, M.E.J.: Community structure in social and biological networks. Proc. Nat. Acad. Sci. **99**(12), 7821–7826 (2002)
5. Fortunato, S.: Community detection in graphs. Phys. Rep. **486**(3), 75–174 (2010)
6. Zeng, X., Wang, W., Chen, C., Yen, G.G.: A consensus community-based particle swarm optimization for dynamic community detection. IEEE Trans. Cybern. **50**(6), 2502–2513 (2020)
7. Lu, M., Zhang, Z., Qu, Z., Kang, Y.: LPANNI: Overlapping community detection using label propagation in large-scale complex networks. IEEE Trans. Knowl. Data Eng. **31**(9), 1736–1749 (2019)
8. Yang, L., Cao, X., He, D., Wang, C. , Zhang, W.: Modularity based community detection wih deep learning. In: Proceedings of International Joint Conference on Artificial Intelligence, pp. 2252–2258 (2016)
9. Newman, M.E.J., Girvan, M.: Finding and evaluating community structure in networks. Phys. Rev. E **69**(2), 026113 (2004)
10. Zhan, Z., Shi, L., Tan, K., Zhang, J.: A survey on evolutionary computation for complex continuous optimization. Artificial Intell. Rev. **55**, 59–110 (2021)
11. Guimerà, R., Sales-Pardo, M., Amaral, L.A.N.: Modularity from fluctuations in random graphs and complex networks. Phys. Rev. E **70**(2), 025101 (2004)
12. Duch, J., Arenas, A.: Community detection in complex networks using extremal optimization. Phys. Rev. E **72**(2), 027104 (2005)
13. Liang, S., Li, H., Gong, M., Wu, Y., Zhu, Y.: Distributed multi-objective community detection in large-scale and complex networks. In: Proceedings of the 2019 15th International Conference on Computational Intelligence and Security (CIS), pp. 201–205 (2019)
14. Bian, K., Sun, Y., Cheng, S., Liu, Z., Sun, X.: Adaptive methods of differential evolution multi-objective optimization algorithm based on decomposition. In: Zhang, H., Yang, Z., Zhang, Z., Wu, Z., Hao, T. (eds.) NCAA 2021. CCIS, vol. 1449, pp. 458–472. Springer, Singapore (2021)
15. Chakraborty, T., Ghosh, S., Park, N.: Ensemble-based overlapping community detection using disjoint community structures. Knowl.-Based. Syst **163**, 241–251 (2019)
16. Shi, Y.: Brain storm optimization algorithm. In: International Conference on Swarm Intelligence, vol. Part I, pp. 303–309 (2011)
17. Estévez, P.A., Tesmer, M., Perez, C.A., et al.: Normalized mutual information feature selection. IEEE Trans. Neural Networks **20**(2), 189–201 (2009)
18. Danon, L., Díaz-Guilera, A., Duch, J., et al.: Comparing community structure identification. J. Stat. Mech: Theory Exp. **2005**(09), P09008 (2005)
19. Arthur, D., Vassilvitskii, S.: K-means++: the advantages of careful seeding. In: Proceedings of the Eighteenth Annual ACM-SIAM Symposium on Discrete Algorithms, pp. 1027–1035 (2007)
20. Newman, M.E.J.: Modularity and community structure in networks. Proc. Natl. Acad. Sci. U.S.A. **103**(23), 8577–8582 (2006)
21. Guimerà, R., Danon, L., Díaz-Guilera, A.: Self-similar community structure in a network of human interactions. Phys. Rev. E **68**(6), 065103 (2003)
22. Gleiser, P., Danon, L.: Community structure in jazz. Adv. Complex Syst. **6**(4), 565 (2003)

23. Lusseau, D., Schneider, K., Boisseau, O.J., Haase, P., Slooten, E., Dawson, S.M.: The bottlenose dolphin community of doubtful sound features a large proportion of long-lasting associations. Behav. Ecol. Sociobiol. **54**(4), 396–405 (2003)
24. Zachary, W.W.: An information-flow model for conflict and fission in small groups. J. Anthropol. Res. **33**(4), 452–473 (1977)
25. Gong, M., Cai, Q., Chen, X., Ma, L.: Complex network clustering by multiobjective discrete particle swarm optimization based on decomposition. IEEE Trans. Evol. Comput. **18**(1), 82–97 (2014)
26. Gong, M., Fu, B., Jiao, L., Du, H.: Memetic algorithm for community detection in networks. Phys. Rev. E **84**(5), 056101 (2011)
27. Pizzuti, C.: GA-Net: a genetic algorithm for community detection in social networks. Proc. Parallel Problem Solving Nat. **5199**, 1081–1090 (2008)
28. Gong, M., Ma, L., Zhang, Q., Jiao, L.: Community detection in networks by using multiobjective evolutionary algorithm with decomposition. Phys. A **391**(15), 4050–4060 (2012)
29. Shi, C., Yan, Z., Cai, Y., Wu, B.: Multi-objective community detection in complex networks. Appl. Soft Comput. **12**(2), 850–859 (2012)
30. Pizzuti, C.: A multiobjective genetic algorithm to find communities in complex networks. IEEE Trans. Evol. Comput. **16**(3), 418–430 (2012)
31. Coello, C., Pulido, G., Lechuga, M.: Handling multiple objectives with particle swarm optimization. IEEE Trans. Evol. Comput. **8**(3), 256–279 (2004)
32. Clauset, A., Newman, M.E.J., Moore, C.: Finding community structure in very large networks. Phys. Rev. E **70**(6), 066111 (2004)
33. Rosvall, M., Bergstrom, C.T.: Maps of random walks on complex networks reveal community structure. Proc. Natl. Acad. Sci. USA **105**(4), 1118–1123 (2008)

Optimization Method of Multi-body Integrated Energy System Considering Air Compression Energy Storage

Chongyi Tian[1,2], Xiangshun Kong[1,2], Yubing Liu[1,2], and Bo Peng[1,2(✉)]

[1] School of Information and Electrical Engineering, Shandong University of Construction, Jinan 250101, China
1539834847@qq.com
[2] ShandongKey: Laboratory of Intelligent Building·Technology, Jinan 250101, China

Abstract. Integrated energy system (IES) with compressed air energy storage (CAES) has significant benefits in energy utilization and gradually becomes a research hotspot, and the problem of how to establish a set of integrated energy optimization operation model in the context of compressed air energy storage needs to be solved urgently. Based on the above background, a master-slave game strategy between energy service providers and load aggregators is proposed. Firstly, this paper introduces the system framework and the mathematical model of the equipment. Secondly, the optimal operation model is established for energy service provider (ESP) and load aggregator (LA) respectively, and two scenarios are set to verify the effectiveness of compressed air energy storage. The results show that the model proposed in this paper can effectively balance the interests of energy service provider and load aggregator and ensure the economy of the system.

Keywords: Energy Service Provider · Load Aggregators · Compressed Air Energy Storage · Optimization

1 First Section

With the increasing demand for energy in China and the "dual carbon" strategy, economic and low-carbon energy has become the trend of future energy development [1, 2]. In this context, integrated energy systems with multi-energy coupling and joint dispatch characteristics have become an important form of efficient and clean energy utilization [3].

Nowadays, the research on the optimal scheduling scheme for IES is a hot issue in integrated energy systems. In the literature [4], an optimal scheduling scheme for multi-agent systems based on the system layer, network layer, energy center layer, and local layer is proposed. For each layer, the economic optimality is the goal, and the simultaneous distributed processing of multiple types of tasks in the scheduling process is realized, which improves the rapidity and accuracy of scheduling. The literature [5] establishes an integrated energy system planning model under fixed investment, solves

H. Zhang et al. (Eds.): NCAA 2023, CCIS 1869, pp. 403–415, 2023.
https://doi.org/10.1007/978-981-99-5844-3_29

the optimal capacity allocation of equipment under fixed investment, integrates various constraints such as fixed investment, equipment output, and energy coupling, and uses the pareto front solution set to achieve an economical, environmentally friendly system. In the literature [6], an integrated energy system combining energy supply and energy conversion is constructed with economy and environment as the starting point, which improves the economy and environmental protection of the system. The literature [7] analyzed the electric boiler equipment modeling problem and the joint electric-thermal scheduling problem from multiple perspectives; the literature [8] proposed various storage electric heating optimization configuration methods considering multiple factors such as park heating constraints, load source uncertainty, and multi-energy coupling. The introduction of the game idea in the optimal scheduling of IES has provided a new direction for the study of optimal scheduling problems, and many scholars have done a lot of research on games in recent years. In the literature [9], a two-stage optimal scheduling strategy for PIES based on robust stochastic model predictive control is proposed by considering the uncertainty of PIES load demand and renewable energy output. The literature [10] further considers the influence of market price uncertainty and constructs a day-ahead optimal dispatching model for PIES in the electricity market environment. Based on this, the literature [11] constructed a PIES two-stage interactive optimization model based on the energy hub model, using the coupling matrix to simplify the coupling and interaction of PIES energy flows. To solve this problem, existing studies generally approximate the model first, and then invoke a commercial solver to solve it. Specifically, the literature [12] reduces the optimal scheduling model of PIES to a mixed-integer linear programming problem by means of linearization techniques and solves it using the CPLEX solver. However, the results obtained from this approximation deviate from the actual situation. Heuristic algorithms, which do not depend on the model properties, are more popularly used in PIES optimal scheduling problems. Specifically, literature [13] proposed an improved genetic algorithm to achieve the solution of the PIES optimal scheduling strategy that balances the interests of energy suppliers, service providers and users; literature [14] used an improved simulated annealing algorithm to solve the NP-hard problem of PIES optimal scheduling; literature [15] based on the idea of two-layer iterative optimization, the objective cascade method combined with the improved particle swarm algorithm to achieve efficient solution of PIES multi-source coordinated optimization. However, the drawback of the heuristic algorithm is that its computational burden grows exponentially with the increase of the problem size.

In the literature [16], the Stackelberg master-slave game model is constructed by combining the auxiliary service model and the economic operation model, and the existence of a unique equilibrium solution of this model in the game process is proved, and finally the model is solved by a two-layer distributed algorithm. The literature [17] proposed a distributed cooperative optimal scheduling scheme for regional integrated energy systems based on the Stackelberg master-slave game model, which embedded the transaction mode and mathematical model of integrated energy systems into the framework of game theory and established a one-master-multi-slave noncooperative game, resulting in a significant improvement of both system and user consumption surplus. Game theory has a certain research base in the direction of share energy storage However, the non-cooperative game model of perfect competition cannot intuitively reflect the value of sharing, and the existence of competition among participants also leads to efficiency loss. In order to further reduce the cost of power consumption in new energy communities, some studies have established a cooperative game model between multiple communities to realize power mutual aid between communities. The cooperative game model is widely used, in which the benefit redistribution mechanism for cooperative alliances is a prerequisite and guarantee for reaching cooperation. In the literature [18], an optimal energy dispatching model for a multi-community optical storage system is proposed, and the hapley value method is used as the benefit redistribution scheme for the system. In the paper [19], a share energy storage model based on cooperative game was developed on the new energy generation side. The paper [20] established a share energy storage capacity planning model based on cooperative game. The paper [21] establishes a share energy storage capacity planning model based on a cooperative game on the new energy generation side, in which energy storage resources are allocated to each power plant while energy storage is shared and based on the marginal revenue created by the participants for the alliance to establish the alliance allocation strategy based on the marginal revenue created by the participants for the alliance. The allocation method based on Shapley's value is based on the marginal contribution to but it lacks the ability to understand the impact of complex factors, such as the volatility of new energy output, on the sharing costs in a practical context. The Shapley value-based allocation method is based on marginal contribution, but it lacks further analysis and research on the impact of complex factors such as the volatility of new energy output on the apportioned cost in the actual context. It does not reflect the actual situation well.

In this paper, we establish an integrated energy operation optimization model based on master-slave game and CAES for a class of IES containing ESP and industrial type, and verify the effectiveness of the proposed model through the analysis of calculation cases. Among them, Sect. 1 introduces the current status of previous developments in this field, Sect. 2 shows the IES architecture and device workflow, Sects. 3 and 4 introduce the device work model and game logic in this paper, and Sect. 5 presents the result analysis.

2 Framework for IES in Industrial Parks

In order to unify and optimize the arrangement, the distributed user group within the IES is equated to a LA. The framework of the IES is shown in Fig. 1. The equipment mainly includes compressed air energy storage (CAES), wind power (WT), photovoltaic (PV), heat exchanger (HE), gas turbine (GT), gas boiler (GB), energy storage (ES), thermal storage tank (TS), electric chiller (ER) and Chillers (AC), loads including electric load (EL), cold load (CL) and heat load (HL).

The system is mainly composed of three parts: energy supply side, energy conversion side and load side. Renewable energy such as PV and WT as well as the power grid are used as power supply units to input electrical energy to the CCHP system. Natural gas is input to the CCHP system as primary energy and is converted into electrical and thermal energy through GT and GB respectively. The process of GT electricity production is accompanied by heat generation, which can b0e supplied to heat users or used to drive AC for conversion into cold energy through waste heat recovery, and CAES and heat release, heat absorption and GT and thermal coupling, and power complementation to achieve the graded utilization of energy.

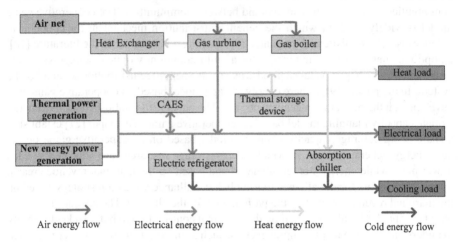

Fig. 1. IES structure diagram

3 Mathematical Models

3.1 Equipment Model

The main equipment model of the system is as follows:

(1) WT.

The relationship between wind power characteristics, mainly output power and wind speed, can be expressed by Eq. (1):

$$P_{WT}(t) = \begin{cases} 0, v(t) \leq v_{in} \, or \, v(t) \geq v_{out} \\ \dfrac{v(t)^3 - v_{in}^3}{v_r^3 - v_{in}^3} P_r, v_{in} \leq v(t) \leq v_r \\ P_r, v_r \leq v(t) \leq v_{out} \end{cases} \quad (1)$$

where, $P_{WT}(t)$ is the output power of the wind turbine at time t; P_r is the rated power; $v(t)$ is the actual outdoor wind speed at time t; v_r is the rated wind speed; v_{in} is the cut-in wind speed; v_{out} is the cut-out wind speed.

(2) PV.

The variation of photovoltaic output power with light intensity and temperature can be expressed by Eq. (2)-(3):

$$P_{PV}(t) = \frac{P_{STC} I(t)[1 + k(T_{PV}(t) - T_r)]}{I_{STC}} \quad (2)$$

where, $P_{PV}(t)$ is the output power of PV at moment t; $I(t)$ is the light intensity at moment t; P_{STC} is the maximum test power under standard test conditions (light intensity I_{STC} is 1000W/m2 and T_r is 25°C); k is the power temperature coefficient; $T_{PV}(t)$ is the temperature of PV power module at moment t, which can be estimated by testing the ambient temperature.

$$T_{PV}(t) = T_0(t) + 0.03I(t) \tag{3}$$

where, $T_0(t)$ is the ambient temperature at the moment t.

(3) GB.

The mathematical model of the gas boiler can be expressed by Eq. (4):

$$P_b(t) = F_b(t)L_{NG}\eta_b \tag{4}$$

where, $P_b(t)$ is the heating power of gas boiler at time t; $F_b(t)$ is the biomass gas volume of gas boiler at time t; L_{NG} is the gas calorific value; η_b indicates the heating efficiency of gas boiler.

(4) AC.

The mathematical model of the absorption chiller can be represented by equation:

$$P_{ac}(t) = Q_{ac\_in}(t)COP_{ac} \tag{5}$$

where, $P_{ac}(t)$ is the power of absorption refrigeration machine supply cooling at moment t; $Q_{ac\_in}(t)$ is the heat input of absorption refrigeration machine at moment t; COP_{ac} is the energy efficiency ratio of absorption refrigeration machine.

(5) CAES.

The mathematical model of CAES is presented in the literature [22].

3.2 ESP Model

Let a day be divided into H time periods, and ESP uses the electricity sale price and heat sale price as strategies, with the corresponding constraints as:

$$\lambda_h^{EG,b} < \lambda_h^{ESP,s} < \lambda_h^{EG,s} \tag{6}$$

$$\gamma_h^{ESP,min} < \gamma_h^{ESP,s} < \gamma_h^{ESP,max} \tag{7}$$

where, $\lambda^{EG,b}_h$, $\lambda^{EG,s}_h$, $\lambda^{ESP,s}_h$ are the grid purchase price and sale price of electricity and the ESP sale price of electricity in the hth time period of the day, respectively; $\gamma^{ESP,s}_h$ is the selling heat price of ESP in the hth time period of the day; $\gamma^{ESP,s}_h$ and γ^{ESP}_{min} Separate upper and lower heat price limits.

The gain of ESP in one day can be expressed as:

$$E_{ESP} = E_{ESP}^{LA,e} + E_{ESP}^{l,e} + E_{ESP}^{l,h} - E_{MT} \tag{8}$$

where, $E^{EG,e}_{ESP}$, $E^{l,e}_{ESP}$ and $E^{l,h}_{ESP}$ denotes the revenue generated by the ESP from trading with grid power, trading with customer-side power, and supplying heat to the customer side in a day, respectively.

Where:

$$E_{MGO}^{l,e} = \lambda_h^{MGO,s} \max\left(L_h^{l,c}, 0\right) \tag{9}$$

$$E_{MGO}^{l,h} = \sum_{h=1}^{H} (u_h L_h^{MGO,h} \gamma_h^{MGO,s}) \tag{10}$$

$$E_{MT} = \sum_{h=1}^{H} E_{MT,h}$$

where, $L^{l,c}{}_h$ is the net electrical load of LA in the hth time period of the day.

4 Two-Tier Game Model

4.1 Objective Function

The ESP first sets the set of electricity and heat purchase price strategies for a day, and the LA then adjusts the electricity and heat loads for each period in real time according to the ESP pricing plan and rationalizes the use of CAES. The game S can be expressed as follows:

$$S = \left\{ \begin{array}{c} (ESA \cup LA); \lambda^{ESP,b}, \lambda^{ESP,s}; \\ \overline{LA_l^s}, \Delta LA^{l,h}, LA^{ESs}; E_{ESP}; E_l \end{array} \right\} \tag{11}$$

where, the micro-network operator ESP is the leader and the LA l is the follower; $\overline{LA_l^S}$ denotes the LA adjusted flexible negative load set in a day set of load policies; $\Delta LA^{l,h}$ denotes the set of thermal energy abatement strategies of LA in a day the set of strategies; $LA^{l,ESS}$ denotes the set of policies for the use of CAES services by LA during a day; E_l denotes the gain of LA in one day; $\lambda^{ESP,b}$ and $\lambda^{ESP,s}$ respectively denotes the strategic set of the ESP one-day electricity purchase price and electricity sale price combined; E_{MSA} denotes the revenue of the micro network ESP in one day.

In the system designed in this study, ESP and LA are independent stakeholders, and ESP maximizes the revenue by optimizing the unit output and energy offer, while LA improves the customer satisfaction by optimizing the customer's energy purchase plan; the energy price provided by ESP affects LA's energy purchase plan, and LA's energy purchase plan affects ESP's energy sale revenue, and there is a game of interests between the two. There is a game of interests between the two.

4.2 Constraints

The constraints of this study are as follows:

(1) Scope of energy exchange between the campus and external power network.

$$P_e^{min} \quad P_e(t) \quad P_e^{max} \tag{12}$$

where, $P^{min}{}_e$ and $P^{max}{}_e$ are the upper and lower limits of the system power purchased from outside the system in unit time period t, respectively.

(2) PV/WT output.

$$P^{i,t}_{PV,min} \leq P^{i,t}_{PV} \leq P^{i,t}_{PV,max} \tag{13}$$

$$P^{i,t}_{WT,min} \leq P^{i,t}_{WT} \leq P^{i,t}_{WT,max} \tag{14}$$

where, $P^{i,t}_{pv,min}$ and $P^{i,t}_{pv,max}$ are the minimum and maximum values of PV output force respectively; $P^{i,t}_{WT,min}$ and $P^{i,t}_{WT,max}$ are the minimum and maximum values of WT output force respectively.

(3) GT output/climbing constraint.

$$P^{i,n,t}_{MT,min} \leq P^{i,n,t}_{MT} \leq P^{i,n,t}_{MT,max} \tag{15}$$

$$-R^{i,n}_{MT,dn} \Delta t \leq P^{i,n,t}_{MT} - P^{i,n,t-1}_{MT} \leq R^{i,n}_{MT,up} \Delta t \tag{16}$$

where, $P^{i,n,t}_{MT,min}$ and $P^{i,n,t}_{MT,max}$ are the minimum and maximum values of GT output force are respectively; $-R^{i,n,t}_{MT,dn}$ and $R^{i,n,t}_{MT,up}$ are climbing/declining power of gas turbine on/off in a period of time, respectively.

(4) Gas boiler output.

$$P_{GB,min} \leq P_{GB} \leq P_{GB,max} \tag{17}$$

where, $P_{GB,min}$ and $P_{GB,max}$ are the minimum and maximum values of GB output force, respectively.

(5) CAES constraint.

$$E_{CASE,min} \leq E_{CASE} \leq E_{CASE,max} \tag{18}$$

$$P_{CASE,ch,min} \leq P_{CASE,ch} \leq P_{CASE,ch,max} \tag{19}$$

$$P_{CASE,dis,min} \leq P_{CASE,dis} \leq P_{CASE,dis,max} \tag{20}$$

where, $E_{CAES,max}$ and $E_{CAES,min}$ are the upper and lower limits of the energy storage state, respectively; $P_{CAES,ch,min}$, $P_{CAES,ch,max}$ and $P_{CAES,dis,min}$, $P_{CAES,dis,max}$ are the minimum and maximum values of CAES charging and discharging, respectively.

5   Analysis of Results

ESP first develops a collection of electricity purchase price and heat purchase price strategies for a day, and LA then adjusts the electricity and heat loads for each time period and reasonably plans the use of CAES services in real time according to the pricing scheme and use of ESP, and solves the process as follows:

(1) Parameters of initial ESP and LA.
(2) Randomly generate the electricity and heat selling prices for m sets of ESPs.
(3) LA returns the remaining electricity and heat to ESP.

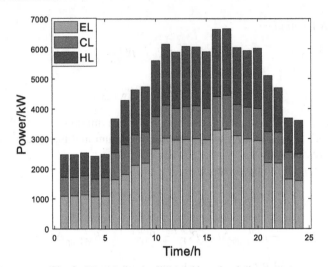

Fig. 2. Electricity, cooling and heat load diagram

(4) ESP calculates the revenue for one day.

(5) Re-generate the heat and electricity sales prices and repeat steps (4)–(5).

In this paper, an industrial park is selected as the application scenario, and the 24h load is shown in Fig. 2.

To further verify the validity of the model, different scenarios are set up as shown in Table 1.

Table 1. Set two scenes to compare

Programs	CAES
A	×
B	√

The proposed model is applied to the above scenarios and programmed using MATLAB. The results of revenue optimization for each party in the above two scenarios are shown in Table 2.

Table 2. ESP and LA gain

Programs	ESP	LA
A	5956.36	1689.23
B	5757.28	1956.46

Among them, the benefit of LA under Scenario B is increased by 267.23 yuan and the benefit of ESP is decreased by 199.08 yuan compared to Scenario A. This is because the customer side is equipped with CAES under Scenario B, which further improves the flexibility of customer side heat demand.. The analysis in Table 2 shows that the introduction of CAES (Scenario 2) significantly increases the benefits of LA compared to ESP (Scenario A) for electricity and heat trading with customers, while the benefits of ESP are abated. Among them, the selling price of electricity with ESP for option A and B are shown in Figs. 3, 4, 5 and 6.

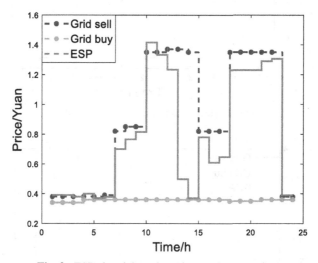

Fig. 3. ESP electricity sales prices under scenario A

Figures 3, 4, 5 and 6 reflect the difference between the ESP electricity and heat sales prices under the two scenarios, which is the result of the participation of Scenario B CAES in the regulation of the electricity and heat loads on the customer side, further changing the game between the two parties. The ESP heat and electricity sales prices are reduced, effectively reducing the cost and increasing the revenue of the LA.

Figure 7 gives the capacity change curve of CAES in one day, CAES keeps increasing in 0:00–8:00, 12:00–16:00 and 23:00–24:00, which coincides with the increase of customers' electric load in these periods under scenario B, i.e., customers charge CAES; while the capacity shows a decreasing trend in the periods of 9:00–11:00 and 7:00–22:00. The trend of decreasing capacity in 9:00–11:00 and 7:00–22:00 corresponds to the decrease of user's electric load in this period, i.e., CAES discharges to the user. The customer can use the ESP price for heat and electricity to plan for a less costly heating solution to improve their profit.

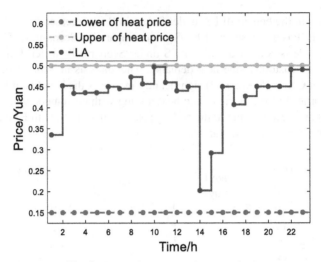

Fig. 4. Heat sales price under scenario A

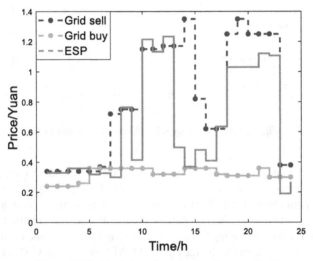

Fig. 5. ESP electricity sales prices under scenario B

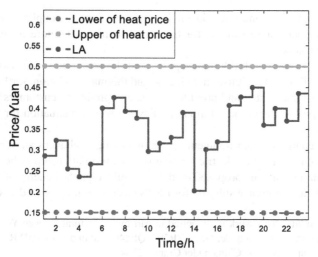

Fig. 6. Heat sales price under scenario B

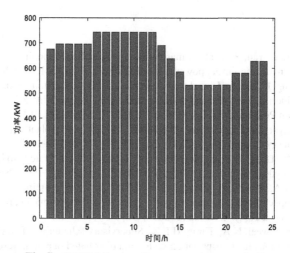

Fig. 7. CAES 24h capacity curve under scenario B

6 Conclusion

The decision-making model with ESP as the leader severely abates the benefits of the user side. In this paper, we optimize the decision-making scheme for the user side in the context of industrial-type integrated energy systems with compressed air energy storage, and the main conclusions are as follows:

Most of the existing studies only consider the upper layer (ESP) of the model with electro-thermal coupling, but the proposed model in this paper takes into account the participation of CAES on the user side, and both the upper and lower layers have electro-thermal. The model proposed in this paper takes into account the participation of user-side

CAES, and both upper and lower layers have electro-thermal coupling, which is more comprehensive and more suitable for the actual scenario of future integrated energy system optimization.

CAES technology is developing rapidly, and the proposed model is conducive to improving the flexibility of customer electric and thermal regulation, further improving the customer-side revenue, and providing decision options for customers participating under CAES, while the proposed model achieves a win-win situation for both LA and ESP revenue.

With the continuous development of customer-side appliances and energy storage services, the customer-side electric heating regulation relationship is becoming more and more complex, and the proposed model can well weigh the benefits of ESP and LA, and also provide a reference solution for ESP's electric heating price decision.

Acknowledgements. This work is supported in part by the Natural Science Youth Foundation of Shandong Province, China under Grant ZR2021QE240, and in part by PhD Research Fund of Shandong Jianzhu University, China under Grant X21040Z.

References

1. Zhang, T., Li, Y., Yan, R., et al.: A master-slave game optimization model for electric power companies considering virtual power plant. IEEE Access **10**, 21812–21820 (2022)
2. Deng, F., Meng, D., Deng, Z., et al.: Study on master-slave game strategy of integrated energy system considering users demand response and energy balance. In: 2022 12th International Conference on Power and Energy Systems (ICPES), pp. 957–961. IEEE (2022)
3. Liu, X.: Bi-layer game method for scheduling of virtual power plant with multiple regional integrated energy systems. Int. J. Electr. Power Energy Syst. **149**, 109063 (2023)
4. Xu, X., Xu, B., Dong, J., et al.: Near-term analysis of a roll-out strategy to introduce fuel cell vehicles and hydrogen stations in Shenzhen China. Appl. Energy **196**, 229–237 (2017)
5. El-Taweel, N.A., Khani, H., Farag, H.E.Z.: Hydrogen storage optimal scheduling for fuel supply and capacity-based demand response program under dynamic hydrogen pricing. IEEE Trans. Smart Grid **10**(4), 4531–4542 (2019)
6. Khani, H., El-Taweel, N.A., Farag, H.E.Z.: Supervisory scheduling of storage-based hydrogen fueling stations for transportation sector and distributed operating reserve in electricity markets. IEEE Trans. Industr. Inf. **16**(3), 1529–1538 (2020)
7. Zhu, H., Zhu, J., Li, S., et al.: Review of optimal scheduling of integrated electricity and heat systems. J. Glob. Energy Interconnection **5**(4), 383–397 (2022)
8. Zhang, J., Chen, W., Zhang, Y., et al.: Bilevel optimal planning method for park regenerative electric heating considering capacity of distribution network and reliable heating. J. Glob. Energy Interconnection **4**(2), 142–152 (2021)
9. Liu, C., Li, R., Yin, Y., et al.: Two-stage optimization for community integrated energy system based on robust stochastic model predictive control. Electr. Power Autom. Equip. **42**(5), 1–7 (2022)
10. Zhou, Y.Z., Wei, Z.N., Sun, G.Q., et al.: A robust optimization approach for integrated community energy system in energy and ancillary service markets. Energy **148**, 1–15 (2018)
11. Bu, F., Tian, S., Fang, F., et al.: Optimal day-ahead scheduling method for hybrid energy park based on energy hub model. Proc. CSU-EPSA **29**(10), 123–129 (2017)

12. Li, P., Yang, S., Wei, C., et al.: Robust stochastic optimization model of the Park's integrated energy system considering impact of energy policy. Electric Power, 1–12 (2022)
13. Chen, Y., Zhu, J., Yang, D., et al.: Research on economic optimization operation technology of park-level integrated energy system based on multi-party interest game. High Voltage Eng. **47**(1), 102–110 (2021)
14. Hamdy, M., Sirén, K., Attia, S.: Impact of financial assumptions on the cost optimality towards nearly zero energy buildings-a case study. Energy Build. **153**, 421–438 (2017)
15. Ma, Y., J Xie, S Zhao, et al.: Multi-objective optimal dispatching of active distribution network considering park-level integrated energy system. Autom. Electr. Power Syst., 1–16 (2022)
16. Zhou, C., Ma, X., Guo, X., et al.: Leader-follower game based optimized operation method for interaction of integrated energy system in industrial park. Autom. Electr. Power Syst. **43**(7), 74–80 (2019)
17. Wang, H., Li, K., Zhang, C., et al.: Distributed cooperative optimal operation strategy of community integrated energy system based on master-slave game. Proc. CSEE **40**(17), 5435–5445 (2020)
18. Zhao, H., Wang, X., Li, B., et al.: Distributionally robust optimal dispatch for multi-community Photovoltaic-energy storage system considering energy sharing. Autom. Electr. Power Syst. **46**(9) (2022)
19. Sun, C., Chen, L., Qiu, X., et al.: A generation-side shared energy storage planning model based on cooperative game. J. Glob. Energy Interconnection **2**(4), 360–366 (2019)
20. Jiang, Y., Zheng, C.: Two-stage operation optimization for a grid-connected wind farm cluster with shared energy storage. Power Syst. Technol., 1–14
21. Xue, W., Yue, C., Feng, L., et al.: Stackelberg game based retailer pricing scheme and EV charging management in smart residential area. Power Syst. Technol. **39**(4), 939–945 (2015)
22. Ma, X., Zhang, C., Li, K., et al.: Optimal dispatching strategy of regional micro energy system with compressed air energy storage. Energy **212** (2020)

Sequential Seeding Initialization for SNIC Superpixels

Jinze Zhang[1], Yanqiang Ding[1], Cheng Li[2(✉)], Wangpeng He[2(✉)], and Dan Zhong[3]

[1] Northwest A&F University, Yangling, China
{3022716521,dingyq}@nwafu.edu.cn
[2] Xidian University, Xi'an, China
licheng812@stu.xidian.edu.cn, hewp@xidian.edu.cn
[3] Northwestern Polytechnical University, Xi'an, China
henryzhongdan@mail.nwpu.edu.cn

Abstract. In this paper, a novel seeding initialization strategy is introduced to simple non-iterative clustering (SNIC) superpixels for further optimizing the performance. First, half the total seeds are initialized on the image plane via the conventional grid sampling. The remaining seeds are then sequentially sampled from the spatial midpoint of a seed pair that shows the closest inter-seed correlation. During the process, a dynamic region adjacency graph is established to indicate the spatial relationship of all seeds. In addition, a novel correlation measurement based on the linear path is proposed to quantitatively describe the inter-seed correlation. Compared with conventional initialization, sequential sampling considers both the complexity of regional context and the regularity of global distribution. Experimental results verify that the proposed strategy could effectively improve the regional describing ability of SNIC superpixels, especially for objects with various granularities.

Keywords: Superpixels · Seeding Initialization · Correlation Measurement

1 Introduction

Superpixel segmentation over-segments an image into semantic sub-regions termed superpixels [1], thus promoting the implementation of several advanced visual tasks. Since it could reconstruct an image with fewer elements compared with pixel-wise representation, superpixel segmentation has gradually become a universal preprocessing step in various advanced visual tasks. Apart from the conventional image processing applications, recently, they are becoming popular in agriculture and monitoring [2–5].

Over the past two decades, extensive studies on superpixels have been put forward to boost their performance. In general, there are several crucial properties within superpixels that surpass conventional image over-segmentation algorithms, including segmentation accuracy, running efficiency, and visual regularity [6]. Accuracy directly affects the final performance of advanced visual tasks. Efficiency decides whether a superpixel algorithm can be deployed in a timing-stringent application. Regularity reveals the appearance satisfaction of superpixels, such as size and shape uniformity.

H. Zhang et al. (Eds.): NCAA 2023, CCIS 1869, pp. 416–427, 2023.
https://doi.org/10.1007/978-981-99-5844-3_30

From the view of property preference, the State-Of-The-Art (SOTA) works in the field of superpixels can be divided into three categories. Simple linear iterative clustering (SLIC) [7] adopts a restricted region inspection strategy to improve the efficiency of k-means clustering-based superpixel generation. It also proposes a joint color-spatial correlation measurement to both control color homogeneity and shape compactness. Inspired by this pioneering work, Simple Non-Iterative Clustering (SNIC) [8] introduces online averaging to substitute the iterative framework. As a result, the label assignment and updating process can be executed in a one-pass region-growing manner. It not only strengthens the region connectivity but avoids a mass of redundant calculations. As reported in [9], these two approaches mainly focus on a balanced trade-off between accuracy and efficiency. Linear Spectral Clustering (LSC) [10] also adopts k-means based clustering to generate superpixels. Different from SLIC, it maps pixels into a ten-dimensional feature space, thus achieving more accurate correlation measurements. The work in [11] maps the image plane to a 2-dimensional manifold to develop the content density. The result superpixels are con-tent-sensitive that could catch more detailed information. In general, superpixel generated from more informative feature space could acquire better performance in accuracy. Based on the assumption that adjacent pixels are likely to be assigned with an identical label, Fast Linear Iterative Clustering (FLIC) [12] overcomes the restricted region inspection in SLIC. In contrast, it performs an active search to update pixels in the image plane efficiently. Preemptive SLIC (PSLIC) [13] re-examines the global termination criteria and then proposes a local interruption strategy to guide the convergence of partial regions. Consequently, an increasing number of homogenous superpixels avoid being updated in the main loop of global iteration, resulting in lower computational costs. In Summary, these two superpixels achieve rapid convergence with a few iterations and therefore reveal superiority in running efficiency.

As mentioned above, there is a large amount of related work on optimizing the linear clustering framework [10–13] proposed in SLIC. Generally, they mainly focus on three aspects: feature optimization, structure optimization, and clustering optimization. Meanwhile, modification on initialization sometimes could yield an immediate return to the performance of superpixels. For example, Dynamic Random Walk (DRW) [14] adopts a new initialization strategy to further improve the performance of superpixels. By averaging the distribution of areas in both the 2D image plane and 3D curved manifold, the strategy could produce seeds that cover the objects as much as possible, especially for small objects and strip areas. In addition, as reported in [15], an ideal initialization is always beneficial to clustering results. Therefore, research on subtler initialization strategy is of great significance for the clustering-based superpixels.

Recently, as the demand for performance has increased, several challenging computer vision applications put forward higher requirements for superpixels. For example, it is necessary for some vision-based environmental monitoring tasks to accurately describe objects with different granularities. In this case, scale adaptation and content sensibility become critical for superpixel-based scene analysis. In other words, since there is seldom uniform visual information in real-world scenarios, superpixels without those two properties may be inferior as a pre-processing tool.

It is also worth noting that SNIC superpixels have already achieved SOTA segmentation performance. In practical applications, the results could be further promoted once it overcomes the following bottlenecks.

- The greedy strategy performed via a priority queue in the online averaging process is prone to premature convergence;
- The label assignment and updating process via a region-growing manner is sensitive to initial clustering centroids.

Towards this goal, a novel seeding strategy is proposed for SNIC in this paper, referred to as Sequential Seeding Initialization (SSI). Rather than regular grid sampling seeds for clustering pixels, it adopts a context-aware approach to sample new seeds. Specifically, half the total seeds are first initialized on the image plane via the conventional grid sampling, which could roughly indicate the richness of regional information. The remaining seeds are then iteratively sampled from the spatial midpoint of a seed pair that shows the closest inter-seed correlation. The spatial relationship of all sampled seeds is modeled by an undirected Region Adjacency Graph (RAG), which could be efficiently updated during the whole SSI process. Meanwhile, the inter-seed correlation is measured by an accumulated joint color and spatial distance, which can be calculated with less computation burden in the concise sequential processing strategy. Therefore, the local concentration of seeds is adaptive to the complexity of regional information. Experimental results show that the proposed SSI strategy is compatible with the non-iterative clustering framework, which facilitates SNIC superpixels to represent image regions with various scales. Accordingly, the non-iterative clustering from these seeds could easily avoid falling into the local optimum trap. Moreover, the optimized superpixels could achieve higher interior homogeneity and better boundary adherence.

2   Preliminaries

As shown in Fig. 1, the process of generating SNIC superpixels mainly consists of two stages. The first stage is the grid sampling to initialize all seeds on the image plane. The second stage is the label assignment for all pixels via non-iterative clustering.

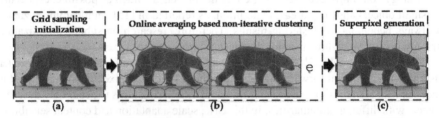

Fig. 1. Schematic diagram of generating SNIC superpixels. (a) Grid sampling initialization of the input image; (b) Dynamic processes of non-iterative clustering; (c) SNIC superpixel results. The red dots and outlines indicate the seeds and boundaries of superpixels, respectively.

2.1 Grid Sampling Initialization

For an image $I = \{I_i\}_{i=1}^{W \times H}$, the 2-dimensional position coordinates in the Euclidean space and the 3-dimensional color intensity in the CIELAB color space of each pixel I_i can be denoted by $\mathbf{p}(I_i) = (x(I_i), y(I_i))$ and $\mathbf{c}(I_i) = (l(I_i), a(I_i), b(I_i))$, where W and H denote the width and the height of I, respectively.

As a seed-demand method, SNIC adopts grid sampling to initialize seeds (i.e., incipient cluster centers) of all superpixels. Firstly, given an image I, it is uniformly partitioned into K rectangular grid cells $\{\Box_k\}_{k=1}^{K}$ with the step $S = \sqrt{W \times H / K}$. In kth grid cell $\Box_k$, the barycenter b_k is selected as a seed with a unique label $L(b_k) = k$. From a spatial quantization perspective, the relation between $\Box_k$ and its seed b_k can be described as follows

$$\Box_k = \left\{ I_i \big| \|\mathbf{p}(I_i) - \mathbf{p}(b_k)\|_2 \le \|\mathbf{p}(I_i) - \mathbf{p}(b_j)\|_2, \forall k \ne j \right\}, \tag{1}$$

where $\| \bullet \|_2$ represents the Euclidean metric (ℓ_2 norm).

In practice, this strategy is simple yet essential to produce approximately the same size superpixels in an image plane, which makes it very popular in many superpixel algorithms [16–18].

2.2 Non-Iterative Clustering

Conventional non-iterative clustering sequentially performs three steps to classify all pixels on the image plane, including pixel inspection, label assignment, and cluster updating.

In pixel inspection, the growing point of cluster Ω_k inspects its 4-neighbor points and calculates the feature distance $D(I_i, b_k)$ from every unlabeled element I_i by

$$D(I_i, b_k) = \sqrt{\|\mathbf{c}(I_i) - \mathbf{c}(b_k)\|_2^2 + \lambda^2 \|\mathbf{p}(I_i) - \mathbf{p}(b_k)\|_2^2}, \tag{2}$$

where λ is a normalized factor to balance color and spatial proximity, which is pre-set to 10 in the conventional SNIC. Then a pixel node $\mathbf{N}(I_i^k) = \left[\mathbf{c}(I_i), \mathbf{p}(I_i), D(I_i, b_k) \right]$ defined for I_i is pushed into a small-root priority queue $\mathbf{Q}$, where I_i^k means that I_i acquires a temporary label k from b_k.

Notice that the pixel inspection begins from all seeds $\{b_k\}_{k=1}^{K}$, and the first growing point in Ω_k is b_k itself. The priority queue sorts the nodes according to the value of feature distance by Eq. (2).

In label assignment, the top-most node $\mathbf{N}(I_i^m)$ with the global minimal feature distance is popped from $\mathbf{Q}$, and the temporary label m from b_m becomes the definite label of I_i (i.e., $L(I_i) = m$).

In cluster updating, the cluster Ω_m absorbs I_i as a new member and then upgrades the center b_m to b'_m as follows

$$\mathbf{c}(b'_m) = \frac{\sum_{I_j \in \Omega_m} \mathbf{c}(I_j) + \mathbf{c}(I_i)}{|\Omega_m| + 1}, \tag{3}$$

$$\mathbf{p}(b'_m) = \frac{\sum_{I_j \in \Omega_m} \mathbf{p}(I_j) + \mathbf{p}(I_i)}{|\Omega_m| + 1}. \tag{4}$$

where $|\Omega_m|$ calculates the area of Ω_m (i.e., number of pixels). Next, I_i becomes the growing point of cluster Ω_m to perform the subsequent pixel inspection operations.

SNIC integrates label assignment and cluster updating to establish an online averaging process, which is repeated until $\mathbf{Q}$ is empty during the subsequent operations.

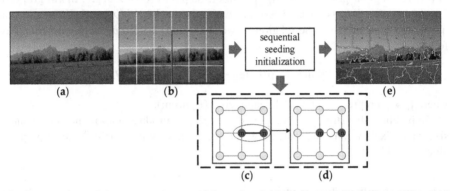

Fig. 2. Workflow of the proposed sequential seeding initialization. (a) Input image; (b) Grid sampling initialization with half the expected number of seeds; (c-d) Zoom-in performance of sequential seeding initialization, wherein the white-marked point denotes a newly sampled seed; (e) Result of SNIC superpixels optimized by the proposed initialization strategy.

3   Methodology

This section introduces the proposed SSI strategy in detail. As shown in Fig. 2, the workflow contains two major steps. Firstly, it initializes half the expected number of seeds via conventional grid sampling. Next, a sequential re-sampling process is performed to allocate the remaining seeds according to the inter-seed correlation. After the two-fold seeding, superpixels could acquire scale adaptation and content sensitivity.

3.1   Sequential Sampling

The goal of SSI is to produce a set of seeds that are beneficial to generate superpixels with context awareness. This target can be interpreted as obtaining partitions in areas of low visual complexity with a larger number of homogeneous pixels, and high complexity with fewer heterogeneous pixels [19]. As mentioned in [6], an effective approach is hierarchical sampling instead of evenly grid distribution. To this end, a sequential framework is proposed to sample all seeds from whole to part.

As shown in Fig. 2, after the global grid sampling, a rough context distribution for all rectangular regions is established on the image plane. Then a set of evenly distributed seeds $\{s_k\}_{k=1}^{K/2}$ are sampled as the initial center of clusters, which is done to establish a rough context distribution for all rectangular regions.

Specifically, the average color $\bar{\mathbf{c}}(\Box_k)$ of all pixels in a $S \times S$ region $\Box_k$ centered at b_k is calculated to represent the richness of context information in kth grid G_k

$$\bar{\mathbf{c}}(\Box_k) = \sum_{I_m \in \Box_k} D_{color}(I_m, b_k) = \sum_{I_m \in \Box_k} \|\mathbf{c}(I_m) - \mathbf{c}(b_k)\|_2. \tag{5}$$

Notice that the lightness channel varies more sharply than the other two intensity channels in the CIELAB color space. In other words, it plays a major role in the color difference and the calculation of $\|\mathbf{c}(I_m) - \mathbf{c}(b_k)\|_2$ can be replaced by $\|l(I_m) - l(b_k)\|_2$ for computing simplicity. Therefore, Eq. (5) can be recast into

$$\bar{\mathbf{c}}(\Box_k) = \sum_{I_m \in \Box_k} \|l(I_m) - l(b_k)\|_2 = \left| \sum_{I_m \in \Box_k} l(I_m) - l(b_k) \times |\Box_k| \right|, \tag{6}$$

where the first term $\sum_{m \in \Box_k} l(I_m)$ can be efficiently calculated by a summed area table (SAT) [20] with a constant complexity, and the value of $|\Box_k|$ is S^2.

Next, the remaining $K/2$ seeds are sampled one by one to adapt to the richness of the regional context, thus reducing the heterogeneity in the potential superpixels. The detailed implementation can be demonstrated as follows:

- An undirected adjacency graph (RAG) $G = (V, E)$ is established to depict the inter-seed correlations, including spatial adjacency and color similarity. A vertex $v_k \in V$ represents a seed b_k and an edge $\omega_{k\leftrightarrow j}$ between v_k and v_j measures the correlation of b_k and b_j, which will be discussed in Sect. 3.2. A big-root priority queue $\mathbf{Q}'$ with is introduced to preserve all edges, which always returns the maximum element that contains the minimum correlation (i.e., the edge with the greatest value) while it is not empty.
- Once the top-most element ω_{mn} is popped from $\mathbf{Q}'$, there is a new seed sampled from the spatial midpoint of the corresponding seed pair (b_m, b_n). That is, a new seed $b_{m|n}$ emerges from $(\mathbf{p}(b_m) + \mathbf{p}(b_n))/2$ to promote the inter-seed correlations.
- During the process, the newly sampled seed $b_{m|n}$ would create new inter-seed correlations with existing seeds, the RAG G is updated by the new vertex $v_{m|n}$ and edges that are then pushed on $\mathbf{Q}'$.

The last two steps are repeated until the number of sampled seeds increases to K.

3.2 Linear Path Correlation

As mentioned above, the proposed SSI adopts a RAG to depict the inter-seed correlations, wherein the spatial adjacency and color similarity of the two seeds should be pre-defined.

In this work, the adjacency of two seeds is simply determined by the spatial relationship. Specifically, when the first $K/2$ seeds are settled via the conventional grid sampling, there is an incipient adjacent relationship between two seeds with a spatial distance of the step S. In the next sequential re-sampling process, when a new seed $b_{m|n}$ is sampled,

Fig. 3. Schematic diagram of the linear path. (a) Grid sampling initialization with half the expected number of seeds; (b) Zoom-in performance of blue rectangle box in (a), the straight line in cyan is the linear path between two adjacent seeds; (c) Zoom-in performance of the linear path in (b), wherein all elements make joint effects on the inter-seed correlation.

a rectangular search region $\square_{m|n}$ is derived to inspect the underlying neighbor seeds. If an existing seed belongs to $\square_{m|n}$, it can be defined as an adjacent element of $\square_{m|n}$ in G. Notice that, $\square_{m|n}$ centered at $b_{m|n}$ with a step of S is similar to a grid cell $\square_k$ in Sect. 2.1.

Another important parameter in SSI is inter-seed similarity, which decides whether a new seed should be sampled in the midpoint of the two seeds. To accurately measure the similarity, a novel correlation measurement based on the linear path is proposed. As shown in Fig. 3, unlike the pixel-cluster correlation distance in Eq. (2), the context information along the linear path is additionally introduced to the distance measurement [6]

$$D'(b_k, b_j) = \sqrt{D^2_{color}(b_k, b_j, \mathcal{P}_{kj}) + \lambda^2 \|\mathbf{p}(b_k) - \mathbf{p}(b_j)\|^2_2}, \qquad (7)$$

where

$$D_{color}(b_k, b_j, \mathcal{P}_{kj}) = \sqrt{\gamma \|\bar{\mathbf{c}}(\square_k) - \bar{\mathbf{c}}(\square_j)\|^2_2 + (1 - \gamma)\frac{1}{|\mathcal{P}_{kj}|}\sum_{i \in \mathcal{P}_{kj}} \|\mathbf{c}(I_i) - \bar{\mathbf{c}}(\square_j)\|^2_2}, \qquad (8)$$

where γ is a parameter to adjust the influence of the additional information on the inter-seed correlation. Mapping the parameters to the RAG, the correlation between two seeds is described as follows

$$\omega_{kj} = \begin{cases} D'(b_k, b_j) & \text{if } \square_k | \square_j \\ 0 & \text{otherwise} \end{cases}, \qquad (9)$$

where $\square_k | \square_j$ represents b_k and b_j is spatially adjacent in G.

By calculating ω_{kj} between b_k and b_j, the information complexity around adjacent seeds b_k and b_j can be roughly described. That is, if $D'(b_k, b_j)$ is great enough, the difference of color distribution in $\square_k$ and $\square_j$ is significant. Therefore, additional seeds are required to generate fine-grained superpixels.

Eventually, an optimized distribution that is content-aware with the exact pre-set number of seeds is established. It is more sensitive to the complexity of regional information than conventional grid sampling and also remains relatively far from the contour. This lays the foundation for efficient convergence without local optimum.

4   Experiments

In this section, the proposed SSI strategy is integrated into SNIC as a substitute for GSI, which forms an improved SNIC superpixel algorithm. The rest of the paper terms it as SSI-SNIC to simplify the expression.

The overall performance is tested on the Berkeley Segmentation Data Set 500 (BSDS500) [21] and evaluated by Stutz's Extended Berkeley Segmentation Benchmark (EBSB) [22]. BSDS500 contains 500 visual images with 481×321 in size, including prospect and close-up landscapes, buildings and vegetation, as well as human and animal photographs. In addition, each image is manually labeled with at least two ground truths. EBSB contains several popular metrics to comprehensively evaluate the superpixel performance. Among them are Boundary Recall (BR), Under-segmentation Error (UE) and Achievable Segmentation Accuracy (ASA), which draw more important in the application of environmental monitoring. Superpixels produced by SNIC are utilized as the baseline.

4.1   Metrical Evaluation

Given $\Omega=\{\Omega_k\}_{k=1}^K$ and $G=\{G_m\}_{m=1}^M$ as superpixel results and the corresponding ground truth, respectively, those metrics are mathematically defined as follows.

BR is the ratio of ground truth boundaries covered by superpixel boundaries

$$\text{BR} = \frac{\sum_{i \in G_b} \Pi\left(\min_{j \in \Omega_b} \|P(I_i) - P(I_j)\|_2 < r\right)}{G_b}, \tag{10}$$

where Ω_b and G_b are the outline pixels in Ω and boundary pixels in G, respectively. The indicator $\Pi()$ returns the logic value of whether the expression is true. The coverage radius r is set to 2 in this paper. A greater BR indicates a closer consistency from the superpixel edge to actual image boundaries.

Table 1. Comparison of two algorithms in terms of boundary recall.

Method	User-expected superpixel number									
	50	100	150	200	250	300	350	400	450	500
SNIC	0.674	0.792	0.842	0.866	0.895	0.906	0.919	0.931	0.938	0.949
SSI-SNIC	0.721	0.798	0.851	0.881	0.898	0.916	0.928	0.935	0.942	0.951

As for BR in Table 1, SSI-SNIC outperforms the conventional SNIC in a wide range of superpixel sizes. The superiority of SSI lies in the dynamical sampling that considers the complexity of image content. For example, this strategy might be executed iteratively in some detailed regions to promote the inter-seed correlation, thus generating sufficient seeds for over-segmentation. In this case, the boundaries of different objects can be captured accurately by superpixel outlines.

UE measures how each superpixel overlaps with only one object. Its value is the ratio of the leaked pixels to the actual segmented pixels, wherein the former refers to the pixels beyond the intersection of the superpixel and the ground truth

$$UE = \frac{\sum_{m=1}^{M} \left(\sum_{\Omega_k | \Omega_k \cap G_m \neq \phi} |\Omega_k| \right) - N}{N}. \tag{11}$$

It utilizes segmentation regions instead of boundaries for measurement in BR, which is negatively correlated with segmentation accuracy.

Table 2 lists the comparison of UE, wherein SSI-SNIC acquires lower results. Despite conventional SNIC holding an acceptable performance on UE, the proposed SSI could further optimize the process of pixel classification, which leads to higher homogeneity within each superpixel. From another perspective, sequential re-sampling on the outcome of grid sampling can be regarded as the beginning of the region sub-division. Therefore, the problem of under-segmentation can be effectively avoided.

Table 2. Comparison of two algorithms in terms of under-segmentation error.

Method	User-expected superpixel number									
	50	100	150	200	250	300	350	400	450	500
SNIC	0.175	0.109	0.089	0.080	0.070	0.068	0.064	0.060	0.057	0.054
SSI-SNIC	0.142	0.106	0.085	0.073	0.068	0.062	0.058	0.056	0.055	0.051

ASA describes the accuracy of segmentation results. It reveals the percentage of the correct segmentation in terms of the ground truth

$$ASA = \frac{\sum_{k=1}^{K} \arg \max_m |\Omega_k \cap G_m|}{\sum_{m=1}^{M} |G_m|}, \tag{12}$$

which also uses regional information to evaluate the performance. A higher ASA value indicates the performance of superpixels in subsequent is more effective and robust. As reported in [22], there is a close relationship between UE and (1-ASA), while they could also implicitly measure boundary adherence.

Table 3 further confirms the availability of the proposed strategy. Similar to BR and UE, with the growing number of superpixels, SSI-SNIC still outperforms the corresponding implementation without any initialization optimization. It also indicates that the SSI-based SNIC superpixels could prompt higher upper bounds of accuracy for subsequent visual tasks.

Table 3. Comparison of two algorithms in terms of achievable segmentation accuracy.

Method	User-expected superpixel number									
	50	100	150	200	250	300	350	400	450	500
SNIC	0.837	0.896	0.914	0.922	0.931	0.934	0.937	0.941	0.943	0.946
SSI-SNIC	0.865	0.897	0.918	0.927	0.931	0.938	0.942	0.943	0.945	0.948

Table 4. Comparison of superpixel number between user expectation and actual generation.

Method	User-expected superpixel number									
	50	100	150	200	250	300	350	400	450	500
SNIC	40	96	150	187	260	294	330	400	442	504
SSI-SNIC	50	100	150	200	250	300	350	400	450	500

In addition to the three metrics listed above, the number of actual generated superpixels is introduced to determine the controllability of the number of superpixels. Table 4 illustrates the number comparison of superpixels between actual generation and user expectation. Since there is no merging post-processing in the non-iterative clustering framework, the amount of result superpixels is equal to the final sampled seeds. Compared with the grid sampling in conventional SNIC, the proposed SSI could generate an exact number of seeds that conforms to the input K. During the grid sampling, a floor operation is exerted on the result S to maintain an identical step in both the horizontal and vertical directions. As a result, the number of grid cells is not always equal to the user's pre-definition. On the contrary, the accumulation of sampled seeds is controlled by the priority queue, which separately adds new seeds from $K/2$. Therefore, the number of seeds in SSI can be exactly determined.

4.2 Visual Comparison

Figure 4 shows the visualized performance of SNIC and SSI-SNIC on BSDS500. Obviously, both two methods could generate regularly shaped superpixels with pleasant visual results. The non-iterative clustering framework inherited from SNIC enables the latter to make a balanced trade-off between spatial compactness and color homogeneity. Additionally, SSI-SNIC superpixels pay more attention to the local details, whose content awareness enables partitions to be adaptive to objects with different scales. Compared with the conventional SNIC, it shows better versatility in complex imaging scenes, such as twig context, low contrast, and small granularities.

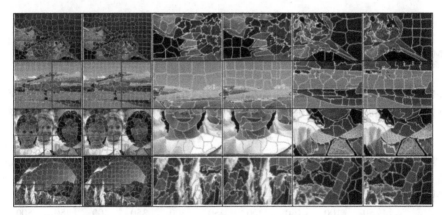

Fig. 4. Visual comparison of SNIC superpixels with and without SSI strategy in alternating columns. Each image is over-segmented by nearly and exactly 200 superpixels, followed by two zoom-in results of local regions enclosed by red rectangles.

5 Conclusion

In this paper, a novel initialization strategy is integrated into the conventional SNIC superpixel algorithm to further improve the overall performance. Compared with the conventional grid sampling that straightforwardly acquires all seeds, the proposed sequential seeding initialization (SSI) iteratively updates the seeding distribution from whole to part. This process is modeled by a dynamic region adjacency graph, which could effectively preserve the spatial relationship and color similarity. Meanwhile, an optimized correlation measurement is designed to accurately describe the similarity of two adjacent seeds. Experimental results show that SSI-embedded SNIC superpixels outperform the conventional SNIC in terms of several metrics on the BSDS500 dataset. In addition, it acquires the controllability of outputting the exact number of superpixels.

Future work will focus on deploying the strategy on several other state-of-the-art superpixels to explore the synergistic performance with their major frameworks.

References

1. Ren, X., Malik, J.: Learning a classification model for segmentation. In: Proceedings of the International Conference on Computer Vision (ICCV), pp. 10–17. IEEE, Nice (2003)
2. Cui, L., et al.: Superpixel segmentation integrated feature subset selection for wetland classification over Yellow River Delta. Environ. Sci. Pollut. Res. **30**, 50796–50814 (2023)
3. Raine, S., Marchant, R., Kusy, B., Maire, F., Fischer, T.: Point label aware superpixels for multi-species segmentation of underwater imagery. IEEE Rob. Autom. Lett. **7**(3), 8291–8298 (2022)
4. Wang, P., Zhang, J., Zhu, H.: Fire detection in video surveillance using superpixel-based region proposal and ESE-ShuffleNet. Multimedia Tools Appl. **82**, 13045–13072 (2023)
5. Fang, Q., Peng, Z., Yan, P., Huang, J.: A fire detection and localisation method based on keyframes and superpixels for large-space buildings. Int. J. Intell. Inf. Database Syst. **16**(1), 1–19 (2023)

6. Li, C., He, W., Liao, N., Gong, J., Hou, S., Guo, B.: Superpixels with contour adherence via label expansion for image decomposition. Neural Comput. Appl. **34**(19), 16223–16237 (2022)
7. Achanta, R., Shaji, A., Smith, K., Lucchi, A., Fua, P., Susstrunk, S.: SLIC superpixels compared to state-of-the-art superpixel methods. IEEE Trans. Pattern Anal. Mach. Intell. **34**(11), 2274–2282 (2012)
8. Achanta, R., Susstrunk, S.: Superpixels and polygons using simple non-iterative clustering. In: Proceedings of the Conference on Computer Vision and Pattern Recognition (CVPR), pp. 4895–4904. IEEE, Honolulu (2017)
9. Zhong, D., Li, T., Dong, Y.: An efficient hybrid linear clustering superpixel decomposition framework for traffic scene semantic segmentation. Sensors **23**(2), 1002 (2023)
10. Chen, J., Li, Z., Huang, B.: Linear spectral clustering superpixel. IEEE Trans. Image Process. **26**(7), 3317–3330 (2017)
11. Liu, Y., Yu, M., Li, B., He, Y.: Intrinsic manifold SLIC: a simple and efficient method for computing content-sensitive superpixels. IEEE Trans. Pattern Anal. Mach. Intell. **40**(3), 653–666 (2018)
12. Zhao, J., Hou, Q., Ren, B., Cheng, M., Rosin, P.: FLIC: fast linear iterative clustering with active search. In: Proceedings of the AAAI Conference on Artificial Intelligence, pp. 7574–7581. AAAI, New Orleans (2018)
13. Neubert, P., Protzel, P.: Compact watershed and preemptive SLIC: on improving trade-offs of superpixel segmentation algorithms. In: Proceedings of the International Conference on Pattern Recognition (ICPR), pp. 996–1001. IEEE, Stockholm (2014)
14. Kang, X., Zhu, L., Ming, A.: Dynamic random walk for superpixel segmentation. IEEE Trans. Image Process. **29**, 3871–3884 (2020)
15. Li, C., Guo, B., Huang, Z., Gong, J., Han, X., He, W.: GRID: GRID resample by information distribution. Symmetry **12**(9), 1417 (2020)
16. Jampani, V., Sun, D., Liu, M.-Y., Yang, M.-H., Kautz, J.: Superpixel sampling networks. In: Ferrari, V., Hebert, M., Sminchisescu, C., Weiss, Y. (eds.) ECCV 2018. LNCS, vol. 11211, pp. 363–380. Springer, Cham (2018). https://doi.org/10.1007/978-3-030-01234-2_22
17. Yuan, Y., Zhu, Z., Yu, H., Zhang, W.: Watershed-based superpixels with global and local boundary marching. IEEE Trans. Image Process. **29**, 7375–7388 (2020)
18. Xiao, X., Zhou, Y., Gong, Y.: Content-adaptive superpixel segmentation. IEEE Trans. Image Process. **27**(6), 2883–2896 (2018)
19. Liao, N., Guo, B., Li, C., Liu, H., Zhang, C.: BACA: Superpixel segmentation with boundary awareness and content adaptation. Remote Sens. **14**(18), 4572 (2022)
20. Viola, P., Jones, M.: Rapid object detection using a boosted cascade of simple features. In: Proceedings of the Conference on Computer Vision and Pattern Recognition (CVPR), pp. 511–518. IEEE, Kauai (2001)
21. Arbelaez, P., Maire, M., Fowlkes, C., Malik, J.: Contour detection and hierarchical image segmentation. IEEE Trans. Pattern Anal. Mach. Intell. **33**(5), 898–916 (2011)
22. Stutz, D., Hermans, A., Leibe, B.: Superpixels: an evaluation of the state-of-the-art. Comput. Vis. Image Underst. **166**, 1–27 (2018)

Dynamic Multi-objective Prediction Strategy for Transfer Learning Based on Imbalanced Data Classification

Danlei Wang[(✉)] and Cuili Yang

Beijing University of Technology, Beijing 100124, China
wangdanlei0801@foxmail.com, clyang5@bjut.edu.cn

Abstract. Prediction strategies based on transfer learning have been proved to be effective in solving Dynamic Multi-Objective Optimization Problems (DMOPs). However, slow running speed impedes the development of this method. To address this issue, this paper proposes a transfer learning method based on imbalanced data classification and combines it with a decomposition-based multi-objective optimization algorithm, referred to as ICTr-MOEA/D. This method combines the prediction strategies based on transfer learning with knee points to save computational resources. In order to prevent the prediction accuracy from being affected by the insufficient of knee points, ELM classifier selects more high-quality points from a large number of random solutions. Moreover, SMOTE resolves the class imbalance problem during the ELM training process. The simulation results demonstrate that ICTr-DMOEA shows good competitiveness.

Keywords: Dynamic multiobjective optimization · Transfer learning · Imbalanced data

1 Introduction

Multi-objective optimization problems (MOPs) refer to a class of problems with multiple objectives that are mutually conflicting [1]. Unlike static MOPs, the Pareto optimal solution (POS) of dynamic multi-objective optimization problems (DMOPs) is not fixed, but dynamically changes over time [2]. Influenced by conditions such as environmental change frequency and degree of change, the optimization objectives, parameters, and constraints of DMOPs may also change dynamically over time [3,4]. These characteristics require dynamic multi-objective optimization algorithms (DMOEA) to quickly converge to the Pareto frontier (POF) after environmental changes. For example, in wastewater treatment, energy saving and water quality improvement are two conflicting objectives [5,6]. The optimization settings of dissolved oxygen and nitrate need to be dynamically adjusted according to the real-time inflow and weather conditions. In addition, in portfolio optimization problems, maximizing profits and minimizing risks are two conflicting objectives, and market prices need to be optimized

H. Zhang et al. (Eds.): NCAA 2023, CCIS 1869, pp. 428–441, 2023.
https://doi.org/10.1007/978-981-99-5844-3_31

in real time according to the running situation [7]. These practical problems are difficult to solve with traditional static multi-objective optimization algorithms, so research on DMOEA has important practical significance.

The optimization objectives of DMOPs usually have complex nonlinear characteristics, which poses a challenge to the design of DMOEA [8]. In recent years, many strategies for solving DMOPs have been proposed. For example, static optimization algorithms are often directly applied to solve DMOPs, and the optimization algorithm is restarted after each environmental change. Although this method is simple to operate, its efficiency is low and it cannot provide guidance for the evolution of the population quickly after environmental changes. Diversity-based methods are also a typical solution to DMOPs, which can prevent evolutionary stagnation caused by local optima during the running process [9–11]. However, blindly increasing diversity is likely to lead the population to evolve in the wrong direction. In addition, memory-based methods store useful information of the past few generations of populations for reuse [12,13]. This method is suitable for periodic changes, but is highly dependent on the diversity of the population.

In the research of DMOEA, it has been found that many DMOPs in real life change in a deterministic way, which means that there is a correlation between the POF or POS in a continuous environment [14]. Therefore, prediction-based methods are becoming increasingly popular. This approach utilizes historical data to build a forecasting model to analyze the fluctuation of DMOPs [15]. However, this method is more effective for environment with stable changes. When the environment undergoes drastic changes, the early historical information often lacks reference value. Using past experience to predict the position of POF in the new environment is a challenge to the research of prediction strategies.

Transfer learning-based prediction strategies, which reuse a small number of historical environments, can effectively deal with DMOPs with drastic environmental changes [16]. However, the large amount of computation and slow running speed hinder the development of this method. In order to tackle this issue, this paper proposes a imbalanced data classification-based transfer learning method (ICTr-DMOEA). This algorithm combines transfer learning strategies with knee points to avoid a large number of low-quality individuals occupying computing resources. Although combining with knee points can improve computational efficiency, due to the insufficient number of turning points, it is unable to depict the movement trend of the entire population, thus affecting the prediction accuracy. Based on this characteristic, a classifier based on extreme learning machine (ELM) [17] is trained to separate more high-quality individuals similar to knee points from a large number of random solutions. In addition, in order to overcome the imbalance between the population and the knee points in the ELM training process, this paper adopts synthetic minority over-sampling technique (SMOTE) [18] to improve the training effect of ELM.

2   Preliminary

2.1   Transfer Component Analysis

Transfer Component Analysis (TCA) [19] is a transfer learning technique which adjusts the marginal probability distribution across different domains by seeking a feature mapping that preserves the internal characteristics of the domains to the greatest extent while reducing the dissimilarity between them. Assume that the source domain $D_s = \{X_{src}, Y_{src}\}$, where X_{src} is the sample data set of the source domain, Y_{src} is the label data set of the source domain. Target domain $D_t = \{X_{tar}\}$, where X_{tar} is the target domain sample data set, the target domain label data is unknown. The data distribution between the source domain and target domain is dissimilar, that is, $P(X_{src}) \neq P(X_{tar})$. The transition component analysis (TCA) postulates the existence of a feature mapping $\bar{\phi}$, which makes the edge probability distribution between two domains in the mapped new feature space as consistent as possible, i.e. $P(\bar{\phi}(X_{src})) \approx P(\bar{\phi}(X_{tar}))$. To solve for this feature mapping, TCA employs Maximum Mean Discrepancy (MMD) to quantify the dissimilarity between the source domain and the target domain.

$$\text{dist}(X_s, X_t) = \left\| \frac{1}{n_1} \sum_{i=1}^{n_n} \bar{\phi}(X_{si}) - \frac{1}{n_2} \sum_{j=1}^{n_2} \bar{\phi}(X_{tj}) \right\|_H^2 \tag{1}$$

where n_1 denotes the number of samples in the source domain and n_2 stands for the number of samples in the target domain; $\| \cdot \|_H$ denotes the Reproducing Kernel Hilbert Space (RKHS) norm; $X_{si} \in X_{src}$, $X_{tj} \in X_{tar}$. By transforming the inner product after expansion into a kernel matrix form, Formula 1 can be further articulated as:

$$\text{dist}(X_{src}, X_{tar}) = \text{tr}(KL) \tag{2}$$

$$K = \begin{bmatrix} K_{s,s} & K_{s,t} \\ K_{t,s} & K_{t,t} \end{bmatrix} \in \mathbb{R}^{(n_1+n_2) \times (n_1+n_2)} \tag{3}$$

where $K(i, j) = \left[\bar{\phi}(x_i)^T \bar{\phi}(x_j) \right]$; and $K_{s,s}$, $K_{t,t}$, $K_{s,t}$, and $K_{t,s}$ represent the kernel matrices of the source domain, target domain, and both domains, respectively; L is the metric matrix, whose calculation expression is:

$$L(i, j) = \begin{cases} \frac{1}{n_1^2} & x_i, x_j \in X_{src}, \\ \frac{1}{n_2^2} & x_i, x_j \in X_{tar}, \\ -\frac{1}{n_1 n_2} & \text{Otherwise.} \end{cases} \tag{4}$$

In order to make the calculation simpler, a low-dimensional matrix $\tilde{W} \in \mathbb{R}^{(n_1+n_2) \times m}$ is defined to map the features to an m-dimensional space. Therefore, the final optimization objective of TCA is:

$$\min_{W} \ \mathrm{tr}\left(W^T KLKW\right) + \mu\,\mathrm{tr}\left(W^T W\right)$$
$$\text{s.t.} \ \ W^T KHKW = I_m \tag{5}$$

where μ is a balancing factor. Finally, by solving Formula 5, the optimal mapping matrix W is obtained to finish the feature data space mapping between the source and target domains.

3   Proposed Methodology

DMOPs requires the optimization algorithm to quickly and accurately converge to the changed POF after the environment changes. It has been demonstrated that the prediction strategy based on transfer learning has a good prediction performance, however, its slow running speed is deemed to be the primary disadvantage. How to improve the running speed while maintaining the prediction accuracy is the focus of this chapter. The knee point is a special point with representative in the population. Combining the knee point with the transfer learning strategy can effectively save the computing resources. In order to ensure the prediction accuracy, the ELM classifier is able to distinguish a considerable amount of high-quality solutions similar to the knee points from a multitude of random solutions, while the synthetic minority oversampling technique is utilized to resolve the class imbalance issue during the training of the ELM classifier.

3.1   Knee Points

In the development of multi-objective optimization algorithms, it is often necessary to find some representative high-quality individuals, which are usually given by the decision-maker according to his own preferences. However, in practice, the decision-maker often does not have prior knowledge to form preferences. In this case, the knee point can be selected as a favorable solution. The knee point is part of the POS. This type of point is intuitively represented as the most concave part of the POF, as shown in Fig. 1. Near the knee point, any change in the objective value of one dimension will lead to a significant change in the objective value of other dimensions. In other words, the solutions at the knee point, if slightly improved in one objective, will cause at least one other objective to degrade significantly. The characteristic of the knee point indicates that the solutions at this position have a higher hypervolume (HV) value [20]. Therefore, from the decision-maker's point of view, if no special preference is specified, the knee point is most likely to be the first choice point to represent the population. In addition, from a theoretical point of view, the selection of knee points can improve the convergence performance of the algorithm.

There are many methods for identifying knee points, such as fitting the curvature of the curve by regression analysis, if the current curvature changes sharply, then this is the knee point. Or judge the knee point by the derivative equation,

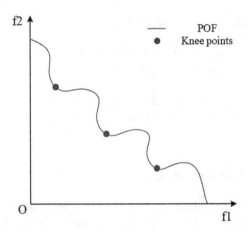

Fig. 1. Illustration of knee points

the derivative equation can measure the change rate of the function, if the first derivative of the function changes, it indicates that this is the knee point. This paper determines the knee point by computing the distance between the point and the extreme line or hyperplane. In the two-objective optimization problem, the distance expression formula is as follows:

$$\text{Dis}(KP, L) = \begin{cases} \frac{|ax_s + by_s + c|}{\sqrt{a^2+b^2}}, & ax_s + by_s + c < 0 \\ -\frac{|ax_s + by_s + c|}{\sqrt{a^2+b^2}}; & \text{otherwise.} \end{cases} \tag{6}$$

where Dis is the distance from the knee point to the line L, a and b are coefficients determined by the line L, and the point with the maximum value is the knee point KP. In three-objective problems, the point with the farthest perpendicular distance from the plane composed of boundary points is defined as the knee point.

In order to enhance the prediction accuracy, the target space is divided into numerous sub-regions, each of which possess its own local knee points. The division method is shown in Fig. 2. If the number of sub-regions is three, take any maximum value A and minimum value B of the objective function, form a line with A and B and divide it into three parts. Find the non-dominated point farthest from the line in each sub-region, these points are the inflection points in each sub-region. When the problem rises to three dimensions, the boundary points form a plane, and the point farthest from the plane in each sub-region is taken as the knee point.

3.2   Data Augmentation Strategy Based on ELM-SMOTE

Traditional gradient learning methods are unsatisfactory in terms of the learning speed when training multi-layer feedforward neural networks (FNN). The

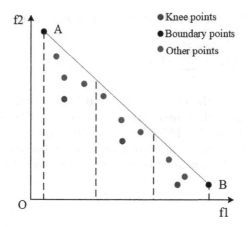

Fig. 2. Sketch map of finding knee point in each subspace

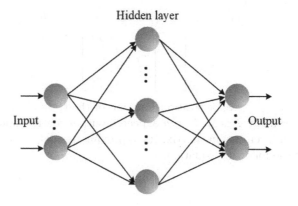

Fig. 3. The architecture of the ELM

connection weights and hidden biases between the input layer and the hidden layer of the ELM can be randomly selected, and the output layer weights can be calculated analytically, thus efficiently avoiding the tedious training process. In comparison to traditional FNN, ELM possesses superior generalization capacity and computational efficiency, and has been widely used in classification problems in different fields [21]. Therefore, ELM is adopted to differentiate high-quality individuals from a vast amount of random solutions, thereby augmenting the accuracy of the prediction model while preserving the operational speed. The basic structure of ELM is shown in Fig. 3. Assuming a set of N training data $D(x_i, t_i)$, $i = 1 \cdots N$ is given, where the corresponding input can be set as $x_i = [x_{i1}, x_{i2}, \cdots x_{in}]^T \in R^n$ and the expected output as $t_i = [t_{i1}, t_{i2}, \cdots t_{im}]^T \in R^m$, and m is the dimension of the output layer, i.e. the number of attributes of the training samples. As shown in Fig. 3, if a FNN with M hidden layer nodes is able to fit all training samples with zero error, it implies that the β_i, w_l and b_l values will satisfy the following equation,

$$f_M(x_j) = \sum_{i=1}^{M} \beta_i \bar{G}(w_i, b_i, x_j) = t_j, j = 1, \ldots, N \tag{7}$$

Here, w_l and b_l denote the input weights and hidden node thresholds, respectively, which are randomly generated in the interval $[-1,1]$; represents the connection weights from the ith hidden layer node to each output node, and $\bar{G}$ is the hidden layer activation function, also known as the feature mapping function. For the convenience of formula derivation, the above equation can be simplified as:

$$H\beta = T \tag{8}$$

where

$$H = \begin{bmatrix} \bar{G}(w_1 \cdot x_1 + b_1) & \cdots & \bar{G}(w_M \cdot x_1 + b_M) \\ \vdots & \vdots & \vdots \\ \bar{G}(w_1 \cdot x_N + b_1) & \cdots & \bar{G}(w_M \cdot x_N + b_M) \end{bmatrix} \tag{9}$$

$$\beta = \begin{bmatrix} \beta_1^T \\ \vdots \\ \beta_M^T \end{bmatrix}, T = \begin{bmatrix} T_1^T \\ \vdots \\ T_N^T \end{bmatrix} \tag{10}$$

where $G(w_i \cdot x_j \cdot b_i)$ is the activation function of the training sample x_j on the ith hidden layer node, T is an $N \times m$ expected output matrix, and H is referred to as the hidden layer output matrix. Its approximate solution $\hat{\beta}$ can be directly calculated by the following equation:

$$\hat{\beta} = H^+ T \tag{11}$$

where H^+ is the generalized inverse of H, and $\hat{\beta}$ is the least-norm least-squares solution of the network. Therefore, the calculation of the output weights can be obtained in one step, reducing the training time of the network. Secondly, the l_2 norm of the output weight matrix β can be effectively constrained, thus ensuring that the ELM algorithm has strong generalization ability.

The historical solutions comprising of knee points are utilized as the training set to acquire the output weights of ELM. During the training process of ELM, there is an apparent class imbalance issue, wherein the number of knee points considerably differs from the size of the complete population. In order to reduce the training error, the synthetic minority oversampling technique is used to process the training data.

Traditional oversampling methods often randomly duplicate some minority class samples to increase the sample size, but this operation can lead to overfitting problems. The Synthetic Minority Oversampling Technique (SMOTE) combines the KNN algorithm with the oversampling method, randomly selecting minority group samples and their k nearest neighbors, and randomly inserting artificial synthesized samples between these samples, as shown in Fig. 4. If the

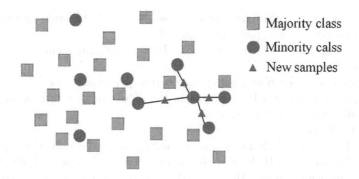

Fig. 4. Schematic of SMOTE

minority class sample is $x_{original}$, an artificial sample x_{new} is generated by the following method between $x_{original}$ and its adjacent sample x_{knn}:

$$x_{new} = x_{\text{original}} + \alpha * (x_{\text{knn}} - x_{\text{original}}).$$ (12)

The process of the data augmentation strategy based on ELM-SMOTE is as follows:

Step 1. For the ELM classifier training set, the estimated turning point $Knee_t$ and randomly generated solutions are selected in the new environment, and the training set is preprocessed using SMOTE;

Step 2. For each individual $x_{original}$ in the estimated turning point set, the distance to other turning points is calculated using the Euclidean distance as the measurement standard.

Step 3. A parameter k is set, and the k nearest neighbors of $x_{original}$ are obtained from the estimated turning point $x_{original}$ and x_{knn}.

Step 4. For each of the neighbors x_{knn}, generate a number of new samples x_{new} respectively with $x_{original}$, merge the new samples into the estimated turning point set to balance the classes of the training set.

Step 5. Train the ELM model to obtain the connection weights O between the hidden layer and the output layer.

Step 6. Generate a great number of random solutions, and use the trained ELM classifier to separate high-quality individuals and $Knee_t$ to form $Knee_{ES}$ together.

3.3   The Process of ICTr-MOEA/D

Based on the above analysis, the flow of ICTr-MOEA/D is as follows:

Input: Dynamic multi-objective optimization function $F_t(\cdot)$; the number of sub-regions p.

Output: POS and POF at different moments.
Step 1. Random initialization of population Pop_0 at $t = 0$;

Step 2. Detecting if the environment has altered, if the environment alters, proceed to Step 4, if not, proceed to Step 3;

Step 3. Using MOEA/D to obtain the POS, POF and Pop_t of the current environment;

Step 4. Calculate the $Knee_t$ set of the current population, use TCA to obtain the mapping relationship W, and obtain the HSK_t after $Knee_t$ is mapped to the high-dimensional Hilbert space;

Step 5. $t = t + 1$;

Step 6. For each individual hk in the HSK_t, find the individual q_k in the current solution space. If the mapping of q_k's objective value is closest to h_k in the high-dimensional space, then q_k is the $Knee_{t+1}$ in the new environment;

Step 7. Obtain $Knee_{ES}$ using ELM-SMOTE;

Step 8. Obtain a complete Pop_{t+1} through Step 1;

Step 9. Determine if the stopping condition is met, if so, stop, otherwise jump to Step 2.

4   Experiments

In order to validate the efficacy of the ICTr-MOEA/D algorithm, the CEC2018 competition on dynamic multiobjective optimization benchmark test suite [22] was used to evaluate the proposed algorithm. The benchmark test functions emulate the dynamic characteristics of multiple real systems, and for this experiment, the nine double-objective optimization problems DF1–DF14 are chosen as the benchmark test problems. The dynamic characteristics of DMOPs are defined as $t = (1/n_t)\,[\tau/\tau_t]$, where τ is the maximum number of iterations, n_t and τ_t represent the severity and frequency of changes, respectively.

4.1   Performance Indicators

In this paper, the following performance metrics were introduced.

Inverted generational distance (IGD) [23] is a comprehensive metric for measuring the convergence and distribution of the Pareto optimal solutions obtained by the algorithm,

$$IGD\,(POF^*, POF) = \frac{\sum_{v \in POF^*} d(v, POF)}{|POF^*|} \tag{13}$$

where $POF*$ denotes the true Pareto optimal front, which is composed of a series of discrete points uniformly distributed on the true front. POF denotes the Pareto optimal front obtained by a certain method. The lower the IGD value, the nearer the solution set attained is to the true front and the more uniform the dispersion is.

MIGD is modified based on IGD. This index reflects the comprehensive performance of the Pareto optimal solution set obtained by a certain method through the IGD statistical results in a period of time, as shown below,

$$MIGD\left(POF^{t*}, POF^{t}\right) = \frac{1}{|T|} \sum_{t \in T} IGD\left(POF^{t*}, POF^{t}\right) \qquad (14)$$

where PF^{t*} and PF^{t} represent the true Pareto optimal front of the problem at time t and the Pareto optimal front obtained by a certain method, respectively, and T is the number of environments. The lower the MIGD value, the nearer the Pareto front is to the true front and the more uniform the dispersion is.

4.2   Compare Algorithms

1. Tr-DMOEA [16]: A DMOEA based on transfer learning maps the POF in the previous environment to the latent space through transfer component analysis (TCA), and these mappings are employed to generate high-quality initial populations. It is compared with the original transfer learning prediction strategy and the proposed algorithm.
2. KT-DMOEA [24]: A transfer learning method based on turning points. A trend prediction model (TPM) is developed to predict turning points, and an imbalanced transfer learning method is used to generate high-quality initial populations from the turning points.
3. D-MOEA/D: D-MOEA/D is adapted from MOEA/D. 10% the population is initialized after environmental changes

The parameters are set as follows,

1. Compare algorithms: The static optimizer for all experiments was MOEA/D. The parameters of Tr-MOEA/D and KT-MOEA/D are set by referring to [16] and [24].
2. Static optimizer: For all double-objective optimization problems, the population size is set to 100, and the dimension of the decision variable is 10.
3. Operating standard: The test questions are executed independently 10 times, with each run consisting of 30 environmental changes.

4.3   Performance on DF Problems

The MIGD results on the DF1–DF14 problems are shown in Table 1. The "+" and "−" marked after the data indicate that the results are better or worse than the proposed algorithms. On DF2, DF3, DF5, DF9, DF12 and DF14, the results of ICTr-MOEA/D are superior to those of other algorithms. On DF1 and DF4, two indicators are better than those of other algorithms. This indicates that ICTr-MOEA/D exhibits excellent dynamic tracking capability in the majority of cases. It is worth noting that ICTr-MOEA/D performs slightly worse on DF6 and DF7. This may be because the complex changes of DF6 and DF7 make the problem unpredictable, resulting in the failure of the prediction model.

Table 1. Mean MIGD values of compared algorithms

Problems	τ_t, n_t	ICTr-MOEA/D	Tr-MOEA/D	KT-MOEA/D	D-MOEA/D
DF1	10,10	0.0969	0.1251(+)	**0.0892(−)**	0.1258(+)
	10,5	**0.1306**	0.1374(+)	0.1290(+)	0.1244(+)
	5,10	**0.1346**	1.1877(+)	0.1423(+)	0.1948(+)
DF2	10,10	**0.0862**	0.0891(+)	0.0912(+)	0.1919(+)
	10,5	**0.0826**	0.0865(+)	0.0907(+)	0.0909(+)
	5,10	**0.1208**	0.1497(+)	0.1324(+)	0.1460(+)
DF3	10,10	**0.3440**	0.3829(+)	0.3653(+)	0.4315(+)
	10,5	**0.3279**	0.3640(+)	0.3996(+)	0.4753(+)
	5,10	**0.4021**	0.4218(+)	0.4210(+)	0.4387(+)
DF4	10,10	0.1294	1.1846(+)	**0.1002(−)**	1.2214(+)
	10,5	**0.9613**	1.1601(+)	0.9814(+)	1.0992(+)
	5,10	**1.2047**	1.3125(+)	1.3268(+)	1.3799(+)
DF5	10,10	**1.2037**	1.6880(+)	1.3124(+)	1.3128(+)
	10,5	**1.2741**	1.6323(+)	1.3370(+)	1.9498(+)
	5,10	**1.2905**	1.8674(+)	1.3056(+)	2.4453(+)
DF6	10,10	2.8022	2.8275(+)	2.4380(+)	**2.1083(−)**
	10,5	3.3125	3.2019(+)	**1.9623(−)**	2.4854(+)
	5,10	3.9088	3.6823(+)	**3.0623(−)**	4.2567(+)
DF7	10,10	3.0082	**2.0342(−)**	2.3246(+)	2.0721(+)
	10,5	2.6495	**2.1843(−)**	2.3097(+)	2.5513(+)
	5,10	4.0378	3.2963(+)	3.1479(+)	**3.1054(-)**
DF8	10,10	0.9702	0.9896(+)	**0.8427(−)**	1.1419(+)
	10,5	0.9223	0.9945(+)	**0.8829(−)**	0.9688(+)
	5,10	0.8956	0.9214(+)	**0.8214(−)**	0.9027(+)
DF9	10,10	**1.6349**	1.6460(+)	1.6791(+)	2.1883(+)
	10,5	**1.4323**	1.4721(+)	1.8026(+)	2.1441(+)
	5,10	**1.8024**	1.8274(+)	1.8648(+)	2.3498(+)
DF10	10,10	**0.1026**	0.1128(+)	0.1980(+)	0.2104(+)
	10,5	0.1732	**0.1543(−)**	0.1952(+)	0.2174(+)
	5,10	**0.1804**	0.1617(+)	0.2104(+)	0.2497(+)
DF11	10,10	**0.1421**	0.1961(+)	0.1478(+)	0.1926(+)
	10,5	0.2024	0.2834(+)	**0.1757(−)**	0.2652(+)
	5,10	0.1920	0.2173(+)	**0.1821(−)**	0.2398(+)
DF12	10,10	**1.1692**	1.1893(+)	1.1774(+)	0.9452(+)
	10,5	**1.1753**	1.1986(+)	1.1820(+)	0.9327(+)
	5,10	**1.2024**	1.2203(+)	1.2164(+)	0.9217(+)
DF13	10,10	1.6449	1.6069(+)	**1.5217(−)**	1.5573(+)
	10,5	**1.4623**	1.5364(+)	1.4972(+)	1.5307(+)
	5,10	**1.6524**	1.7590(+)	1.6736(+)	1.7634(+)
DF14	10,10	**0.9049**	0.9314(+)	0.9181(+)	0.9328(+)
	10,5	**0.9413**	0.9426(+)	0.8765(+)	0.9417(+)
	5,10	**0.9524**	1.1024(+)	0.9864(+)	0.9986(+)

Table 2. Mean MIGD values of different algorithms with $n_t = 10$ and $\tau_t = 10$

Problems	ICTr-MOEA/D	D-MOEA/D	ICTr-NASGII	D-NASGII
DF1	0.0969	0.1258	0.1089	0.1324
DF2	0.0862	0.1919	0.0925	0.2123
DF3	0.3440	0.4315	0.3726	0.4612
DF4	0.1294	1.2214	0.1347	1.2421
DF5	1.2037	1.3128	1.3042	1.3564
DF6	2.8022	2.1083	2.9162	3.0178
DF7	3.0082	2.0721	3.2166	2.5677
DF8	0.9702	1.1419	0.9964	1.2312
DF9	1.6349	2.1883	1.7396	2.3124
DF10	0.1026	0.2104	0.1529	0.2201
DF11	0.1421	0.1926	0.1507	0.2016
DF12	1.1692	0.9452	1.1706	1.1024
DF13	1.6449	1.5573	1.6501	1.5927
DF14	0.9049	0.9328	0.9281	0.9564

To ascertain the feasibility of integration of the ICTr-based prediction model with other MOEAs, MOEA/D was substituted with NSGAII (hereafter referred to as ICTr-NSGAII). The original NSGAII was modified to DNSGAII suitable for environmental changes, in which 10% the population individuals were randomly initialized when the environment changes. The experimental results are shown in Table 2. It is evident that for various static algorithms, the algorithms incorporating ICTr are evidently superior to the random initialization methods, that is, ICTr-MOEA/D and ICTr-NSGAII are obviously better than D-MOEA/D and NSGAII. This observation suggests that the outstanding performance of the proposed algorithm is largely attributed to its forecasting strategy, rather than a particular static optimization algorithm.

5 Conclusion

In order to address the issue of slow running speed of prediction strategies based on transfer learning, this paper proposes a DMOEA based on ICTr-MOEA/D. Firstly, a turning point is introduced into the dynamic multi-objective optimization algorithm based on transfer learning, avoiding the problem of excessive computational resources occupied by low-quality individuals. Secondly, a classification model based on ELM is trained to separate more high-quality individuals from a large number of random solutions, in order to converge more accurately to the changed POF. Finally, the data processing method based on SMOTE solves the class imbalance problem in the ELM training process. The benchmark experiments demonstrate that the proposed algorithm attains desirable results in the majority of experiments, especially in the cases of predictable problems. Therefore, ICTr-MOEA/D has obvious advantages in solving DMOPs.

Acknowledgement. This work was supported by the National Key Research and Development Project under Grant 2021ZD0112002, the National Natural Science Foundation of China under Grants 61973010, 62021003 and 61890930-5.

References

1. Xie, Y., Qiao, J., Wang, D., Yin, B.: A novel decomposition-based multiobjective evolutionary algorithm using improved multiple adaptive dynamic selection strategies. Inf. Sci. **556**, 472–494 (2021)
2. Xie, H., Zou, J., Yang, S., Zheng, J., Ou, J., Hu, Y.: A decision variable classification-based cooperative coevolutionary algorithm for dynamic multiobjective optimization. Inf. Sci. **560**, 307–330 (2021)
3. Hu, Y., Zheng, J., Zou, J., Yang, S., Ou, J., Wang, R.: A dynamic multi-objective evolutionary algorithm based on intensity of environmental change. Inf. Sci. **523**, 49–62 (2020)
4. Sun, J., Gan, X., Gong, D., Tang, X., Dai, H., Zhong, Z.: A self-evolving fuzzy system online prediction-based dynamic multi-objective evolutionary algorithm. Inf. Sci. **612**, 638–654 (2022)
5. Han, H., Zhang, L., Liu, H., Yang, C., Qiao, J.: Intelligent optimal control system with flexible objective functions and its applications in wastewater treatment process. IEEE Trans. Syst. Man Cybern. Syst. **51**(6), 3464–3476 (2021)
6. Qiao J., Zhang W.: Dynamic multi-objective optimization control for wastewater treatment process. Neural Comput. Appl. **29**(11), 1261–1271 (2018)
7. Xu, J., Luh, P., White, F., Ni, E., Kasiviswanathan, K.: Power portfolio optimization in deregulated electricity markets with risk management. IEEE Trans. Power Syst. **21**(4), 1653–1662 (2006)
8. Zhou, A., Jin, Y., Zhang, Q.: A population prediction strategy for evolutionary dynamic multiobjective optimization. IEEE Trans. Cybern. **44**(1), 40–53 (2014)
9. Yang, S.: Genetic algorithms with memory and elitism-based immigrants in dynamic environments. Evol. Comput. **16**(3), 385–416 (2008)
10. Daneshyari, M., Yen, G.: Dynamic optimization using cultural based PSO. In: Proceedings of IEEE Congress of Evolutionary Computation, pp. 509–516, June 2011
11. Shang, R., Jiao, L., Ren, Y., Li, L., Wang, L.: Quantum immune clonal coevolutionary algorithm for dynamic multiobjective optimization. Soft. Comput. **18**(4), 743–756 (2014)
12. Nakano, H., Kojima, M., Miyauchi, A.: An artificial bee colony algorithm with a memory scheme for dynamic optimization problems. In: Proceedings of IEEE Congress of Evolutionary Computation, pp. 2657–2663, May 2015
13. Sahmoud, S., Topcuoglu, H.: Sensor-based change detection schemes for dynamic multi-objective optimization problems. In: Proceedings of IEEE Symposium Series on Computational Intelligence, pp. 1–8 (2017)
14. Cao, L., Xu, L., Goodman, E., Bao, C., Zhu, S.: Evolutionary dynamic multiobjective optimization assisted by a support vector regression predictor. IEEE Trans. Evol. Comput. **24**(2), 305–319 (2020)

15. Hatzakis, I., Wallace, D.: Dynamic multi-objective optimization with evolutionary algorithms: a forward-looking approach. In: Proceedings of GECCO, pp. 1201–1208 (2006)
16. Jiang, M., Huang, Z., Qiu, L., Huang, W., Yen, G.: Transfer learning-based dynamic multiobjective optimization algorithms. IEEE Trans. Evol. Comput. **22**(4), 501–514 (2018)
17. Yang, C., Nie, K., Qiao, J.: Design of extreme learning machine with smoothed L0 regularization. Mobile Networks Appl. **25**(1), 2434–2446 (2020)
18. Chawla, N., Bowyer, K., Hall, L.: SMOTE: synthetic minority over-sampling technique. AI Access Found. **2002**(1) (2002)
19. Pan, S., Tsang, I., Kwok, J., Yang, Q.: Domain adaptation via transfer component analysis. IEEE Trans. Neural Networks **22**(2), 199–210 (2011)
20. Zou, F., Yen, G., Tang, L.: A knee-guided prediction approach for dynamic multiobjective optimization. Inf. Sci. **509**, 193–209 (2020)
21. Huang, G., Zhu, Q., Siew, C.: Extreme learning machine: theory and applications. Neurocomputing **70**(1–3), 489–501 (2006)
22. Jiang, S., Yang, S., Yao, X., Tan, K., Kaiser, M., Krasnogor, N.: Benchmark functions for the CEC 2018 competition on dynamic multiobjective optimization, Newcastle University, Newcastle upon Tyne, U.K., Report (2018)
23. Xu, D., Jiang, M., Hu, W., Li, S., Pan, R., Yen, G.: An online prediction approach based on incremental support vector machine for dynamic multiobjective optimization. IEEE Trans. Evol. Comput. **26**(4), 690–703 (2022)
24. Jiang, M., Wang, Z., Hong, H., Yen, G.: Knee point-based imbalanced transfer learning for dynamic multiobjective optimization. IEEE Trans. Evol. Comput. **25**(1), 117–129 (2021)

A Twin Learning Evolutionary Algorithm for Capacitated Vehicle Routing Problem

Hanshi Yang[1], Xin Zhang[2], Liang Feng[3], Zhou Wu[1(✉)], and Choujun Zhan[4]

[1] College of Automation, Chongqing University, Chongqing, China
a17703779@163.com, zhouwu@cqu.edu.cn
[2] College of Artificial Intelligence, Tianjin Normal University, Tianjin, China
ecemark@tjnu.edu.cn
[3] College of Computer Science, Chongqing University, Chongqing, China
brightfengs@gmail.com
[4] College of Electrical and Computer Engineering, Nanfang College Guangzhou, Guangzhou, China
zchoujun2@gmail.com

Abstract. Capacitated vehicle routing problem (CVRP) is challenging or even impossible to be solved. Most studies solve similar CVRPs without prior knowledge. In order to use the knowledge obtained by solving similar historical CVRPs to solve the target CVRP, a twin learning evolutionary algorithm (TLEA) is proposed. An autoencoder is used to search a twin CVRP which is similar to the target CVRP. If not found, a graph filter is used to construct a twin CVRP. Further, the initial solution of the target CVRP is obtained by learning a mapping from the twin CVRP to the target CVRP. Finally, a traditional meta-heuristic algorithm embedded in the TLEA to improve the quality of the initial solution. To evaluate the performance of TLEA, comprehensive studies have been conducted with the CVRP benchmarks, against the traditional meta-heuristic algorithms and a famous learning based algorithm, confirming the efficacy of TLEA for CVRP.

Keywords: Capacitated vehicle routing problem · Twin learning · Knowledge transfer

1 Introduction

CVRP is one of the most important researches in transportation, supply chain management and other research fields. There are three main categories of existing CVRP methods: deterministic algorithms, meta-heuristic algorithms and learning based algorithms. Deterministic algorithms, e.g., dynamic programming and branch-bound methods [1,2]. They can precisely solve small-scale CVRP, but as the nodes of CVRP increases, they show low efficiency. In order to solve large scale CVRP, meta-heuristic algorithms are proposed and have been studied extensively. Meta-heuristic algorithms, e.g., genetic algorithm (GA) and ant colony algorithm

H. Zhang et al. (Eds.): NCAA 2023, CCIS 1869, pp. 442–454, 2023.
https://doi.org/10.1007/978-981-99-5844-3_32

[3,4]. They can search acceptable solutions efficiently for CVRP with hundreds of nodes. However, meta-heuristic algorithms ignore the knowledge obtained by solving similar historical tasks. In order to learn the knowledge from historical tasks to improve the efficiency of solving CVRP, learning based algorithms are widely studied. Learning based algorithms, e.g., graph neural network (GNN) and deep reinforcement learning (DRL) [5,6]. They can use knowledge learned from related historical tasks to quickly generate an acceptable solution for the CVRP. However, the time cost of model training is relatively large.

There are some recent studies for CVRP.

(1) Meta-heuristic algorithms: For large scale CVRP, the clustering methods are used to divide the consumer nodes into a number of clusters that suits the number of vehicles and their capacity. And the optimal route of clusters are connected as the result [7]. Evolutionary multi-task optimization (EMTO) is a new general framework that simultaneously optimizes each task, aided by the implicit knowledge between tasks. For example, multifactorial evolutionary algorithm (MFEA) optimizes multiple tasks simultaneously and performs knowledge transfer between them [8].

(2) Learning based algorithms: Recent research in learning based algorithms has focused on two categories: learning construction heuristics and learning improvement heuristics [9]. Learning construction heuristics, e.g., GNNs, attention mechanisms and sequence representations, which could produce a solution sequentially by extending a partial tour. Learning improvement heuristics, which learn a behavior or rule to improve solutions. For example, train a DRL to learn the search behavior of 2-opt algorithm [10].

However, most methods have neglected similar historical CVRPs, few historical experience and knowledge could be utilized to solve the problem. Evolutionary transfer optimization (ETO) is a new framework that aims to improve the search efficiency of evolutionary algorithms (EAs) by utilizing knowledge from historical problems or ongoing problems. Osaba et al. [11] proposed a multifactorial cellular genetic algorithm for CVRP. Through positive knowledge transfer between CVRPs to enhance the searching ability of cellular genetic algorithm. Feng et al. [12] proposed an explicit evolutionary multitasking for CVRP. By learning sparse mapping matrices between two CVRPs to achieve explicit knowledge transfer. However, most of related methods assume that the source task is similar to the target task. If they are not similar, these methods would show incapability. Based on the above analysis, Wu et al. [13] proposed a twin learning framework. An autoencoder is used to efficiently find a twin task in history which is similar to the target task. Then, the best solution found in the twin task will be transferred by a mapping to the target task. If there is no twin task, a novel construction method based on graph filters is used to construct a small-scale twin task. Inspired by [13], this paper proposed a twin learning evolutionary algorithm for CVRP. It is the first application of twin learning framework to VRP family of problem.

The rest of the paper is organized as follows. The definition of CVRP and related research for CVRP are given in Sect. 2. Section 3 represents general twin

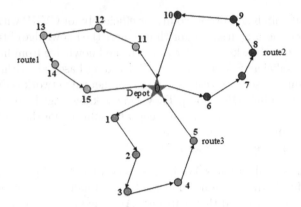

Fig. 1. The graph of CVRP

learning framework. The details of the twin learning evolutionary algorithm for CVRP is described in Sect. 4. In Sect. 5, Analysis of experimental results. The conclusion is given in Sect. 6.

2 Preliminaries

2.1 CVRP

The CVRP with $n+1$ nodes is defined on a graph $G(V, E)$, where V is a set of nodes with $V = \{0, 1, 2, \cdots, n\}$ and E is a set of arcs with $E = \{(i, j) \in V^2, i \neq j\}$. Node 0 represents the depot, while others represent consumers with certain demands. A simple version about graph of CVRP is shown in Fig. 1. The purpose is to find a trip schedule such that the cost of servicing all consumers is minimized. It must be guaranteed that the total customer demand serviced by a particular vehicle less than or equal to its capacity. The formulations in terms of mathematical operation are shown in Eq. (1) and Eq. (2):

$$\min \sum_{i,j \in E} c_{ij} x_{ij} \tag{1}$$

$$\sum_{i,j \in 1,\cdots,n, i \neq j} x_{ij} = 1 \tag{2}$$

where c_{ij} is the cost of travel over arc $(i, j) \in A$. Equation (2) represents there is a route between consumer node i and consumer node j, and each node can be accessed only once.

2.2 Meta-Heuristic Algorithms for CVRP

Variable neighborhood search (VNS) [14] is an effective algorithm based on local search for CVRP. The search procedure of VNS includes variable neighborhood descent (VND) and shaking. For an initial solution S_0, uses the specified neighborhood action rule to calculate m neighborhoods $N_k = \{N_1, N_2, \cdots, N_m\}$ for

S_0. Then perform shaking that uses a local search operator such as 2-opt to improve the N_k, if find a better solution S_t than S_0 in the N_k, $S_0 = S_t$, until all neighborhoods have been searched. GA is a famous single task meta-heuristic algorithm. When using GA to solve CVRP, the combination of all nodes could be regarded as a chromosome. Crossover operator, mutation operator and selection operator are used to improve the chromosomes. MFEA is a famous EMTO algorithm which is inspired by the multi-factorial optimization. It solves multiple CVRPs simultaneously and transfer knowledge between them.

2.3   Learning Based Algorithms for CVRP

Learning construction heuristics based on deep learning model (LCH-DLM) is a well-known learning based algorithm. And a large number of randomly generated CVRP instances are used to train the LCH-DLM. When solving the CVRP, the trained LCH-DLM can generate an acceptable solution quickly. LCH-DLM such as the graph attention network (GAT) [15]. The 2D graph of CVRP node is passed into the encoder of GAT to encode the input features. Then, the decoder outputs the probabilities of unselected nodes. Finally, the unselected nodes are subsequently selected according to the probability distribution by greedy search. The reinforcement learning algorithms are used to train the GAT.

3   Twin Learning Framework

This section introduces the general twin learning framework. Specifically, an autoencoder is used to search a twin task in history for the target task firstly. If there is no twin task, generate a simple twin task by the graph filter. Moreover, a mapping is used to transfer knowledge from the twin task to get initial solutions for the target task. Finally, a traditional meta-heuristic algorithm is used to improve the quality of the solutions. Details of matching and construction twin task and learning mapping are given below.

3.1   Twin Task Matching

An autoencoder [16] is used to obtain feature vectors. The simple structure of the autoencoder is shown by Fig. 2. The autoencoder is consist of the encoder and decoder.

In particular, for the 2D combinatorial optimization problem (COP), we get instance images of the same size firstly. Then get the binary matrix $T \in R^{d \times 1}$ of the image matrix as the input vector for the encoder, as shown in Eq. (3):

$$h = \sigma(wT + b) \tag{3}$$

where the $h \in R^{d' \times 1}$ is the hidden layer vector, $w \in R^{d' \times d}$ is a weight matrix, $b \in R^{d' \times 1}$ is the bias vector, σ is the logistic sigmoid, as shown in Eq. (4):

$$\sigma = 1/(1 + e^{-T}) \tag{4}$$

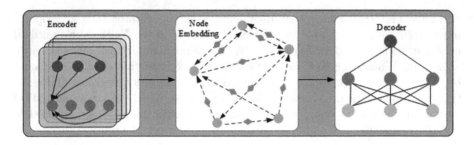

Fig. 2. The structure of autoencoder

Through the decoder, the T is reconstructed as $T' \in R^{d \times 1}$, as shown in Eq. (5):

$$T' = \sigma(w' h + b')$$ (5)

where $w \in R^{d \times d'}$, $b \in R^{d \times 1}$. The training objective function is defined as follows to update parameters w, w', b, b'.

$$\min_{w,w',b,b'} \sum_{r=1}^{m} \frac{1}{2m} \left\| T^{(r)} - T'^{(r)} \right\|^2$$ (6)

where m is the number of training samples.

The encoder outputs the feature vector h. Suppose there are two vectors h_A and h_B, the cosine distance between them is denoted as Eq. (7):

$$dist(A, B) = 1 - cos(h_A, h_B) = \frac{\|h_A\|_2 \cdot \|h_B\|_2 - h_A \cdot h_B}{\|h_A\|_2 \cdot \|h_B\|_2}$$ (7)

The smaller the $dist(A, B)$ is, the more similar A and B are. In this paper, set $d_{min} = 0.05$, if $dist(A, B) <= 0.05$, A and B are similar.

3.2   Twin Task Construction

If all the tasks in historical task library are not similar to the target task, use a graph filter [17] to construct a twin task. In particular, for a certain 2D COP, its C nodes can be defined as Eq. (8):

$$P = \{p_i = (x_i, y_i) | i = 1, \ldots, C\}$$ (8)

Firstly, k-nearest neighbor (kNN) [18] is used to get the neighbors $\delta_i = \{\delta_{i,1}, \ldots, \delta_{i,k}\}$ of each node p_i.

Then, calculate the weight $W_{i,j}$ between p_i and p_j as follows:

$$W_{i,j} = \begin{cases} exp(-\frac{\|p_i - p_j\|_2^2}{\eta^2}), & if \; j \; \in \; \delta_i \\ 0, & otherwise \end{cases}$$ (9)

Further, construct a diagonal matrix D as follows:

$$D_{i,i} = \sum_{i=1}^{C} W_{i,i} \tag{10}$$

Moreover, a filter is used to get the sampling probability of each point. More details about construct a twin task can be found in [13].

3.3   Mapping Strategy

After matching twin tasks, transfer the knowledge from the twin task to the target task by learning a mapping M. In particular, for the 2D COP, the objective function of learning map from the twin task T_s to the target task T_t is defined as Eq. (11):

$$\min_{M} ||T_s \times M - T_t||_F + \lambda ||G \odot M||_F \tag{11}$$

Where T_s is a $2 \times n_s$ coordinate matrix and T_t is a $2 \times n_t$ coordinate matrix. M is a $n_s \times n_t$ matrix. F is the Frobenius norm and $\odot$ denotes the dot product. In this paper $\lambda = 0.2$. G is a $n_s \times n_t$ matrix and each element g_{ij} is defined as Eq. (12):

$$g_{ij} = (e_{max}^j - e_{ij})/(e_{max}^j - e_{min}^j) \times e_{ij} \tag{12}$$

where e_{ij} is the distance between the ith node in the T_s and jth node in the T_t. e_{min}^j represents the shortest distance between all the nodes in the T_s to the jth node in the T_t. e_{max}^j represents the longest distance between all the nodes in the T_s to the jth node in the T_t. In the M, the weight of the most similar node in the T_s for node in the T_t is largest. Due to use the M depending on specific COP. How to use the M to construct solutions for CVRP is detailed in Sect. 4.

4   Twin Learning Evolutionary Algorithm for CVRP

This section introduces TLEA for CVRP. Specifically, match the twin CVRP for target CVRP firstly. Secondly, construct feasible solutions for target CVRP.

4.1   Match the Twin CVRP for Target CVRP

The procedure of matching the twin CVRP for target CVRP is the same as Subsect. 3.1 and a little different from Subsect. 3.2. In particular, after getting the twin CVRP, we need to add the depot coordinate to the twin CVRP. A sample version considers matching the twin CVRP for target CVRP (Fig. 3).

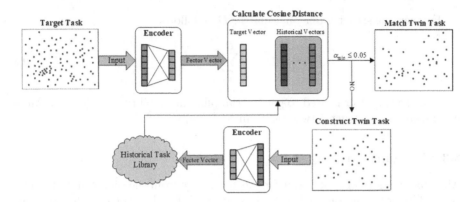

Fig. 3. The flowchart of matching the twin CVRP for target CVRP

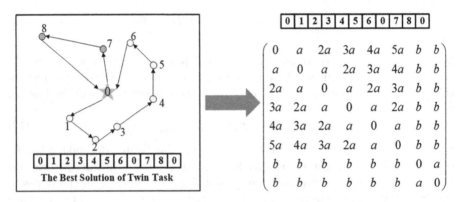

Fig. 4. Illustration of the D_{new}

4.2   Construct Feasible Solutions for CVRP

The procedure of learning mapping for CVRP is a little different from Subsect. 3.3. In particular, before learning the mapping, we need to delete the depot coordinate from the twin CVRP and the target CVRP. After getting the mapping, we need to construct feasible initial solutions for the target CVRP. The process of constructing feasible initial solutions for the target CVRP is detailed below.

In particular, construct a $n_s \times n_s$ distance matrix D_{new} according to the best solution found in the twin task for all customers of the twin task firstly. D_{new} as shown in Fig. 4. a is a small real number and b is a large number. D_{new} is to guarantee the customers served by a common vehicle to be close to each other and away from customers served by a other vehicles. Then, get the new customers representations T_s^{new} of the twin CVRP through the multidimensional scaling with D_{new} [19]. Further, get the new customers representations T_t^{new} of the target CVRP via $T_t^{new} = T_s^{new} \times M$. Finally, use greedy search algorithm and T_t^{new} to construct a feasible initial solution S_M for the target CVRP. To improve the quality of S_M, introduce meta-heuristic algorithms to optimize it.

5 Experiment on CVRP Benchmarks

In this section, six CVRP instances from the CVRP benchmarks with different number of consumers and node distribution are used to verify the performance of the TLEA for CVRP. In particular, E-n101-k8 is designed by Christofides et al. [20]. P-n101-k4 is designed by Augerat et al. [21]. X-n176-k26, X-n143-k7, X-n115-k10 and X-n167-k10 are designed by Uchoa et al. [22]. In this paper, suppose the CVRP library has only six CVRP instances. In Table 1, notice that the cosine distance of the optimal CVRP matched by X-n176-k26 and X-n167-k10 are greater than 0.05, so we reconstruct the twin CVRPs of 50% of the samples of these two tasks, they were named as X-n176-k26-50%, X-n167-k10-50% and their vehicle amount are set to 50% of the original task. The cosine distance between X-n176-k26-50% and X-n176-k26 is 0.013. The cosine distance between X-n167-k10-50% and X-n167-k10 is 0.04. The experimental settings and results are presented and discussed.

Table 1. Similarity of CVRPs

	E-n101-k8	P-n101-k4	X-n176-k26	X-n143-k7	X-n115-k10	X-n167-k10
E-n101-k8	0	0.046	0.089	0.048	0.037	0.093
P-n101-k4	0.046	0	0.057	0.053	0.042	0.056
X-n176-k26	0.089	0.057	0	0.052	0.054	0.183
X-n143-k7	0.048	0.053	0.052	0	0.022	0.181
X-n115-k10	0.037	0.042	0.054	0.022	0	0.131
X-n167-k10	0.093	0.056	0.183	0.181	0.131	0

5.1 Experimental Configuration

We implement the local search algorithm VNS for solving CVRP, which is labeled as EA1. Further, the single task evolutionary algorithm GA is introduced, which is named as EA2. MFEA is introduced as multi-task solver. Finally, we implement adaptive large neighborhood search (ALNS) [23], which is named as EA3.

The search operators of MFEA are kept the same as EA2 for fair comparison. TLEA adopt EA3 as the basic optimizer. The details are given below.

1) EA1
   (a) The perturbation phase : $0.1 \times N$
2) EA2
   a) Population size: 100
   b) Probability of cross: 0.8
   c) Probability of mutate: 0.1

3) MFEA
   (a) Population size: 200
   (b) Probability of mutate: 0.1
   (c) Random mating probability: 0.1
4) EA3
   (a) Degree of destruction: 0.05
   (b) Max string removals: 2
   (c) Max string size: 12
5) TLEA
   (a) Solver: EA3

5.2   Results and Discussion

In this subsection, we compare the results obtained by TLEA and other algorithms.

Solution Quality. Table 2 shows results obtained by EA1, EA2, MFEA, EA3 and TLEA over 20 independent runs. It can be observed that TLEA obtained best solution qualities on all instances, such as "X-n115-k10", "X-n143-k7", "X-n167-k10" and "E-n101-k8". The results of MFEA are better than EA2 shows that knowledge transfer between similar tasks can facilitate all tasks solving. EA1 and EA3 achieved better mean values than EA2 and MFEA. This is because EA1 and EA3 have stronger search ability than EA2.

Table 2. Solution quality of EA1, EA2, MEFA, EA3 and TLEA on CVRP instances

| | EA1 | EA2 | EA3 | MFEA | | TLEA | |
Test case	Mean	Mean	Mean	Similar Task	Mean	Twin Task	Mean
E-n101-k8	893	993	844	X-n115-k10	966	X-n115-k10	**836**
P-n101-k4	726	839	708	X-n115-k10	822	X-n115-k10	**696**
X-n115-k10	14259	21336	13582	X-n143-k7	17091	X-n143-k7	**13207**
X-n143-k7	18637	21797	17326	X-n115-k10	19760	X-n115-k10	**16117**
X-n167-k10	24832	27999	23006	X-n167-50%	26875	X-n167-50%	**21697**
X-n176-k26	50122	301141	50713	X-n176-k26-50%	204481	X-n176-k26-50%	**49908**

Search Efficiency. The convergence profiles of TLEA, MFEA, EA1, EA2 and EA3 are presented in Fig. 5. The y-axis denotes the averaged fitness value in log scale, while the x-axis denotes evaluation times. Since the initial solutions of EA1, EA2, MFEA and EA3 are all randomly generated, it is not conducive to comparison. Therefore, use EA2 to create feasible solutions as the initial solutions of the respective algorithm.

From Fig. 5, it can be noticed that the initial fitness value of EA1, EA2, MFEA and EA3 is very large. Because they start searching from the zero prior knowledge of target CVRP. Since the initial solution of the target CVRP is generated from the best solution of the twin CVRP, TLEA obtains a good convergence and high quality initial solution for the target CVRP. We can find that on "X-n167-k10" and "X-n176-k26", the convergence speed are relatively fast at beginning. Because the scale of twin CVRPs are smaller, some nodes in initial solution are unordered and through a simple search, the solution can be quickly converged. On "E-n101-k8", "P-n101-k4", "X-n115-k10", and "X-n143-k7", the convergence speed are slow. Because their twin CVRPs are larger, most nodes in initial solution are ordered.

In Table 3, we study the performance of TLEA and E-GAT [5] on CVRP benchmarks. E-GAT uses knowledge obtained by solving historical CVRPs to solve a new CVRP, which is similar to TLEA. The results of E-GAT obtained from [5]. In [5], the E-GAT is trained 100 epochs by using instances randomly generated from the unit square. For fair comparison, the fitness evaluation times

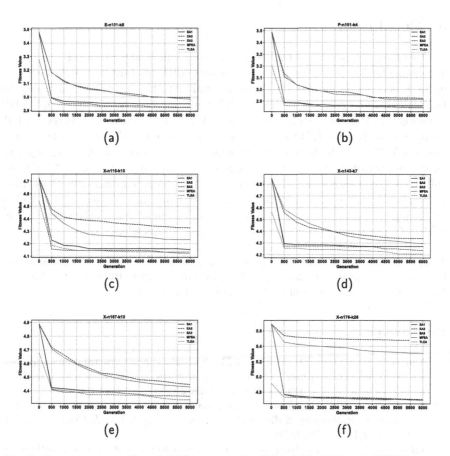

Fig. 5. Convergence profiles of TLEA versus the EA1, EA2, EA3 and MFEA on test cases. (a) E-n101-k8. (b) P-n101-k4. (c) X-n115-k10. (d) X-n143-k7. (e) X-n167-k10. (f) X-n176-k26.

for TLEA is set to 100 and TLEA uses EA1 as the base optimizer. The results obtained by TLEA over 10 independent runs. It can be noticed that TLEA outperforms E-GAT in most instances.

Table 3. Solution quality of TLEA and E-GAT

Test case	TLEA	E-GAT	Optimal values
P-n101-k4	**748**	820	681
X-n106-k14	**27300**	28085	26362
X-n134-k13	12878	**12342**	10916
X-n148-k46	**45834**	50462	43448
X-n157-k13	**17602**	20273	16876

6   Conclusion

In this paper, we propose a twin learning evolutionary algorithm utilizing useful knowledge buried in the historical similarity CVRP for target CVRP. As shown in the experiments, compared with traditional meta-heuristic algorithms and the famous learning based algorithm, TLEA performs best on the most of CVRP instances. TLEA can generate high quality initial solutions for target CVRP. Compared with most ETO algorithms, we consider the similarity between the source CVRP and the target CVRP, which can effectively inhibit the negative knowledge transfer. If there is no source CVRP similar to the target CVRP, we will construct a twin CVRP. However, the knowledge contained in the CVRP has various forms, e.g., search direction, search track and landscape, etc. And in this paper we only did the solution transfer from the source CVRP to the target CVRP. In the future, we will study other forms of knowledge transfer.

Acknowledgements. The work is supported by the National Key Research and Development Program of China (2021YFF0500903, 2022YFE0198900), the National Natural Science Foundation of China (52178271, 52077213).

References

1. Marinelli, M., Colovic, A., Dell'Orco, M.: A novel dynamic programming approach for two-echelon capacitated vehicle routing problem in city logistics with environmental considerations. Transp. Res. Procedia **30**, 147–156 (2018)
2. Tüzemen, A., Yildiz, Ç.: A solution proposal to vehicle routing problem with integer linear programming: a distributor company sample. Int. J. Contemp. Econ. Adm. Sci. **9**(1), 46–78 (2019)
3. Filipec, M., Skrlec, D., Krajcar, S.: An efficient implementation of genetic algorithms for constrained vehicle routing problem. In: SMC 1998 Conference Proceedings. 1998 IEEE International Conference on Systems, Man, and Cybernetics (Cat. No. 98CH36218), vol. 3, pp. 2231–2236. IEEE (1998)

4. Mazzeo, S., Loiseau, I.: An Ant Colony algorithm for the capacitated vehicle routing. Electron. Notes Discrete Math. **18**, 181–186 (2004)
5. Lei, K., Guo, P., Wang, Y., Wu, X., Zhao, W.: Solve routing problems with a residual edge-graph attention neural network. Neurocomputing **508**, 79–98 (2022)
6. Li, J., et al.: Deep reinforcement learning for solving the heterogeneous capacitated vehicle routing problem. IEEE Trans. Cybern. **52**(12), 13572–13585 (2021)
7. Son, D.V.T., Tan, P.N.: Capacitated vehicle routing problem—a new clustering approach based on hybridization of adaptive particle swarm optimization and grey wolf optimization. In: Aljarah, I., Faris, H., Mirjalili, S. (eds.) Evolutionary Data Clustering: Algorithms and Applications. AIS, pp. 111–128. Springer, Singapore (2021). https://doi.org/10.1007/978-981-33-4191-3_5
8. Gupta, A., Ong, Y.S., Feng, L.: Multifactorial evolution: toward evolutionary multitasking. IEEE Trans. Evol. Comput. **20**(3), 343–357 (2015)
9. Kim, M., Park, J., et al.: Learning collaborative policies to solve NP-hard routing problems. Adv. Neural. Inf. Process. Syst. **34**, 10418–10430 (2021)
10. da Costa, P., Rhuggenaath, J., Zhang, Y., Akcay, A., Kaymak, U.: Learning 2-Opt heuristics for routing problems via deep reinforcement learning. SN Comput. Sci. **2**, 1–16 (2021)
11. Osaba, E., Martinez, A.D., Lobo, J.L., Laña, I., Del Ser, J.: On the transferability of knowledge among vehicle routing problems by using cellular evolutionary multitasking. In: 2020 IEEE 23rd International Conference on Intelligent Transportation Systems (ITSC), pp. 1–8. IEEE (2020)
12. Feng, L., et al.: Explicit evolutionary multitasking for combinatorial optimization: a case study on capacitated vehicle routing problem. IEEE Trans. Cybern. **51**(6), 3143–3156 (2020)
13. Wu, J., Yang, H., Zeng, Y., Wu, Z., Liu, J., Feng, L.: A twin learning framework for traveling salesman problem based on autoencoder, graph filter, and transfer learning. IEEE Trans. Consum. Electron. (2023)
14. Amous, M., Toumi, S., Jarboui, B., Eddaly, M.: A variable neighborhood search algorithm for the capacitated vehicle routing problem. Electron. Notes Discrete Math. **58**, 231–238 (2017)
15. Zhang, Y., Wang, J., Zhang, Z.: Edge-based formulation with graph attention network for practical vehicle routing problem with time windows. In: 2022 International Joint Conference on Neural Networks (IJCNN), pp. 01–08. IEEE (2022)
16. Hu, K., Liu, H., Zhan, C., Tang, Y., Hao, T.: Learning knowledge graph embedding with a bi-directional relation encoding network and a convolutional autoencoder decoding network. Neural Comput. Appl. **33**(17), 11157–11173 (2021). https://doi.org/10.1007/s00521-020-05654-4
17. Zeng, Y., Wu, Z., Liu, J., Feng, L.: Point cloud simplification based on decomposed graph filtering. In: 2020 IEEE 18th International Conference on Industrial Informatics (INDIN), vol. 1, pp. 725–728. IEEE (2020)
18. Kramer, O., Kramer, O.: K-nearest neighbors. In: Dimensionality Reduction with Unsupervised Nearest Neighbors, vol. 51, pp. 13–23. Springer, Cham (2013). https://doi.org/10.1007/978-3-642-38652-7_2
19. Carroll, J.D., Arabie, P.: Multidimensional scaling. In: Measurement, Judgment and Decision Making, pp. 179–250 (1998)
20. Christofides, N., Eilon, S.: An algorithm for the vehicle-dispatching problem. J. Oper. Res. Soc. **20**(3), 309–318 (1969)

21. Augerat, P., Naddef, D., Belenguer, J., Benavent, E., Corberan, A., Rinaldi, G.: Computational results with a branch and cut code for the capacitated vehicle routing problem (1995)
22. Poggi, M., Uchoa, E., et al.: New exact algorithms for the capacitated vehicle routing problem (2014)
23. Pisinger, D., Ropke, S.: A general heuristic for vehicle routing problems. Comput. Oper. Res. **34**(8), 2403–2435 (2007)

Research on Full-Coverage Path Planning Method of Steel Rolling Shop Cleaning Robot

Tengguang Kong[1,2(✉)], Huanbing Gao[1,2(✉)], and Xiuxian Chen[1,2]

[1] School of Information and Electrical Engineering, Shandong Jianzhu University,
Jinan 250101, China
1104457815@qq.com, gaohuanbing2004@sdjzu.edu.cn
[2] Shandong Key Laboratory of Intelligent Building Technology, Jinan 250101, China

Abstract. As a large indoor unstructured environment, the steel rolling shop has many unknown factors that greatly affect the full-coverage cleaning path planning of the cleaning robot. In this paper, a round-trip full-coverage path planning algorithm is improved by using the breadth-first search (BFS) algorithm combined with the improved A* algorithm to complete the full-coverage cleaning task of the steel rolling shop by the cleaning robot. First, the obstacles in the raster map are inflated using the expansion strategy to prevent the cleaning robot from colliding with the obstacles; second, the map is traversed with full coverage by the BFS algorithm; finally, the round-trip full-coverage algorithm is used for path planning based on the traversal information, and when it enters the dead zone location, the improved A* algorithm is used to find the unplanned nodes and plan an optimal path to escape from the dead zone. The simulation results show that the improved round-trip full-coverage path planning algorithm is more efficient than the traditional full-coverage path planning algorithm, and the proposed method can solve the complex problems in unstructured environment and complete the full-coverage cleaning task of the steel rolling shop cleaning robot.

Keywords: Cleaning robot · Complete coverage path planning · Breadth-first search · A* algorithm

1 Introduction

Since entering the 21st century, the majority of scientific and technological workers in the process, equipment, products and other aspects of technological innovation, and gradually solve the major key technologies and common technical problems that limit the development of steel rolling technology, to promote the leapfrog development of the steel rolling industry. However, the increase in steel rolling output is accompanied by the generation of large amounts of dust. In the face of large amounts of coal dust and iron oxide dust in the workshop, in order to ensure the normal operation of the production line, the dust must be cleaned in a timely manner to complete the cleaning task of covering the entire area of the steel rolling workshop. Cleaning robots are widely used in full-coverage operations in the fields of intelligent cleaning, farming operations, and military detection because of their capabilities of environmental perception, behavioral decision-making and motion control [1]. In this paper, we study the full-coverage path planning problem for cleaning robots in large indoor unstructured environments, and use the full-coverage path planning algorithm to reduce the path repetition rate and improve the efficiency of cleaning robots.

The complete coverage path planning algorithm, abbreviated as CCPP problem, refers to obtaining a shortest path that travels everywhere except for obstacles in a certain area or space. In recent years, scholars at home and abroad have conducted a lot of research around the problem of full-coverage path planning for robots in specific work areas [2, 3]. Choi et al. [4] proposed an online full-area coverage algorithm based on an inward spiral approach and restricted inverse distance conversion; Kapanoglu et al. [5] used the template method combined with genetic algorithms for path planning to achieve a full-coverage path planning problem for robots; Kai Li [6] used the West-Move First algorithm to achieve local area coverage, and used the improved A* algorithm to plan a smooth and obstacle-free path from the dead point to the backtracking point. Hao Zongbo et al. [7] proposed the internal spiral coverage algorithm ISC (Internal Spiral Coverage); Qiu Xuena et al. [8] improved the bio-inspired neural network algorithm and applied it to full coverage of paths in uncertain dynamic environments.

The above-mentioned mobile robot path planning studies are very inspiring for the development of full-region coverage path planning algorithms, but all of their algorithms have some shortcomings in today's mobile robot working environment with increasing complexity.

For example, the algorithm has high repetition rate of covering paths in unstructured environments with multiple types of obstacles, low search efficiency, etc. To address the problems of the above algorithms, this paper proposes an improved round-trip full coverage path planning algorithm using breadth-first search (BFS) algorithm, which first covers the full area using breadth-first search algorithm based on the unstructured environment information in the steel rolling shop, and then performs full area path planning by round-trip path planning algorithm based on the environment information obtained from full area coverage, and when entering the dead zone position, the improved A* algorithm is used to escape from the dead zone position, and finally the full-area coverage task is completed, thus realizing the cleaning task of the cleaning robot.

2 Brief Description of the Algorithm Principle

2.1 Map Environment Modeling

In large unstructured environments, it is necessary to create environment maps in order for the cleaning robots to achieve a full area traversal effect. The commonly used map environment modeling can be broadly classified as raster maps, topological maps, viewable maps [8]. Since the raster map is binary in nature, it is more intuitive to represent the coverage area and obstacle areas compared with other map building methods. Therefore, this paper uses the raster map method to partition the environment space of the steel rolling shop and build a raster map. When the raster map is established, the robot is often considered as a mass point, and in order to prevent the robot edge from collision, the obstacle is expanded according to the shortest distance between the center of the cleaning robot and the edge of the body, and the obstacle that is less than one raster is calculated by one raster, and the unit length of the cell in the raster map is set according to the working range of the cleaning robot. The results of environment rasterization and obstacle expansion are shown in Fig. 1, A, B1, C, D, E, F and G denote the obstacles after expansion treatment; B2 denotes the initial state of obstacle B1 before expansion treatment.

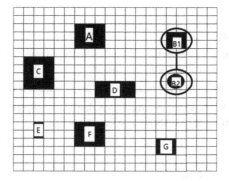

Fig. 1. Raster map obstacle expansion treatment

2.2 Breadth-First Search Algorithm

This section focuses on full-coverage traversal of the map environment, and the traversal order planning is based on the breadth-first search algorithm hereafter referred to as BFS. The BFS algorithm is one of the best-known search algorithms in graph theory that finds an exhaustive coverage path to travel. With the input of a graph G (V, E) consisting of a vertex set V and an edge set E and a search starting point, the BFS algorithm first visits the search starting point, then visits the unvisited neighbors of the search starting point $v_1, v_2, v_3, \ldots, v_n$, and then visits the unvisited neighbors of each vertex in the order of $v_1, v_2, v_3, \ldots, v_n$, until The algorithm ends when all the vertices in the graph that have paths to the starting point are visited [9]. The traversal connectivity graph is shown in Fig. 2.

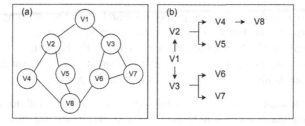

Fig. 2. Traversing the connectivity diagram (a) Connected graph (b) Traversal sequence graph

In the BFS algorithm, since all the vertices of the graph may be connected to other vertices, to avoid going back to the vertices that have been visited before, an auxiliary array visited[n] is set to mark each vertex that has been visited. Table 1 shows the algorithm steps.

Table 1. BFS algorithm step

Steps	Content
Input	Connected Graph G
Step 1	Select the starting cell v, set the value of visited[v] to true, insert it into the traversal list and mark it as visited, then put v into the queue
Step 2	If the queue is not empty, the head element is queued out and set to u
Step 3	Check all neighbors w of u in turn, if the value of visited[v] is false, visit w and set the value of visited[w] to true, then put w into the queue
Step 4	Repeat steps 2 and 3 above until all cells in the connectivity diagram have been accessed
Output	Full-region traversal list

2.3 A* Algorithm

The A* algorithm is a heuristic search algorithm [10], which guides the search direction of the algorithm by means of a heuristic function. It is often applied to find the optimal solution of a path under a static map, which searches in the state space according to the cost function to select a shortest path from the starting point to the target point. The formula for the cost function of the conventional A* algorithm is expressed in Eq. 1 as:

$$f(n) = g(n) + h(n) \tag{1}$$

where f(n) is the estimated cost from the initial state to the target state via state n, g(n) is the actual cost from the initial state to state n in the state space, and h(n) is the estimated cost from state n to the target state. Expressing the estimated cost of the optimal path from state n to the target state in terms of h*(n), then h(n) is chosen in three broad ways as follows:

(1) h(n) < h*(n), in this case, more points are searched, the search range is large, and the efficiency is low. But the optimal solution can be obtained.
(2) h(n) = h*(n), in which case the search efficiency is the highest.
(3) h(n) > h*(n), the number of points searched is small, the search range is small, and the efficiency is high, but the optimal solution is not guaranteed.

For the path search problem, the commonly used ranging methods are Euclidean distance, Manhattan distance, Chebyshev distance, etc. In this paper, according to the actual cleaning needs of cleaning robots, only the four-directional A* algorithm is considered here. The so-called four-direction means that it can move up and down, left and right, but does not allow diagonal movement. Therefore, the exploration distance cost is calculated using Manhattan distance to mark the absolute axis distance sum of two points on the standard coordinate system. The Manhattan distance in the 2-dimensional plane is the distance between two points on the vertical axis plus the distance on the horizontal axis, and the formula is shown below:

$$d(x, y) = |x_1 - x_2| + |y_1 - y_2| \tag{2}$$

where d(x, y) for Manhattan distance, (x1, y1) and (x2, y2) correspond to the starting and ending coordinate positions, respectively.

The principle of A* algorithm path planning starts from the starting point, updates the neighboring nodes of the current node in turn, and uses the neighboring node with the lowest weight as the new starting node until it traverses to the target node [11]. A path to escape the dead zone is obtained by solving the A* algorithm, avoiding obstacles to find nearby uncovered nodes, and continuing the path planning based on the coverage rules, but this path has the problem of many turns and low search efficiency, so the path needs to be optimized [12].

3   Improved Algorithm for Full Coverage Path Planning in Clean Areas

3.1   Improved A* Obstacle Avoidance Path Planning

Since the cleaning robot needs to cover all areas to complete the floor cleaning task, it is easy in entering the dead zone area when performing full coverage path planning. Therefore, A* algorithm is chosen to plan the optimal path and escape from the dead zone location. And the path planned by the traditional A* algorithm has problems such as the number of turns and low search efficiency, so this paper optimizes the traditional A* algorithm according to the needs of the cleaning robot. The constraints in the traditional A* algorithm are only obstacles and the shortest path, and the estimated cost h(n) for the best path from state n to the target state in the traditional A* algorithm is optimized, and the weight of h(n) is increased on the basis of this coefficient Weights, and control the search direction of A* by setting the coefficient size to improve the speed of searching uncovered nodes. The improved A* algorithm estimation function is:

$$f(n) = g(n) + w(n) \cdot h(n) \tag{3}$$

The improved A* algorithm increases the weight coefficient w(n), w(n) can influence the evaluation value, reduce the number of expansion nodes in the search process, and improve the search efficiency in the path search process by flexibly adjusting w(n) according to the actual condition of the cleaning robot's cleaning environment.

For the above-mentioned A* algorithm path planning for fleeing dead-end areas is theoretically optimal path, and in the actual cleaning robot to avoid obstacles to continue cleaning operations will produce too many turns, increasing the possibility of robot collision, so to avoid unnecessary turns and collisions in the premise of ensuring the optimal path to optimize the corner processing. The specific idea is shown below:

(1) Firstly, the index value of the current parent node is used to calculate the node where the straight line is expected to be expanded. If the parent node position information is L, it means that the current parent node is obtained from the expansion of the point on the left, then the desired expansion straight line node is on the right of the parent node. If the cost of the current expanded node is equal to the cost of the desired expanded node, the optimization correction is made, otherwise the correction is skipped for the next step.

(2) If the current expansion node is the same as the node of the desired expansion line, otherwise skip the correction and proceed to the next step.

The steps of the improved A* algorithm are shown in Table 2:

Table 2. Steps of improved A* algorithm

Steps	Content
Input	Create raster maps
Step 1	Inflation of unstructured obstacles to obtain a new map model
Step 2	Optimal path planning using improved A* algorithm
Step 3	Optimization of path corners
Step 4	Generate a path to the optimal escape zone
Output	Output continuous and unobstructed optimal path

First, according to the above, a 20 × 20 raster map is established, the expansion strategy does expansion on obstacles, and the starting and target points are set in the maps with few obstacles and many obstacles, respectively, and the expansion nodes are marked in the expansion process, where black represents obstacles, green represents the starting position, yellow represents the end point, and the black straight line represents the generated optimal path [13]. The w(n) weights are flexibly adjusted according to the actual needs of the cleaning robot.

Based on the above algorithm ideas, simulation experiments were carried out using Matlab, and the comparison between the traditional A* algorithm Fig. 3(a) and the improved A* algorithm Fig. 3(b) in the simulation results shows that when w(n) takes a suitable value, the improved A* algorithm is significantly reduced in the number of expansion nodes compared to the traditional A*, and the search rate is more advantageous compared to the traditional A*, and the corners of the actual cleaning robot cleaning path are optimized when avoiding obstacles and planning the shortest path after entering the dead-end area, which reduces unnecessary turns and avoids the possibility of collision in the cleaning process.

3.2 Improved Round-Trip Full-Coverage Path Planning Algorithm

The path planning without environment model is divided into random full-area coverage and planned full-area coverage path planning, and the planned full-area coverage

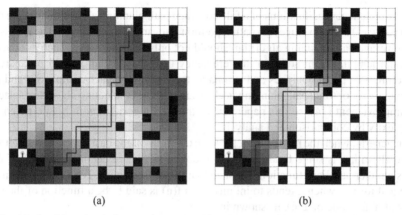

(a)   (b)

Fig. 3. A* algorithm simulation results (a) Traditional A* algorithm (b) Improved A* algorithm

path planning method has many practical applications and mature technology, which generally includes Fig. 4(a) round-trip path planning and Fig. 4(b) internal spiral path planning, and the cleaning rules of round-trip path planning are: first, the cleaning robot is placed in a corner of the room; then, along a Then, it travels in a certain direction; when it encounters an obstacle, it adopts an obstacle avoidance strategy, and when it encounters a wall, it moves a body to turn around and continue to travel, and so it moves back and forth to traverse the whole environment [8]. However, due to the lack of information from the environment model, the path obtained by the round-trip path planning algorithm has disadvantages such as low coverage and low planning efficiency. Therefore, in this paper, based on the traditional round-trip path planning algorithm, the BFS algorithm is used to traverse the cleaning area of the cleaning robot with full coverage in advance, and then use the rules of the round-trip path planning algorithm combined with the traversal information to plan the full-coverage path of the cleaning area through the map model state information obtained after the traversal.

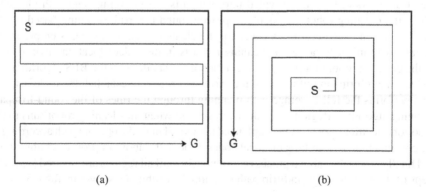

(a)   (b)

Fig. 4. Traditional full-coverage path planning (a) Round-trip path planning (b) Spiral path planning

The technical specifications of a full-coverage path planning algorithm include three main parts: (1) area coverage, (2) path repetition rate, and (3) total travel. The merits of a full-coverage path planning algorithm should not only consider issues such as coverage and path length, but in addition, the efficiency of the algorithm should be analyzed. Therefore, this paper analyzes and compares the algorithm space complexity and time complexity for optimal path planning in clean areas while conducting algorithm selection and improvement [14], so this paper selects the BFS algorithm with high efficiency in complexity by comparison and analysis.

In general, when the number of repeated executions of the basic operation statements of the algorithm is some function of the problem size, here denoted by T(n); if there exists another auxiliary function f(n) such that the limit value of T(n)/ f(n) is a constant not equal to zero when n tends to infinity, then f(n) is said to be a function of the same order of magnitude of T(n), as shown in Eq. 4:

$$T(n) = O(f(n)) \tag{4}$$

Then the above equation O (f(n)) is called the asymptotic time complexity of the algorithm, or the time complexity for short.

Regarding the storage space requirement of the algorithm, similar to the time complexity of the algorithm, we adopt the asymptotic space complexity as a measure of the storage space required by the algorithm, referred to as the space complexity, which is also a function of the problem size n, as shown in Eq. 5:

$$S(n) = O(f(n)) \tag{5}$$

The proposed breadth-first search algorithm has a time complexity of $O(n + e)$ when stored using the adjacency matrix table and a space complexity of $O(n)$ when borrowed from the queue, where n is the number of elements and e is the number of scanned nodes, making the BFS algorithm more efficient in practical cleaning robot cleaning applications.

Through the above analysis, it is known that the full-coverage path planning of the cleaning robot is mainly divided into two parts: area coverage and path planning. When the cleaning robot adopts the improved round-trip path planning based on the BFS algorithm proposed in this paper for the cleaning task, its model is empty in the initial environment because the cleaning robot is in the indoor location environment, so the cleaning robot first performs the region traversal through the BFS algorithm and collects the information of nodes in the traversal process to establish the node list, and then performs the full-coverage path planning through the rules of the round-trip path planning algorithm. Region for path coverage, and when the cleaning robot runs to the dead zone position, it obtains the unplanned nodes and finds the optimal path according to the cost function by searching the local nodes through the above improved A* algorithm, and finally repeats the above process continuously until all regions are covered [15]. The steps of the BFS-based round-trip path planning algorithm are shown in Table 3:

Table 3. Steps of complete coverage path planning algorithm

Steps	Content
Input	Raster map model
Step 1	Set the initial position and start full-coverage traversal using the BFS algorithm
Step 2	Create PL table to store traversal node information
Step 3	Full-coverage planning based on traversal results, using the priority order of motion direction to make a turn when an obstacle is encountered until it enters the dead zone position
Step 4	Determine the location of uncovered nodes using the improved A* algorithm and generate the optimal path to escape the dead zone
Step 5	Loop through steps 3 ~ 4 until the PL table is empty, then terminate the loop
Output	Output continuous and unobstructed full coverage path

4 Simulation and Data Analysis

To verify the feasibility of the above algorithm when applied to a cleaning robot for cleaning tasks in a steel rolling shop, the full-coverage path planning algorithm was simulated and analyzed using environment modeling via Matlab to create a 20 × 20 raster map [16]. Figure 5 shows the simulation comparison results obtained in any two unstructured indoor environments. In Fig. 5, green represents the custom starting position, yellow represents the end position, blue represents the repeating nodes, and black straight lines indicate the full-coverage planning path.

In this paper, the performance of the full-area coverage algorithm is analyzed by simulation comparison, and the technical metrics used in the analysis are mainly four metrics such as the total length of the full-area coverage path of the cleaning robot, the number of repeated nodes, the path repetition rate, and the area coverage rate [6]. A comparison of the simulation analysis of the two algorithms is shown in Table 4.

The simulation results show that the proposed algorithm has fewer repetitive nodes and lower repetitive coverage rate than the traditional round-trip path planning algorithm in the same map environment, and thus the total length of the path is shorter, so the algorithm can enable the cleaning robot to achieve 100% coverage of the cleaning area in the steel rolling shop and complete the dust cleaning task in the shop in a complex unstructured environment.

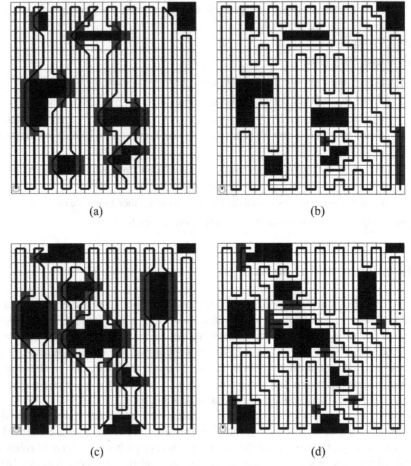

(a)

(b)

(c)

(d)

Fig. 5. Simulation results of path planning for full area coverage (a) Simulation result of traditional algorithm 1 (b) Simulation result of this paper 1 (c) Simulation result of traditional algorithm 2 (d) Simulation result of algorithm in this paper 2

Table 4. Simulation results data comparison

	Traditional Algorithms		Algorithm in this paper	
	Figure 4(a)	Figure 4(c)	Figure 4(b)	Figure 4(d)
Total path length	420.3	406.3	362.0	347.0
Number of repeating nodes	53	62	7	16
Repeat coverage percentage	12.6%	15.3%	1.9%	4.6%
Coverage percentages	99.75%	98.19%	100%	100%

5 Conclusion

In this paper, for the cleaning robot sweeping road task in the complex unstructured environment of steel rolling shop, the traditional round-trip path planning algorithm is used to search for slow speed, high repeat coverage rate of path and poor obstacle avoidance effect, etc. The BFS algorithm is used to traverse the whole region and then use the round-trip path planning algorithm motion rules to cover the whole region through the four directions of front, back, left and right, and when entering the dead zone When entering the dead zone position, the improved A* algorithm is used to avoid obstacles and then escape from the dead zone position, finally realizing the path coverage of the whole region. According to the above simulation result 1 and simulation real result 2, the path length is reduced by 13.8% and 14.6%, and the number of repeated nodes is reduced by 86.8% and 74.2%, respectively, compared with the traditional algorithm. For the full-coverage path planning problem of the cleaning robot in the steel rolling shop, the full-coverage traversal speed full problem is solved by the BFS algorithm, and the full-coverage path planning is performed in the unstructured environment under the round-trip path planning algorithm combined with the optimized A* algorithm, and the simulation results show that the algorithm in this paper has good stability, and the repetition rate is greatly reduced compared with the traditional algorithm and the coverage rate can reach 100%. Coverage. It enables the cleaning robot to better adapt to the complex unstructured rolling mill environment and complete the cleaning task of the cleaning area.

Acknowledgments. The above work was partially supported by the Natural Science Foundation of Shandong Province (ZR2022MF267): Research on the method of road condition identification and friction estimation in autonomous driving. We would also like to acknowledge the support from the National Natural Science Foundation of China (No. 61903227).

References

1. Xu, B., Xu, M., Chen, L.-P., et al.: Review on coverage path planning algorithm for intelligent machinery. Comput. Meas. Control **24**(10), 1–5+53 (2016)
2. Tang, D.-L., Yuan, B., Hu, L., et al.: Complete coverage path planning method for oil tank inspection wall climbing robot. Chin. J. Eng. Des. **25**(3), 253–261 (2018)
3. Li, W.-L., Zhao, D.-H.: Complete coverage path planning for mobile robot based on grid method and neuronal. Mach. Des. Manuf. **8**, 232–234 (2017)
4. Choi, Y.H., Lee, T.K., Baek, S.H., et al.: Online complete coverage path planning for mobile robots based on linked spiral paths using constrained inverse distance transform. In: Proceedings of IEEE/RSJ International Conference on Intelligent Robots and Systems, pp. 5788–5793. IEEE, Missouri (2009)
5. Kapanoglu, M., Ozkan, M., Yazici, A., et al.: Pattern-based genetic algorithm approach to overage path planning for mobile robots. Springer, Berlin (2009)
6. Li, K., Chen, Y.F., Jin, Z.Y., et al.: A full coverage path planning algorithm based on backtracking method. Comput. Eng. Sci. **41**(7), 1227–1235 (2019)
7. Hao, Z.-B., Hong, B.-R., Huang, Q.-C.: Study of coverage path planning based on grid-map. Appl. Res. Comput. **24**(10), 56–58 (2007)

8. Qiu, X.-N., Liu, S.-R., Yu, J.-S., et al.: Complete coverage path planning of mobile robots biologically inspired neural network and heuristic template approach. Pattern Recogn. Artif. Intell. **19**(01), 122–128 (2006)
9. Zhao, X.-D., Bao, F.: Survey on cleaning robot path planning algorithm. J. Mech. Electr. Eng. **30**(11), 1440–1444 (2013)
10. Xu, Q.-Z., Han, W.-T., Chen, J.-S., et al.: Optimization of breadth-first search algorithm based on many-core platform. Comput. Sci. **46**(01), 314–319 (2019)
11. Gao, F., Zhou, H., Yang, Z.-Y.: Global path planning for surface unmanned vessels based on improved A* algorithm. Appl. Res. Comput. **37**(S1), 120–121+125 (2020)
12. Wang, M.-C.: Path Planning of Mobile Robots Based on A* algorithm. Shenyang University of Technology, Shenyang (2017)
13. Wang, H.-W., Ma, Y., Xie, Y., et al.: Mobile robot optimal path planning based on smoothing A* algorithm. J. Tongji Univ. (Nat. Sci.) **38**(11), 1647–1650+1655 (2010)
14. Tang, B., Zhang, D., Peng, T.-L., Tang, D.-W.: Research on path planning of mobile robot based on improved A* algorithm. Mech. Manage. Dev. **38**(01), 74–77 (2023)
15. Liu, X.-D., Yu, R.-F., Li, P.-C., et al.: Research on time complexity of randomized quick select algorithm. Comput. Digit. Eng. **46**(02), 256–259+280 (2018)
16. He, L.-L., Liu, X.-L., Huang, T.-Z., et al.: Research on the complete-coverage path planning algorithm of mobile robot. Mach. Des. Manuf. **361**(03), 280–284 (2021)

Deep Reinforcement Learning Method of Target Hunting for Multi-agents with Flocking Property

Yujiao Dong[1], Jin Cheng[1(✉)] (iD), and Qiqi Wang[2]

[1] School of Electrical Engineering, University of Jinan, Jinan 250022, China
`cse_chengj@ujn.edu.cn`
[2] Qilu Hospital of Shandong University, Jinan 250012, China

Abstract. A multi-agent deep deterministic policy gradient method is proposed for the target hunting problem in this paper. The behaviours of target hunting and flocking are trained with the action experiences under random environment while the multi-agents explore the environment randomly. A danger zone model is designed to construct a reward function module for the hunting behaviour of the target, and an improved artificial potential energy function is designed to impel the agents to evolve to the flocking state. To evaluate the proposed method, experiment is simulated with gym. The experimental results show that the multi-agent hunting of dynamic targets with flocking property is successfully realized.

Keywords: Target Hunting · Multi-agent Deep Deterministic Policy Gradient · Danger Zone Model · Artificial Potential Energy

1 Introduction

Research on multi-agents has clearly shown that the collective ability of a group is becoming increasingly important, surpassing the significance of individual abilities. Individuals give feedback to the group system, which makes group system become more and more intelligent. In modern military warfare [1], where a single military force is much lower than a group military force, multiple drones cooperate against a single drone, which can improve survival. Individuals in a group can work together to achieve more efficient tasks. In practical applications of modern agriculture [2], multiple agricultural robots can irrigate and seed a large area according to the planned farmland path. Deep reinforcement learning(DRL) algorithm is also applied to the practical application of multi-robots. The main feature of deep reinforcement learning [3] is to integrate the two learning methods of deep learning and reinforcement learning, such as extracting corresponding pixel feature parameters from a given large amount of image data, and passed into the convolutional neural network for image classification. A multi-robot cooperative algorithm based on improved double neural network structure is proposed in the literature [4], which can take the relative angle between robots as

H. Zhang et al. (Eds.): NCAA 2023, CCIS 1869, pp. 467–478, 2023.
https://doi.org/10.1007/978-981-99-5844-3_34

the input of convolutional neural network to train the behaviour of robots. In [5], a path optimization method for unmanned aerial vehicles a behavioural threat framework are designed according to the actual environment, and then a grey wolf algorithm model with constraints is established to control the unmanned aerial vehicles.

Target hunting has become an emerging research hotspot in the past 5 years. A local cooperative control method based on output feedback linearization is proposed in [6]. The relative distance information and angle information between robots and neighbouring nodes and targets, as well as the robot's own azimuth information ware used to design a cooperative controller to tail the target. In [7], aiming at the target encirclement problem of robot system, a target hunting control method based on reinforcement learning is proposed, and the effectiveness of target hunting control based on reinforcement learning method is verified by simulation.

With the growing maturity of the flocking technology, the multi-agent flocking technology has been applied to the target capture design. Multiple-agents keep the state of the cluster moving to hunt the target, which greatly improves the group strength in the confrontation and reduces their own losses and improves the efficiency of target capture. In [8], Pandey et al. utilized image extraction technology to extract image data from videos. They manually annotated the shapes of birds in flight within the images. The annotated image features are fed into a neural network to update the strategy, enabling multi-agents to learn the optimal formation of birds and achieve a swarm state. In [9], the performance of the traditional state-action-reward-state'-action'(SARSA) algorithm was evaluated in flocking control, which founds that the slow search action of the agent in the early stage of the training process can easily lead to unsatisfactory results in the later stage. At the same time, the Deep Q-network (DQN) algorithm is evaluated, showing that the greedy algorithm is used to estimate the action based on the maximum Q value, which is prone to overestimate the action and cause excessive deviation. In [10], an experience pool storage mechanism was proposed to improve the performance of the agent in dynamic random exploration. During training, the model can extract the required data from the experience pool at any time, avoiding the delay in the training process. The insufficient and unreasonable use of the empirical database has a great impact on the convergence of the model, which can easily lead to the omission of a large amount of edge data in the experience pool. Cao et al. introduced a high-value experience priority playback mechanism in [11], which boosts the importance of high-value experiences to facilitate multiple learning for the agent. This approach allows the agent to fully leverage high-value experiences to compensate for any deficiencies in learned experience. This way significantly increases the learning rate in the task, so that the multi-agents can execute the flocking hunting strategy with the optimal action of exploration. In [12], a multi-agent deep deterministic policy gradient (MADDPG) algorithm with shared state parameters was proposed for the research of behaviour confrontation of multi-agents. It improves the optimization of pursuit actions by transferring the state parameters of pursuers to each other, so that the escapes can be successfully rounded up in a small-scale training ground. To enhance the targeted hunting of agents in a continuous environment, a Gym environment suitable for reinforcement learning algorithms is utilized to construct a multi-agent learning framework [13]. This framework enables seamless interaction between agents and the environment, facilitating the acquisition of necessary neural network input information for model training.

The contents of this paper are as follows. Firstly, the deep deterministic policy gradient (DDPG) algorithm model of multi-agent is established, enabling agents to learn better target hunting actions in flocking form by training the model. Then, the dangerous area model is established to make the agents learn the hunting strategy, and the improved artificial potential energy function is employed to make the multi-agent target hunting in flocking state. Finally, the target hunting behaviour of multi-agent is simulated by using Gym environment, and the experimental results and conclusions are obtained.

2   Reinforcement Learning Process

In reinforcement learning, the mapping relationship between the learning state and behaviour of agents is designed to maximize reward. The markov decision process(MDP) of the agent is shown in Fig. 1.

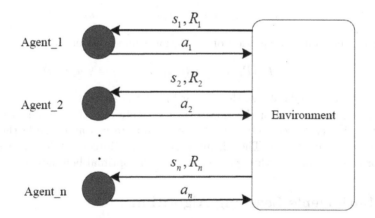

Fig. 1. The MDP process of agents.

At first, the agents do not know which action to select, and thus they can only explore different behaviours through repeated trials to identify the one that yields the highest reward. MDP can simplify the reinforcement learning model. The probability of agent to select an action a in state s and moves to the next state $s^{'}$ is expressed as $P^a_{ss^{'}}$. The probability of the agent for the next state is only related to the previous state, which is expressed as:

$$P^{a_t}_s = E(s_{t+1} = s^{'} | s_t = s, a_t = a) \tag{1}$$

In addition, the probability of action a taken by the agent in state s is only related to state s, and the adopted strategy is:

$$\pi(a \mid s) = P(a_t = a \mid s_t = s) \tag{2}$$

The value function of agent state at time t is

$$v_\pi(s) = E_\pi \sum_{i=t}^{n-1} (\zeta^{i-t} R_{i+1} \mid s_t = s) \tag{3}$$

where ζ is the discounted value. The selected actions should pay attention to the value of actions that are not only considering the rewards at the current moment, but also paying attention to the follow-up delay rewards. If agents want to get the maximum return action value, it needs to combine the sum of continuous state rewards made by the agent.

The action value function is defined as:

$$q_\pi(s) = E_\pi \sum_{i=t}^{n-1} (\zeta^{i-t} R_{i+1} \mid s_t = s, a_t = a) \tag{4}$$

There is a correlation between the state s_t time t and the state s_{t+1} time $t+1$, and it can be concluded that:

$$v_\pi(s) = E_\pi(R_{t+1} + \zeta v_\pi(s_{t+1}) \mid s_t = s) \tag{5}$$

Therefore, the Bellman equation of the action value function is:

$$q_\pi(s, a) = E_\pi(R_{t+1} + \zeta q_\pi(s_{t+1}, a_{t+1}) \mid s_t = s, a_t = a) \tag{6}$$

The above formula shows that the value of the current action is composed of the attenuation ratio of the reward in this state and the subsequent action reward, which can effectively get a greater value return and enable the agent to select a better action. The ultimate goal of reinforcement learning can be transformed into finding the optimal value of the optimal bellman equation.

3   Multi-agents Learning Algorithm

In order to make multi-agents learn flocking hunting action, MADDPG algorithm is used to train agents. According to need of hunting the target environment, the algorithm adopts the policy of overall training and distributed control of agents. Algorithm model is shown in Fig. 2.

The main component of the algorithm lies in the actor-critic framework structure, in which the actor structure makes the next exploration of the environment according to the current state of the agent. The critic structure evaluates the actions and states obtained by the agent exploration to obtain the action value. Based on the action value and loss function, the model performs back propagation and updates the neural network parameters, and obtains the better action

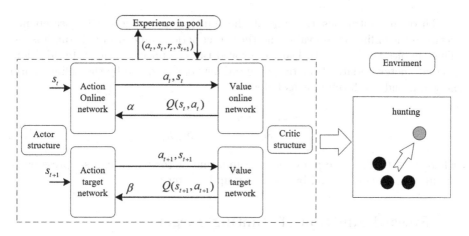

Fig. 2. MADDPG algorithm framework.

of rounding up the target. The model then references the action value and loss function again to perform back propagation and further update the neural network parameters, resulting in an enhanced hunting action for the target.

Figure 2 is essentially an agent's action optimization training process. The action online network takes the current state s as input and generates the corresponding action a_t for the current time. And the value online network evaluates the online action and the state s to obtain the value $Q(s_t, a_t)$. The action online network updates the neural network parameters α through back propagation based on the obtained action value. The corresponding formula for the loss gradient is as follows:

$$\bigtriangledown_J(\alpha) = \frac{1}{n}\sum_{j=1}^{n}(\bigtriangledown_a Q(s_t, a_t, \alpha)|_{s=s_t, a=a_t} \bigtriangledown_\alpha \pi_\alpha(s)|s = s_t) \tag{7}$$

The action target network utilizes the state s at the next moment to calculate the next action s_{t+1} of the agent, and the value of the value target network is:

$$y_t = R_t + \zeta Q(s_{t+1}, a_{t+1}, \beta_{t+1}) \tag{8}$$

where β_t is the value online network parameter.

The loss function equation of the value online network is:

$$L = \frac{1}{n}\sum_{j=1}^{n}(y_t - Q(s_t, a_t, \beta_t))^2 \tag{9}$$

where n is the number of experiences taken from the experience pool during the training process.

Through continuous training of the multi-agent models, the discrepancy between the online action value and the target action value gradually diminishes. This enables the agents to learn improved target hunting actions. In addition, the parameter update of the target neural network is copied by the online neural network, and the formula is as follows:

$$\alpha_{t+1} = m\alpha_t + (1 - m)\alpha_t$$
$$\beta_{t+1} = m\beta_t + (1 - m)\beta_t \tag{10}$$

where $m \in (0, 1)$ is the parameter of updating factor which can suppress overfitting cooperation caused by too many update steps.

4   Reward Function of Hunting Target

The multi-agent model evaluates the loss value of actions according to the reward function module, and then feeds back to the system to update the strategy of hunting the target.

4.1   Reward Function of Danger Zone Model

In an unknown environment, a danger perception zone model is established for pursuers and escapers. Within a certain range, the escaper can perceive the presence of the pursuer and take corresponding actions to escape. Additionally, the pursuer can also sense the presence of the escapers within its sensing area in order to encircle and capture them. This allows for the establishment of a target escape mechanism. As shown in Fig. 3.

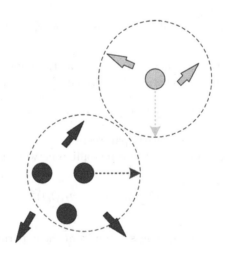

Fig. 3. The danger zone model.

In the above picture, the red circle represents the escaper, the blue circle represents the pursuer, the arrow represents the action of the agent, and the dotted line represents the perception range. Each agent has its own sensing area. In the picture, agents explore randomly in unknown areas, and do not make pursuit or escape actions.

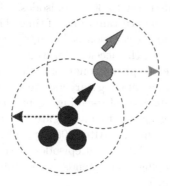

Fig. 4. The danger zone model.

In Fig. 4, escapers and pursuers sense each other and make corresponding actions. The reward equation for both the escaper and the pursuer is set to:

$$rew_1 = \omega_1 \parallel p_{escaper} - p_{pursuer} \parallel_2 \tag{11}$$

In the above formula, p is the position of the agent in the unknown environment, and ω_1 is the proportional size of the reward.

4.2 Establishment of Flocking Reward Function

In order to achieve the flocking effect of agents in hunting moving targets, an improved artificial potential energy reward function is used to maintain the distance between pursuers, namely:

$$rew_2 = \omega_2 \left(\frac{H}{\parallel d_{pp} \parallel^2} + log_2 \parallel d_{pp} \parallel^2 \right) \tag{12}$$

where H is the proportionality coefficient, and d_{pp} is the distance between pursuers.

5   Simulation Results and Analysis

The experiment of hunting moving the target is simulated in Gym environment. Firstly, the map is constructed and simulated in a map with a scale of 10 cm*10 cm. The pursuers and escapers set the initial positions. Then, the learning rate of the model is set to 0.01, and the step size is selected to be 80. The number of training rounds is set to 30000 episodes, which can effectively make agents learn better actions in a certain range of time. The maximum speed of the escaper is 15 cm/s and the acceleration is 15 cm/s^2, while the maximum speed of the pursuer is 10 cm/s and the acceleration is 20 cm/s^2.

To better demonstrate the phenomenon of multi-agents maintaining a flocking behaviour while hunting moving targets, we set the number of pursuing agents to three and the number of escaping agents to one. This configuration allows the three pursuing agents to investigate the clustering pattern of the escaping agent in the environment. After 2000 episodes of training, the rewards of agents tend to be stable, After 2000 episodes of training in the multi-agent model, the reward of escaping agents tends to be stable, and the reward curve of escape0 is shown in Fig. 5.

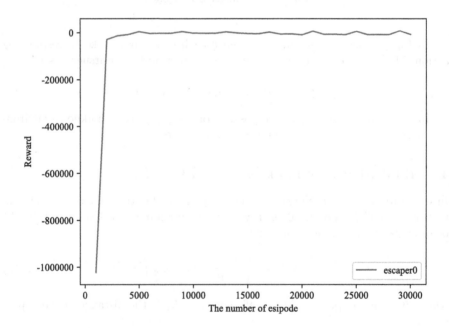

Fig. 5. The reward for escape0.

The reward changes of pursuing agents is shown in Fig. 6. The pursuing agents need to constantly update its own behaviour strategy because the escaping agent is a moving target. After 7000 episodes of training, so the pursuing agents keep stable, but there is a slight fluctuation.

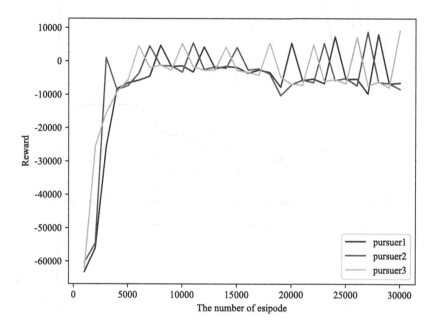

Fig. 6. The reward for pursuers.

The difference in distance between escape0 and pursuer1 is shown in Fig. 7. It shows the distance between escaper0 and pursuer1 in the process of hunting the moving target, which indicates that the tacit behaviour relationship between the escaping agent and the pursuing agents can be formed. After moving about 65cm, the agents begin to maintain an appropriate displacement to move, which proves the success of the danger zone model designed.

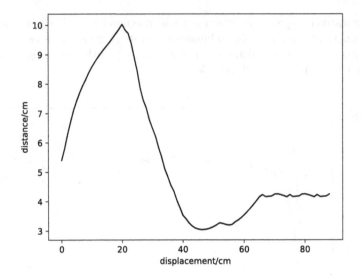

Fig. 7. The difference in distance between escape0 and pursuer1.

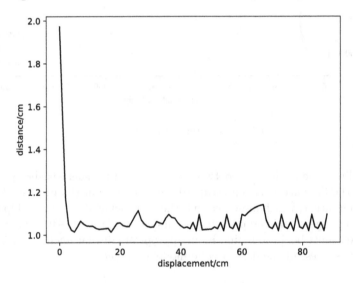

Fig. 8. The difference in distance between pursuer1 and pursuer2.

Figure 8 shows the difference of displacement between pursuer1 and pursuer2. After the pursuers move about 5cm, the pursuing agents begin to keep a proper distance to move, which shows that the improved artificial potential function can enable the pursuers to move in a flocking form.

Figure 9 shows the image of the pursuit trajectory, where the purple lines represent the trajectories of the escapee and the red, green, and brown lines represent the trajectories of the three pursuers. After the escapee moves about 2cm, the pursuer pursues at an appropriate distance, and the effect is relatively stable.

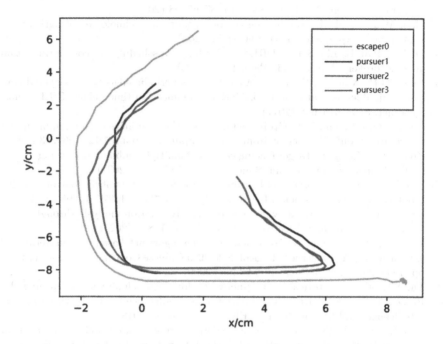

Fig. 9. The chase trajectories.

6   Conclusion

This paper adopts the deep reinforcement learning MADDPG method to address the challenge of hunting moving targets. During the hunting process, establishing risk perception is crucial for enabling agents to make appropriate pursuit and escape actions based on distance judgement. Additionally, the algorithm incorporates an improved artificial potential energy function to maintain a flocking formation among agents while hunting moving targets. This research endeavour enhances the cooperative capabilities of multi-agents and enables them to accomplish more intricate tasks. Artificial intelligence algorithm is used to study the pursuit of target, which will have great influence on agriculture and military fields. In future research, the algorithm performance and application value will be further improved.

Acknowledgement. This work was supported by National Nature Science Foundation under Grant 61203335.

References

1. Li, Y., Han, W., Wang, Y.Q.: Deep reinforcement learning with application to air confrontation intelligent decision-making of manned/unmanned aerial vehicle cooperative system. IEEE Access **8**, 67887–67898 (2020)
2. Mao, W.J., et al.: Research progress on synergistic technologies of agricultural multi-robots. Appl. Sci. **11**(4), 1448–1482 (2021)
3. Li, J.X., Pan, L., Liu, S.J.: A DRL-based online VM scheduler for cost optimization in cloud brokers. World Wide Web, 1–27 (2023)
4. Duan, G.P., Xiao, F., Wang, L.: Asynchronous periodic edge-event triggered control for double-integrator networks with communication time delays. IEEE Trans. Cybernet. **8**(2), 675–688 (2017)
5. Xu, C., Xu, M., Yin, C.J.: Optimized multi-UAV cooperative path planning under the complex confrontation environment. Comput. Commun. **162**, 196–203 (2020)
6. Kou, L.W., Xiang, J.: Target fencing control of multiple mobile robots using output feedback linearization. Autom. Sinica. **48**(5), 1285–1291 (2022)
7. Fan, Z.L., et al.: Reinforcement learning method for target hunting control of multi-robot systems with obstacles. Int. J. Intell. Syst. **37**(12), 11275–11298 (2022)
8. Pandey, B., et al.: Towards video based collective motion analysis through shape tracking and matching. Electron. Lett. **56**(17), 881–884 (2020)
9. Li, M.L., et al.: A new reinforcement learning algorithm based on counterfactual experience replay. In: Proceedings of 2020 39th Chinese Control Conference (CCC), pp. 1994–2001. IEEE, Shenyang, China (2020)
10. Yu, W., et al.: Historical best Q-networks for deep reinforcement learning. In: Proceedings of 2018 IEEE 30th International Conference on Tools with Artificial Intelligence (ICTAI), pp. 6–11. IEEE, Volos, Greece (2018)
11. Cao, X., et al.: High-value prioritized experience replay for off-policy reinforcement learning. In: Proceedings of 2019 IEEE 31st International Conference on Tools with Artificial Intelligence (ICTAI), pp. 1510–1514. IEEE, Portland, USA (2019)
12. Hu, C.Y.: A confrontation decision-making method with deep reinforcement learning and knowledge transfer for multi-agent system. Symmetry **12**(4), 631–655 (2020)
13. Cruz, D., Cruz, J.A., Lopes Cardoso, H.: Reinforcement learning in multi-agent games: open AI gym diplomacy environment. In: Moura Oliveira, P., Novais, P., Reis, L.P. (eds.) EPIA 2019. LNCS (LNAI), vol. 11804, pp. 49–60. Springer, Cham (2019). https://doi.org/10.1007/978-3-030-30241-2_5

Design of Particle Swarm Optimized Fuzzy PID Controller and Its Application in Superheat Degree Control

Xingkai Zhao$^{(\boxtimes)}$, Haoxiang Ma, Hao Sun, Dong Wu, and Yuanxing Yang

School of Information and Electrical Engineering, Shandong Jianzhu University, Jinan, China
916810848@qq.com

Abstract. Heat pump system operation is often accompanied by strong coupling, large time lag and nonlinear characteristics, strong coupling will lead to more perturbations in the operation of the system, and even produce large changes in system conditions, These characteristics present a great challenge for control systems, for these characteristics of the superheat degree control system, this paper uses fuzzy PID to control the superheat degree, but the controller production cycle is long, the quantization factor does not match the system and other problems seriously affect the fuzzy PID control accuracy and development. Due to the introduction of fuzzy algorithm, the adjustment of PID initial parameters becomes more complicated, and the "Z-N method" commonly used in industry and other methods of human adjustment are no longer applicable. In this paper, we optimize the design of fuzzy PID with particle swarm algorithm, train the fuzzy quantization factors K_e and K_{ec} with high matching to the system for heat pump system, and solve the problem of selecting three initial parameters K_p, K_i, K_d, of fuzzy PID, which greatly shorten the fabrication cycle of fuzzy PID controller and improve the control accuracy. And the control characteristics of the optimized fuzzy PID controller are analyzed under MATLAB/Simulink platform. The results show that the control performance of the optimized fuzzy PID controller is significantly improved, and the long and time-consuming fabrication cycle of the fuzzy PID controller is greatly improved.

Keywords: Fuzzy control · Quantization factor · Particle swarm algorithm · Overheating degree · Global optimization search

1 Introduction

Heat pump system superheat control is a nonlinear system with large time lag and strong coupling. There are many subsystems such as hot water cycle, refrigeration cycle, solution cycle, etc., and the systems are coupled with each other, so it increases the complexity and control difficulty of heat pump system, especially in the control of evaporator outlet superheat degree, which becomes especially complicated.

Many scholars have applied fuzzy PID to complex control systems in various fields [1–3], and it has been confirmed by many scholars that it has obvious advantages in

© The Author(s), under exclusive license to Springer Nature Singapore Pte Ltd. 2023
H. Zhang et al. (Eds.): NCAA 2023, CCIS 1869, pp. 479–490, 2023.
https://doi.org/10.1007/978-981-99-5844-3_35

complex system control, but for multi-coupled complex systems, to develop a fuzzy quantization factor that matches the system well requires multiple experiments, a complex reasoning process and a long formulation period, and there are blindness and errors in human tuning of the parameters Once the fuzzy rules, quantization factors and other parameters are determined based on expert experience, the variability and strong coupling of heat pump system superheat will seriously affect the control effect of the controller. And the introduction of fuzzy algorithm will increase the difficulty of PID tuning, industry commonly used "Z-N method" and other manual tuning methods are no longer applicable, obviously, this will increase the production cycle of fuzzy PID, labor and material resources, but also become a fuzzy PID difficult industrial large-scale use of the problem [4–6].

To solve the above problems, this paper applies particle swarm algorithm to optimize the design of fuzzy PID controller, and applies it to two types of parameter seeking in fuzzy PID, namely K_e, K_{ec} and K_p, K_i, K_d. It solves the problems of long production cycle of fuzzy controller and tuning of PID controller, greatly shortens the production cycle of fuzzy PID controller, and at the same time can give full play to the performance of fuzzy PID controller. Performance, which provides ideas for the wide promotion of fuzzy PID in industrial control.

2 Description of the Controlled Object

2.1 Description of Experimental Setup and Superheat

Compression heat pump system consists of compressor, condenser, electronic expansion valve, evaporator and other core components, which is a widely used refrigeration/heat method by driving the compressor by electric energy to compress the work vapor to the condensing pressure corresponding to the condensing temperature. And the evaporator in the heat pump system is the key component for making cold, and the refrigerant in the evaporator is heat exchanged through phase change, and the process of phase change will be accompanied by the generation of gas-liquid two-phase state and the two-phase zone and superheat zone, and the heat exchange efficiency of the two end areas is very different, and in order to ensure the heat exchange efficiency, the two-phase zone is theoretically required to be as large as possible, and the superheat zone is as small as possible [7–9]. The refrigerant continues to absorb heat during the phase change, and the temperature does not increase after reaching the saturation temperature, and continues to absorb heat and becomes higher when the phase change is completed, i.e., after the refrigerant enters the superheat zone. The difference between the evaporator outlet temperature and the saturation temperature in the evaporator tube is the superheat degree. The expression is as follows:

$$T_{esh} = T_{vo} - T_e \qquad (1)$$

In the above equation, T_{esh} is the superheat degree, T_{vo} is the evaporator outlet temperature, and T_e is the saturation temperature inside the evaporator.

2.2 Control System Structure

For the heat pump system coupling between the various aspects of the system resulting in complex and variable situation, the traditional PID algorithm is obviously difficult to achieve accurate control, the author used fuzzy algorithm to modify the parameters of the PID controller in real time to achieve the purpose of improving the adaptive capacity of the controller. During the experiment, it is found that the selection of fuzzy quantization factor has great influence on the effect of fuzzy algorithm, and the introduction of fuzzy algorithm also makes the adjustment of the initial parameters of PID controller complicated, the above-mentioned parameters are often used in the industry to adjust the parameters of the human method, and its limitations are obvious.

After the study, the particle swarm algorithm has advantages in multi-parameter global optimization, and it is applied to the design of the optimized fuzzy PID controller. On the one hand, the design of the fuzzy controller is optimized, and the global optimization of the two fuzzy quantization factors is carried out; on the other hand, the problem of complicated initial value rectification of PID parameters after the introduction of the fuzzy algorithm is solved, and the optimization of the whole fuzzy PID controller production process is completed. It is verified that the optimized fuzzy PID can well handle the control problems of such complex systems as heat pumps. The overall structure of the controllers shown in Fig. 1, and the design process and simulation verification structure of the controller will be developed in detail later.

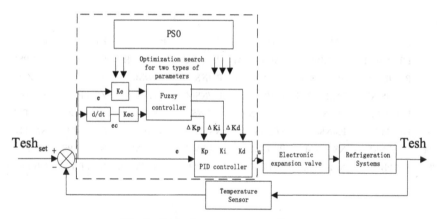

Fig. 1. Control system structure diagram.

3 Implementation of the Control Algorithm

3.1 Implementation of Fuzzy PID Algorithm

According to the actual working conditions of the existing heat pump system, the fuzzy PID structure was selected as a series structure to improve the controller adaptive capability in order to adapt to the strongly coupled and large time lag heat pump system. The fuzzy module adopts a two-dimensional control with two inputs and three outputs, and the fuzzy module inputs are the error between the set value of the superheat degree and the actual value E and the error change rate E_c, respectively. Based on the summary and reasoning of a large amount of data from the heat pump system simulation experiments and considering the actual situation of the simulation system, the author sets the output of the fuzzy controller as the increments of the three parameters K_p, K_i, K_d of the PID algorithm.

The actual range of the input variable error and error rate of change of the fuzzy PID is its basic domain, let the basic domain of the error and error rate of change be $[-e_{max}, e_{max}]$ and $[-ec_{max}, ec_{max}]$ respectively. The quantities in the basic theoretical domain are the exact quantities. The fuzzy theoretical domains of the error and the rate of change of the error in the input variables are set as $\{-m, -m+1, \cdots, 0, \cdots, m-1, m\}$, $\{-n, -n+1, \cdots, 0, \cdots, n-1, n\}$. Accordingly, the basic domain of the output is set to $[-u_{max}, u_{max}]$ and the fuzzy domain is set to $\{-l, -l+1, \cdots, 0, \cdots, l-1, l\}$. In general, to avoid runaway phenomena, it is required that the fuzzy set can completely cover the theoretical domain.

The fuzzy rules are described in the language form of "", and finally 49 fuzzy control rules are designed. The fuzzy controller makes decisions according to the fuzzy rules, and the control rules are shown in Table 1.

Table 1. Fuzzy control rules.

E	E_c						
	NB	NM	NS	Z	PS	PM	PB
NB	PB/NB/PM	PB/NB/PB	PM/NM/PM	PM/NS/Z	PS/NS/PB	PS/Z/PM	Z/Z/PM
NM	PB/NB/NS	PB/NB/NM	PM/NM/NS	PS/NS/NS	PS/NS/Z	Z/Z/NS	PS/Z/Z
NS	PM/NS/NM	PM/NM/NS	PM/NS/NS	PS/NS/Z	Z/Z/PS	Z/PS/PS	NS/PS/Z
Z	PM/NM/NM	PM/NM/Z	PS/NS/NS	Z/Z/NS	NS/PS/Z	NM/PS/Z	NM/PM/Z
PS	PS/NM/NM	PS/NS/NS	Z/Z/NS	NS/PS/Z	NS/PS/Z	NM/PM/PS	NM/PB/Z
PM	NS/Z/NM	NS/Z/NM	PS/Z/NM	NS/PM/PS	NM/PM/PS	NM/PS/PS	NB/PB/PB
PB	Z/Z/PB	NM/Z/PM	NM/PS/NB	NM/PM/Z	NM/PM/PS	NB/PB/PM	NB/PB/PB

3.2 Determination of Quantifiers and Theoretical Domains

In order to enable a smooth fuzzification non-defuzzification process, the input and output variables must be converted from the basic domain to the corresponding fuzzy domain, the input variables are multiplied by the corresponding quantization factors. The quantization factor K_e corresponding to the input error and the quantization factor K_{ec} corresponding to the error rate of change are determined by Eq. (2) (3), respectively:

$$k_e = \frac{2m}{e_H - e_L} \tag{2}$$

$$k_{ec} = \frac{2n}{ec_H - ec_L} \tag{3}$$

Defuzzification is the conversion of the fuzzy quantities output by the fuzzy algorithm into specific values that directly change the PID parameters, with the output corresponding to a scaling factor determined by the following equation:

$$K_{\Delta p} = \frac{u_{kpH} - u_{kpL}}{2l_p} \tag{4}$$

$$K_{\Delta i} = \frac{u_{kiH} - u_{kiL}}{2l_i} \tag{5}$$

$$K_{\Delta d} = \frac{u_{kdH} - u_{kdL}}{2l_d} \tag{6}$$

In the above equation, e_H, e_{cH} and e_L, e_{cL} are the upper and lower limits of error and error rate of change, respectively; $u_{kpH}, u_{kiH}, u_{kdH}$ and $u_{kpL}, u_{kiL}, u_{kdL}$ indicate the upper and lower limits of output, respectively.

The accuracy of quantization factors K_e, K_{ec} has a great impact on the control performance of the controller. From Eq. (2) (3), we can see that the fuzzification formulae are obtained from their corresponding fuzzy theoretical domain and the actual theoretical domain by certain changes, so this paper applies the particle swarm algorithm to the selection of quantization factors to complete the selection of quantization factors and the selection of fuzzy theoretical domain and actual theoretical domain.

The scale factor also affects the magnitude of the output value, but its influence on the control effect of the fuzzy controller is smaller than that of the quantization factor. In the following parameter optimization, its influence on the controller will be downplayed or ignored in order to avoid unnecessary coupling effects caused by too many optimization targets.

3.3 System Model Identification

According to many scholars, the transfer function of the superheat degree of the heat pump system should be a first order plus pure hysteresis model with the model assumption of Eq. (7).

$$Y(s) = \frac{K}{Ts + 1} e^{-\tau s} U(S) \tag{7}$$

In the above equation, parameters K, T, τ are the parameters to be identified by the model and are the input signals, $U(S)$ is the input signal. After obtaining the data on the relationship between the electronic expansion valve opening and the superheat degree of the heat pump system, the data are processed and the first-order plus pure hysteresis model is identified by MATLAB using the least squares method, and the three parameters as follows: $K = -9.18, T = 6.22, \tau = -4.45$, The data model obtained by identification is:

$$Y(s) = \frac{-9.18}{6.22s + 1} \cdot e^{-4.45s} \tag{8}$$

The model obtained from the identification was brought into the simulation system for simulation and compared with the curve of the actual superheat variation of the system, and the results are shown in Fig. 2.

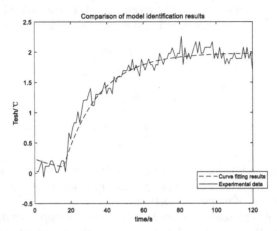

Fig. 2. Comparison of fitted curves.

4 Design of Particle Swarm Optimized Fuzzy PID Controller

4.1 Introduction of Pso Optimization Algorithm

Particle Swarm Optimization (PSO) is a bionic optimization method [10–12]. At any $t + 1$ moment, the velocity and position update equations for the j dimension of the i particle in the swarm are Eq. (9) (10):

$$v_{ij}(t + 1) = \omega v_{ij}(t) + c_1 r_1 \left(\text{pbest}_{ij}(t) - x_{ij}(t) \right) + c_2 r_2 \left(\text{gbest}_j(t) - x_{ij}(t) \right) \quad (9)$$

$$x_{ij}(t + 1) = x_{ij}(t) + v_{ij}(t + 1) \quad (10)$$

where ω is the inertia weight; $v_{ij}(t)$ is the j dimensional velocity component of particle i at evolution to generation t; $x_{ij}(t)$ is the j dimensional position component of particle i at evolution to generation t; $pbest_{ij}(t)$ is the j dimensional individual optimal position $pbest_i$ component of particle i at evolution to generation t; $gbest_j(t)$ is the j dimensional component of the optimal position $gbest$ of the whole particle population at evolution to generation t; c_1, c_2 are the acceleration factors; r, r_2 are the random numbers within [0, 1].

After several experiments, the initial values of the two aforementioned quantization factors K_e, K_{ec} and the three parameters K_p, K_i, K_d of the fuzzy PID are regarded as five particles. Determine the algorithm population size $N = 100$, particle dimension $j = 5$, acceleration constants $c_1 = 2$, $c_2 = 2$, Maximum number of iterations $n = 50$ and minimum fitness value $m = 1$, If the search does not work well, a limit boundary can be set for the particles to be searched separately to achieve a more accurate search.

In order to obtain the optimal quantization factor, the Integrated Time Absolute Error (ITAE) performance index is used as the optimization-seeking objective function.

4.2 Optimization Results

The optimal values of the two quantization factors K_e, K_{ec} of the fuzzy PID controller and the initial values K_p, K_i, K_d of the three parameters of the fuzzy PID are finally obtained after the iterative optimization search of the particle swarm algorithm., $K_e = 0.9664$, $K_{ec} = -1$, $K_p = -10$, $K_i = -0.1358$, $K_d = -0.9516$.

We analyze the difference between the fuzzy PID controller optimized by particle swarm algorithm and the fuzzy PID controller using the "trial-and-error" method of parameter tuning. The experimental results are used to compare the control performance indexes of the two controllers. The K_p value obtained by the "trial-and-error" method is slightly larger, and there are many factors affecting the adjustment time, and the slightly smaller K_d value is probably the biggest factor leading to the poor control effect. The fuzzy PID controller optimized by the particle swarm algorithm, however, coordinates the relationship between system speed and stability by optimizing multiple target parameters as a whole, resulting in the lowest system adaptation value and better control effect. The comparison effect graph with each performance index is shown in Fig. 3 and Table 2.

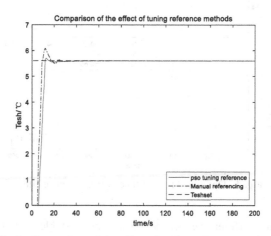

Fig. 3. Comparison of the effect of tuning reference methods.

Table 2. Comparison of performance indicators.

	Rise time $t_r(s)$	Adjustment time $t_s(s)$	Peak time $t_p(s)$	Overshoot amount $\sigma(\%)$	Overshoot amount ess
Manual referencing	16	39	20	13.91	0.83
PSO	18	31	21	3.5	0.28

5　Control Effect Verification

5.1　Simulation Verification

In order to verify the application effect of the controller, the optimized fuzzy PID controller is introduced into the subject compression heat pump simulation system, and two rounds of experimental tests are conducted on the controller: the given value tracking experiment and the controller anti-interference ability experiment.

The given value tracking experiment is the heat pump simulation system with the original superheat set value of 5.6 °C on the basis of stable operation, in the following figure at 400 s to change the superheat set value of 7.6 °C, observe the system operating state and the controller control effect, the control effect graph is shown in Fig. 4.

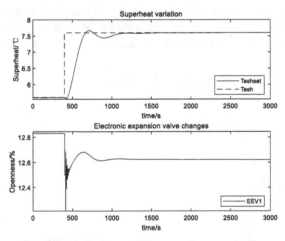

Fig. 4. Optimized fuzzy PID controller control effect.

Controller anti-interference experiment is also in the stable operation of the system, by changing the disturbance of the compressor frequency change, in order to verify the processing ability of the controller on the compressor disturbance, after the evaporator superheat at 7.6 °C to maintain stable operation, respectively, the introduction of the compressor 10, −10 Hz interference, anti-interference experimental results are shown in Fig. 5, Fig. 6.

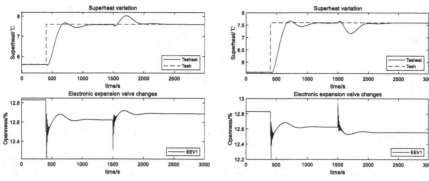

Fig. 5. Anti-jamming capability test of the optimized controller (−10 Hz).

Fig. 6. Anti-jamming capability test of the optimized controller (−10 Hz).

Through the above two rounds of simulation experiments, it is proved that the controller can make the superheat degree converge to the target value quickly by controlling the opening of the electronic expansion valve after the superheat degree setting value is changed, and at the same time, it can ensure that the system overshoot is small and the system operation is stable. In the test of anti-disturbance ability of introducing compressor disturbance, the system does not show violent vibration no matter the compressor increases or decreases the frequency, and it can always converge to the set value of superheat quickly and smoothly.

5.2   Experimental Validation

The controller tested through simulation experiments is added to the experimental platform of the subject group, and the control program is called through LabVIEW and added to the main LabVIEW program of the system. Experimental verification is also tested through two rounds of experiments.

After the stable operation of the subject heat pump experimental platform, the superheat degree is 5.6 °C at this time, the compressor frequency is 50 Hz, the electronic expansion valve opening pulse number is 120. Change the set value of superheat degree to 7.6 °C at 30 s in the following figure, the system controlled quantity response curve is shown in Fig. 7, and the system control quantity response curve is shown in Fig. 8.

As can be seen from Fig. 7, after the set value of superheat degree is changed, the electronic expansion valve will respond in time to adjust the opening degree under the control of the control algorithm, so that the superheat degree will quickly approach the set value, and the system can be restored to a stable working condition in a short time.

Change the system load and increase the disturbance to test the performance. In this experiment, the compressor frequency is increased by 10 Hz at 30 s in the following figure for the first experiment after the system is running stably, and then the compressor frequency is reduced by 20 Hz for the second experiment, and the set value of superheat degree is still 5.6 °C, and the automatic control program of superheat degree is started by LabVIEW program. The experimental results are shown in Fig. 9 and Fig. 11, and the response curve of system control quantity is shown in Fig. 10 and Fig. 12.

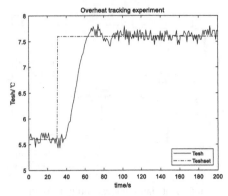

Fig. 7. System controlled quantity response curve.

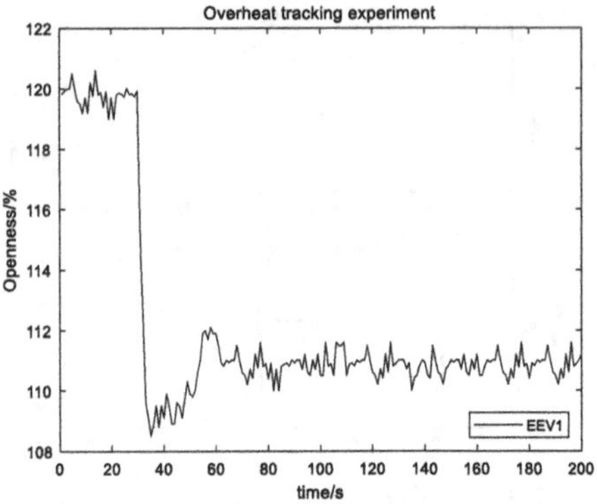

Fig. 8. System control volume response curve.

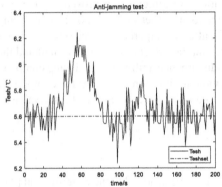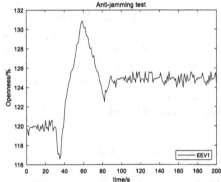

Fig. 9. Experimental effect of superheat control (+10Hz)

Fig. 10. Electronic expansion valve opening response

5.3 System Analysis

Due to the limitation of the accuracy of the temperature and pressure sensors, the experimental data are more volatile than the actual magnitude. However, we can observe from the controller given value tracking experiment that the system superheat can be stabilized within the range of 0.2 °C above and below the preset 5.6 °C when no interference of compressor change is introduced manually, which ensures the stable operation of the system, and when the set value of superheat is changed, the controller can act quickly and make the superheat close to the set value quickly by adjusting the opening of electronic expansion valve quickly; in the controller In the anti-interference experiment, after introducing the interference of compressor frequency change, the system superheat value will be affected by the coupling and change, no matter the compressor frequency increases or decreases, the controller can quickly respond and affect the superheat value by changing the opening of the electronic expansion valve, so that the system superheat can quickly return to the set value.

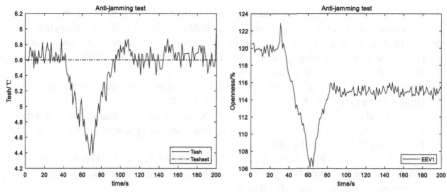

Fig. 11. Experimental effect of superheat control (−20 Hz)

Fig. 12. Electronic expansion valve opening response

6 Conclusion

The optimized fuzzy PID controller shows excellent control effect in complex system control, which is significantly better than traditional PID controller in terms of overshoot, stabilization time and regulation time, and the precise parameters also let the controller play a more excellent control effect. After the introduction of particle swarm optimization algorithm, it not only greatly simplifies the process of making the controller to shorten the production cycle, but also makes the controller have excellent control effect and stability with each quantization factor that matches well with the system. The author believes that the particle swarm optimized fuzzy PID controller can undertake the control work of most complex systems and has good prospects for industrial applications.

References

1. Wang, H.-K., Zhao, L.-Y.: Application of improved predictive fuzzy PID control in pneumatic valve positioning system. Comb. Mach. Tools Autom. Mach. Technol. **587**(01), 142–146 (2023). https://doi.org/10.13462/j.cnki.mtmtamt.2023.01.032
2. Liu, Y., Liu, D.: Design of composite structure greenhouse system based on fuzzy PID technology. Agric. Mech. Res. **45**(12), 45–49 (2023). https://doi.org/10.13427/j.cnki.njyi.2023.12.001
3. Yan, C., Zhao, Y., Hou, P., Xie, Y.: Design of fuzzy PID control system for gravity compensation device of suspended air cushion. Mech. Des. Res. **38**(05), 142–147 (2022). https://doi.org/10.13952/j.cnki.jofmdr.2022.0257
4. He, L.-Y.: Analysis and simulation of the effect of fuzzy control system parameter selection on system performance. Mach. Tools Hydraul. (05), 361–364 (2008)
5. Chen, W.Y., Chen, Z.J., Zhu, R.Q., Wu, Y.Z.: Control algorithm of electronic expansion valve to regulate evaporator superheat. J. Shanghai Jiaotong Univ. (Chin. Ed.) **08**, 1228–1232 (2001)
6. Liu, H., Liu, M.: Layer. Simulation study based on fuzzy PID in boiler temperature control system. Autom. Instrum. **33**(04), 20–25 (2018). https://doi.org/10.19557/j.cnki.1001-9944.2018.04.005
7. Liu, B., Li, M.: Research on particle swarm rectification fuzzy PID control of pressure in the drying section of paper machine. Chin. J. Pap. Mak. **32**(04), 42–46 (2017)

8. Niu, R., Chen, X., Wang, Z., et al.: Numerical simulation of heat pump driven moisture regulating evaporator. J. Central South Univ. **26**(08), 2197–2213 (2019). (in English)

9. Liu, H., Ma, X., Wang, H., et al.: Optimization study of solar heat pump combined with wind power heat storage heating system. J. Sol. Energy **43**(10), 104–112 (2022). https://doi.org/10.19912/j.0254-0096.tynxb.2021-0419

10. Kong, X., Xu, X., Zhang, P., et al.: Operation control strategy of direct expansion solar heat pump heating system. J. Agric. Eng. **38**(08), 38–44 (2022)

11. Wang, Y., Ma, C., Wang, C., Zhang, H., Fan, S.: Characteristics analysis and optimization design of bridge crane based on improved particle swarm optimization algorithm. J. Low Freq. Noise Vib. Act. Control **42**(1) (2023)

12. Oyewola, O.M., Petinrin, M.O., Labiran, M.J., Bello-Ochende, T.: Thermodynamic optimisation of solar thermal Brayton cycle models and heat exchangers using particle swarm algorithm. Ain Shams Eng. J. **14**(4) (2023)

A Multi-style Interior Floor Plan Design Approach Based on Generative Adversarial Networks

Xiaolong Jia[1], Ruiqi Jiang[1], Hongtuo Qi[2], Jiepeng Liu[3], and Zhou Wu[1(✉)]

[1] College of Automation, Chongqing University, Chongqing, China
202213021012@stu.cqu.edu.cn, {jiang_ruiqi,zhouwu}@cqu.edu.cn
[2] Smart City Research Institute, Chongqing University, Chongqing, China
hitqht@163.com
[3] College of Civil Engineering, Chongqing University, Chongqing, China
liujp@cqu.edu.cn

Abstract. Artificial intelligence is reshaping the process of interior floor plan design, making it more intelligent and automatic. Due to the increasing demand for interior design, this paper proposes a dual-module approach to achieve multi-style floor plan design, which adopts the pix2pix series models, a type of Generative Adversarial Networks (GAN). The proposed approach consists of two modules to generate semantic label images from design sketches and multi-style design drawings from label images. To this end, the key area layout and style texture information are separated through extracting and coding areas' label. Afterwards, the experimental comparison between dual-module and single-module generation validates the superiority of proposed dual-module generation approach. Finally, the dual-module approach with label coding is verified in multi-style interior floor plan generation experiments. Based on the proposed style evaluation method, the stylization indicators of generated results exceed 0.8, which further denotes the multi-style generation ability of the proposed dual-module approach.

Keywords: Interior floor plan design · Generative adversarial networks · Multi-style

1 Introduction

With the rapid development of urbanization and living quality, the housing demands have been universally elevated, which causes an increasing demand for interior design. In actual interior floor plan design, the overall interior structure would be analyzed and each spatial area should be divided reasonably. Then, the furniture need to be arranged in each room areas. However, the entire floor plan design process is high human resource consumption, and multi-style interior floor plan design further increases the workload of designers [1]. Therefore, an efficient and intelligent multi-style interior design method is urgently needed.

H. Zhang et al. (Eds.): NCAA 2023, CCIS 1869, pp. 491–506, 2023.
https://doi.org/10.1007/978-981-99-5844-3_36

Artificial intelligence provides feasibility for reducing manual workload and improving design efficiency, which makes intelligent design possible [2–4]. In recent years, some researchers have adopted certain neural networks in interior design [5,6]. Through analyzing the training and processing of neural networks, Huang, et al. found that neural networks could recognize and generate interior building drawings [5]. However, this study only visually verified the processing capability of neural networks, the quality of generated designs have not been further analyzed and evaluated. Based on deep learning, Chaillou trained three generative models to handle specific generative tasks, achieving the entire process from house outlines to furniture layouts [6]. However, the blurriness of generated results and the lack of multi-style designs have limited practical application of the study. Therefore, existing interior design research could accomplish a certain degree of generative design, but specific generative effects and frameworks are still not perfect.

None of the above indoor design methods have considered multi-style generation, and we first analyzed multi-style image generation methods. Existing multi-style generation methods are mostly based on Generative Adversarial Networks(GAN) for image style transfer [7–13]. Some methods [8–10,13] map the specific image information to the intermediate latent space to obtain the latent code of corresponding image, and control corresponding levels of features by manipulating latent code in different dimensions, ultimately achieving multi-style generation. Additionally, methods [14–16] learn different attribute codes and inject the desired style attributes into the generator to obtain corresponding outputs. These studies have shown excellent performance in multi-style implementation of face or landscape images. However, since the detailed design principles need to be learned for interior floor plan design, the preservation and generation of position information for specific elements is extremely crucial. Therefore, for interior plan design images, the "content" and "style" may not be accurately coded. Additionally, multi-style design in interior design requires ensuring local stylization of each area, which makes it more difficult to find universally representative style images, so the above multi-style implementation methods are not suitable in interior floor plan design.

Compared to traditional convolutional neural networks, using GAN for image generation has better precision and realism [17–22]. In [22], a shear wall generation method based on GAN is proposed. The method learns from existing shear wall design files and the generated results are highly consistent with engineer's designs. Therefore, we also use GAN as the basis for interior design. In this study, a novel dual-module generative method based on Generative Adversarial Networks is used in interior multi-style floor plan design. The overall structure is mainly composed of two parts, which we refer to as Module 1 and Module 2. Among them, the learning of design concepts such as spatial layout is mainly reflected in the training process of module 1, while module 2 realizes multi-style generative design. Compared to previous studies, this method could achieve superior results in the generation of interior floor plan design, and the designs are not limited to single style. By learning from existing interior design schemes, the intelligent design method could achieve generative design from wall sketches to design drawings. Therefore, the method greatly reduces workload of manual design and further improves efficiency of interior design.

This paper describes the establishment of datasets and evaluation indicators in Sect. 2. Then, Sect. 3 introduces two generative adversarial network algorithms, pix2pix and pix2pixHD, and designs an overall dual-module generation framework. Next, Sect. 4 conducts algorithm comparison experiments between pix2pix and pix2pixHD, and the superiority verification experiment of dual-module generation method. Moreover, The dual-module generation experiment further prove the capability of the proposed framework for multi-style design. Finally, Sect. 5 presents the conclusions of the study.

2 Datasets and Evaluation Metrics

2.1 Datasets Pre-processing

The process of automatic interior plan design is an image-to-image generation process. Therefore, the preparation of datasets must first be considered. At present, there is no open source corresponding datasets available, so it is necessary to collect suitable interior design drawings to make datasets. To this end, more than 200 indoor design images were collected from internet, which include two main styles: modern minimalist style and European classical style (defined as Class0 and Class1, respectively). These images were obtained from real decorated housing from websites such as https://www.lianjia.com and https://www.anjuke.com, which have high reference value.

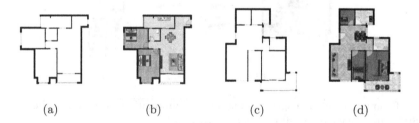

(a)	(b)	(c)	(d)

Fig. 1. Examples of design sketches and design drawings. (a) Sketch of class0. (b) Drawing of class0. (c) Sketch of class1. (d) Drawing of class1.

Due to problems such as inconsistent sizes, low clarity, and watermark obstruction in downloaded images, pre-processing was required to improve usability and consistency. Specifically, the following steps were taken: 1)Remove watermark: Use the methods of blurring filter and morphological transformation in the *OpenCV* library. 2)Enhance image quality: the contrast, sharpness and color balance of images were automatically adjusted using online platforms to make images clearer. 3)Unify the resolution of images: the resolution were transformed into 1024×1024 using batch processing tool in *Photoshop*. After the pre-processing, we obtained 200 high-quality interior design drawings with uniform sizes. Then, the wall lines in design drawings are extracted to obtain the design sketches. The results of the design drawings and design sketches are shown in Fig. 1.

After above datasets creation, we applied image enhancement operations such as rotation and mirroring to increase the quantity of data, so as to prevent over-fitting and improve the model's generalization ability. In the end, we obtained 748 sets of data images for training. Due to the low quality of directly generated design drawings and inability to achieve multi-style design [6], we added label image as the intermediate state for dual-module generation model. In order to segment design information, we require that the label image must contain key spatial layout information for interior design. Based on this premise, the rule for labeling is to represent different areas with different RGB colors, and the final label image is composed of multiple RGB color blocks. The principle of labeling is as follows: the areas to be labeled are divided into spatial areas and furniture placement. The spatial areas mainly include living rooms, bedrooms, kitchens, etc., and corresponding furniture also exists in each spatial area. Since the furniture is directly located in relevant areas, in order to maximize the recognition, we also differentiate the RGB values of labels to the greatest extent. For example, the RGB value of bedroom area is R: 255, G: 0, B: 0, and the RGB value of bed is R: 0, G: 255, B: 255. The specific labeling information is shown in Fig. 2.

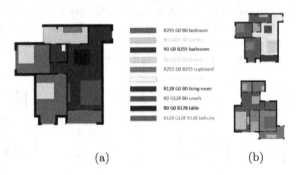

(a)   (b)

Fig. 2. Examples of label images. (a) The specific information on standard label image. (b) Style label image of class0(up) and class1(down).

2.2   Evaluation and Metrics

After training to obtain the corresponding model, it is crucial to conduct reasonable evaluation. For the generated results, key aspects of evaluation rest on quantifying the difference between generated images and target images. Existing evaluation methods are mainly based on questionnaire surveys and computer vision synthetic images evaluation. Due to the unique nature of interior design, especially multi-style design, the survey results are not unique for different research objects. Therefore, the questionnaire survey method lacks objectivity to certain extent. Additionally, traditional computer vision image evaluation methods are relatively complex.

FID(Fréchet Inception Distance) has widespread applications in evaluating the quality of generative models and the diversity of generated images, particularly in the context of GANs [23]. FID is a metric used to evaluate the difference of distribution between generated model and real data, which is based on the Inception network and Fréchet distance. The Inception network is a deep learning model used to extract feature representations from images. The Fréchet distance is a statistical measure, which could quantify the difference between two probability distributions. Mathematically, FID is is calculated using Equation (1).

$$FID = \|\mu_G - \mu_R\|^2 + \text{Tr}\left(C_G + C_R - 2\sqrt{C_G \cdot C_R}\right) \tag{1}$$

where, μ_G and μ_R represent the feature of the generated image distribution G and real image distribution R respectively, while C_G and C_R represent the covariance matrices of G and R respectively. First, the generated images and real images are mapped to feature space using the Inception network. Then, these feature representations are used to calculate the Fréchet distance. A lower FID value denotes that the generated images are closer to the distribution of real data, indicating better performance of the model.

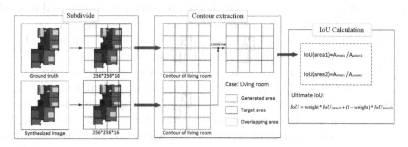

Fig. 3. The detailed steps of the IoU evaluation

IoU(Intersection over Union) has the advantages of simple principles, small calculation amounts, and the ability to evaluate generation accuracy. It was also selected to evaluate the rationality of structural design in the generation design of shear walls [22]. Accordingly, we adopt IoU to measure the consistency between the generated results and target images, and the detailed steps of the IoU evaluation are shown in Fig. 3. First, the 1024×1024 images is divided into 16×256×256 sub-images to improve the accuracy of the contour detection algorithm. Secondly, since the HSV(Hue, Saturation, and Value) color space is more intuitive, we choose to extract specific color block elements in HSV space. Finally, the contours of images are extracted through the contour detection algorithm, so as to obtain the total intersection corresponding to each area, and the IoU of each area is calculated using Equation (2).

$$IoU_{(\text{area})} = \frac{A_{\text{inter(area)}}}{A_{\text{union(area)}}} \tag{2}$$

where $_{area}$ represents specific color block areas, etc., A_{inter} represents the intersection area between generated image and target images, and A_{union} represents the union area, calculated as $A_{union} = A_{generation} + A_{target} - A_{inter}$, $A_{generation}$ and A_{target}, which respectively represent the areas of generated area and target area. The final IoU calculation formula is Equation (3):

$$IoU = weight \times IoU_{(area1)} + (1 - weight) \times IoU_{(area2)} \qquad (3)$$

where $_{area1}$ includes spatial areas such as living room, bedroom, kitchen, etc., $_{area2}$ includes furniture such as sofa, bed, etc., and $weight$ represents weighting coefficient. Since for the result that the position of furniture does not completely coincide, but the generated position is reasonable, we still consider such generation meets the requirements. Therefore, we appropriately reduced its weight and ultimately selected $weight = 0.6$.

3   Overall Design Framework

3.1   Pix2pix and Pix2pixHD

In the GAN algorithm, the pix2pix and pix2pixHD algorithms both have high performance in image generation [24, 25]. Therefore, in this study, we used these algorithms for training. The pix2pix algorithm is an image-to-image generative model based on conditional Generative Adversarial Networks [18], which could process image data with resolution of 256×256. The generator adopts a U-Net [26] structure with skip connections based on encoder-decoder, thereby preserving shared information and improving the quality of generation. Calculating loss on small images could significantly amplify the capturing ability of texture details and effectively enhance the discriminator's attention to generated details. Accordingly, the discriminator adopts the PatchGAN structure, which divides image into multiple small patches and then uses the average of outputs as the overall result. Furthermore, according to existing research [27, 28], mixing the GAN objective loss with traditional loss functions can yield better results. The pix2pix algorithm also utilizes this advantage. Its objective function adds L1 loss to ensure the images in the original domain and target domain are as close as possible.

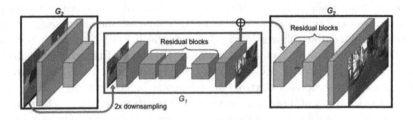

Fig. 4. The generator structure of pix2pixHD

The pix2pixHD algorithm is a high-resolution image-to-image generation model, which has multi-scale generators and discriminators. As shown in Fig. 4, the pix2pixHD generator has two sub-networks, the global generator $G1$ and the local enhancer $G2$. The front-end network extract features on input images, and the down-sampling result and the up-sampling result of global generator are used as the back-end input concurrently. Afterwards, the sampling network restores the resolution of global generator output and improves the details and clarity. Moreover, the discriminator directly operates on high-resolution images requires high computing requirements. To solve the problem, pix2pixHD uses three discriminators with the same network structure. The discriminators need to be trained at different scales, and the loss function needs to be calculated for each scale. Therefore, so as to stabilize the training process, feature matching loss based on the discriminator is added to the loss function of the pix2pixHD model. Since the VGG [29] network has better perceptual loss ability, it is used instead of L1 loss in pix2pixHD model as a pre-trained feature extractor to calculate the quality of the generated high-resolution details.

3.2 Dual-Module Structure

The design of dual-module structure is mainly to solve two problems: (1) poor quality of results directly generated by single module, and (2) existing models cannot achieve multi-style design [5,6]. Based on the adversarial training mechanism, the probability distribution of specific pixel values is learned, so as to master interior floor plan design skills such as space division and texture addition. Firstly, in the training of module 1 and module 2, we adopted consistent GAN algorithm to simplify the model complexity. The corresponding trained models are defined as sketch2label and label2draw. Since the generation task of module 1 is to layout the specific areas and furniture positions based on design sketches, the corresponding training datasets consists of sketches and standard label images. Module 2 is responsible for generating multi-style interior designs from label images. Therefore, We use two sets of label images corresponding to specific style designs in training, which is the core to multi-style generation. With the trained model, multi-style design drawings can be obtained by using different label images as input, even for the same house. The complete model structure is shown in Fig. 5.

Meanwhile, for the purpose of improving the connectivity between two modules, we have added the process of image fine-tuning and style selection. Since the generation process of sketch2label is mainly based on the probability distribution of pixel point spatial positions, there may be noise interference and irregular color blocks in generated label images when there are multiple reasonable layouts. To address these issues, we perform a certain processing on generated label images of sketch2label module. Here, we treat generated label images as probability density images of spatial regions. The distribution of RGB values represents the maximum probability distribution for the corresponding location of furniture or areas. Based on the probability distribution range, we calculate the probability center of specific furniture placement and then directly divide it into appropriate furniture labels.

In addition, to eliminate the impact of noise, after completing the standardization of furniture labels, we use the characteristics of black lines to divide areas, and quickly fill each room with region-related semantics by means of corrosion. Through this method, the label images generated by sketch2label module could be easily and efficiently standardized. The specific algorithm workflow of image fine-tuning is shown in Algorithm 1.

As the generated and standardized result image is single type of label images, we have added a custom "style selector" after the image fine-tuning process. By resetting the RGB values of each area block, we obtain the input label images for module 2. We refer to this process as style selection process. The overall design workflow is shown in Algorithm 2.

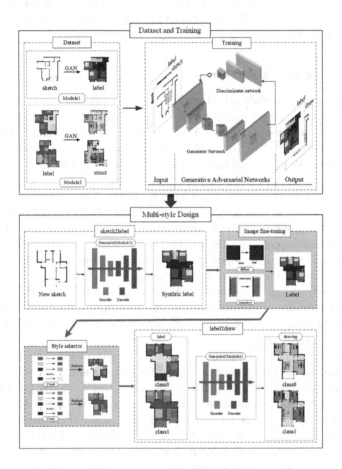

Fig. 5. The structure of dual-module approach

4 Experiment

4.1 Algorithm Comparison and Selection

Based on the pre-processed datasets and dual-module generation implementation method proposed above, we compared the generation abilities of the pix2pix and pix2pixHD algorithms in interior floor plan design. For the purpose of supporting the evaluation, we compared the two models for the task of generating label images from design sketches. It is worth noting that since the pix2pix model can only generate images with resolution of 256×256, we appropriately scaled the existing 1024×1024 images before using them as the pix2pix model's datasets. While ensuring that the number of learning iterations was sufficient and did not lead to over-fitting, we ultimately determined that the optimal number of iterations was 350.

Algorithm 1. The normalization process of generated label images

---

Input: Semantic label images generated by sketch2label module.
Output: Normalized semantic label images.
1: Calculate the color types K in generated label images;
2: **for** int i=1 to K **do**
3:    **if** i is the preset standard color **then**
4:      **if** i is the standard color of furniture **then**
5:        Taking the neighboring group as statistical unit, count the number of group pixels in ith color distribution;
6:        Select the largest number of group pixels and calculate the center point;
7:        The corresponding furniture area is directly divided through the position of center point;
8:        Set furniture area to standard RGB color and calculate the area contour;
9:      **else**
10:        The color is standard area color and record the color;
11:      **end if**
12:    **else**
13:      Set the pixels to general color to be processed and wait for filling;
14:    **end if**
15: **end for**
16: Based on the contour of furniture, the general color is corroded outward to obtain normalized semantic label images.

As the performance of the pix2pix and pix2pixHD algorithms is affected by hyper-parameters, it is essential to analyze these parameters for obtaining optimal performance. For the pix2pix algorithm, as the loss function adds L_1 loss on the basis of cGAN, and two loss functions have different impacts on the local features and global clarity of the generated images, it is meaningful to experimentally analyze the weight parameter $\lambda L_1(L_{1weight}/L_{GANweight})$ of the loss function. On the other hand, the pix2pixHD algorithm uses the feature matching loss L_{FM} instead of L_1 loss to stabilize the training process. Therefore,

---

Algorithm 2. Workflow of multi-style interior floor plan design

---

Input: Design sketches.

Output: Specified style design drawings.

1: Input the design sketches into sketch2label module to generate semantic label images;
2: Normalize the generated semantic label images according to Algorithm 1;
3: Choose the appropriate style corresponding requirements;
4: Input the modified semantic label images into label2draw module;
5: Generate the final interior design drawings.

---

in pix2pixHD algorithm, the quality of generated images is affected by adjusting the weight parameter $\lambda L_{FM}(L_{FMweight}/L_{GANweight})$. To this end, we conducted four sets of parameter adjustment experiments and the generated results are shown in Fig. 6.

It is obvious that the experimental results of pix2pix and pix2pixHD are similar. When $\lambda L_1 = 0$ or $\lambda L_{FM} = 0$, neither can generate the label images correctly. After increasing the weight parameters, the quality of the generated results is significantly improved in accuracy and clarity. In addition, the design results of different hyper-parameters of two algorithms were quantitatively evaluated, and the evaluation results are shown in Fig. 7. Apparently, The evaluation results are consistent with the perceptual results in Fig. 6. Moreover, the image generation quality and structural design quality of pix2pixHD are significantly better than pix2pix, with small design dispersion and high stability. The pix2pixHD with the optimal parameters achieves an IoU of over 0.7, indicating that its floor plan layout is very reasonable. Therefore, in the generation design of the dual-module, pix2pixHD was ultimately chosen as the core algorithm, and the parameter λL_{FM} was chosen as 10.

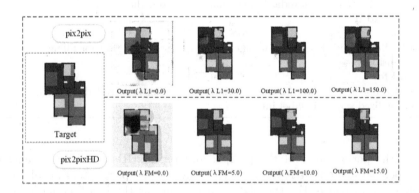

Fig. 6. Label images generated by pix2pix and pix2pixHD

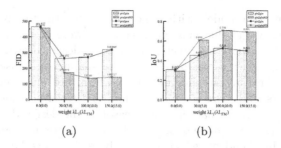

(a)   (b)

Fig. 7. Comparisons between pix2pix and pix2pixHD with various parameters. (a) FID evaluation results. (b) IoU evaluation results.

4.2 The Experimental Comparison of Single-Module and Dual-Module Approach

In previous section, we compared the generative performance of algorithms and ultimately chose pix2pixHD as the core algorithm. Therefore, in this section, we validate the intelligent interior design process based on this algorithm. In the experiment, the learning rate of the pix2pixHD model is 0.0002, using the Adam optimizer with momentum parameter of $\beta_1 = 0.5$ and weight parameter of $\lambda L_{FM} = 10$. Using pix2pixHD algorithm with above parameters, we compared and validated the superiority of the dual module generation. Firstly, the use of GAN algorithm to achieve indoor floor plan generation design was implemented in [6], and we define this method as a single-module generation method, which directly generating design drawings from design sketches. Since this method can only generate single-style design drawings, the comparison process is limited to single-style floor plan design. For the training model of single-module generation, the datasets used is 328 sketches and drawing diagrams of Class0. To make comparison process more objective, class1 is removed from the datasets of dual-module generation model. Therefore, the final dual-module generation uses only Class0 single-style sketches, labels, and drawings. In addition, for the generated results of models, since there is no intuitive measurement standard for design drawings, here we directly evaluate the design results using visual perception.

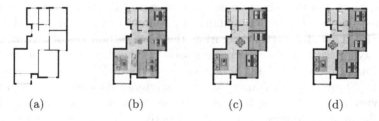

(a)   (b)   (c)   (d)

Fig. 8. Generation results of two models. (a) The input of models. (b) Generation result of single-module. (c) Generation result of dual-module. (d) Target design drawing.

After training two models with same parameters, the test results of single-module generation and dual-module generation were compared as shown in Fig. 8. It is obvious that the direct generation of design drawings from design sketches is below expectations under the same amount of datasets and scale of neural network. Area overlapping and blurred furniture occurs in single-module generation results, and there is a phenomenon that toilet position layout cannot be generated. We believe that the main reason is the design drawings contains both layout design concept and details of furniture texture, and the two are not completely independent, which ultimately leads to overlapping areas. The dual-module design uses independent label images as the intermediate state, which convert the position of core area blocks into corresponding spatial layout information. Accordingly, the design drawings generated indirectly through the dual-module approach are more excellent in terms of texture and design details. The experiment fully validated the superiority and necessity of dual-module generation in interior floor plan design.

4.3   Multi-style Interior Floor Plan Design

Above, we have validated that the dual-module approach has better design quality. Therefore, in multi-style generation design, we continued to add the class1 datasets to dual-module approach for complete generation process verification. The corresponding output of sketch2label module is shown in Fig. 9(a). It is apparent that the module has high accuracy in area division, while the layout of furniture has a certain flexibility. In addition, because the generation of GAN is based on the probability distribution of datasets, a small amount of irregular color blocks and interference noise also represents low likelihood of generation. After the generated label image is processed by custom image fine-tuner and style selector, the processing results are shown in Fig. 9(b). This kind of noise is greatly filtered out, making the label images conform to the input specifications of label2draw module.

Then, the corresponding design drawings are generated using the stylized label images, and the generated results are shown in Fig. 9(c). It is evident that the proposed dual-module approach could quickly and conveniently generate corresponding style floor plan designs, and the designs have a high visual perception effect. Then, we continued to test the multi-style generation for individual area and furniture, and the mix-style design is shown in Fig. 9(d). According to the generated results, this model preliminarily achieve mix-style design. But compared with the uniform style generation, there is a certain loss of texture details in mix-style design. We consider the reason is that there is no such mix-style design image in training dataset. In subsequent research, we could design a digital twin [30] process to enhance this type of interior floor plan design process.

In view of the above stylized generations, we utilize clustering method to verify the multi-style generation ability. Firstly, we adopt the method of converting design images into grayscale histograms to reduce feature dimensions. Afterwards, we obtain the center points of two styles respectively through clustering method, and calculate the distance between feature vector of genera-

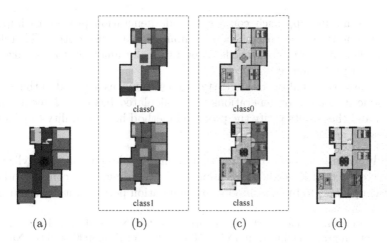

(a)   (b)   (c)   (d)

Fig. 9. Experimental results of dual-module approach. (a) Label image generated by sketch2label. (b) Standardized and stylized label images. (c) Multi-style design drawings generated by label2draw. (d) Mix-style design drawings.

tions and clustering center. Based on the distance reflecting the degree of similarity between images and two styles, we ultimately quantify the results as probability values P_{class0} and P_{class1} that the generations belong to the two styles, respectively. The style indicators for generations are as follows: class0 ($P_{class0} = 0.867, P_{class1} = 0.133$); class1 ($P_{class0} = 0.167, P_{class1} = 0.833$); mix-style ($P_{class0} = 0.536, P_{class1} = 0.464$). According to test indicators, the style indicators of class0 and class1 have reached more than 0.8, indicating that the model have excellent multi-style generation ability. In addition, the style indicator of mix-style design is moderate, which verifies the rationality and correctness of the style assessment methods.

5   Conclusions

In this paper, we propose a multi-style interior floor plan design approach to reduce the workload of manual design. The method consists of two modules, which respectively accomplish the tasks of generating layout label images and adding texture information, and finally generate multi-style design drawings. Moreover, reasonable evaluations were adopt in area layout and multi-style design, and the evaluation results verify the reliability and rationality of the dual-module approach. This multi-style automatic design method minimizes the need for human resources and greatly improves the efficiency of interior design. The conclusions drawn are as follows:

1. The label image separates the layout and texture information of floor plan designs, which allows the model to be trained correspondingly, thereby improving the generation performance. The idea can be leveraged in other generative design tasks as well.

2. The dual-module approach could subdivide generation process to improve learning ability at the cost of more computational consumption. Therefore, the number of subdivisions needs to be rationally analyzed to optimize the generation performance.
3. The quantitative analysis of multi-style generation was adopted in this study. The style indicators of generations were calculated based on k-means algorithm, and the results verify the proposed method has the ability to generate multi-style designs.

This study could generate overall multi-style interior design, but the stylization of specific objects still needs to be improved. Accordingly, we consider adding digital twin process to further develop the generation performance in the future.

Acknowledgements. The work is supported by the National Key Research and Development Program of China (2021YFF0500903,2022YFE0198900) and the National Natural Science Foundation of China (52178271,52077213).

References

1. Ahlin, E.M.: A mixed-methods evaluation of a hybrid course modality to increase student engagement and mastery of course content in undergraduate research methods classes. J. Crim. Justice Educ. **32**(1), 22–41 (2020)
2. Tafraout, S., Bourahla, N., Bourahla, Y., Mebarki, A.: Automatic structural design of RC wall-slab buildings using a genetic algorithm with application in BIM environment. Autom. Construct. **106**, 102901 (2019)
3. Djenouri, Y., Hatleskog, J., Hjelmervik, J., Bjorne, E., Utstumo, T., Mobarhan, M.: Deep learning based decomposition for visual navigation in industrial platforms. Appl. Intell., 1–17 (2022)
4. Zhang, M., Kadam, P., Liu, S., Kuo, C.C.J.: GSIP: green semantic segmentation of large-scale indoor point clouds. Pattern Recogn. Lett. **164**, 9–15 (2022)
5. Huang, W., Zheng, H.: Architectural drawings recognition and generation through machine learning. In: Proceedings of the 38th Annual Conference of the Association for Computer Aided Design in Architecture, Mexico City, Mexico, pp. 18–20 (2018)
6. Chaillou, S.: ArchiGAN: artificial intelligence x architecture. In: Yuan, P.F., Xie, M., Leach, N., Yao, J., Wang, X. (eds.) Architect. Intell., pp. 117–127. Springer, Singapore (2020). https://doi.org/10.1007/978-981-15-6568-7_8
7. Li, X, et al.: Image-to-image translation via hierarchical style disentanglement. In: Proceedings of the IEEE/CVF Conference on Computer Vision and Pattern Recognition, pp. 8639–8648 (2021)
8. Huang, X., Liu, M.-Y., Belongie, S., Kautz, J.: Multimodal unsupervised image-to-image translation. In: Ferrari, V., Hebert, M., Sminchisescu, C., Weiss, Y. (eds.) ECCV 2018. LNCS, vol. 11207, pp. 179–196. Springer, Cham (2018). https://doi.org/10.1007/978-3-030-01219-9_11
9. Patashnik, O., Wu, Z., Shechtman, E., Cohen-Or, D., Lischinski, D.: StyleCLIP: text-driven manipulation of styleGAN imagery. In: Proceedings of the IEEE/CVF International Conference on Computer Vision, pp. 2085–2094 (2021)

10. Karras, T., Laine, S., Aila, T.: A style-based generator architecture for generative adversarial networks. In: Proceedings of the IEEE/CVF Conference on Computer Vision and Pattern Recognition, pp. 4401–4410 (2019)
11. Karras, T., Laine, S., Aittala, M., Hellsten, J., Lehtinen, J., Aila, T.: Analyzing and improving the image quality of styleGAN. In: Proceedings of the IEEE/CVF Conference on Computer Vision and Pattern Recognition, pp. 8110–8119 (2020)
12. Karras, T., et al.: Alias-free generative adversarial networks. Adv. Neural. Inf. Process. Syst. **34**, 852–863 (2021)
13. Zhu, J.Y., et al.: Toward multimodal image-to-image translation. In: Advances in Neural Information Processing Systems, vol. 30 (2017)
14. Li, X., et al.: Attribute guided unpaired image-to-image translation with semi-supervised learning. arXiv preprint arXiv:1904.12428 (2019)
15. Romero, A., Arbeláez, P., Van Gool, L., Timofte, R.: SMIT: stochastic multi-label image-to-image translation. In: Proceedings of the IEEE/CVF International Conference on Computer Vision Workshops (2019)
16. Yu, X., Chen, Y., Liu, S., Li, T., Li, G.: Multi-mapping image-to-image translation via learning disentanglement. In: Advances in Neural Information Processing Systems, vol. 32 (2019)
17. Goodfellow, I., et al.: Generative adversarial networks. Commun. ACM **63**(11), 139–144 (2020)
18. Mirza, M., Osindero, S.: Conditional generative adversarial nets. arXiv preprint arXiv:1411.1784 (2014)
19. Dai, B., Fidler, S., Urtasun, R., Lin, D.: Towards diverse and natural image descriptions via a conditional GAN. In: Proceedings of the IEEE International Conference on Computer Vision, pp. 2970–2979 (2017)
20. Bao, J., Chen, D., Wen, F., Li, H., Hua, G.: CVAE-GAN: fine-grained image generation through asymmetric training. In: Proceedings of the IEEE International Conference on Computer Vision, pp. 2745–2754 (2017)
21. Lu, Y., Wu, S., Tai, Y.W., Tang, C.K.: Image generation from sketch constraint using contextual GAN. In: Proceedings of the European Conference on Computer Vision (ECCV), pp. 205–220 (2018)
22. Liao, W., Lu, X., Huang, Y., Zheng, Z., Lin, Y.: Automated structural design of shear wall residential buildings using generative adversarial networks. Autom. Construct. **132**, 103931 (2021)
23. Heusel, M., Ramsauer, H., Unterthiner, T., Nessler, B., Hochreiter, S.: GANs trained by a two time-scale update rule converge to a local nash equilibrium. In: Advances in Neural Information Processing Systems, vol. 30 (2017)
24. Isola, P., Zhu, J.Y., Zhou, T., Efros, A.A.: Image-to-image translation with conditional adversarial networks. In: Proceedings of the IEEE Conference on Computer Vision and Pattern Recognition, pp. 1125–1134 (2017)
25. Wang, T.C., Liu, M.Y., Zhu, J.Y., Tao, A., Kautz, J., Catanzaro, B.: High-resolution image synthesis and semantic manipulation with conditional GANs. In: Proceedings of the IEEE Conference on Computer Vision and Pattern Recognition, pp. 8798–8807 (2018)
26. Ronneberger, O., Fischer, P., Brox, T.: U-Net: convolutional networks for biomedical image segmentation. In: Navab, N., Hornegger, J., Wells, W.M., Frangi, A.F. (eds.) MICCAI 2015. LNCS, vol. 9351, pp. 234–241. Springer, Cham (2015). https://doi.org/10.1007/978-3-319-24574-4_28
27. Mao, X., Li, Q., Xie, H., Lau, R.Y., Wang, Z.: Multi-class generative adversarial networks with the l2 loss function. arXiv preprint arXiv:1611.04076 5, 1057–7149 (2016)

28. Zhang, Z., et al.: On loss functions and recurrency training for GAN-based speech enhancement systems. arXiv preprint arXiv:2007.14974 (2020)
29. Simonyan, K., Zisserman, A.: Very deep convolutional networks for large-scale image recognition. arXiv preprint arXiv:1409.1556 (2014)
30. Zhu, Z., Liu, C., Xu, X.: Visualisation of the digital twin data in manufacturing by using augmented reality. Procedia Cirp **81**, 898–903 (2019)

Linear Model-Based Optimal VVC Intercoding Rate Control Scheme

Xuekai Wei[1], Mingliang Zhou[2(✉)], and Xingran Liao[3]

[1] State Key Laboratory of Internet of Things for Smart City and Department of Electrical and Computer Engineering, University of Macau, Macao 999078, China
xuekaiwei2-c@my.cityu.edu.hk
[2] School of Computer Science, Chongqing University, Chongqing 400044, China
mingliangzhou@cqu.edu.cn
[3] Department of Computer Science, City University of Hong Kong, Kowloon 999077, Hong Kong, China
xingrliao2-c@my.cityu.edu.hk

Abstract. In this paper, we propose a rate control (RC) scheme with a linear model in versatile video coding (VVC) intercoding, our method introduces a compromise global Lagrange multiplier to minimize the distortion induced under group-of-pictures (GOP)- and frame-level bit budget constraints. To obtain the optimal solution, the corresponding problem is transformed into a convex optimization problem. Then, the Karush-Kuhn-Tucker (KKT) conditions are used to obtain the optimal quantization parameter (QP). The experimental results show that our RC algorithm achieves better coding efficiency, a better RC effect, and higher subjective quality than the default algorithm in VVC Test Model (VTM) 17.0 and other state-of-the-art algorithms.

Keywords: R-D model · rate control · linear · optimal · convex optimization

1 Introduction

With the development of future mainstream 4K, immersive and high-dynamic-range (HDR) videos on various mobile devices and applications, people's lives and industrial manufacturing have been significantly facilitated [1]. Versatile video coding (VVC) is a new-generation video coding standard that has been developed based on high-efficiency video coding (HEVC) [8]. Rate control (RC) plays an important role in actual video applications, especially video communication applications with real-time requirements [22]. The R-lambda RC algorithm has been widely acknowledged for its superior RC accuracy and high video quality, making it the preferred choice in HEVC [14]. The algorithm has been integrated into the present VVC Test Model (VTM) platform. Li *et al.* introduced an enhanced R-lambda RC method in [13], which has since been adopted in VTM. [12] proposed using convolutional neural network (CNN) in VTM intracoding RC method formulation. The proposed method yielded an average Bjøntegaard

© The Author(s), under exclusive license to Springer Nature Singapore Pte Ltd. 2023
H. Zhang et al. (Eds.): NCAA 2023, CCIS 1869, pp. 507–517, 2023.
https://doi.org/10.1007/978-981-99-5844-3_37

distortion rate reduction under the All Intra configuration compared to the RC method in VTM 17.0. JVET-L0241 [16] introduced an adaptive lambda ratio estimation approach under the random access (RA) configuration. Additionally, a quality dependency factor (QDF) obtained from the skip ratio is proposed in [15] by Liu *et al.*. The QDF was integrated into the frame-level bit allocation process to improve the RC performance [21].

However, the aforementioned models based on the R-lambda algorithm in VVC have the following problems [5].

First, the RC principle in VVC inherits H.265/HEVC and slightly modifies it to deal with the increasing number of skipped coding blocks [2]. In addition, RC has been further extended to screen content coding [2] and all intracoding [2] scenes. In VVC, lambda is determined prior to determining the quantization parameter (QP) [4]. If the video coding scheme in VVC becomes too complex, the QP and lambda parameters will no longer have a fixed relationship.

Second, model-based RC presents many problems, which also restrict the ability to achieve the target bitrate through adjustment. Second, although an RC method based on a Lagrange multiplier can be used to accurately control the bitrate, the order of the coding tree unit (CTU) sequence in VVC, which is a zig-zag sequence, cannot achieve the best quality; thus, a global optimization method for RC in VVC is urgently needed [10].

Furthermore, with the application of novel coding tools in VVC, the rate-distortion (R-D) model has been further investigated, resulting in a new R-D model that can be used to determine the encoding parameters in RC. On this basis, we derive an RC method that conforms to the characteristics of VVC [9].

The contributions of this work are listed below.

First, we propose an optimized RC algorithm based on a linear R-D model. First, by introducing the linear R-D model, the RC strategy can be expressed as an optimization problem, that is, determining the best optimization (assigning the best QP to each frame and CTU). Specifically, we propose a global R-D optimization formation, which is constrained by the total bitrate. Then, the optimal QP can be obtained by incorporating the linear R-D model.

Second, to obtain the optimal QP, the corresponding problem is transformed into a convex optimization problem. Then, the Karush-Kuhn-Tucker (KKT) conditions are used to solve the problem. Therefore, compared with other RC methods, the RC algorithm proposed in this paper can achieve better performance and RC performance.

2   Proposed Linear R-QP Model

R-D optimization is an important feature of video coding. The size of the bitstream output by the encoder and the distortion of the reconstructed video are important indicators when evaluating video coding quality. A recently developed Lagrange multiplier-based RC method can control the bitrate in a highly accurate manner; however, the order of the CTU sequence in VVC, which is a zig-zag

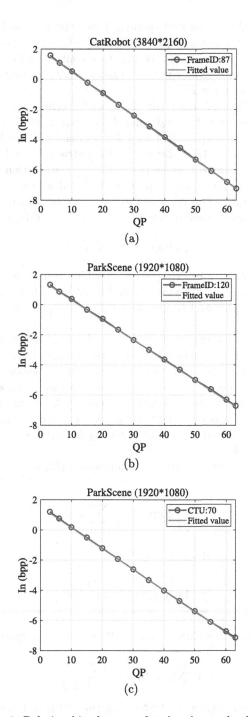

Fig. 1. Relationships between fitted and actual values.

sequence, cannot achieve the best quality. Thus, the proposal of a global optimization method for achieving optimal RC in VVC is urgently needed. To this end, the application of novel VVC coding rate control method investigates the R-D relationship, resulting in a new R-D model that can be used to determine the encoding parameters in RC.

With the adoption of more advanced VVC technology, the relationships among the RD characteristics, QP and λ become more flexible. Although λ plays an important role in pattern-based decision making, the λ effect on output distortion and bitrate is still quite vague. In contrast, the QP affects both the mode decision and quantization results, and the quantization results dominate the distortion and bitrate. This inspires us to build a new analysis framework combining R and Q to better capture the internal relationship between the two. We use a typical linear model to describe the corresponding relationship [18]:

$$\ln \text{bpp} = c \times \text{QP} + d, \tag{1}$$

where bpp is the number of bits per pixel, which can be expressed as bpp = $R/(f \times M \times N)$, where R and f are a predefined bitrate and a frame rate, respectively. M and N are the width and height of the frame, respectively. c and d are the model parameters.

Outlined below is the experiment conducted using the IBBB encoding structure to investigate the correlation between bpp and QP, in which the VTM 17.0 [3] default encoder configuration was utilized. While a higher-order approximation may result in a higher R^2 value, we contend that a linear approximation suffices in practical applications. Multiple sequences were employed in the experiment to validate our model's effectiveness. As depicted in Fig. 1, our proposed model demonstrated a remarkable fit, yielding favorable outcomes at both the CTU and frame levels.

The arithmetic-geometric-averages relationship [19] is considered.

$$\sqrt[L]{\prod_{i=1}^{L} \text{bpp}_i} \leq \frac{1}{L} \sum_{i=1}^{L} \text{bpp}_i \tag{2}$$

It can be further rewritten as

$$\sum_{i=1}^{L} \ln(\text{bpp}_i) \leqslant L \ln \left(\frac{\text{bpp}_c}{L} \right). \tag{3}$$

The mean square error (MSE) and the PSNR have the following relationship [20]:

$$\text{PSNR} = 10 \times \lg \left(\frac{255^2}{\text{MSE}} \right). \tag{4}$$

The relationship between the PSNR and the QP is as follows [7]:

$$\text{PSNR} = -\alpha \, \text{QP} + \beta, \tag{5}$$

where α and β are model parameters, which vary with the video content. We selected some representative sequences to verify this, where the horizontal axis represented the value of the QP and the vertical axis represented the PSNR.

We can obtain

$$\lg\left(\frac{255^2}{\text{MSE}}\right) = -\frac{\alpha}{10}\,\text{QP} + \frac{\beta}{10}. \tag{6}$$

The above formula can be transformed into

$$\text{MSE} = \frac{255^2}{10^{-\frac{\alpha}{10}\text{QP} + \frac{\beta}{10}}} = 255^2 \times 10^{\frac{\alpha}{10}\text{QP} - \frac{\beta}{10}} \triangleq 10^{a\,\text{QP}+b}, \tag{7}$$

where a and b are model parameters: $a = \alpha/10$ and $b = -\beta/10$.

By transforming the constrained problem

$$\{\text{QP}_1^*,\ \text{QP}_2^*, ...,\ \text{QP}_L^*\} = \arg\min \sum_{i=1}^{L} 10^{a\,\text{QP}_i + b},$$

$$s.t. \quad \sum_{i=1}^{L}(c_i \times \text{QP}_i + d_i) \le L\ln\frac{\text{bpp}_c}{L}, \tag{8}$$

into an unconstrained problem, a Lagrange multiplier u is introduced:

$$Z = \sum_{i=1}^{L}\left(10^{a_i\,\text{QP}_i + b_i}\right) + u\left(L\ln\frac{\text{bpp}_c}{L} - \sum_{i=1}^{L}(c_i \times \text{QP}_i + d_i)\right), \tag{9}$$

The bpp_c represents the bpp value associated with a given coding bit budget. In frame-level RC, bpp_c indicates the bits per pixel for the current group of pictures (GOP). On the other hand, bpp_c (CTU-level RC) denotes the bits per pixel for the current frame. The model parameters for the ith coding unit are represented by a_i, b_i, c_i, and d_i.

3   Proposed RC Scheme

3.1   Optimized Solution

The utility vector is noted as $U_m = (\text{QP}_1^m, \text{QP}_2^m, ..., \text{QP}_L^m)$, $m \in [0, M]$, which can be represented as a possible utility-values combinations, where M is the number of possible utility combinations. We define the minimum utilities for all coding units as $U_* = (\text{QP}_1^*, \text{QP}_2^*, ..., \text{QP}_L^*)$. We prove that the utility set $U = (U_1, U_2, ..., U_M)$ is nonempty and bounded and that the set of feasible utilities U is convex, which implies that it is possible to achieve optimal RC performance.

The Lagrangian theory presents that if the minimum value u in Eq. (9) can be obtained at $U_* = [\text{QP}_1, \text{QP}_2', \cdots, \text{QP}_L^*]$, then $U*$ represents an optimal solution to Eq. (9).

Furthermore, when minimizing convex functions and differentiable functions on convex sets using Eq. (9), the KKT conditions are a set of necessary and

Table 1. PSNR comparison between our method and other methods.

Class	VTM 17.0 with RC as ref.		FixQP as ref.		Li et al. [11] as ref.		Gao et al. [6] as ref.		Mao et al. [17] as ref.		Zhou et al. [23] as ref.	
	BD -BR	BD -PSNR	BD -BR	BD -PSNR	BD -BR	BD -PSNR	BD -BR	BD -PSNR	BD -BR	BD -PSNR	BD -BR	BD -PSNR
Class A1	-4.40	0.20	-0.60	0.04	-3.90	0.18	-3.20	0.13	-3.55	0.16	-2.50	0.09
Class A2	-4.50	0.21	-0.50	0.03	-3.90	0.17	-3.20	0.12	-3.55	0.15	-2.50	0.08
Class B	-4.10	0.19	-0.20	0.01	-4.20	0.19	-2.90	0.10	-3.55	0.15	-2.20	0.07
Class C	-3.80	0.17	-0.30	0.02	-3.60	0.17	-2.80	0.10	-3.20	0.14	-1.90	0.07
Class D	-4.60	0.20	-0.10	0.01	-4.20	0.19	-3.30	0.13	-3.75	0.16	-2.30	0.08
Class E	-4.50	0.21	-0.10	0.01	-4.10	0.18	-3.30	0.13	-3.70	0.16	-2.30	0.08
Average	-4.32	0.20	-0.30	0.02	-3.98	0.18	-3.12	0.12	-3.55	0.15	-2.28	0.08

Table 2. RC errors induced at the sequence level and frame level.

Class	VTM 17.0 RC		Li et al. [11]		Gao et al. [6]		Mao et al. [17]		Zhou et al. [23]		Ours	
	Sequence (%)	Frame (%)	Sequence (%)	Frame (%)	Sequence (%)	Frame (%)	Sequence (%)	Frame (%)	Sequence (%)	Frame (%)	Sequence (%)	Frame (%)
Class A1	0.06	1.15	0.05	1.03	0.05	1.12	0.05	1.10	0.05	1.11	0.04	0.96
Class A2	0.05	1.14	0.05	1.05	0.06	1.11	0.05	1.02	0.06	1.07	0.04	0.93
Class B	0.03	2.23	0.05	1.92	0.03	2.11	0.02	1.68	0.03	1.90	0.01	1.53
Class C	0.06	1.51	0.06	1.31	0.06	1.58	0.05	1.33	0.06	1.46	0.04	1.11
Class D	0.08	2.35	0.06	2.22	0.09	2.37	0.08	2.22	0.09	2.30	0.07	2.11
Class E	0.09	5.41	0.10	4.68	0.12	4.21	0.09	3.93	0.11	4.07	0.06	3.72
Average	0.06	2.30	0.06	2.04	0.07	2.08	0.06	1.88	0.06	1.98	0.04	1.73

sufficient optimality conditions for solving optimization problems to ensure the optimality of $U*$.

The optimal solution can be obtained by the point that satisfies KKT conditions orresponding to the global minimum:

$$
\begin{cases}
a_i > 0, \ i = 1, ..., L, & (i) \\
b_i > 0, \ i = 1, ..., L, & (ii) \\
c_i > 0, \ i = 1, ..., L, & (iii) \\
d_i > 0, \ i = 1, ..., L, & (iv) \\
\text{QP}_i > 0, \ i = 1, ..., L, & (v) \\
\nabla Z_{\text{QP}_i, u} = 0, & (vi) \\
u \left(L \ln \dfrac{\text{bpp}_c}{L} - \sum_{i=1}^{L} (c_i \times \text{QP}_i + d_i) \right) = 0. & (vii)
\end{cases}
\tag{10}
$$

where, QP_i can be obtained as follows.

$$
\text{QP}_i = \frac{\lg \left(\dfrac{u c_i}{\ln 10 a_i 10^{b_i}} \right)}{a_i},
\tag{11}
$$

where

$$u = 10^{\left(\dfrac{\phi}{\sum_{i=1}^{L} \frac{c_i}{a_i}}\right)}, \tag{12}$$

$$\phi = L \ln \frac{\mathrm{bpp}_c}{L} + \left(\sum_{i=1}^{L} \left(\frac{c_i}{a_i} \lg \left(\frac{\ln 10 a_i 10^{b_i}}{c_i} \right) - d_i \right) \right). \tag{13}$$

Model Parameter Updating - Based on the strategy proposed in [15], we update the model parameters based only on the previous frame at the same level and the previous frame in the current GOP in this paper.

To estimate the model parameters for RC, the previous frame and the corresponding frame in the previous GOP are utilized in frame-level RC. Similarly, in CTU-level RC, the CTU of the previous frame and the CTU at the corresponding position in the previous GOP are employed to estimate the model parameters.

The model parameters, denoted by a, b, c, and d, are typically used to model the relationship between the bit rate and the distortion in RC. The initial values of these parameters are commonly set based on previous experimental results or some reasonable assumptions. In this case, the initial values of a, b, c, and d are set to -0.12, 1.7, -1.5, and 25, respectively. These initial values are obtained by averaging the fitted values of a, b, c, and d to provide a reasonable starting point for the optimization process.

The proposed RC scheme is summarized as follows. In our method, we consider RC only at the frame level and at the CTU level. For GOP-level RC, we follow the method used in the VVC reference software (VTM 17.0) [3]. The details of our method are as follows. First, the GOP-level bitrate is determined [3]; then, the frame-level QP is determined based on the GOP-level bitrate using Eq. (11). Unlike the traditional method, which first determines lambda and subsequently determines the QP, our method uses a global approach to obtain the QP value. Similar to frame-level RC, we follow the traditional strategy of VTM 17.0 for frame-level bitrate allocation [3], and the CTU-level QP is then determined based on the frame-level bitrate using Eq. (11). Performing bitrate allocation at the CTU level is not needed in our algorithm. Finally, the model parameters are updated. Note that except for the above steps, all other steps (bit allocation at the GOP level and frame level, *etc.*) are identical to those of the VTM RC algorithm [3].

4   Experimental Results

To verify the superiority of the proposed RC optimization method, we conducted experiments using the VTM 17.0 VVC reference software [3]. The original VTM 17.0 benchmark [3] was also employed for testing using an R-lambda model-based RC scheme. For comparison, we implemented three state-of-the-art RC schemes from references [6,11,23], denoted by "Li et al. [11]", "Gao et al. [6]" and "Zhou *et al.* [23]". Mao *et al.* [17] is a method specifically designed for VVC. We applied these methods on the VTM 17.0 platform for comparison purposes. We tested according to the standard common test condition (CTC) [3]. As an additional reference, FixQP was also tested. FixQP encodes videos multiple times using different fixed QP sequences to achieve the optimal encoding performance. In most cases, FixQP offers good R-D performance and smoothness.

In the comparisons, we used the PSNR metric for quality evaluation purposes. The coding performances and their differences were assessed using the Bjøntegaard delta bitrate (BD-Rate) [3] and Bjøntegaard delta PSNR (BD-PSNR) [3] metrics. The BD-PSNR is the differential integral of the PSNR within a set bitrate range based on the fitted R-D curves, while the BD-Rate is the differential integral of the bitrate within a set PSNR range based on the fitted R-D curves.

These values reflect the average quality improvements achieved in terms of the coding performance at the same bitrate and the average rate-saving percentage at the same quality, respectively. The total number of frames was 300; however, because some of the sequences consisted of fewer than 300 frames, we used the actual number of frames as the benchmark for those sequences and enabled the adaptive ratio bit allocation. 128×128 is the CTU size. It is worth noting that setting the target bitrate is a crucial step in RC. The target bitrate needs to be chosen appropriately to achieve a balance between video quality and bitrate allocation. To determine the target bitrate, we compressed the same sequence using a non-RC VVC coding method with fixed QP values of 22, 27, 32, and 37, respectively. The actual bitrates obtained from these compression runs were then used as the target bitrates in our RC experiments. We conducted experiments under the LDB configuration. The rest of the settings were identical to those of the VTM [3].

4.1   Performance Evaluations

The R-D performance results obtained by the different methods are evaluated. According to Tables 1, the proposed R-D performance improvements yielded was 4.3% (baseline is VTM 17.0) and 2.3% (baselines are SOTA methods). Our method performed well because it obtained the QPs directly and determined the frame-level and CTU-level QPs from a global perspective. We also compared our method with FixQP. The experimental results show that the performance of our method was very close to that of FixQP.

4.2   Bitrate Accuracy

The RC errors Er of the different algorithms are calculated as follows:

$$\mathrm{Er} = \frac{|R_{tar} - R_{act}|}{R_{tar}} \times 100\%. \tag{14}$$

Here, R_{act} and R_{tar} are the actual and target bitrates, respectively. According to Table 2, the average control accuracies of the default RC algorithm in VTM

17.0, the SOTA algorithms and our algorithm were within 0.06%, 0.06%, 0.07%, 0.06%, 0.06%, and 0.04%, respectively, in terms of the sequence-level bitrate. The accuracies of the default RC algorithm in VTM 17.0, the SOAT algorithms and our algorithm were within 2.30%, 2.04%, 2.08%, 1.88%, 1.98%, and 1.73%, respectively, in terms of the frame-level bitrate. The accuracies of the three state-of-the-art algorithms considered for comparison purposes differed slightly, but all approaches exhibited high levels of control accuracy, outperforming the VTM 17.0 algorithm.

Nevertheless, Our method slightly outperformed all other methods, which can be fundamentally attributed to the implementation of the global optimization approach during QP determination.

5   Conclusions

In this paper, we propose a RC scheme with a linear model in VVC intercoding. First, we fully consider the features of VVC and directly obtain the QP values based on the bitrate instead of determining lambda based on a linear model. Second, our method introduces a compromise global Lagrange multiplier to minimize the distortion induced under GOP- and frame-level bit budget constraints. Finally, the proposed RC algorithm is implemented on the latest VVC test platform, VTM 17.0. The experimental results show that our RC algorithm achieves better coding efficiency, a better RC effect and higher subjective quality than VTM 17.0 and other SOTA methods. Only interframe RC is discussed in this paper; intraframe RC will be the focus of future research. Moreover, it should be noted that our method addresses only frame-level and CTU-level bit allocation.

References

1. Bonnineau, C., Hamidouche, W., Fournier, J., Sidaty, N., Travers, J.F., Déforges, O.: Perceptual quality assessment of HEVC and VVC standards for 8k video. IEEE Trans. Broadcast. **68**(1), 246–253 (2022). https://doi.org/10.1109/TBC.2022.3140710
2. Bross, B., et al.: Overview of the versatile video coding (VVC) standard and its applications. IEEE Trans. Circuits Syst. Video Technol. **31**(10), 3736–3764 (2021). https://doi.org/10.1109/TCSVT.2021.3101953
3. Chen, J., Ye, Y., Kim, S.: Algorithm description for versatile video coding and test model 17.0 (VTM 17.0). In: Joint Video Exploration Team (JVET) Meeting by teleconference (2022)
4. Choi, H., Nam, J., Yoo, J., Sim, D.: Improvement of the rate control based on pixel-based URQ model for HEVC (JCTVC-i0094). In: Joint Collaborative Team on Video Coding (JCT-VC) 9th Meeting, Geneva, CH (2012)
5. Choi, H., Nam, J., Yoo, J., Sim, D.: Rate control based on unified RQ model for HEVC (JCTVC-h0213). In: Joint Collaborative Team on Video Coding (JCT-VC) 8th Meeting, San Jose, US (2012)

6. Gao, W., Kwong, S., Jia, Y.: Joint machine learning and game theory for rate control in high efficiency video coding. IEEE Trans. Image Process. **26**(12), 6074–6089 (2017)

7. Gao, W., Kwong, S., Jiang, Q., Fong, C.K., Wong, P.H.W., Yuen, W.Y.F.: Data-driven rate control for rate-distortion optimization in HEVC based on simplified effective initial QP learning. IEEE Trans. Broadcast. **65**(1), 94–108 (2019). https://doi.org/10.1109/TBC.2018.2865647

8. Huo, J., Du, H., Li, X., Wan, S., Yuan, H., Ma, Y., Yang, F.: Unified cross-component linear model in vvc based on a subset of neighboring samples. IEEE Transactions on Industrial Informatics pp. 1–1 (2022). https://doi.org/10.1109/TII.2022.3151746

9. Karczewicz, M., Wang, X.: Intra frame rate control based on satd (JCTVC-m0257). In: Joint Collaborative Team on Video Coding (JCT-VC) 13th Meeting, Incheon, KR (2013)

10. Li, B., Li, H., Li, L., Zhang, J.: Rate control by r-lambda model for HEVC (JCTVC-k0103). In: Joint Collaborative Team on Video Coding (JCT-VC) 11th Meeting, Shanghai, CN (Oct 2012)

11. Li, S., Xu, M., Wang, Z., Sun, X.: Optimal bit allocation for CTU level rate control in HEVC. IEEE Trans. Circuits Syst. Video Technol. **27**(11), 2409–2424 (2017)

12. Li, Y., Liu, D., Chen, Z.: Ahg9-related: CNN-based lambda-domain rate control for intra frames (JVET-m0215). In: Joint Video Exploration Team (JVET) 13th Meeting, Marrakech, MA (2019)

13. Li, Y., Chen, Z., Li, X., Liu, S.: Rate control for VVC (JVET-k0390). In: Joint Video Exploration Team (JVET) 11th Meeting, Ljubljana, SI (2018)

14. Li, Z., et al.: Adaptive rate control for h.264. J. Visual Communication and Image Representation **17**, 376–406 (2006). https://doi.org/10.1016/j.jvcir.2005.04.004

15. Liu, Z., Chen, Z., Li, Y., Wu, Y., Liu, S.: Ahg10: quality dependency factor based rate control for VVC (JVET-m0600). In: Joint Video Exploration Team (JVET) 13th Meeting, Marrakech, MA (2018)

16. Liu, Z., Li, Y., Chen, Z., Li, X., Liu, S.: Ahg10: Adaptive lambda ratio estimation for rate control in VVC (JVET-l0241). In: Joint Video Exploration Team (JVET) 12th Meeting, Macao, CN (2018)

17. Mao, Y., Wang, M., Wang, S., Kwong, S.: High efficiency rate control for versatile video coding based on composite cauchy distribution. IEEE Trans. Circuits Syst. Video Technol. **32**, 1–1 (2021). https://doi.org/10.1109/TCSVT.2021.3093315

18. Sanchez, V.: Rate control for HEVC intra-coding based on piecewise linear approximations. In: 2018 IEEE International Conference on Acoustics, Speech and Signal Processing (ICASSP). pp. 1782–1786 (2018). https://doi.org/10.1109/ICASSP.2018.8461970

19. Wang, M., Ngan, K.N., Li, H.: Low-delay rate control for consistent quality using distortion-based lagrange multiplier. IEEE Trans. Image Process. **25**(7), 2943–2955 (2016)

20. Yu, Y., Wang, H., Chen, P., Zhang, Y., Guo, Z., Liang, R.: A new approach to external and internal fingerprint registration with multisensor difference minimization. IEEE Trans. Biometrics Behav. Identity Sci. **2**(4), 363–376 (2020). https://doi.org/10.1109/TBIOM.2020.3007289

21. Zhang, J., Wang, M., Jia, C., Wang, S., Ma, S., Gao, W.: Scalable intra coding optimization for video coding. IEEE Trans. Circuits Syst. Video Technol. **32**, 7092–7106 (2022). https://doi.org/10.1109/TCSVT.2022.3174214

22. Zhang, Y., Kwong, S., Zhang, G., Pan, Z., Yuan, H., Jiang, G.: Low complexity hevc intra coding for high-quality mobile video communication. IEEE Trans. Industr. Inf. **11**(6), 1492–1504 (2015). https://doi.org/10.1109/TII.2015.2491646
23. Zhou, M., et al.: SSIM-based global optimization for CTU-level rate control in HEVC. IEEE Trans. Multimedia **21**(8), 1921–1933 (2019)

Detection and Analysis of Hanging Basket Wire Rope Broken Strands Based on Mallat Algorithm

Dongyue Luo, Yanping Li[✉], Zhiheng Luo, and Chunxue Han

School of Information and Electrical Engineering, Shandong Jianzhu University, Jinan 250101, China

Liyanping0531@126.com

Abstract. To reduce the occurrence of high-altitude hanging basket fall accidents, it is crucial to develop a method for detecting broken strands in the wire rope. This can be achieved by utilizing electromagnetism to detect the leakage of magnetic induction lines at damaged sections of the wire rope. The magnetic dipole model is employed to determine the magnetic induction strength of the axial, radial, and circumferential components at the defect. The analysis of the magnetic induction strength of the leakage field is conducted, and Hall elements are utilized to measure the magnetic induction strength. An information acquisition and processing system based on the STM32 microcontroller is established. To obtain more accurate and stable detection data, the signal is processed using wavelet transform. The Mallat decomposition algorithm and reconstruction algorithm are employed to determine the soft threshold for signal processing, ultimately identifying wire rope damage. The proposed wire rope detection method effectively detects broken strands, as supported by theoretical analysis and experimental results.

Keywords: wire rope · magnetic leakage detection · Mallat algorithm

1 Introduction

In the process of working at height gondolas, incidents of high working basket falls caused by broken strands of wire ropes are common. Preventing such high-altitude fall incidents caused by broken wire ropes is crucial for outdoor safety.

During the construction process, the strength of the wire rope is reduced due to long-term load use and collisions with mechanical equipment. This can lead to failure, wear, and wire breakage. If these issues are not detected and repaired in a timely manner, major construction accidents are more likely to occur. Unfortunately, the damage to the metal cross-sectional area of the wire rope on construction sites often happens internally and remains invisible to the human eye [1]. Furthermore, there is a lack of detailed detection methods, making it impossible to quantitatively assess the extent of the damage.

Nondestructive Testing (NDT) offers various methods for wire rope defect detection, including image recognition, ultrasonic detection, and magnetic flux leakage Hall element detection [2].

H. Zhang et al. (Eds.): NCAA 2023, CCIS 1869, pp. 518–532, 2023.
https://doi.org/10.1007/978-981-99-5844-3_38

Image recognition analyzes defect characteristics by extracting image features, but is limited to surface inspection and may face challenges in image acquisition.

Ultrasonic detection relies on echo analysis and is suitable for static wire ropes, but has limitations in operational settings and faces challenges such as field clutter and low signal-to-noise ratios.

Magnetic flux leakage Hall element detection is widely used and effective for wire rope inspection. It utilizes the principle of magnetic flux leakage to detect defects through Hall sensors, providing precise defect identification [3].

Considering feasibility, operational challenges, and theoretical maturity, this study adopts the magnetic flux leakage Hall element detection method for real-time inspection of hoisting wire ropes, surpassing limitations of other techniques. This approach promises reliable and efficient wire rope defect detection.

2 Magnetic Field Distribution Detection

The Hall element leakage detection method is commonly used for nondestructive testing of wire ropes. In this method, the wire rope is magnetized to saturation. At the defect, there will be leakage of the axial and radial components of the magnetic induction lines [4]. According to Coulomb's law of magnetic charge, the magnetic field strength at the defect can be calculated. The magnetic dipole model, depicted in Fig. 1, is used in this method. The diameter of the wire rope is denoted as 'd', and the width of the wire rope at the defect is denoted as '2δ'. The magnetic charges $+Q$ and $-Q$ are positioned at P1 ($-l$, 0) and P2 ($+l$, 0), respectively. The distance between point P and the two magnetic charges is represented by r1 and r2.

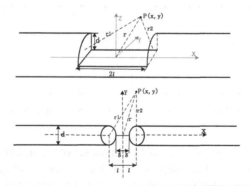

Fig. 1. Magnetic dipole model diagram

Let's denote B_{1x} and B_{2x} as the components of the magnetic charge $+Q$ and the magnetic charge $-Q$ along the x-axis at point P, respectively. Similarly, let B_{1y} and B_{2y} represent the components of the two magnetic charges along the y-axis at point P. The magnetic field strength $B_p(r)$ at point P can be calculated as follows:

$$B_p(r) = B_{1x} + B_{2x} + B_{1y} + B_{2y} \tag{1}$$

The axial component B_x and radial component B_y at the defect can be calculated as follows:

$$B_x = B_{1x} + B_{2x} = \frac{Q}{4\pi\mu_0} \left\{ \frac{x-l}{\left[(x-l)^2 + y^2\right]^{3/2}} - \frac{x+l}{\left[(x+l)^2 + y^2\right]^{3/2}} \right\} \quad (2)$$

$$B_y = B_{1y} + B_{2y} = \frac{Q}{4\pi\mu_0} \left\{ \frac{y}{\left[(x-l)^2 + y^2\right]^{3/2}} - \frac{y}{\left[(x+l)^2 + y^2\right]^{3/2}} \right\} \quad (3)$$

Figures 2 and 3 depict the relationship between the detection position's distance from the center of the two magnetic charges (x, y), the length of the defect l, and the magnetic charge Q [5], and the magnetic field strengths of the axial and radial components at the defect, respectively. Figure 2 provides an illustration of how the magnetic field strength of the axial component varies with the detection position's coordinates along the x-axis and y-axis. Similarly, Fig. 3 shows how the detection position's coordinates along the x-axis and y-axis influence the magnetic field strength of the radial component. These figures demonstrate the correlation between the detection position, the defect parameters, and the resulting axial and radial magnetic field strengths.

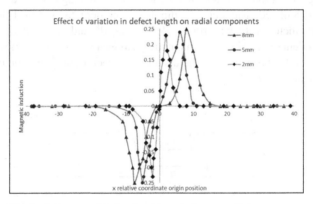

Fig. 2. Influence of defect length variation on radial component

Upon comparing the radial and axial components of the defective leakage field, it becomes evident that the radial component is more effective in reflecting the characteristics of the defective leakage field compared to the axial component. This discrepancy arises due to the structural composition of the wire rope, which consists of multiple steel rope twist strands combined together. During the excitation process, the wire rope generates regular wave noise caused by the twisting strands. The wave signal possesses significant energy and exhibits a direction consistent with the axial direction. In contrast, the leakage signal of the radial component can more accurately capture the primary characteristics of the leakage signal by overcoming the influence of the axial direction wave signal. Thus, the radial component is better suited for detecting and analyzing the defective leakage field in the wire rope [6].

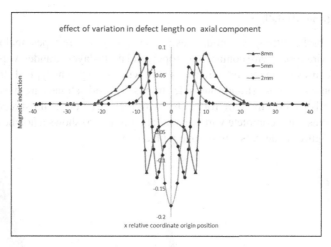

Fig. 3. Influence of defect length variation on axial component

3 Wire Rope Defect Detection Design

The wire rope defect detection device, depicted in Fig. 4, consists of a permanent magnet, Hall sensor, and STM32 core microcontroller. The device operates as follows: Magnetization Device: The permanent magnet is responsible for fully magnetizing the wire rope under inspection, ensuring that it reaches saturation during the excitation process. Acquisition and Data Processing Module: This module converts the leakage magnetic field signal from the wire rope into an analog signal. The analog signal is then further converted into a digital signal that can be interpreted by the microcontroller. Microcontroller: The STM32 core microcontroller receives the digital signal from the acquisition module. It processes the signal through various algorithms and denoising techniques to determine the extent of damage or defects in the wire rope. The device utilizes the magnetization, acquisition, and data processing capabilities to detect and analyze the wire rope's leakage magnetic field signal. By applying denoising techniques and establishing a damage threshold, the device can assess the severity of wire rope defects.

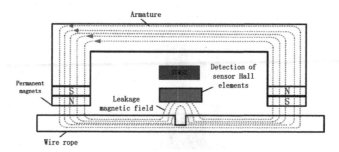

Fig. 4. Wire rope defect detection device

3.1 Wire Rope Modeling

The hanging basket lifting system utilizes various types of wire ropes, including triangular strand wire ropes, wire contact wire ropes, and multi-layer stranded wire ropes [7]. This paper focuses on analyzing wire ropes with wire contact. The typical structure of a wire rope consists of core strands and side strands. The side strands are twisted around the core strands to form the wire rope. This wire rope structure is illustrated in Fig. 5. Figure (a) represents a complete wire rope model, Figure (b) shows a notched wire rope model, and Figure (c) depicts a wire rope with a gap.

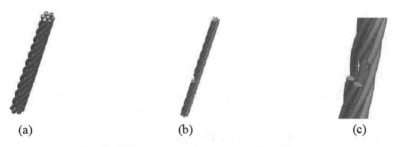

(a)   (b)   (c)

Fig. 5. Different state wire rope modeling diagram

3.2 Excitation Device

Permanent magnet excitation has become the preferred method for leakage detection due to its advantages, such as not requiring a power supply and offering flexible assembly options based on the desired environment, even though it lacks the ability to control magnetization intensity. In this study, we utilize permanent magnets to magnetize the wire rope in the high working basket. The excitation device employs Nd-Fe-B rare earth permanent magnets as the magnetization material, while industrial pure iron (Fe) is selected as the soft magnet [8]. Deep excitation of the wire rope is performed to ensure the leakage of magnetic susceptibility at the wire rope defects. The wire rope model is imported into Ansys software in IGS format, and the permanent magnet modeling is depicted in Fig. 6.

Fig. 6. Permanent magnet model diagram

3.3 Leakage Field Simulation

In Ansys simulation software, the color of magnetic induction lines varies from light to dark, representing different magnetic induction strengths. Based on the simulation results, it can be observed that the internal magnetic induction strength of a wire rope without a notch is higher compared to a wire rope with a notch. Additionally, by analyzing the density of magnetic induction lines outside the wire rope, it is evident that the density of magnetic induction lines is higher at the fracture of the wire rope compared to a section without any fractures. This suggests that there is magnetic field leakage at the fracture. The simulation results are depicted in Fig. 7.

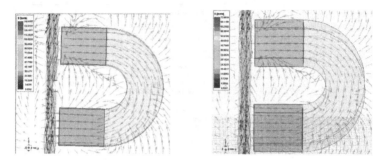

Fig. 7. Leakage field simulation results

3.4 Signal Acquisition Module

To comprehensively and accurately detect wire rope damage signals, multiple measures are implemented: Detection Distance and Hall Element Group: The sensor's detection distance is set at 4mm, and three Hall elements are grouped together to detect the axial, radial, and circumferential components of the wire rope. Multiple Hall element groups are evenly placed around the wire rope for comprehensive coverage [9]. Distance from the Wire Rope: To prevent wear caused by relative movement, a specific distance is maintained between the wire rope's surface and the Hall element group. Improved Detection Accuracy: To enhance accuracy, as many Hall element groups as possible are utilized. In this study, six groups of Hall element groups are strategically positioned based on the characteristics of the wire rope and the excitation device. The placement of the multiplexed sensors is depicted in Fig. 8.

3.5 Signal Processing Module

The signal processing hardware design block diagram of the defect detection system is illustrated in Fig. 9. The Hall sensor employed in the system utilizes a high sensitivity A1302 sensor. The static output voltage of the Hall sensor is approximately 2.492–2.508 V when there is no external magnetic induction. [10] Since the output can reach 4.5–4.7 V in extreme cases, the input pin threshold of the AD7606 chip at the backend is set to 2.5 V. Therefore, a high-precision resistor is required to perform a 1:1 voltage divider processing on the Hall sensor's output signal [11].

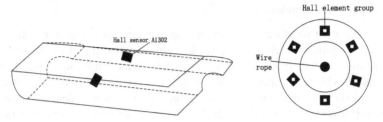

Fig. 8. Multiplex sensor location map

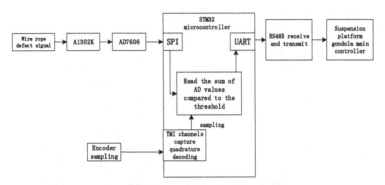

Fig. 9. Defect detection system signal processing hardware setup

Due to the limitation of the A1302 Hall sensor, which has only 4 sampling channels and cannot perform synchronous sampling [12], the AD7606 conversion chip from ADI is chosen. This conversion chip offers 16-bit conversion accuracy and allows for 8-channel simultaneous sampling. To achieve high precision signal acquisition, an independent reference voltage is required for the AD7606. The internal reference voltage of the AD7606 ranges from 2.49 V to 2.505 V. To meet this requirement, the TL431 voltage regulator chip from TI is selected. It provides a stable reference voltage with low noise output and consistent temperature characteristics.

For the core processing module of the signal processing part, the STM32F103CBT6 microcontroller is selected. This chip operates at a main frequency of 72M, providing faster computing and processing speeds. Additionally, it offers a wide range of peripheral resources, enhancing the functionality of the wire rope defect detection system.

4 Wire Rope Defect Signal Processing

Due to the twisting and wrapping of multiple wires in the wire rope, the leakage magnetic signal from fine damage is relatively weak [13]. This poses a challenge in detecting low-frequency damage signals embedded within the strand wave signal, which ultimately affects the accuracy of defect signal identification [6].

Wavelet transform provides an accurate extraction and amplification of signal feature points in the time domain. It effectively captures the essential information and detailed characteristics of the defect signal by decomposing the high-frequency components and highlighting the detailed information. Through reconstruction, it combines the basic information and detailed features of the signal. This approach overcomes the limitations of traditional filters.

The wavelet function $\psi_{(a,b)}$ is translated and scaled to obtain a sequence of functions, resulting in a set of wavelet bases. This set of wavelet bases is characterized by the fluctuations of the functions being concentrated near the origin, while the attenuation of the functions in other regions is zero.

$$\psi_{a,b}(t) = \frac{1}{\sqrt{|a|}} \psi\left(\frac{t-b}{a}\right), a, b \in R, a \neq 0 \tag{4}$$

In the continuous wavelet transform, the scale parameter "a" controls the level of refinement in the analysis, while the displacement parameter "b" shifts the wavelet's center position to the location of interest, allowing for targeted analysis. The continuous wavelet transform of a signal $f(t)$ in the $L^2(R)$ space is expressed as:

$$W_f(a, b) \leq f, \psi_{a,b} \geq \frac{1}{\sqrt{|a|}} \int f(t) \psi\left(\frac{t-b}{a}\right) dt \tag{5}$$

By discretizing the continuous wavelet transform, we can obtain discrete wavelet groups. In this case, discretization involves using binary scale parameter "a" and translation parameter "b" such that $a = 2^{-j}$ and $b = k2^{-j}$, where j and k are integers. The resulting discrete wavelet is:

$$\Psi_{2^{-j},k2^{-j}}(t) = 2^{\frac{j}{2}} \Psi\left(2^j t - k\right) \tag{6}$$

At this point, for any function or signal $f(t)$, we can achieve the wavelet level expansion as follows:

$$f(t) = \sum_{k=-\infty}^{+\infty} \sum_{j=-\infty}^{+\infty} A_{k,j} \Psi_{k,j}(x) \tag{7}$$

where the $A_{k,j}$ combined coefficient represents the contribution of the function $f(t)$ to the discrete wavelet orthogonal basis at the corresponding scale and translation parameters.

4.1 Mallat Algorithm

Multi-resolution analysis, also known as multi-scale analysis, serves as the foundational theory for signal analysis and reconstruction. It considers the squared integrable signal $f(t)(f(t) \in L^2(R))$ as the limit of a series of successive approximations. Each approximation is obtained by convolving $f(t)$ with a low-pass signal $\varphi(t)$ and scaling $\varphi(t)$ step by step [14]. The signal's multi-resolution analysis, involves iterative decomposition of the low-frequency components, enhancing their resolution. This mapping of different resolution components to distinct frequency bands is achieved through the Mallat decomposition algorithm and reconstruction algorithm [15], represented as follows:

$$\begin{cases} A_0[f(t)] = f(t) \\ A_j[f(t)] = \sum_k H(2t - k)A_{j-1}[f(t)] \\ D_j[f(t)] = \sum_k G(2t - k)A_{j-1}[f(t)] \end{cases} \tag{8}$$

$$A_j[f(t)] = 2\left\{ \sum_k h(t - 2k)A_{j+1}[f(t)] + \sum_k g(t - 2k)D_{j+1}[f(t)] \right\} \tag{9}$$

In the equations, Aj and Dj represent the wavelet coefficients of the high-frequency and low-frequency parts, respectively, at the jth layer. H and G correspond to the low-pass and high-pass filters used in the wavelet decomposition [16]. The h and g filters are the wavelet low-pass and high-pass reconstruction filters.

For signal noise reduction, the criteria are smoothness and similarity, and an appropriate threshold value should be selected. The soft thresholding method addresses the issue of edge jitter in the denoised signal caused by rejecting wavelet coefficients using a hard threshold. This method results in a smoother reconstructed signal. The expression for soft thresholding is as follows:

$$\tilde{W}_{j,k} = \begin{cases} sign(W_{j,k})(|W_{j,k}| - \lambda) & |W_{j,k}| \geq \lambda \\ 0 & |W_{j,k}| \geq \lambda \end{cases} \tag{10}$$

4.2 Machine Learning Based Wire Rope Defect Detection

The artificial neural network is a mathematical model that processes information through interconnected neurons [17]. Neurons are the fundamental units of neural networks, and Fig. 10 depicts the model structure.

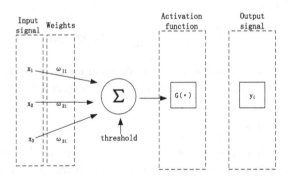

Fig. 10. Neural network model

RBF neural networks have fewer layers, making them faster to train [18]. They utilize the local approximation property of RBF, simplifying computation. Only neurons close to the input value respond, and weights w do not require parameter adjustments [19] (Fig. 11).

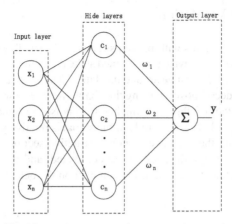

Fig. 11. Basic structure of RBF neural network

Among them, w_{ki} represents the weight connecting the hidden layer node to the output layer node. $R(x)$ represents the radial basis function, and the Gaussian function is commonly chosen as the radial basis function. Equation 11 represents the expression of the radial basis function, where ci represents the center point of the Gaussian function.

$$R(x_p - c_i) = \exp(-\frac{1}{2\sigma^2}\|x_p - c_i\|^2) \tag{11}$$

Figure 12 depicts the complete process of implementing an RBF neural network. The unknown parameters in the process are the center vector c_i, the constant δ_i in the Gaussian function, and the weight w. The algorithm begins by employing the K-means clustering algorithm to determine the center vector c_i. Subsequently, the K-nearest neighbor (KNN) algorithm is applied to compute the value of δ_i Finally, the weight w is obtained using the least squares method.

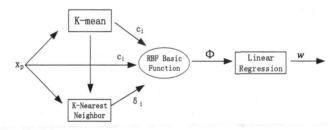

Fig. 12. The overall process of RBF neural network implementation

In this study, the wire rope is utilized as the test sample to investigate the effects of different damage levels at evenly spaced locations. Multiple tests are conducted on the damaged samples. The experimental process involves initial screening of representative samples to identify relevant signals, followed by data sampling of the selected samples. Finally, the sampled signals are subjected to feature value extraction. The degree of damage is varied, and multiple sets of sample values are extracted accordingly.

5 Experimental Validation

In this test, wire rope samples with broken strands are used. To extract the signal data multiple times, the detection device is positioned at the location of the defect. Several tests are conducted within a 1.2 m range around the defect. The original signal is obtained by performing a 3-layer decomposition using the db5 wavelet. The wavelet coefficients are then processed using the soft threshold method to denoise the signal [20]. The denoised signal is shown in Fig. 13. The amplitude of the signal is relatively small and resembles a sine wave, indicating the presence of non-destructive sections of the rope affected by external interference. Multiple tests were conducted, and it was observed that the interference signal did not have a significant impact on the detection results.

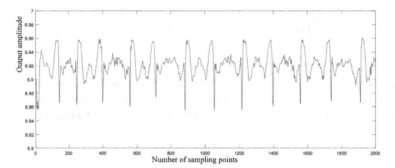

Fig. 13. Wire rope broken strand denoising signal diagram

After the denoising process, the waveform of the wire rope defect signal exhibits a certain pattern with the variation of sampling points. At this stage, the smaller amplitude parts of the waveform, which resemble interference or sine wave patterns, are discarded. The focus is on extracting the more prominent features of the defect signal waveform, as shown in Fig. 14.

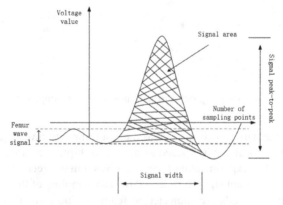

Fig. 14. Defect signal waveform diagram

From the figure, it can be observed that when detecting wire rope defects, the signal exhibits a sudden change, with a significantly higher amplitude compared to the strand wave signal. The amplitude change is more prominent with an increasing number of broken strands in the wire rope. Therefore, the peak of the signal can be used to indicate the degree of damage, and the width of the signal wavelength within a certain range is proportional to the degree of damage.

To account for the diversity of wire rope damage in practical operations and avoid relying solely on sample data, a 4*31SW+FC-8.3 mm steel wire was chosen for experiments. A total of 20 sets of experimental data were selected, as shown in Table 1. Out of these, 11 sets were used for training samples, while 9 sets were used for test verification. The traditional BP neural network and RBF neural network were employed to train the wire rope defect strand breaking signal.

Table 1. Extraction table of the eigenvalues of the damage signal

Test serial number	Wire rope diameter (mm)	Wire diameter (mm)	Lift height (mm)	Wave width (M)	crest (V)	Actual number of broken wires
1	8.3	0.09	4	0.3009	0.039	2
2	8.3	0.09	4	0.3006	0.038	2
3	8.3	0.09	4	0.3374	0.036	2
4	8.3	0.09	4	0.2630	0.035	2
5	8.3	0.09	4	0.4236	0.113	3
6	8.3	0.09	4	0.4612	0.096	3
7	8.3	0.09	4	0.3374	0.086	3
8	8.3	0.09	4	0.4878	0.092	3
9	8.3	0.09	4	0.5464	0.113	3
10	8.3	0.09	4	0.5440	0.093	3
11	8.3	0.09	4	0.4780	0.097	3
12	8.3	0.09	4	0.4236	0.162	3
13	8.3	0.09	4	0.4464	0.099	3
14	8.3	0.09	4	0.3484	0.183	4
15	8.3	0.09	4	0.3780	0.186	4
16	8.3	0.09	4	0.4176	0.172	4
17	8.3	0.09	4	0.3972	0.179	4
18	8.3	0.09	4	0.3164	0.165	4
19	8.3	0.09	4	0.3394	0.159	4
20	8.3	0.09	4	0.4146	0.188	4

The test results for the 11 data sets are presented in Figs. 18 and 19. In these figures, the solid blue line represents the true results, while the red dashed line represents the predicted results. The left figure illustrates the comparison between the model's predicted values and the true values. On the right, the graph displays the model's loss function.

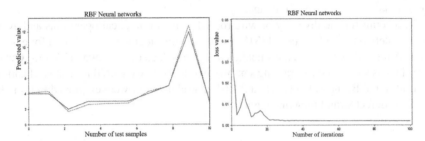

Fig. 15. RBF model for wire rope broken strand detection

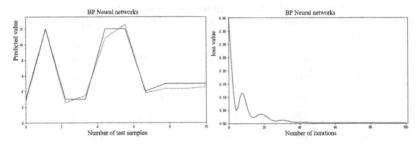

Fig. 16. BP model for wire rope broken strand detection

From Figs. 15 and 16, it is evident that the RBF model outperforms the BP model in terms of approximation accuracy and convergence speed. The RBF model exhibits smaller prediction errors, with a convergence value of 0.0019812735 for the test results and a training error of 0.3646402156521739. These findings validate the high accuracy of the quantitative defect identification model. The experimental results demonstrate the suitability of the proposed method for wire rope broken strand detection.

6 Conclusion

This paper presents a highly effective and accurate wire rope break detection method utilizing the magnetic leakage Hall element as the detection sensor. By establishing a magnetic dipole model and analyzing the axial and radial components of the magnetic leakage signal using Ansys simulation software, the optimal arrangement of Hall elements is determined for precise detection. The developed wire rope break detection system, based on the STM32F103CBT6 microcontroller and equipped with the AD7606 analog-to-digital conversion chip, ensures reliable data sampling. In this study, we employ the Mallet algorithm for wavelet denoising to extract crucial information from the collected signals. The denoised data, containing valuable features, undergoes further processing using a meticulously chosen set of characteristic quantities. Subsequently, a powerful RBF neural network is trained to achieve quantitative identification of wire rope defects with remarkable precision. The effectiveness and reliability of this detection method are demonstrated through successful determination of wire rope strand break information. The proposed approach represents a significant breakthrough in wire rope break detection, providing improved performance and practical applicability. Its exceptional accuracy and reliability make it a promising solution for real-world applications.

References

1. Yao, Y., Li, G., Zhang, X., et al.: Research on wavelet denoising method based on soft threshold in wire rope damage detection. In: 2020 IEEE 8th International Conference on Computer Science and Network Technology (ICCSNT). IEEE (2020)
2. Liu, S., Sun, Y., Jiang, X., Kang, Y.: A review of wire rope detection methods. sensors and signal processing techniques. J. Nondestruct. Eval. **39**(4) (2020)
3. Zhang, J., Peng, F., Chen, J.: Quantitative detection of wire rope based on three-dimensional magnetic flux leakage color imaging technology. IEEE Access **PP**(99), 1 (2020)
4. Zhou, Z., Liu, Z.: Fault diagnosis of steel wire ropes based on magnetic flux leakage imaging under strong shaking and strand noises. IEEE Trans. Ind. Electron. **68**(3), 2543–2553 (2020)
5. Kim, J.-W., Park, S., et al.: Magnetic flux leakage-based local damage detection and quantification for steel wire rope non-destructive evaluation. J. Intell. Mater. Syst. Struct. (2018)
6. Zhao, W., Xue, T., Fan, C., et al.: Research on online detection method of wire rope damage of mine hoist. Min. Mach. **50**(06), 22–26 (2022)
7. Lin, W., Liu, L.J.: Non-destructive testing of ship steel wire rope based on electromagnetic principle. In: 2019 6th International Conference on Information Science and Control Engineering (ICISCE). IEEE (2019)
8. Sun, Y., Liu, S., He, L., et al.: A new detection sensor for wire rope based on open magnetization method
9. Kim, J.W., Park, S.: Magnetic flux leakage sensing and artificial neural network pattern recognition-based automated damage detection and quantification for wire rope non-destructive evaluation. Sensors **18**(2), 109–114 (2018)
10. Li, X., Yu, L., Chen, L., et al.: Analysis of winding configurations and end magnetic leakage shielding in joint motors. In: 2019 22nd International Conference on Electrical Machines and Systems (ICEMS) (2019)
11. Ma, W.J.: Non-destructive testing system of steel wire rope based on STM32F1. Electromech. Eng. Technol. **49**(06), 28–30 (2020)
12. Lihuan, X., Yuanfeng, H., Wenze, D., et al.: Implementation of a high-precision data acquisition system based on AD7606. Comput. Knowl. Technol. **13**(20), 210–214 (2017)
13. Wang, H., Tian, J., Li, X., et al.: Inspection of mine wire rope using magnetic aggregation bridge based on magnetic resistance sensor array. IEEE Trans. Instrum. Meas. **PP**(99), 1 (2020)
14. Dautov, I.P., Zerdem, M.S.: Wavelet transform and signal denoising using Wavelet method. In: 2018 26th Signal Processing and Communications Applications Conference (SIU). IEEE (2018)
15. Zhong, X., Chen, K.: Research on accurate identification system for small defects of wire rope based on leakage detection mechanism. Comput. Meas. Control **28**(01), 195–199 (2020)

16. Jie, T., Shan, S.: Research on improved particle swarm optimized wavelet threshold for mining wire rope damage signal processing method. Coal Eng. **52**(04), 103–107 (2020)
17. Narendra, K.G., Sood, V.K., Khorasani, K., et al.: Application of a radial basis function (RBF) neural network for fault diagnosis in a HVDC system. In: International Conference on Power Electronics, pp. 177–183. IEEE (1998)
18. Nordgard-Hansen, E., Falconer, S., Grasmo, G.: Remaining useful life estimation of HMPE rope during CBOS testing through machine learning. Ocean Eng. **238** (2021)
19. Ma, C.L., Tao, J., Hu, J.L., et al.: Research on signal processing of power with the method of improved wavelet denoising. In: 2016 5th International Conference on Computer Science and Network Technology (ICCSNT) (2016)
20. Liu, G., Huang, Y., Zeng, Z., Cao, Y., Zhao, E., Xing, C.: Research on image denoising method based on wavelet thresholding. J. Xinxiang Coll. **38**(03), 42–47 (2021)

PointAF: A Novel Semantic Segmentation Network for Point Cloud

Tianze Chen[1], Xuhong Wang[2], Dongsheng Li[3], Jiepeng Liu[2], and Zhou Wu[1(✉)]

[1] College of Automation, Chongqing University, Chongqing, China
zhouwu@cqu.edu.cn
[2] The Hong Kong University of Science and Technology, Hong Kong, China
liujp@cqu.edu.cn
[3] School of Civil Engineering, Chongqing University, Chongqing, China
lds@ust.hk

Abstract. Point cloud semantic segmentation is a crucial problem in computer vision, which aims to assign semantic labels to each point in a point cloud. However, the sparsity and irregularity of point cloud data pose significant challenges to achieving accurate segmentation. Unlike traditional image classification tasks, each point in a point cloud not only has location information but also other feature information that must be considered. To address this issue, this paper proposes a novel point cloud semantic segmentation algorithm called PointAF. The proposed method utilizes soft projection operation during downsampling to better combine neighborhood information, an attention mechanism to achieve an adaptive offset effect, and residual connections to solve the problem of gradient disappearance. Experimental results show that the proposed method achieves great performance, with an mIoU of 70.6% and OA of 90.2% on the S3SDIS dataset, as well as mIoU of 69.2% and mACC of 70.1% on the ScanNetV2 dataset. The proposed method demonstrates great potential in point cloud semantic segmentation and may have practical applications in areas such as autonomous driving, robot navigation, and augmented reality.

Keywords: Soft projection · Density adaptation · Point cloud segmentation

1 Introduction

In recent years, significant progress has been made in many fields through the combination of deep learning methods. Semantic segmentation, which is widely used in deep learning, can be categorized into 2D image semantic segmentation and 3D point cloud semantic segmentation [1–3]. While 2D image semantic segmentation has made significant advancements in both academic and engineering fields, 3D semantic segmentation started later and has not yielded as fruitful results. Point cloud semantic segmentation, which is one of the key tasks in 3D scene understanding, holds great importance for various application fields such as autonomous driving, robot navigation, and augmented reality. However, point cloud semantic segmentation also encounters challenges such as the discreteness, dis-order, inhomogeneity, and noise interference of point cloud data.

H. Zhang et al. (Eds.): NCAA 2023, CCIS 1869, pp. 533–545, 2023.
https://doi.org/10.1007/978-981-99-5844-3_39

Therefore, the core issue of point cloud semantic segmentation research lies in effectively extracting useful information from the original point cloud data and parsing the semantic information at different locations.

Point cloud semantic segmentation is the process of assigning a specific semantic label to each point in the point cloud, enabling the segmentation of objects and giving them a specific meaning. In comparison to semantic segmentation in 2D images, semantic segmentation of 3D point clouds poses many challenges. 2D images are structured and can be directly processed using CNN. However, point cloud data is disorderly, interactive, and transformationally independent, making it unsuitable for direct CNN processing. To extend deep learning methods to point cloud processing, various methods have been developed. These methods aim to overcome the challenges posed by point cloud data and enable effective and accurate semantic segmentation.

At present, point cloud semantic segmentation is mainly divided into point-based methods [4–7], grid-based methods [8, 10] and multi-view methods [11–13]. These methods have good rotation and translation invariance, but they have high computational complexity and are difficult to deal large-capacity point cloud data. Grid-based methods transform point cloud data into regular grid structures, such as voxels [8] or octrees [9]. Then, convolutional neural networks are used to extract network features and classify each grid. These methods have high computational efficiency and scalability, but they may lose some detail information and require additional space to store the grid structure. Multi-view methods are simple in concept. It projects the 3D data onto a 2D plane, such as a sphere or cylinder, and then applies image semantic segmentation techniques to extract features of the projected image and classify each pixel. However, this method introduces projection errors, and consistency between different views must be considered.

In this study, we focus on point-based methods and propose a novel semantic segmentation network for point clouds called PointAF. The network consists of two main modules. The first module is the adaptive fusion module, which utilizes local soft projection operation combined with context to obtain key points, and self-attention to fuse the key point information, perform overall adaptive offset, and finally map to high-dimensional information space by MLP. The second module is a feature mapping module, which incorporates a residual channel to suppress gradient disappearance and extends the dimension to obtain a more comprehensive and accurate point feature representation. Finally, the same Feature Propagation module as PointNet++ is utilized to generate the final semantic segmentation results. More detailed information will be covered in Sect. 3.

2   Related Work

2.1   DL in Point Cloud

With CNN [14] and Transformer [15] gaining great success in computer vision and natural language processing, there has significant interests in utilizing CNN or Transformer models for analyzing 3D data. Due to the irregular and sparse characteristics of point cloud, some methods used for image processing can't be directly applied. Early efforts primarily revolve around employing CNN or Transformer architectures with regular representations of 3D geometric data, such as 3D voxels. These approaches involve converting point clouds into uniform voxels within a predefined resolution and utilizing volumetric convolution techniques. However, more recent research has delved into novel designs of local aggregation operators specifically tailored for point clouds. These advancements aim to process point sets more efficiently while minimizing the loss of intricate details.

PointNet [4] introduced a groundbreaking approach by directly processing raw point clouds. It incorporates per-point multi-layer perceptrons (MLPs) that elevate each point from its coordinate space to a higher-dimensional feature space. A global pooling operation then consolidates the information into a representative feature vector, which is subsequently mapped to the object class of the input point cloud using fully connected (FC) layers. PointNet++ [5] is an extension of PointNet. It uses a hierarchical neural network structure that can effectively process point clouds of different densities while extracting local and global features of point clouds. PointCNN [6] is a deep learning model for point cloud segmentation. It uses a novel convolution algorithm that demonstrates efficient capability in performing convolution operations on point clouds, leading to commendable performance in classification and segmentation tasks. DGCNN [16] is a deep learning model for point cloud classification and segmentation. It uses a neighborhood graph-based dynamic graph construction method that can effectively capture the local and global structural features of point clouds. Point Transformer [17] is a deep learning model for point cloud classification and segmentation. It uses attention mechanisms in the Transformer model to efficiently extract and process features, achieving great result in point cloud segmentation tasks.

2.2 Attention Mechanism

Neural networks have limited storage capacity, which is proportional to the number of neurons and the network's complexity. Increasing the amount of stored information leads to a higher number of neurons and network complexity, resulting in a geometric increase in network parameters. However, computers have finite computational resources, requiring an effective balance between network complexity and expressive power to make the most of available resources.

To improve neural network efficiency and train using limited computing resources, researchers have introduced attention mechanisms [18] inspired by the biological brain's attention mechanism. The attention mechanism filters out irrelevant information through top-down information selection, focusing on a small part of useful information.

The essence of the attention mechanism is an addressing process that involves two steps: attention distribution calculation and information weighted average calculation. To compute the attention value, the key vector sequence is used to calculate the attention distribution, followed by a weighted average calculation with the value vector sequence, given a query vector.

The upcoming section incorporates the utilization of self-attention. Self-attention is a mechanism employed in deep learning models to assess the significance of different segments within the input sequence when making predictions. It enables the model to focus on the most relevant information in the input and adaptively adjust its attention weights based on the contextual understanding of the input. The self-attention mechanism calculates a score by computing the dot product between the query vector and the key vectors. Subsequently, the resulting scores are normalized through a softmax function, yielding attention weights that sum up to one. These attention weights are then employed to weigh the value vectors, generating a weighted sum that effectively represents the input information.

2.3 Point Cloud Segmentation

PointNet [4], proposed by Qi et al., is the first neural network model capable of training directly on point clouds. Prior to this, point clouds are often converted into voxel or multi-view representations, but these indirect representations can-not directly reflect the internal characteristics of point clouds. In order to over-come the disorder of point cloud, PointNet adopts Max pooling operation, which does not depend on the order of input data. However, directly adopting Max pooling may lead to feature loss. Therefore, PointNet also uses a Multi-layer Perceptron (MLP) to map the features into a high-dimensional space, where the max-pooling operation is used to minimize the risk of feature loss. Semantic segmentation network depends on both local and global features, so when PointNet performs semantic segmentation of scene point cloud, it needs to concatenate local and global features, and then uses MLP to reduce the dimension and output the semantic segmentation results.

Since PointNet directly extracts global features from the input point cloud as a whole and lacks the correlation information between local points, Qi et al. [5] further proposed a hierarchical feature learning network PointNet++. PointNet++ implements hierarchical learning of the input point cloud through a sequence of collection abstraction layers is employed, wherein each layer comprises a sampling layer, a grouping layer, and a PointNet layer. In the sample layer, PointNet++ use FPS from the input to get sampling points, these sampling points constitute the center of mass of grouping layer set. The grouping layer adopts a fixed neighborhood radius division, which can group the sampled point cloud into multiple grouped point sets. Since the density of points in the local neighborhood is different, the number of neighborhoods of each point must be the same. In the PointNet layer, the network inputs the PointNet in the unit of grouped point set, and finally uses the Max pooling operation to obtain the local point features.

The PointCNN model proposed by Li et al. [6] successfully constructs a 3D convolution operator on the point cloud, so as to realize the feature learning of the point cloud. In order to simulate the convolution operation on the point cloud, PointCNN model proposes a Conv layer, which can calculate the convolution of the k nearest neighbors of the calculated point and overcome the disorder of the point. The skeleton of the model is based on the Conv-DeConv architecture. PointCNN is introduced into the depth of the expansion of the concept of convolution can maintain network, at the same time maintain the growth rate of receptive field, make deep Conv DeConv representative point in the network can be easily observed that the input point cloud's overall shape. Through the process of inverse convolution, representative points with high-dimensional features actively propagate information to their neighboring points. Lastly, by effectively classifying the points based on their distinct features, the segmentation data can be accurately recognized, enabling the achievement of semantic segmentation for the given data.

In order to directly process large-scale point clouds, Hu et al. [19] proposed RandLA-Net. In the model, a random sampling method is proposed to sample the point cloud. However, random sampling may lose feature points in the point cloud. Therefore, the authors propose a local feature fusion module to solve this problem. The local feature fusion module acts on each point in the point cloud in parallel. The module consists of three substructures: a local spatial encoder, an attention-based pooling layer, and a dilated residual module. The main process is as follows. First, the network feeds the point cloud into a shared MLP layer to extract the features of the points and then passes through. In the coding layer, random sampling was used to reduce the number of point clouds and the local feature fusion module was used for feature aggregation. In the decoding layer, then the features of the points in the point cloud are propagated and upsampled. Finally, RandLA-Net utilizes three fully connected layers and a Dropout layer to predict the semantic label for each point, ultimately obtaining the final semantic segmentation result.

The Stratified Transformer proposed by Lai et al. [20] designs a continuous two-layer Transformer module with special aggregated local information, which is composed of a standard multi-head self-attention pattern and a feed forward neural network. Before each Transformer module and feedforward neural net-work, layer normalization operations are empirically added to the modules. It uses point cloud Coordinate information and Color information as input, the encoder, decoder structure. In this network architecture, there are multiple combinations of down-sampling layers and Transformer modules, in which the down-sampling layer of the first layer is replaced by a point embedding module to aggregate the local neighborhood feature.

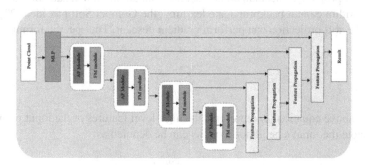

Fig. 1. The overall architecture of PointAF

3   Method

Compared to current popular transformers, the PointAF model uses an encoder-decoder structure that consists of an AS module and an FM module. This overall architecture is depicted in Fig. 1. In Sect. 3.1, we will introduce the AF module, and in Sect. 3.2, we will cover the FM module. Additionally, we will use the same structure as PointNet++ for feature propagation (Fig. 2).

3.1 AF Module

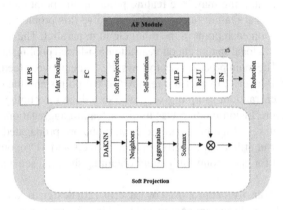

Fig. 2. The specific composition of PointAF module

The AS module is similar to the hierarchical abstraction process of PointNet++, but different from PointNet++ using FPS in the sampling process, this paper uses a simplified version of PointNet to generate downsampling points and uses soft projection operation [21] to better accumulate neighborhood information, and this operation can be backward learning and will be continuously optimized in the iterative training process.

The soft projection mechanism is essentially an attention mechanism, but since the KNN algorithm cannot backpropagate learning, the Gumbel Softmax idea [22] is used to make the soft projection can perform gradient descent. The specific formula is given below:

$$z = \sum_{p_i \in N_k(q_i)} w_i \cdot p_i \tag{1}$$

In the above equation, q_i represents the high-level features of the input point cloud, $N_k(\cdot)$ denote the input's neighborhood, w_i can be denoted as:

$$w_i = \frac{e^{dist_i^2/t}}{\sum_{j=1}^{k} e^{dist_j^2/t}} \tag{2}$$

where t is a hyperparameter which's role is to control the size of the weight w_i and $dist_i$ denotes the distance between the neighborhood and the center point.

Since the general KNN cannot adaptively adjust the neighborhood size according to the density, and if the fixed neighborhood radius is set like PointNet++, it is not conducive to the debugging of the final result. So we use density-adaptive K nearest neighboring (DAKNN) [23] to adaptively change the size of number of neighbors.

In soft projection, DAKNN is used to adaptively calculate the number of neighborhoods according to the density of high-dimensional features of each down-sampled point, and then Gumbel Softmax is used to aggregate neighborhood information to achieve adaptive offset.

Then, the features are grouped by the distance of the high-dimensional feature space, and the features in each group are subjected to the self-attention of advanced features. We use five MRB modules to map and output the obtained features that combine high-level semantic information with low-level location information, and finally obtain the output results by Reduction.

3.2 FM Module

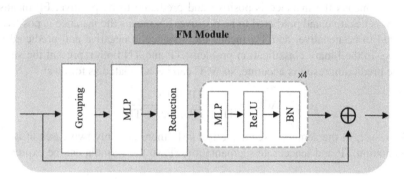

Fig. 3. The specific composition of PointAF module

In the FM module, firstly, in order to alleviate the gradient disappearance problem when the network is too deep, a residual connection is added between the input and output of the module [24]. The features are then grouped, here with the same grouping rules as in the AF module. After the grouping operation is completed, the single-layer MLP maps the neighborhood features. Different from other algorithms, we used four MRBS to map the features after Reduction to achieve better performance. Finally, the input of the module is added with the mapped features to get the final output of the module (Fig. 3).

4   Experiment

We tested the performance of on S3DIS [25] and ScanNetV2 [26] respectively, and we used the approach that is same to PointNext during the data preprocessing phase. In training PointAF, we use CrossEntropy loss with label smoothing, where we use AdamW optimizer for backpropagation and an initial learning rate 0.001, and weight decay 10–4, with Cosine Decay. And the GPU is RTX A5000(24 GB).

4.1   Evaluation Criteria

In the process of neural network model training, the training effect usually needs to be evaluated by calculating the Accuracy (ACC) and Loss during training. For semantic

Table 1. Confusion matrix for binary classification

	Positive	Negative
True	True Positive (TP)	False Positive (FP)
False	False Negative (FN)	False Negative (TN)

segmentation tasks, there is a separate evaluation metric called mean intersection over Union (mIoU). These metrics are described below (Table 1):

The above table represents the confusion matrix for a binary classification problem, where TP means the instance is positive and predicted to be positive, FP means the instance is negative and predicted to be positive, FN means the instance is positive and predicted to be negative, and TN means the instance is negative and predicted to be negative. In the binary classification problem, TP and TN both represent the situation that the prediction result is accurate, so ACC can be calculated as follows:

$$ACC = \frac{TP + TN}{TP + FP + FN + TN} \tag{3}$$

However, in the semantic segmentation task, there are only two cases of accurate segmentation TP and inaccurate segmentation FP, so the formula can be expressed as follows.

$$ACC = \frac{TP}{TP + FP} \tag{4}$$

mIoU is a common metric for semantic segmentation tasks. It first computes the IoU of each class and then computes the mean of the number of classes, which is calculated as follows:

$$IoU = \frac{TP}{TP + FP + FN} \tag{5}$$

$$mIoU = \frac{1}{N} \sum_{i=1}^{N} IoU_i \tag{6}$$

By definition, its practical meaning is the ratio of the intersection of the predicted value and the true value and the union of the predicted value and the true value. A larger mIOU indicates better performance of the trained model and more accurate prediction results for all categories of objects.

The loss function is generally calculated by cross-entropy, which can be calculated from the labels obtained by semantic segmentation and Ground Truth (GT).

4.2 3D Segmentation

S3DIS (Stanford Large-Scale 3D Indoor Spaces) is a challenging benchmark consisting of six Large indoor areas, 271 rooms, and a total of 13 semantic categories Standard Area5 cross-validation results for S3DIS are reported in Table 2.

We first used Area5 of S3DIS as test data and other areas as training data. The results are shown in the following Table 2.In the Table 2, PointAF achieves the best performance in terms of mIoU and ACC compared to others.

In the subsequent experiment, we measured the stability of the algorithm by using the method of 6-fold cross-validation, and the final results were shown in Table 3.

Table 2. Results on S3DIS Area5 for semantic segmentation

Method	mIOU (%)	mACC (%)
PointNet	41.1	–
PointNet++	53.5	83.0
PointCNN	57.3	85.9
RandLA-Net	–	–
PointAF	**70.6**	**90.2**

Table 3. Results on S3DIS 6-Fold for semantic segmentation

Method	mIoU (%)	mACC (%)
PointNet	47.6	78.5
PointNet++	54.5	81.0
PointCNN	65.4	88.1
RandLA-Net	70.0	88.0
PointAF	**74.5**	**73.9**

ScanNetV2 is another well-known large-scale segmentation dataset, containing 3D indoor scenes of various rooms with 20 semantic categories. We follow the publicly available training, validation, and test shards with 1201, 312, and 100 scans, respectively. As shown in Table 4, PointAF gets the best results when compared with other algorithms.

Table 4. Results on ScanNetv2 for semantic segmentation

Method	mIOU (%)	mACC (%)
Pointnet	–	–
Pointnet++	53.4	55.7
PointCNN	–	45.8
RandLA-Net	65.6	64.5
PointAF	**69.2**	**70.1**

To illustrate the superiority of PointAF, we will visualize the result of point cloud semantic segmentation in the S3DIS dataset in Fig. 4.

Ground truth   **Predicted result**

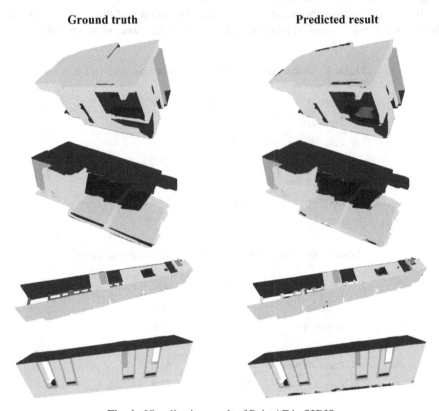

Fig. 4. Visualization result of PointAF in S3DIS

In Fig. 5, we compared the visualization results of other algorithms with PointAF algorithm. It is obvious from the figure above that PointAF's semantic segmentation capability still has some advantages compared with other algorithms.

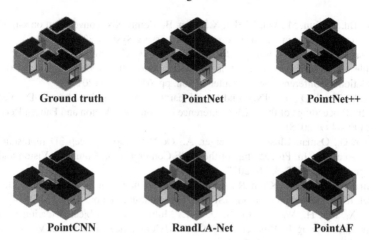

Fig. 5. Comparison results of other algorithms

5 Conclusion

In this paper, we propose PointAF, a point cloud semantic segmentation algorithm based on encoder-decoder architecture design. PointAF solves the problem of insufficient integration of neighborhood information by designing AF module, and in order to avoid over-integration of neighborhood information and ignore global information, it maps into high-dimensional feature space by FM module. The balance between global information and local information is achieved. Finally, the verification on three data shows that the algorithm has strong generalization ability and powerful performance.

Acknowledgements. The work is supported by the National Key Research and Development Program of China (2021YFF0500903, 2022YFE0198900), the National Natural Science Foundation of China (52178271, 52077213).

References

1. Choy, C., Gwak, J., Savarese, S.: 4D spatio-temporal convnets: Minkowski convolutional neural networks. In: Proceedings of the IEEE/CVF Conference on Computer Vision and Pattern Recognition, pp. 3075–3084 (2019)
2. Graham, B., Engelcke, M., Van Der Maaten, L.: 3D semantic segmentation with submanifold sparse convolutional networks. In: Proceedings of the IEEE Conference on Computer Vision and Pattern Recognition, pp. 9224–9232 (2018)
3. Graham, B., Van der Maaten, L.: Submanifold sparse convolutional networks. arXiv preprint arXiv:1706.01307 (2017)
4. Qi, C.R., Su, H., Mo, K., Guibas, L.J.: PointNet: deep learning on point sets for 3D classification and segmentation. In: Proceedings of the IEEE Conference on Computer Vision and Pattern Recognition, pp. 652–660 (2017)
5. Qi, C.R., Yi, L., Su, H., Guibas, L.J.: PointNet++: deep hierarchical feature learning on point sets in a metric space. In: Advances in Neural Information Processing Systems, 30 (2017)

6. Li, Y., Bu, R., Sun, M., Wu, W., Di, X., Chen, B.: PointCNN: convolution on x-transformed points. In: Advances in Neural Information Processing Systems, 31 (2018)

7. Thomas, H., Qi, C.R., Deschaud, J.E., Marcotegui, B., Goulette, F., Guibas, L.J.: KPConv: flexible and deformable convolution for point clouds. In: Proceedings of the IEEE/CVF International Conference on Computer Vision, pp. 6411–6420 (2019)

8. Zhou, Y., Tuzel, O.: VoxelNet: end-to-end learning for point cloud based 3D object detection. In: Proceedings of the IEEE Conference on Computer Vision and Pattern Recognition, pp. 4490–4499 (2018)

9. Riegler, G., Osman Ulusoy, A., Geiger, A.: OctNet: learning deep 3D representations at high resolutions. In: Proceedings of the IEEE Conference on Computer Vision and Pattern Recognition, pp. 3577–3586 (2017)

10. Ochmann, S., Vock, R., Klein, R.: Automatic reconstruction of fully volumetric 3D building models from oriented point clouds. ISPRS J. Photogramm. Remote Sens. **151**, 251–262 (2019)

11. Chen, X., Ma, H., Wan, J., Li, B., Xia, T.: Multi-view 3D object detection network for autonomous driving. In: Proceedings of the IEEE Conference on Computer Vision and Pattern Recognition, pp. 1907–1915 (2017)

12. Dai, A., Nießner, M.: 3DMV: joint 3D-multi-view prediction for 3D semantic scene segmentation. In: Ferrari, V., Hebert, M., Sminchisescu, C., Weiss, Y. (eds.) ECCV 2018. LNCS, vol. 11214, pp. 458–474. Springer, Cham (2018). https://doi.org/10.1007/978-3-030-01249-6_28

13. Gao, Z., Wang, D.Y., Xue, Y.B., Xu, G.P., Zhang, H., Wang, Y.L.: 3D object recognition based on pairwise multi-view convolutional neural networks. J. Vis. Commun. Image Represent. **56**, 305–315 (2018)

14. Krizhevsky, A., Sutskever, I., Hinton, G.E.: Imagenet classification with deep convolutional neural networks. Commun. ACM **60**(6), 84–90 (2017)

15. Vaswani, A., et al.: Attention is all you need. In: Advances in Neural Information Processing Systems, 30 (2017)

16. Phan, A.V., Le Nguyen, M., Nguyen, Y.L.H., Bui, L.T.: DGCNN: a convolutional neural network over large-scale labeled graphs. Neural Netw. **108**, 533–543 (2018)

17. Zhao, H., Jiang, L., Jia, J., Torr, P.H., Koltun, V.: Point transformer. In: Proceedings of the IEEE/CVF International Conference on Computer Vision, pp. 16259–16268 (2021)

18. Bahdanau, D., Cho, K., Bengio, Y.: Neural machine translation by jointly learning to align and translate. arXiv preprint arXiv:1409.0473 (2014)

19. Hu, Q., et al.: RandLA-Net: efficient semantic segmentation of large-scale point clouds. In: Proceedings of the IEEE/CVF Conference on Computer Vision and Pattern Recognition, pp. 11108–11117 (2020)

20. Lai, X., et al.: Stratified transformer for 3D point cloud segmentation. In: Proceedings of the IEEE/CVF Conference on Computer Vision and Pattern Recognition, pp. 8500–8509 (2022)

21. Lang, I., Manor, A., Avidan, S.: SampleNet: differentiable point cloud sampling. In: Proceedings of the IEEE/CVF Conference on Computer Vision and Pattern Recognition, pp. 7578–7588 (2020)

22. Jang, E., Gu, S., Poole, B.: Categorical reparameterization with gumbel-softmax. arXiv preprint arXiv:1611.01144 (2016)

23. Chen, R., Yan, X., Wang, S., Xiao, G.: DA-Net: dual-attention network for multivariate time series classification. Inf. Sci. **610**, 472–487 (2022)

24. He, K., Zhang, X., Ren, S., Sun, J.: Deep residual learning for image recognition. In: Proceedings of the IEEE Conference on Computer Vision and Pattern Recognition, pp. 770–778 (2016)

25. Landrieu, L., Simonovsky, M.: Large-scale point cloud semantic segmentation with superpoint graphs. In: Proceedings of the IEEE Conference on Computer Vision and Pattern Recognition, pp. 4558–4567 (2018)
26. Dai, A., Chang, A.X., Savva, M., Halber, M., Funkhouser, T., Nießner, M.: ScanNet: richly-annotated 3D reconstructions of indoor scenes. In: Proceedings of the IEEE Conference on Computer Vision and Pattern Recognition, pp. 5828–5839 (2017)

Accurate Detection of the Workers and Machinery in Construction Sites Considering the Occlusions

Qian Wang, Hongbin Liu, Wei Peng, and Chengdong Li[✉]

Shandong Key Laboratory of Intelligent Buildings Technology, School of Information and Electrical Engineering, Shandong Jianzhu University, Jinan 250101, China
{liuhongbin19,pengwei19,lichengdong}@sdjzu.edu.cn

Abstract. Traditional supervision of safety in construction sites is usually carried out manually, which may be inefficient and lacks real-time capability. In recent years, scholars have conducted the researches that utilize computer vision methods to perceive workers and machinery information to enhance safety supervision. However, due to complex backgrounds and occlusion issues in construction sites, the detectors still prone to generate missed detections. In order to automatically and accurately identify workers and machinery for accident prevention in construction sites, this paper proposes a novel object detection algorithm. In this method, Mosaic and MixUp data augmentation techniques are employed to obtain more training samples. An improved loss function is adopted to be the detection head to enable more accurate detection of the occluded objects. Experimental results show that the proposed algorithm exhibits superior performance in terms of accuracy and inference speed compared to the baseline detectors. It particularly performs better in detecting occluded objects, with an average precision improvement of 2.41%.

Keywords: Deep learning · Computer vision · Safety supervision · Smart construction

1 Introduction

In recent years, with the popularity of big data and the development of deep learning methods, traditional construction sites have been gradually improved in terms of informationization and digitization. As an important branch of deep learning, computer vision technology has also been widely used in smart construction sites, including equipment condition monitoring [1–3], construction progress monitoring [4], resource management and allocation [5], and safety supervision [6,7]. Compared to traditional construction sites, smart construction sites can better ensure worker safety and effectively safeguard the property of construction companies. Additionally, by utilizing cameras installed at the construction site, real-time monitoring of worker conduct, machine operation,

© The Author(s), under exclusive license to Springer Nature Singapore Pte Ltd. 2023
H. Zhang et al. (Eds.): NCAA 2023, CCIS 1869, pp. 546–560, 2023.
https://doi.org/10.1007/978-981-99-5844-3_40

and material placement can be carried out. This can assist in managing and maintaining order at the construction site, resolving the difficulty in supervising construction sites.

The advent of computer vision techniques in the past decade has given rise to a burgeoning research area exploring the use of image-based approaches for detecting objects on construction sites. Y. Xiang et al. proposed an improved YOLOv5 algorithm for automated detection of workers and machinery in construction sites [8]. Using bidirectional feature pyramid networks and RepVGG blocks, the accuracy of the algorithm was improved by 2.12%. H. Wang et al. devised a multi-scale object detection algorithm to address the challenge of detecting objects of varying sizes in construction sites, which often results in suboptimal detection performance [9]. It was completed through an adaptive data enhancement strategy and a location channel. Experimental results showed that this method could improve detection accuracy, particularly for small objects. Researchers have also focused on the study of mobile object tracking in construction sites. Y. Guo et al. proposed a road construction object tracking algorithm from the perspective of UAV(Unmanned Aerial Vehicle), based on orientation-aware bounding boxes [10]. By utilizing orientation-aware bounding boxes to organize the temporal relationship between various spatial variables. The tracking algorithm obtain excellent results. B. Xiao et al. proposed a specialized algorithm for tracking construction machinery at night in low-light conditions [11]. They utilized deep learning techniques to enhance the night-time monitoring video, and then performed tracking on the construction machinery. As a result, they achieved a 21.7% improvement in robustness.

The environment of construction sites is known for its intricate and diverse nature. Thus, objects within captured images to be frequently obstructed by the surroundings, leading to potential oversights and false detections. Scholars have undertaken extensive research in addressing this issue of occlusion. E. Chian et al. discovered that during the detection of building perimeters, some barricades are lost due to obstructions [12]. In order to automatically detect missing barricades, they proposed two different detectors. X. Wang et al. hypothesized that the problem of missed and inaccurate detections arises from the imbalance between positive and negative samples. Additionally, they suggested that the excessive redundancy of bounding boxes can lead to the failure of detecting occluded objects [13]. They proposed to address this problem by utilizing a Sample Balance-Based Regression Module. However, the effectiveness of this method in detecting overlapping workers and machinery need further improvement. H. Chen et al. proposed the use of Context-Guided data Augmentation, lightweight CNN, and proximity detection techniques to address the issue of occlusion [14]. However, the generated images might have the potential to deviate from the real-world distribution. And the quality of training samples also need improvement.

In order to enhance safety supervision and enable automatic detection of workers and machinery in complex and dynamic construction sites, this paper proposes an algorithm capable of accurately detecting occluded objects in construction sites. To augment the training data with diverse samples, Mosaic and

MixUp data augmentation methods are employed. Subsequently, the YOLOv7 object detection model is trained. To address the challenge of detecting occluded objects, this paper utilizes improved object overlap evaluation parameters. These enables the proposed method to accurately detect occluded objects. Experimental results demonstrate that the proposed detector exhibits superior performance in detecting occluded objects, with an average precision of 86.19%, representing a 2.41% improvement.

2   Methodology

Due to the complexity of construction sites, workers and machinery may overlap and obstruct each other during the construction process. This can result in blind spots in surveillance, and may pose safety hazards. To solve these problems, we propose a novel method in this paper. We replace the evaluation metric in the original YOLOv7 algorithm for obtaining more accurate object detection results. This approach is suitable for evaluating the matching between predicted bounding boxes and real bounding boxes in object detection. The flowchart of the proposed algorithm is shown in Fig. 1. The main steps are as follows.

(1) Image samples are selected from the MOCS benchmark for data enhancement, which includes the Mosaic and MixUp methods to enrich the dataset and accelerate training speed.
(2) The pre-processed images are input into the YOLOv7 network backbone for training.
(3) When calculating the overlap of the detection objects, the original IoU (Intersection over Union) is replaced by DIoU (Distance IoU) and CIoU (Complete IoU). The improved detection results are output when the loss function exceeded the threshold. Below, we will give the details for such steps.

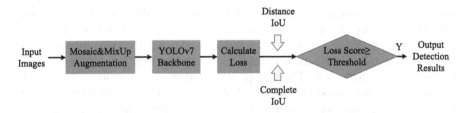

Fig. 1. The flowchart of the proposed detection method

2.1   Data Augmentation

The Mosaic data augmentation technique was first introduced in the YOLOv4 paper [18]. Its fundamental idea is to randomly crop four images and then stitch them together into a single image for training purposes. Firstly, four images are randomly selected from the dataset. Secondly, each image is randomly cropped.

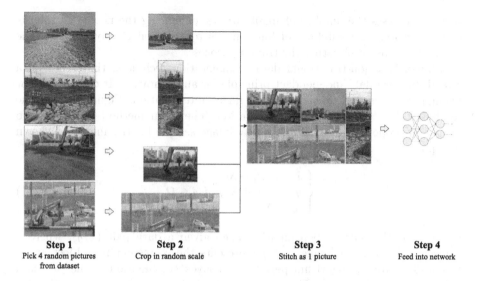

Step 1
Pick 4 random pictures
from dataset

Step 2
Crop in random scale

Step 3
Stitch as 1 picture

Step 4
Feed into network

Fig. 2. The sample of Mosaic data augmentation

Then, these cropped images are arranged in a mosaic formation using random positioning to form a new image, which is subsequently inputted into the neural network for training. Figure 2 shows the results of data after Mosaic augmentation. The benefits of this approach lie in the random usage of four images, random scaling, and random distribution of the spliced images, which significantly enriches the background of the detection dataset. Specifically, the random

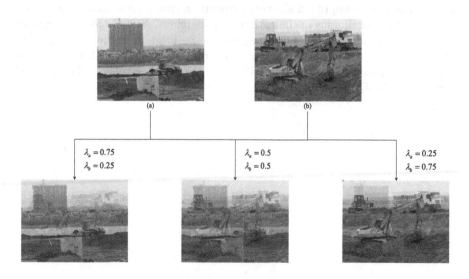

Fig. 3. The result of MixUp data augmentation

scaling increases the number of small objects, enhancing the network's robustness. Moreover, the stitching of four images together effectively increases the batch size, thus accelerating the training process.

MixUp is a non-traditional data augmentation technique that relies on a straightforward data-independent principle for augmentation. It constructs new training samples and labels by using linear interpolation. Figure 3 shows the results of data after MixUp augmentation. The main method is to mix two images in proportion to generate a new image sample for training. It is shown in Eq. 1.

$$\begin{cases} \tilde{x} = \lambda_a x_i + \lambda_b x_j \\ \tilde{y} = \lambda_a y_i + \lambda_b y_j, i, j \in \Omega \\ \lambda_a + \lambda_b = 1 \end{cases} \tag{1}$$

where Ω is the dataset benchmark. When given a data pair (x, y), where x epresents the original image and y represents the corresponding object annotation. The MixUp algorithm performs a convex combination of the two data pairs (x_i, y_i) and (x_j, y_j). The result pair $(\tilde{x}, \tilde{y})$ is a new training sample that has undergone MixUp. At this time, the ground truth of the two original images are also located on the same image. Like the Mosaic method, this can also enrich and expand the dataset and speed up the training process.

2.2   YOLOv7

You Only Look Once(YOLO) is a one-stage object detection algorithm that takes the entire image as input and directly regresses the object's position coordinates and classification probabilities through a neural network [15–18]. The benefit of this approach is an improved algorithm execution speed while maintaining accuracy. Through continuous optimization by researchers, YOLO has gone through many iterations, with YOLOv7 being one of the most promising versions [19]. YOLOv7 offers various network sizes, which differ in the size of their parameter count. Our proposed method is based on the YOLOv7-x version. The network structure of YOLOv7 is shown in Fig. 4.

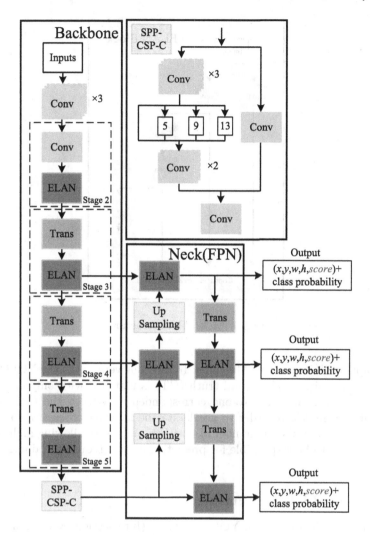

Fig. 4. YOLOv7 network structure

Compared to previous versions, YOLOv7 has the following advantages.

(1) SPP-CSP-C Structure: YOLOv7 introduces a special SPP (Spatial Pyramid Pooling) structure with a CSP (Cross-Stage-Partial) module to enlarge the receptive field. The CSP module has a large residual path that can assist in feature extraction and optimization. The SPP-CSP-C structure is shown in upper right of Fig. 4.

(2) In both the Backbone and FPN, ELAN is utilized for feature extraction, resulting in a more closely interconnected model as compared to previous YOLO versions. The structure of ELAN is shown in Fig. 5.

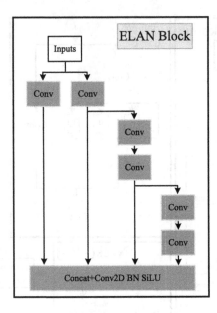

Fig. 5. ELAN block structure

(3) Adaptive Multi-Positive Matching: In previous versions of the YOLO series, during training, each ground truth (GT) is matched with only one predict box, resulting in a one-to-one correspondence between the GTs and the positive samples. In order to accelerate the training process of the model, YOLOv7 increases the number of positive samples by allowing multiple predict boxes to be responsible for predicting each GT during training.

2.3   Improved IoU Loss Function

The evaluation index used in YOLOv7 is IoU (Intersection over Union). It is a popular metric to measure the overlap of predicted and ground-truth bounding boxes, as shown in Eq. 2.

$$IoU = \frac{|A \cap B|}{|A \cup B|} \tag{2}$$

And the IoU loss function can be defined as Eq. 3.

$$\mathscr{L}_{IoU} = 1 - \frac{|A \cap B|}{|A \cup B|} \tag{3}$$

In the YOLO network, the current bounding box is judged whether it needs to be regressed based on whether the GT is in the predict box. Therefore, there may be situations where there is no intersection. According to the definition, the value of IoU is 0, which cannot reflect the distance or overlap between two boxes. Moreover, since the loss value is also 0, there is no gradient backpropagation, and learning training cannot be performed. To improve this situation, some scholars have proposed the use of GIoU(Generalized IoU) [20] as a replacement for the calculation of the indicator instead of IoU, as shown in Eq. 4.

$$\mathscr{L}_{GIoU} = 1 - IoU + \frac{|C - A \cup B|}{|C|} \tag{4}$$

where C is the smallest box covering A and B.

The GIoU metric not only focuses on the overlapping area but also takes into account other non-overlapping regions. Compared to IoU, it better reflects the degree of overlap between the two. However, when the GT and predict box are in a containment relationship, GIoU will degrade to IoU. Additionally, GIoU still heavily relies on IoU, resulting in large errors in the horizontal and vertical directions, thus slowing down the convergence rate. To address this issue, some researchers proposed DIoU (Distance IoU) by introducing the minimum bounding box [21]. The penalty term that maximizes the overlapping area is modified to minimize the standardized distance between the center points of the two bounding boxes, thereby speeding up the convergence of the loss. Its penalty factor as shown in Eq. 5.

$$\mathscr{R}_{DIoU} = \frac{\rho^2(\mathbf{a}, \mathbf{b})}{c^2} \tag{5}$$

And the DIoU loss function can be defined as Eq. 6.

$$\mathscr{L}_{DIoU} = 1 - IoU + \frac{\rho^2(\mathbf{a}, \mathbf{b})}{c^2} \tag{6}$$

In intuitive terms, if the distance between the center points of two bounding boxes is far, then the value of DIoU will be lower, providing a more accurate assessment of the matching situation between the predicted and true bounding boxes. The schematic diagram of calculating $\mathcal{L}_{DIoU}$ is shown in Fig. 6.

The black box is A, while the blue box is B. Red line c is the diagonal length of the smallest enclosing square covering two boxes, and purple line is the distance of central points of two boxes. Although DIoU can improve the accuracy of object detection when objects overlap, it does not take into account the aspect ratio of the bounding box, which is one of the three essential elements of bounding box regression. To better fit real-world scenarios, a new metric called CIoU (Complete IoU) has been proposed. Its penalty factor as shown in Eq. 7.

$$\mathscr{R}_{CIoU} = \frac{\rho^2(\mathbf{a}, \mathbf{b})}{c^2} + \alpha v \tag{7}$$

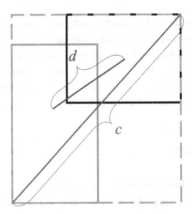

Fig. 6. The schematic diagram of calculating $\mathscr{L}_{DIoU}$

where $\alpha = \frac{v}{(1-IoU)+v}$ is the trade-off parameter, and $v = \frac{4}{\pi^2}\left(\arctan\frac{w_b}{h_b} - \arctan\frac{w}{h}\right)^2$ is the consistency of aspect ratio.

When calculating the CIoU loss, in addition to considering the gradient of v as in the case of DIoU, it is also necessary to consider the gradient of the aspect ratio. In the case where the length and width values are in the range of $[0,1]$, the value of $w^2 + h^2$ for the object box is usually very small, which can lead to gradient explosion. To ensure stable convergence, $\frac{1}{w^2+h^2}$ is replaced with 1 in the implementation. $\mathcal{L}_{CIoU}$ is shown in Eq. 8.

$$\mathcal{L}_{CIoU} = 1 - IoU + \frac{\rho^2(\mathbf{a},\mathbf{b})}{c^2} + \alpha v \tag{8}$$

In intuitive terms, if the predicted bounding box deviates in direction from the true bounding box, the CIoU value will be lower and thus provide a more accurate evaluation of the match between the predicted and true bounding boxes.

3   Experimental Results and Discussion

3.1   Metric Evaluation

The score of each bounding box is calculated according to the method described in Sect. 2.3. A predicted box is considered valid only if its score is above a certain threshold, which is typically set to 0.5. Additionally, the detector outputs the category with the highest confidence score for each box. To further filter out lower-quality boxes, another threshold for the confidence score, called Non-Maximum Suppression (NMS), is used. Any boxes with scores below this threshold are removed. A detection is considered a true positive (TP) only if the predicted bounding box is both valid and correctly labeled. Any detection that fails to meet either of these conditions is considered a false positive (FP).

Conversely, a false negative (FN) arises when a ground truth bounding box is not detected by any bounding box that has sufficient overlap, as determined by predetermined thresholds. Two metrics can be calculated using TP, FP, and FN, as in Eqs. 9 and 10.

$$Precision = \frac{TP}{TP + FP} \tag{9}$$

$$Recall = \frac{TP}{TP + FN} \tag{10}$$

The metric Precision(P) calculates the proportion of correct positive predictions out of all positive predictions. Recall(R) calculates the proportion of positive predictions out of all actual positive instances. To fully evaluate detectors under various thresholds, the concept of Average Precision (AP) is introduced as the evaluation metric. The P-R curve is plotted with different precision and recall values, and the area under the curve is calculated as the AP value, as shown in Eq. 11.

$$AP_i = \int_0^1 p(r)dr, i = 1, 2, \ldots, n \tag{11}$$

By computing the AP values for each category and averaging them, we obtain the mean average precision (mAP), as shown in Eq. 12. To fully evaluate the detector's performance under different thresholds, we vary the score threshold and obtain the mAP values at different indices. Specifically, we compute the mAP@50 when the score threshold is set to 0.5 and mAP@75 when the score threshold is set to 0.75.

$$mAP = \frac{\sum AP_i}{n}, i = 1, 2, \ldots, n \tag{12}$$

3.2   Data Preparation

To evaluate the performance of the new computation metric on the proposed method, the MOCS dataset baseline was selected for experiments [22]. The dataset contains 41,608 images and has the advantages of being large-scale, realistic, and diverse in categories compared to other existing construction-related object detection datasets.For this research, 17,463 images from the MOCS dataset were selected as the training and validation set. The dataset was divided into a 9:1 ratio, resulting in 15,716 images for training and 1,747 images for validation.

3.3   Experimental Setting

All of the experimental tasks are performed on a 12th Gen Intel(R) Core(TM) i9-12900K 3.20 GHz, 64G RAM, NVIDIA GeForce RTX 3090, Windows 10 64bit computer. CUDA 11.0 and cuDNN 8.0.4 are used on the GPU to accelerate model training. The model is implemented on Pytorch v1.7.1 deep learning framework. The input images to the model are resized to 416×416, and the batch size is set to 16.

To conduct a just and equitable comparative experiment, all models employ the improved hyperparameters described in the original paper [17,23]. Furthermore, due to the relatively large number of network parameters, training the model from scratch may result in overfitting. Consequently, the weights of the model that were trained on the COCO dataset are used to fine-tune the initialization pre-training weights. The SGD optimizer is employed throughout the entire training phase, with a momentum factor of 0.937. To prevent overfitting, weight decay is implemented in the algorithm with a value of 0.0005. The initial learning rate is set to 0.001 and is reduced by 10% each time the evaluation index stops improving. The learning rate is decreased using the cosine annealing method, which initially increases the learning rate and then decreases it. This method enables the model to smoothly search for optimal weights with a large learning rate and then gradually converge, ultimately achieving a well-fitted model.

In this paper, there is a 50% chance of using Mosaic data augmentation for each step, and a 50% chance of further applying MixUp to the enhanced images. However, because the enhanced training images deviate from the real-world distribution, in order to prevent overfitting, data augmentation is only used in the first 70% of the training.

3.4   Experimental Results and Analysis

Partial P-R curve of proposed method is shown in Fig. 7. In this figure, precision and recall are inversely related in the same model, meaning that when one is high, the other is low. This results in low accuracy when the confidence threshold is high.

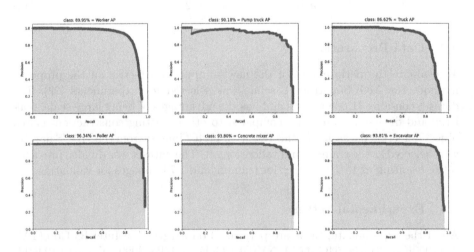

Fig. 7. Partial P-R curve of the proposed method

Some results of different calculation indicators when detection occluded objects are shown in Fig. 8. The first row shows the detection results using IoU as the evaluation metric. It can be observed that all occluded objects in the three images were not detected. The second and third rows show the detection results using DIoU and CIoU as the evaluation metrics, respectively. It can be seen that all occluded objects were successfully detected using both evaluation metrics, with high confidence scores.

Table 1. Comparison of object detection results of different algorithms

Method	mAP@50	mAP@75	Inference speed (FPS)
SSD300 & IoU	59.28	38.03	21.28
YOLOv3 & IoU	65.59	41.66	**27.03**
Mask R-CNN & IoU	**74.89**	**55.98**	7.66
YOLOv7-x & IoU	83.78	66.16	42.34
YOLOv7-x & DIoU	85.36	**68.68**	**42.57**
YOLOv7-x & CIoU	**86.19**	68.60	42.51

Through experiments, the performance index results of different detectors are shown in Table 1. From this table, we can observe that the proposed model in this paper has shown improved performance in all metrics. Compared to the previous three detectors, YOLOv7-x has achieved around a 10% increase in both

Fig. 8. Partial results of different calculation indicators

metrics. Furthermore, after changing the loss calculation method, mAP has been further improved. The DIoU calculation method has achieved the best performance with a 2.52% increase in mAP@75, while the CIoU calculation method has achieved the best performance with a 2.41% increase in mAP@50. Additionally, the inference speed of YOLOv7 has reached the fastest 42fps, ensuring real-time detection, which is better than the other three detectors.

4   Conclusion

To address the issue of detecting occluded objects in construction sites, this paper proposes an accurate object detection algorithm. Firstly, Mosaic and MixUp are employed as data augmentation techniques to enrich the training data and improve the training effectiveness. Then, the YOLOv7-x model is selected to ensure high overall inference speed, ensuring real-time object detection. Finally, the calculation method of the IoU metric is modified, enabling the model to achieve better detection results for overlapping and occluded objects. Through experiments on a large dataset MOCS, the proposed method demonstrates superior detection performance for occluded objects, ensuring that objects are not lost. These two improvements enhance the ability of real-time automated detection of workers and machinery in smart construction sites. Future work will focus on addressing the issue of false detections caused by varying scales of workers and machinery bounding boxes. The main approach is to take into account the aspect ratio of predicted bounding boxes for objects, with the aim of further improving detection accuracy.

Acknowledgment. This study is partly supported by the National Natural Science Foundation of China (Nos. 62076150, 62173216), the Taishan Scholar Project of Shandong Province (No. TSQN201812092), the Key Research and Development Program of Shandong Province (No. 2021CXGC011205, 2021TSGC1053), and the Innovation Team of "Intelligent control and Energy Efficiency Management of Building Equipments" (No. 202228039).

References

1. Guo, Y., Cui, H., Li, S.: Excavator joint node-based pose estimation using lightweight fully convolutional network. Autom. Constr. **141**, 104435 (2022)
2. Wang, D., et al.: Vision-based productivity analysis of cable crane transportation using augmented reality–based synthetic image. J. Comput. Civ. Eng. **36**(1), 04021030 (2022)
3. Assadzadeh, A., Arashpour, M., Li, H., Hosseini, R., Elghaish, F., Baduge, S.: Excavator 3D pose estimation using deep learning and hybrid datasets. Adv. Eng. Inf. **55**, 101875 (2023)
4. Wang, Y., Xiao, B., Bouferguene, A., Al-Hussein, M., Li, H.: Vision-based method for semantic information extraction in construction by integrating deep learning object detection and image captioning. Adv. Eng. Inf. **53**, 101699 (2022)

5. Chen, C., Gu, H., Lian, S., Zhao, Y., Xiao, B.: Investigation of edge computing in computer vision-based construction resource detection. Buildings **12**(12), 2167 (2022)
6. Kong, T., Fang, W., Love, P.E., Luo, H., Xu, S., Li, H.: Computer vision and long short-term memory: learning to predict unsafe behaviour in construction. Adv. Eng. Inf. **50**, 101400 (2021)
7. Zhai, P., Wang, J., Zhang, L.: Extracting worker unsafe behaviors from construction images using image captioning with deep learning–based attention mechanism. J. Constr. Eng. Manag. **149**(2), 04022164 (2023)
8. Xiang, Y., Zhao, J., Wu, W., Wen, C., Cao, Y.: Automatic object detection of construction workers and machinery based on improved YOLOv5. In: Proceedings of the 2022 International Conference on Green Building, Civil Engineering and Smart City, pp. 741–749 (2022)
9. Wang, H., Song, Y., Huo, L., et al.: Multiscale object detection based on channel and data enhancement at construction sites. Multimedia Syst. **29**, 49–58 (2023). https://doi.org/10.1007/s00530-022-00983-x
10. Guo, Y., Xu, Y., Li, Z., et al.: Enclosing contour tracking of highway construction equipment based on orientation-aware bounding box using UAV. J. Infrastruct. Preserv. Resil. **4**, 4 (2023). https://doi.org/10.1186/s43065-023-00071-y
11. Xiao, B., Lin, Q., Chen, Y.: A vision-based method for automatic tracking of construction machines at nighttime based on deep learning illumination enhancement. Autom. Construct. **127**, 103721 (2021)
12. Chian, E., Fang, W., Goh, Y.M., Tian, J.: Computer vision approaches for detecting missing barricades. Autom. Construct. **131**, 103862 (2021)
13. Wang, X., Wang, H., Zhang, C., He, Q., Huo, L.: A sample balance-based regression module for object detection in construction sites. Appl. Sci. **12**(13), 6752 (2022)
14. Chen, H., Hou, L., Zhang, G. K., Wu, S.: Using context-guided data augmentation, lightweight CNN, and proximity detection techniques to improve site safety monitoring under occlusion conditions. Saf. Sci. **158**, 105958 (2023)
15. Redmon, J., Divvala, S., Girshick, R., Farhadi, A.: You only look once: vnified, real-time object detection. In: Proceedings of the IEEE Conference on Computer Vision and Pattern Recognition, pp. 779–788(2016)
16. Redmon, J., Farhadi, A.: YOLO9000: better, faster, stronger. In: Proceedings of the IEEE Conference on Computer Vision and Pattern Recognition, pp. 7263–7271 (2017)
17. Redmon, J., Farhadi, A.: YOLOv3: An incremental improvement. arXiv preprint arXiv:1804.02767(2018)
18. Bochkovskiy, A., Wang, C.Y., Liao, H.Y.M.: YOLOv4: optimal speed and accuracy of object detection. arXiv preprint arXiv:2004.10934(2020)
19. Wang, C.Y., Bochkovskiy, A., Liao, H.Y.M.: YOLOv7: trainable bag-of-freebies sets new state-of-the-art for real-time object detectors. In: Proceedings of the IEEE/CVF Conference on Computer Vision and Pattern Recognition, pp. 7464–7475 (2023)
20. Janardan, R., Lopez, M.: Generalized intersection searching problems. Int. J. Comput. Geom. Appl. **3**(01), 39–69 (1993)
21. Zheng, Z., Wang, P., Liu, W., Li, J., Ye, R., Ren, D.: Distance-IoU loss: faster and better learning for bounding box regression. In: Proceedings of the AAAI Conference on Artificial Intelligence. vol. 34, No. 07, pp. 12993–13000 (2020)

22. Xuehui, A., Li, Z., Zuguang, L., Chengzhi, W., Pengfei, L., Zhiwei, L.: Dataset and benchmark for detecting moving objects in construction sites. Autom. Construct. **122**, 103482 (2021)

23. Liu, W., et al.: SSD: single shot MultiBox detector. In: Leibe, B., Matas, J., Sebe, N., Welling, M. (eds.) ECCV 2016. LNCS, vol. 9905, pp. 21–37. Springer, Cham (2016). https://doi.org/10.1007/978-3-319-46448-0_2

Research on Crack Identification of Highway Asphalt Pavement Based on Deep Learning

Xiuxian Chen[1,2]([✉]), Huanbing Gao[1,2]([✉]), and Tengguang Kong[1,2]

[1] School of Information and Electrical Engineering, Shandong Jianzhu University, Jinan 250101, China
1114266260@qq.com, gaohuanbing2004@sdjzu.edu.cn
[2] Shandong Key Laboratory of Intelligent Building Technology, Jinan 250101, China

Abstract. Highway pavement cracks are the main factors affecting traffic safety, among which asphalt pavement cracks are the main research object. Therefore, the ability to detect cracks accurately and quickly becomes an important research object in pavement identification. In this paper, a method of crack detection for highway asphalt pavement is presented using deep learning. First, the Retinex image enhancement algorithm is applied to the dimmer and lower contrast images in the dataset, so that a brighter and higher contrast image dataset can be obtained. Secondly, by introducing the yolov5 algorithm and classifying the data set cracks and traffic signal lines, 1500 datasets are trained and validated with 200 validation sets. The whole training model was evaluated with mAP (mean Average Precision) and P-R curve as evaluation index. The final training result shows that the crack recognition rate is 86.7%, the ground traffic line is 91.3%, and the mAP is stable at about 0.8. Therefore, the identification algorithm designed in this paper can meet the requirements of crack detection, and has a high accuracy, which has a guiding significance for the maintenance and protection of pavement.

Keyword: Deep learning · Retinex image enhancement · Yolov5 algorithm · mAP Value

1 Foreword

With the input and use of a large number of highways, traffic increases year by year, many highways have been damaged to varying degrees. Surface crack is an early disease of asphalt pavement, but over time, surface crack will accelerate cracking, resulting in large cracks, network cracks, and so on, and eventually lead to pits and even structural damage, so cracks have a great impact on the safety and comfort of driving [1]. Most non-collision traffic accidents are mainly caused by the driver's misjudgement of the road ahead, as well as cracks and depressions on the road surface, which cause the driver to lose control of the vehicle and cause traffic accidents. Especially for high-speed and overhead pavement, the cracks and navigation lines existing in the pavement can not be distinguished in time because of the fast speed of the automobile, so it is easy to form hidden dangers affecting driving safety. Therefore, the repairing of highway cracks becomes the most important task, and crack detection has become the primary task of repairing cracks. Pavement crack detection is an important index to measure pavement [2]. Timely detection and protection of pavement cracks can prevent them from forming larger cracks, thereby improving the service life and safety of the pavement. Pavement cracks are often evaluated manually. Although this method is the simplest, it requires high labor intensity and intensive time arrangement [3], resulting in traditional manual detection that is time-consuming, subjective, and unable to ensure the personal safety of the detection personnel [4]. Therefore, traditional manual inspection cannot meet the requirements of rapid and accurate crack detection.

With the development and innovation of emerging technologies such as artificial intelligence and image processing, deep learning has been widely used in various fields [5]. Using deep learning methods can independently learn the characteristics of cracks, and can learn more comprehensive and accurate feature information about cracks, which is more conducive to improving the accuracy of crack detection [6], thereby achieving the goals of objectivity and fairness, high detection efficiency, and low detection costs. Currently, target detection methods based on deep learning mainly focus on single stage target detection, which extracts image features through one stage based on frame regression [7]. Due to the need to focus on real-time and rapidity in the detection and recognition process, the detection accuracy is low [8]. However, for images with simple details and clear targets, detection can still maintain a high accuracy rate. The representative algorithm is mainly Yolo algorithm [9–11]. In this paper, Yolov5 algorithm is used for highway pavement crack detection to achieve efficient, rapid, and accurate detection requirements.

2 Data Collection and Processing

Highway pavement is mostly divided into asphalt pavement and cement pavement. Because the asphalt pavement is more flat and hard, there is no bump and more comfortable in the driving process, so it becomes the preferred structure of highway pavement. Therefore, the data set used in this study are all the picture sets of highway asphalt pavement cracks. Because the cracks on the asphalt pavement do not differ much from the color of the pavement itself, they often coincide with the white line of traffic on the pavement, and there are gaps at the links of two high-speed road segments, and the traffic signal line is washed out by long sunshine and rain water, which causes the paint discoloration and produces the pattern cracks, all of which can be regarded as miss-check and mis-check during the training process. Therefore, the main classification objects of this study are pavement cracks, traffic indicators and links of highway sections.

The pavement crack dataset used in this paper is provided by Shandong Communications Design Institute. There are 2,000 pictures with a size of 3933. × 2068. The dataset is provided by the Smart Road Detection Vehicle, which is equipped with a laser camera that is vertical to the ground, so the angles taken by the dataset are all vertical to the ground. As shown in Fig. 1. Although the pictures are black and white images, they are clear on cracks and other classification features, which ensures that the detection and recognition algorithm in this study improves the recognition accuracy.

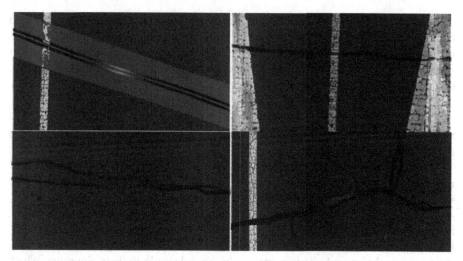

Fig. 1. Asphalt pavement photographed by highway

3 Retinex Image Enhancement

Data set images are mostly black-and-white images, and most of the cracks are black bars, which are difficult to distinguish from the color of ground asphalt. Therefore, image enhancement method is used to process the images. The main purpose of image enhancement is to process the image noise and adjust the image brightness to maximize the sharpness of the target object and make the training of recognition model more effective and accurate in order to extract the feature data of the target object better when the illumination intensity distribution of the image is not uniform or the intensity extreme is large [12]. In this paper, Retinex algorithm image enhancement technology is selected.

Retinex algorithm is a common image enhancement method. The basic principle of Retinex algorithm is that the color of object observed by human is determined by the ability of object to reflect long (red), medium (green), short (blue) light. The observed color of object is not affected by the change of light intensity and has consistency. The main principle model is shown in Fig. 2.

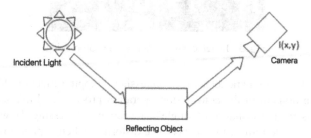

Fig. 2. Principle model of Retinex algorithm

From the model, we can see that the final image we see is illuminated on the target object by the incident light, and the final image is obtained by the reflection effect. This model can be expressed by a formula, that is, the total amount of the resulting image is multiplied by the component of the incident light and the component reflected from the object. As formula 1.

$$I(x, y) = L(x, y) \times R(x, y) \tag{1}$$

where $I(x, y)$ is the original image, $R(x, y)$ is the reflection component, and $L(x, y)$ is the incoming light.

This paper uses the adaptive single-scale Retinex algorithm, which is the most widely used Retinex algorithm. The main purpose is to eliminate the effect of sunlight intensity on the sharpness of the collected image. The color channel of the image is processed, and a Gaussian filter function is selected to enhance the channel. As shown in Fig. 3, an image of the original road surface is selected. It can be seen that because of the black and white image, it is difficult to distinguish the specific shape of the black strip cracks, while the second one is processed by a single-scale Retinex algorithm, which is significantly different from the first one. The strip cracks can have a clear contrast. Although the overall distortion is present, it does not affect the subsequent processing function for recognition. The specific expression for single-scale Retinex is shown in Formula 2.

$$r_i(x, y) = \log(R_i(x, y)) \tag{2}$$

$$\log(R_i(x, y)) = \log(\frac{I_i(x, y)}{L_i(x, y)}) = \log(I_i(x, y)) - \log(I_i(x, y) * G(x, y)) \tag{3}$$

where $*$ denotes a convolution operation, i represents the ith color channel, $G(x, y)$ is a Gaussian surround function, as shown in Formula 3.

$$G(x, y) = \frac{1}{2\pi\sigma^2} e^{(-\frac{x^2+y^2}{2\sigma^2})} \tag{4}$$

Fig. 3. Image contrast after Retinex processing

Among δ The scale parameter of the Gaussian wrapping function directly affects the size of the domain in the convolution operation process, and this parameter also directly affects the performance of single scale retinex processing. However, overall, although single scale retinex may experience distortion when applied to color images with more details, this method is widely used in image enhancement and can effectively improve the contrast of the black and white images studied in this paper. Therefore, the application of single scale retinex algorithm can meet the basic needs of this study.

4   Implementation of Yolov5 Algorithm

Yolo (You Only Look Once, Yolo) refers to the ability to identify the category and location of objects in a graph with only one browse. Yolov5 is a model used for object detection. In this paper, the latest Yolov5 algorithm for image recognition in deep learning is used to detect pavement cracks in the asphalt road image dataset of Shandong Expressway, and the results are compared and analyzed. The Yolo series of algorithms utilize a suitably sized M × The grid of N is covered in the input image, and the content in each small grid is labeled separately. This algorithm is implemented by single convolution, and most of the calculation steps are shared between small grids, making the detection efficiency and running speed of this algorithm high.

In actual pavement crack detection, the pavement background and target cracks to be detected are complex and variable, making it difficult to complete the detection of crack targets through general abstract features. Moreover, traditional detection algorithms require a large number of weights to be trained, resulting in a long training time. However, Yolov5 algorithm can extract the rich features of the same crack target, and can quickly complete the detection of pavement cracks, Therefore, Yolov5 algorithm is adopted as the target detection algorithm for pavement cracks.

4.1   Yolov5 Network Architecture

The network structure of Yolov5 is mainly composed of four parts: input, Backone, Neck, and Prediction, as shown in Fig. 4. This network structure is mainly the Yolov5s network structure, which is the network structure with the smallest depth and the smallest width of the feature mAP (mean Average Precision) in the Yolov5 series. It is suitable for small and medium-sized images and has a faster training time.

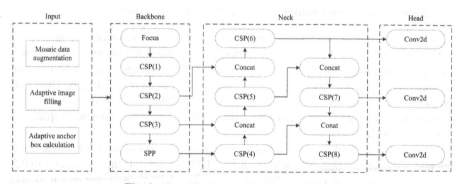

Fig. 4. Network structure diagram of yolov5s

The Input of Yolov5s mainly adopts three methods: Mosaic data augmentation, adaptive anchor box calculation, and adaptive image scaling. Mosaic data augmentation mainly applies to the size and layout of images; Adaptive anchor box calculation mainly involves setting the optimal anchor box value for different datasets; Adaptive image scaling mainly involves adding a small amount of black edges to images of different sizes to set them to the appropriate size. The three methods are mainly aimed at improving monitoring effectiveness and reducing computational complexity.

The main function of the Backbone network is for feature extraction, applying the Focus structure and CSP structure. The Focus structure mainly involves cutting images, obtaining four images based on neighboring down sampling, and increasing the number of RGB color channels to 12, As shown in Fig. 5. Finally, the concatenated images are convolved. The main function of CSP is to perform convolution operations and concatenate with another part, which can reduce computational complexity.

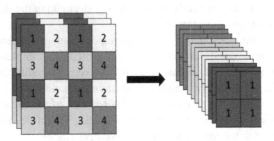

Fig. 5. Focus slicing operation

The Neck end adopts the CSP2 structure in the CSPnet design to enhance the ability of network feature fusion. Head object detection is mainly used for object detection on the feature pyramid, using a multi-level feature fusion method. The feature mAP output by Backbone is reduced in channel number and scaled through a convolution module, and finally fused with the feature mAP.

4.2 Improvement and Evaluation Index of Yolov5 Algorithm

In order to improve the accuracy of training, this article performs enhancement processing on the image and classifies and distinguishes cracks in the image. In the original training results, the Yolov5 algorithm will mistakenly detect cracks in the ground traffic signal lines or road junctions of high-speed roads, as shown in Fig. 6. Therefore, in improving the samples in the training set, careful classification is used to distinguish between signal lines and ground connections, expand the number of samples, and increase the period of epoch.

This article has trained and evaluated the expressway crack dataset. The initial learning rate is 0.01, the weight attenuation rate is 0.005, and the IoU threshold is set to 0.5. The pre training model used is Yolov5s. By setting the training cycle epoch to 200, the training and verification process on the expressway pavement crack dataset can be obtained, as shown in Fig. 7. You can see the box during the training process. The loss curve represents the error between the prediction frame and the standard frame, that is, the magnitude of the difference between the predicted position of the detected target and the position of the target itself. The smaller the value, the more complete and accurate the prediction, and it can be seen that the final value remains around 0.02. And cls_ The loss curve indicates whether the anchor frame during the training process and the calibration

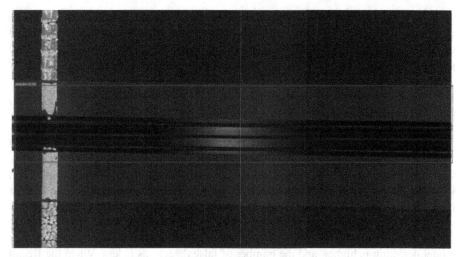

Fig. 6. Original Yolov5s training model error detection image

of the classification carried out in this article are accurate. That is, this article classifies the detected targets into two categories: cracks and traffic signal lines. The higher the detection accuracy of these two categories during the training process, the smaller the error loss, From the chart, it can be seen that the error loss is already less than 0.01 and close to zero, indicating that better detection results have been achieved by adding data sets and classifying them.

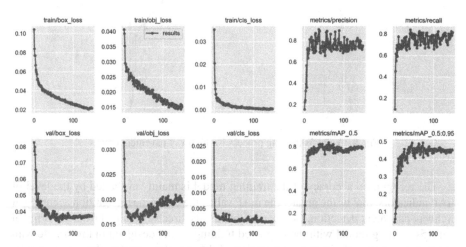

Fig. 7. Yolov5s training process for highway pavement cracks

During the training process, both Precision and Recall can evaluate the quality of the training model in this article. It can be seen that with the continuous increase of the epoch cycle, Precision and Recall still maintain around 0.7 to 0.8 and gradually converge in the late epoch, basically maintaining around 0.8, although their amplitude will fluctuate significantly. Therefore, for this paper, Precision such as Eq. 4, Recall such as Eq. 5, and various average accuracy mAP are used as evaluation indicators for the training model in this article and used to evaluate the real-time performance of the model.

$$precision = \frac{TP}{TP + FP} \tag{5}$$

$$recall = \frac{TP}{TP + FN} \tag{6}$$

where TP is the number of positive cases predicted and the true value is also positive; FP is the number of predicted positive cases and actual negative cases; FN is the number of negative cases predicted and actually positive cases. The P-R curve formed by the two can also be seen as one of the important indicators of the accuracy of the entire training model, as shown in Fig. 8. When the Precision is greater, the target accuracy of model detection is higher, and the number of false detections is less. When the Recall is greater, it can be seen as the number of detected errors is higher, and the number of missed detections is less.

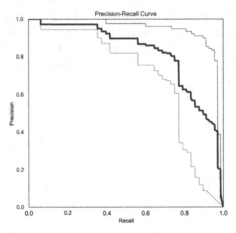

Fig. 8. Yolov5s training process for highway pavement cracks

The recognition accuracy of the training model is mainly evaluated by the level of mAP indicators, that is, the greater the mAP value, the better the recognition accuracy of highway pavement detection. Comparing the results directly obtained from the original Yolov5s training model with the improved training model designed in this article with additional data sets, the optimization effect of the training model in this article can be visually displayed from the mAP training graph, as shown in Fig. 9.

The yellow curve shows the original Yolov5s training model, and the blue curve shows the improved training model. It can be seen that both final mAP values have stabilized at 0.8, but the original mAP has stabilized at 0.8 only when the epoch is around 50, while the improved training model has stabilized at 0.8 when the epoch is 10,

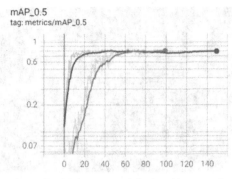

Fig. 9. Comparison of mAP between original and improved training models

Through this comparison, it can be clearly seen that increased data sets and classification can maintain a high mAP even with reduced epoch cycles of training. By comparing the evaluation index tables of the original and improved training models, as shown in Table 1, it can be confirmed from the Recall index that although the original training model can accurately identify cracks, too low a Recall value can easily lead to too high a missed detection rate and is not suitable for the application of high-speed pavement crack identification. However, the improved training model maintains good mAP, Precision, and Recall values, it proves the feasibility and effectiveness of the improved training model in the experimental process.

Table 1. Training Model Evaluation Index Table

Training model	mAP	Precision	Recall
Original Yolov5s	0.793	0.894	0.663
Improved yolov5s	0.80	0.782	0.803

5 Experimental Results and Analysis

This article conducts training and testing on 1500 datasets of highway pavement cracks, and uses 200 images for inspection and testing. Three representative experimental results are selected as shown in Fig. 10.

These three images clearly demonstrate that the Retinex image enhancement for images with low contrast can effectively identify and detect road cracks and traffic signal lines. Through classification training, the dataset has been expanded. Compared with Fig. 5, when the road junction should have been mistakenly detected as a pavement crack, the improved Yolov5s training model has well detected the signal line and did not mistakenly detect the high-speed road junction as a crack, ensuring a more accurate recognition rate.

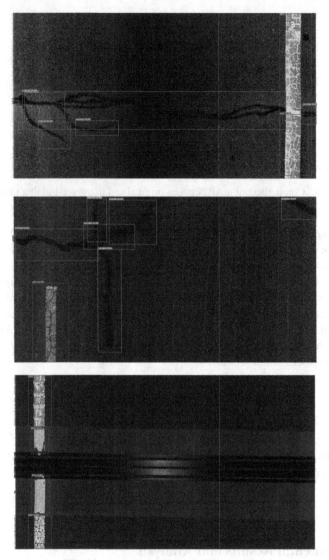

Fig. 10. Experimental results after training

Finally, through the statistics of 200 detection images, the crack recognition rate is 86.7%, and the ground traffic line recognition rate is 91.3%. After training, an evaluation table of crack and traffic signal line detection results can be compiled, as shown in Table 2, both cracks and traffic signal lines have a high recognition rate, and both the false detection rate and the missed detection rate remain at a low level. Compared to cracked ground, traffic signal line recognition has a higher success rate, but with the increase of data sets and their classification error detection rate, it can effectively identify road cracks, meeting the experimental requirements of this article.

Table 2. Evaluation Form of Crack and Traffic Signal Line Detection Results

Class	Recognition Rate	Noise Factor	Loss
Cracks	86.7%	3.5%	4.2%
Traffic signal line	91.3%	1.5%	1.9%

6 Conclusion

In order to meet the accuracy rate of highway pavement crack recognition, this article adopts an expanded classification of the data set, and uses Retinex image enhancement algorithm to greatly improve the contrast of the crack details in the image. Using Yolov5s as the training framework, it realizes the training of the logarithmic data set and ultimately obtains a high recognition rate, which can be applied to the scene of highway pavement crack recognition.

Acknowledgement. This work was partially supported by the Supported by the Natural Science Foundation of Shandong Province (ZR2022MF267): Research on Road Surface Condition Recognition and Friction Estimation Methods in Autonomous Driving. We also wish to acknowledge the support of National Natural Science Foundation of China under Grant (Nos. 61903227).

References

1. Tang, Y.: Research on Intelligent Identification Algorithm of Asphalt Pavement Crack Based on Convolutional Neural Network. Southwest Jiaotong University (2021)
2. Chen, L.: Research on Pavement Crack Detection Based on Deep Learning. Nanjing University of Posts and Telecommunications (2021)
3. Wang, H.: Research on Road Crack Detection Method Based on Machine Learning. Shenyang University of Architecture (2021)
4. Li, Y., Xu, X., Li, S.: Research on intelligent recognition method of asphalt pavement disease image based on deep learning. Transp. Technol. (2022)
5. Le Cun, Y., Bengio, Y., Hinton, G.: Deep learning. Nature **521**, 436–444 (2015)
6. Xu, K.: Research on Pavement Crack Detection and Extraction Method Based on YOLOv5s. Chang'an University (2022)
7. Wang, H., Song, M., Cheng, C.: Road target detection method based on improved YOLOv5 algorithm. J. Jilin Univ. (Eng. Ed.), 1–9 (2023)
8. Wu, R., Bai, Y., Han, J.: A method for detecting and identifying highway pavement damage based on deep learning. Comput. Simul. **40**(01), 208–212 (2023)
9. Li, X.: Highway Pavement Crack Detection Based on Deep Learning. Huazhong University of Science and Technology (2021)
10. Yu, J.: Research on Highway Pavement Disease Detection Method Based on YOLO. East China Jiaotong University (2022)
11. Cheng, L.: Research on Pavement State Image Recognition Technology Based on Deep Learning. Shanghai University of Engineering and Technology (2020)
12. Zhang, W.: Research on Road Surface Image Crack Detection Algorithm Based on Deep Learning. Harbin Institute of Technology (2022)

13. Guo, C.: Research on Asphalt Pavement Disease Detection and Evaluation Methods Based on Deep Learning. Shijiazhuang Railway University (2022)
14. Zhao, X.: Research and Implementation of a Pavement Pits Detection System Based on Deep Learning. Ningxia University (2020)

Author Index

Printed in the United States
by Baker & Taylor Publisher Services